INTELLIGENT SYSTEMS

Technology and Applications

VOLUME II

Fuzzy Systems,
Neural Networks,
and Expert Systems

Edited by
Cornelius T. Leondes

INTELLIGENT SYSTEMS

Technology and Applications

VOLUME II

Fuzzy Systems, Neural Networks, and Expert Systems

CRC PRESS

Boca Raton London New York Washington, D.C.

Library of Congress Cataloging-in-Publication Data

Intelligent systems : technology and applications / edited by Cornelius T. Leondes.
      p. cm.
      Includes bibliographical references and index.
      Contents: v. 1. Implementation techniques -- v. 2. Fuzzy systems, neural networks, and expert systems -- v. 3. Signal, image, and speech processing -- v. 4. Database and learning systems -- v. 5. Manufacturing, industrial, and management systems -- v. 6. Control and electric power systems.
      ISBN 0-8493-1121-7 (alk. paper)
      1. Intelligent control systems. I. Leondes, Cornelius T.

TJ217.5 .I5448 2002
629.8--dc21   2002017473

Visit the CRC Press Web site at www.crcpress.com

© 2003 by CRC Press LLC

No claim to original U.S. Government works
International Standard Book Number 0-8493-1121-7
Library of Congress Card Number 2002017473
Printed in the United States of America 1 2 3 4 5 6 7 8 9 0
Printed on acid-free paper

Foreword

Intelligent Systems: Technology and Applications is a significant contribution to the artificial intelligence (AI) field. Edited by Professor Cornelius Leondes, a leading contributor to intelligent systems, this set of six well-integrated volumes on the subject of intelligent systems techniques and applications provides a valuable reference for researchers and practitioners. This landmark work features contributions from more than 60 of the world's foremost AI authorities in industry, government, and academia.

Perhaps the most valuable feature of this work is the breadth of material covered. Volume I looks at the steps in implementing intelligent systems. Here the reader learns from some of the leading individuals in the field how to develop an intelligent system. Volume II covers the most important technologies in the field, including fuzzy systems, neural networks, and expert systems. In this volume the reader sees the steps taken to effectively develop each type of system, and also sees how these technologies have been successfully applied to practical real-world problems, such as intelligent signal processing, robotic control, and the operation of telecommunications systems. The final four volumes provide insight into developing and deploying intelligent systems in a wide range of application areas. For instance, Volume III discusses applications of signal, image, and speech processing; Volume IV looks at intelligent database management and learning systems; Volume V covers manufacturing, industrial, and business applications; and Volume VI considers applications in control and power systems. Collectively this material provides a tremendous resource for developing an intelligent system across a wide range of technologies and application areas.

Let us consider this work in the context of the history of artificial intelligence. AI has come a long way in a relatively short time. The early days were spent in somewhat of a probing fashion, where researchers looked for ways to develop a machine that captured human intelligence. After considerable struggle, they fortunately met with success. Armed with an understanding of how to design an intelligent system, they went on to develop useful applications to solve real-world problems. At this point AI took on a very meaningful role in the area of information technology.

Along the way there were a few individuals who saw the importance of publishing the accomplishments of AI providing guidance to advance the field. Among this small group I believe that Dr. Leondes has made the largest contribution to this effort. He has edited numerous books on intelligent systems that provide a wealth of information to individuals in the field. I believe his latest work discussed here is his most valuable contribution to date and should be in the possession of all individuals involved in the field of intelligent systems.

Jack Durkin

Preface

For most of our history the wealth of a nation was limited by the size and stamina of the work force. Today, national wealth is measured in intellectual capital. Nations possessing skillful people in such diverse areas as science, medicine, business, and engineering, produce innovations that drive the nations to a higher quality of life. To better utilize these valuable resources, intelligent systems technology has evolved at a rapid and significantly expanding rate to accomplish this purpose. Intelligent systems technology can be utilized by nations to improve their medical care, advance their engineering technology, and increase their manufacturing productivity, as well as play a significant role in a very wide variety of other areas of activity of substantive significance.

Intelligent systems technology almost defines itself as the replication to some effective degree of human intelligence by the utilization of computers, sensor systems, effective algorithms, software technology, and other technologies in the performance of useful or significant tasks. Widely publicized earlier examples include the defeat of Garry Kasparov, arguably the greatest chess champion in history, by IBM's intelligent system known as "Big Blue." Separately, the greatest stock market crash in history, which took place on Monday, October 19, 1987, occurred because of a poorly designed intelligent system known as computerized program trading. As was reported, the Wall Street stockbrokers watched in a state of shock as the computerized program trading system took complete control of the events of the day. Alternatively, a significant example where no intelligent system was in place and which could have, indeed no doubt would have, prevented a disaster is the Chernobyl disaster which occurred at 1:15 A.M. on April 26, 1987. In this case the system operators were no doubt in a rather tired state and an effectively designed class of intelligent system known as "backward chaining" EXPERT System would, in all likelihood, have averted this disaster.

The techniques which are utilized to implement Intelligent Systems Technology include, among others:

Knowledge-Based Systems Techniques
EXPERT Systems Techniques
Fuzzy Theory Systems
Neural Network Systems
Case-Based Reasoning Methods
Induction Methods
Frame-Based Techniques
Cognition System Techniques

These techniques and others may be utilized individually or in combination with others.

The breadth of the major application areas of intelligent systems technology is remarkable and very impressive. These include:

Agriculture	Law
Business	Manufacturing
Chemistry	Mathematics
Communications	Medicine
Computer Systems	Meteorology
Education	Military
Electronics	Mining
Engineering	Power Systems
Environment	Science
Geology	Space Technology
Image Processing	Transportation
Information Management	

It is difficult now to find an area that has not been touched by Intelligent Systems Technology. Indeed, a perusal of the tables of contents of these six volumes, *Intelligent Systems: Technology and Applications*, reveals that there are substantively significant treatments of applications in many of these areas.

Needless to say, the great breadth and expanding significance of this field on the international scene requires a multi-volume set for an adequately substantive treatment of the subject of intelligent systems technology. This set of volumes consists of six distinctly titled and well-integrated volumes. It is appropriate to mention that each of the six volumes can be utilized individually. In any event, the six volume titles are:

1. Implementation Techniques
2. Fuzzy Systems, Neural Networks, and Expert Systems
3. Signal, Image, and Speech Processing
4. Database and Learning Systems
5. Manufacturing, Industrial, and Management Systems
6. Control and Electric Power Systems

The contributors to these volumes clearly reveal the effectiveness and great significance of the techniques available and, with further development, the essential role that they will play in the future. I hope that practitioners, research workers, students, computer scientists, and others on the international scene will find this set of volumes to be a unique and significant reference source for years to come.

Cornelius T. Leondes
Editor

About the Editor

Cornelius T. Leondes, B.S., M.S., Ph.D., Emeritus Professor, School of Engineering and Applied Science, University of California, Los Angeles, has served as a member or consultant on numerous national technical and scientific advisory boards. Dr. Leondes served as a consultant for numerous Fortune 500 companies and international corporations. He has published over 200 technical journal articles and has edited and/or co-authored more than 120 books. Dr. Leondes is a Guggenheim Fellow, Fulbright Research Scholar, and IEEE Fellow, as well as a recipient of the IEEE Baker Prize award and the Barry Carlton Award of the IEEE.

Contributors

Shigeo Abe
Kobe University
Kobe, Japan

Rafael Alcalá
University of Granada
Granada, Spain

Senen Barro
University of Santiago de
    Compostela
Santiago, Spain

Alberto Bugarín
University of Santiago de
    Compostela
Santiago, Spain

Jorge Casillas
University of Granada
Granada, Spain

W.J. Chen
Nanyang Technological
    University
Singapore

Oscar Cordón
University of Granada
Granada, Spain

Toshio Fukuda
Nagoya University
Nagoya, Japan

Francisco Herrera
University of Granada
Granada, Spain

Roberto Iglesias
University of Santiago de
    Compostela
Santiago, Spain

Ficzko Jelena
University of Ljubljana
Ljubljana, Slovenia

Virant Jernej
University of Ljubljana
Ljubljana, Slovenia

Chia-Feng Juang
National Chung Hsing
    University
Taichung, Taiwan, R.O.C.

Abraham Kandel
University of South Florida
Tampa, Florida

Kazuo Kiguchi
Saga University
Saga, Japan

Mraz Miha
University of Ljubljana
Ljubljana, Slovenia

Manuel Mucientes
University of Santiago de
    Compostela
Santiago, Spain

Zimic Nikolaj
University of Ljubljana
Ljubljana, Slovenia

Jong Joon Park
Seokyeong University Sangbuk-Ku
Seoul, Korea

Chai Quek
Nanyang Technological
    University
Singapore

Carlos V. Regueiro
University of A Coruña
A Coruña, Spain

Ying Tan
University of Science and
    Technology of China
Heifei, P.R. China

W.L. Tung
Nanyang Technological University
Singapore

Keigo Watanabe
Saga University
Saga, Japan

K.Y. Michael Wong
Hong Kong University
    of Science and Technology
Hong Kong

Si Wu
Sheffield University
Sheffield, U.K.

Ming Zhou
Indiana State University
Terre Haute, Indiana

Contents

Contents

Volume I: Implementation Techniques

Contents

Contents

Volume IV: Database and Learning Systems

Contents

Contents

Volume VI: Control and Electric Power Systems

1

Neural Network Techniques and Their Engineering Applications

Ming Zhou
Indiana State University

1.1 Introduction

One of the areas in artificial intelligence (AI) that has grown fast and gained popular engineering applications during the last several decades is the artificial neural network (ANN, or simply "neural network;" both terms are used interchangeably in this chapter). Researchers and practitioners in various industries, institutions, and government agencies have applied this powerful technique in solving numerous application problems, ranging from system/process control and automation, reliability/failure predication, and signal/image processing, to financial planning and marketing analysis, and manufacturing process planning and optimization.[2–5,11,14,16,25–27] The primary reason for the wide recognition and application of ANN is due to its simplicity of implementation and effectiveness in solving difficult problems.

The development of ANN was based on the needs for solving large-scale complex problems, and inspired by the successful functions of biological systems, particularly human brains in this regard.[11,21] There are two main characteristics in terms of information processing by human neural systems: parallelism and nonlinearity. Parallelism refers to the human brain's ability of processing

0-8493-1121-7/03/$0.00+$1.50
© 2003 by CRC Press LLC

information through a large number of neurons working simultaneously. Nonlinearity refers to the human's ability to analyze the complex relationship between input and output of complex systems. Parallelism and nonlinearity are two indispensable requirements in developing solution methods for many practical problems due to the complexity involved and the needs for processing massive amounts of information in real time.

The crux of solving many engineering problems lies in the modeling and analysis of the relationship between a set of input variables and one (or more) output variable. There are two basic modeling approaches: analytical modeling and empirical modeling. Analytical models are usually mathematical models developed based on so-called "first principles," i.e., the fundamental scientific laws that govern underlying systems/processes. Although an analytical approach can provide exact or "closed-form" solutions, it is often limited by problem complexity. In reality, analytical models are either too difficult to be constructed or useless due to the lack of "first principle" knowledge and unrealistic assumptions.

Empirical modeling, on the other hand, develops models based on the input/output data collected through experiment or production runs. It estimates the unknown relationship between input and output by investigating important characteristics embedded in the data. Empirical methods can be further divided into two types according to modeling tools: statistical models and neural networks. Statistical models use a defined or predetermined form (e.g., polynomial form) and parameterization (i.e., they are fixed in parametric form). Neural networks, on the other hand, use a network of interconnected "neurons" (computing units) that can learn the input/output relationship of a system through proper training with sampled data. The key difference here is that problem solving is viewed (by a neural network) as a learning problem, i.e., the problem-solving knowledge is learned by a model, not built into the model by experts.

ANNs have been particularly successful in predicting the behavior of large complex and nonlinear systems with a large number of variables.[4,14,16,25] An ANN model takes a set of inputs and produces one or more output, and does not require any pre-assumptions on logical or analytical form between the input and output variables. Its mapping capability is obtained through the architecture of the network and the training of its parameters with experimental data. By examining the patterns between input data and their corresponding outputs, a neural network gains the knowledge on system dynamics, and becomes able to use this knowledge to predict a system's output for any given inputs. These features make a neural network an ideal model for predicting the behavior of a complex engineering system. For example, ANNs have been used successfully in pattern recognition and data mining to capture hidden patterns among raw data, and represent complex relationships between sets of variables.[11,14]

The ability to learn efficiently and effectively the system dynamics, and to represent highly nonlinear relationship between input/output variables has also made neural network a popular tool in process control and automation.[9,14,18] When properly designed and trained, a neural network is capable of learning any process dynamics and making accurate predictions on process output based on the knowledge learned. This mechanism has been used for process design and on-line or off-line control purposes. ANNs are often considered more robust than other models since they are not restricted by the structure of input/output representation, and do not make (explicitly) parametric assumptions on data distribution.

The purpose of this chapter is to provide a brief introduction on the fundamentals of ANN to engineers or other professionals who are interested in, but new to, this field. The emphasis is given to the basic concepts and properties of the technique, and some typical applications in engineering systems. Although the ideas of ANN were inspired by the study of biological behavior of the human brain, the subject is treated from an engineering perspective in this chapter. From this point of view, a neural network is considered a parallel computational machine. In Section 1.2, fundamental concepts and principles of ANN are presented to help readers understand what ANN is and how it works.

Section 1.3 introduces readers to a number of typical engineering applications of ANN technique. A general constructive procedure for design and implementation of ANN applications is provided in Section 1.4. Some recent advancement in the field, particularly combining ANN with fuzzy logic and genetic algorithm, is discussed in Section 1.5. A final conclusion is given in Section 1.6 to close the chapter.

1.2 Fundamental Concepts and Principles

1.2.1 What Is an ANN?

From machine learning point of view, an ANN is a computational network, consisting of a number of interconnected information processing units (neurons), which is able to learn and represent the unknown dependency relationship between a set of input variables and a set of output variables of a system or process. For instance, consider a process with five input variables $(X_1, \ldots, X_5)$ and one output variable Y. The true functional relationship between inputs and output, $Y = F(X)$, is usually unknown. A neural network can be used in this case to estimate the true dependency $F(X)$, i.e., to model or approximate the relationship so that:

$Y \approx ANN(X)$ (read as: Y is approximately equal to the neural network output) (See Figure II.1.1). The approximation capability of a neural network is obtained through training the network with samples of input/output data, *not* by solving a predefined analytical system of equations.

Neural networks are often classified into two main categories according to their architectural characteristics: *multilayer feed-forward networks* and *recurrent networks*. Multilayer feed-forward networks are also commonly referred to as multilayer perceptrons (MLPs). The architecture of a MLP can be represented by a directed acyclic graph (DAG) as shown in Figure II.1.1. Each neuron (or node) is connected only to the neurons in the next layer. There is no connection between neurons in the same layer.

Structurally, a MLP consists of four parts: an input layer containing a number of input nodes; an output layer with one or more output nodes; a set of intermediate or "hidden" layers with a number of hidden neurons; and a set of "links" that connect the neurons between layers (e.g., the network in Figure II.1.1 has one input layer with five input nodes; one hidden layer with four hidden neurons; and one output layer with one output neuron). The input layer contains a set of sensory units (input nodes) that are interfaced with the external world to obtain the data and information from an outside environment. The neurons in hidden layers are computational units that perform nonlinear transformation (mapping) between inputs and outputs. The connection links transfer signals between neurons in different layers. To control the effects (or intensity) of the signals for desired mapping, a weight is associated with each link. These link weights can be adjusted through a training process to improve learning results, and are commonly referred to as the "free parameters" of a neural network.

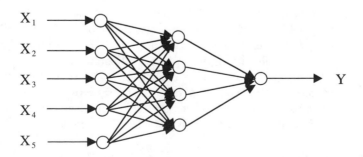

FIGURE II.1.1   A multilayer feed-forward neural network.

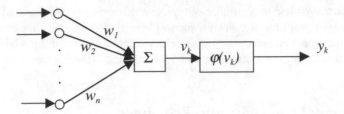

FIGURE II.1.2   The model of a single hidden neuron.

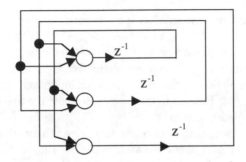

FIGURE II.1.3   A recurrent neural network of 3 neurons.

Given proper design and training, a neural network is capable of approximating any nonlinear function, say $y = f(x)$, where x and y are vector input and output. This means that when an input vector x is presented to the network, it will produce an output vector y^* that is very close to the desired output y, i.e., $y^* \approx y = f(x)$. This mapping capability is mainly due to the nonlinear transfer functions (also called activation functions) of a neural network's hidden neurons. Usually these nonlinear transfer functions are *sigmoidal functions* of the following form:[6,11]

$$y_k = \varphi(v_k) = \frac{1}{1 + e^{-v_k}} \quad \text{with} \quad v_k = \sum_j w_{jk} y_j \qquad \text{(II.1.1)}$$

where y_k = output value of a neuron k; and v_k = activation value (i.e., net internal activity level) of neuron k. Note that v_k is simply a weighted sum of the output signals from the neurons in previous layer (see Figure II.1.2), and $\varphi(v)$ is a nonlinear activation or transfer function that produces the output y_k.

MLPs are probably the most popularly used neural networks among engineering applications. They have been commonly used for solving difficult predictive modeling problems such as regression and classification.[11,19]

Unlike MLPs, recurrent neural networks (RNN) have "feedback loops" so that the output of a neuron can be fed back to the inputs of other neurons in the same or previous layer (Figure II.1.3). The feedback loops (usually denoted by a unit-delay function z^{-1}) cause the network to display a nonlinear dynamic behavior. By allowing the feedback of node-output (activation), the dynamics of such neural network becomes an important property and affects the functional behavior of the network in signal/information processing. The functions of recurrent neural networks are quite different from feedforward networks. For instance, this type of neural network has been used to solve many difficult optimization problems. More detailed discussion is given later in Section 1.3.

1.2.2   How Can an ANN Learn?

For a neural network to be able to learn, several requirements must be satisfied. For instance, *a set of data* must be adequate and available for training the network; *an (network) architecture* must be designed properly; *a learning/training strategy* must be specified to guide the training process; and *a performance measure* needs to be defined to evaluate the results or quality of learning. A learning strategy provides procedural instructions (an algorithm) for training a neural network. There are mainly two types of learning strategies: *supervised learning* and *unsupervised learning*. In supervised learning, a "teacher" is needed, i.e., predetermined target signals are used in the training process to compare with the actual signals produced by a neural network to determine the approximation errors for improvement. With unsupervised learning, the network is self-organized to respond to the characteristics of underlying input patterns. We will focus our discussion on supervised learning in this chapter.

During a (supervised) learning process, an ANN takes a sample of input signals (training data) and produces a corresponding output signal(s). The produced output is then compared with desired output (target signals) to compute an approximation error (i.e., performance measure). The network parameters (e.g., link weights) are then adjusted accordingly to reduce the error. The training process iterates until the performance error is decreased to an acceptable level. For supervised learning, the purpose of training a network is to minimize the approximation error defined as the difference between predicted outputs and target outputs. In general, we define a performance measure P for network learning as the average sum of squared errors (ASE) over a set of N training samples, i.e.:

$$P = ASE = \frac{1}{N} \sum (T_i - D_i)^2 = \frac{1}{N} \sum_{i=1}^{N} e_i^2 \qquad (\text{II.1.2})$$

where D_i and T_i are predicted output value and the target output value, respectively. The changes made to the network's adjustable weights should be in proportion to the reduction of approximation errors to minimize P through training iterations. Such an approach is usually called "**error correction rule.**" Another form of performance measure often used for analyzing learning algorithms is the *instantaneous sum of squared errors* defined as:

$$P = \frac{1}{2} \sum_{k \in O} e_k^2 \qquad (\text{II.1.3})$$

The error $e_k = t_k - y_k$, where t_k and y_k are target and predicted output values. The neuron k is an output neuron, and set O includes all neurons in the output layer.

Backpropagation (BP)[3,11,19,21] has been widely used for several decades as a basic learning algorithm for training feed-forward neural networks. The idea behind this method is to make larger changes to those link weights that will result in larger reduction of approximation error. During a backpropagation iteration, there are two passes of signals. First, a set of input signals is presented to the network and propagated forward to the output layer (forward pass) to produce output signals. Then the errors computed at the output layer are propagated backward through the network to adjust link weights according to an error-correction rule (backward pass). The process stops when a predetermined error goal (e.g., acceptable level of error) is achieved.

By changing weights in the direction of *steepest descent* with respect to approximation error, backpropagation algorithm tries to minimize this error and improve network performance. In backpropagation, the change of a link weight w_{jk} (the link from neuron j to neuron k), say Δw_{jk}, is made according to the so-called *delta rule*, usually in the form: $\Delta w_{jk} = \eta \delta_k y_j$. In this definition, η is a proportional constant called learning rate; y_j is the output of (a previous) neuron j; and δ_k represents the local gradient of error signals (i.e., indicating the direction of maximum reduction of error at a local point) with respect to the weight (i.e., partial derivative $\partial P / \partial w_{jk}$).

A more formal description on the mathematical derivation of backpropagation algorithm is given below. From previous discussion, we know that a weight change Δw_{jk} is made in proportion to the instantaneous change of error performance function (i.e., the gradient of P given in formula (II.1.3)) and in the direction of gradient descent. That is,

$$\Delta w_{jk} \propto \frac{\partial P}{\partial w_{jk}} \ or \ \Delta w_{jk} = \eta \frac{\partial P}{\partial w_{jk}} \tag{II.1.4}$$

Apply the chain-rule:

$$\frac{\partial P}{\partial w_{jk}} = \frac{\partial P}{\partial y_k} \frac{\partial y_k}{\partial w_{jk}}$$

but

$$\frac{\partial y_k}{\partial w_{jk}} = \frac{\partial [\varphi(v_k)]}{\partial w_{jk}} = \frac{\partial y_k}{\partial v_k} \frac{\partial v_k}{\partial w_{jk}} = \varphi'(v_k)y_j$$

and

$$\frac{\partial P}{\partial y_k} = \frac{\partial P}{\partial e_k} \frac{\partial e_k}{\partial y_k} = -e_k$$

therefore,

$$\Delta w_{jk} = -\eta \frac{\partial P}{\partial w_{jk}} = \eta e_k \varphi'(v_k)y_j = \eta \delta_k y_j \tag{II.1.5}$$

where δ_k = local gradient of neuron k, and $\delta_k = e_k \varphi'(v_k)$.

A pseudocode form of backpropagation algorithm for batch mode training of neural networks can be presented as follows:

Begin

```
Specify an error goal ε;
Select a learning rate η;
While (error > ε)
{
For each input xᵢ(0 ≤ i ≤ m);
{
   //perform a forward pass
   Compute output Oi;              //Oᵢ is predicted output
   Calculate error eᵢ = (Ti − Oi);    //Tᵢ is desired/target output

   //perform a backward pass
   Compute local gradient δ for all neurons;
   Calculate weight corrections Δw for all links using delta rule;
}
Sum weight corrections for all inputs x;
Update the weights of all links;
}
```

End.

As mentioned earlier, the purpose of training a neural network is to minimize the approximation error, i.e., forcing the network to approximate or predict as close as possible at data points used in the training phase. However, a neural network that can fit almost perfectly to the training data points may not fit (or fit poorly) with other input signals that are not used to train the network (e.g., "unseen" data points). This type of error is called *generalization error*, i.e., it measures a neural network's ability to generalize the knowledge learned through training to make satisfactory predictions with unseen data inputs. To reduce generalization error, designers often use a technique called "validation"[11,19] to improve the design and performance of a trained neural network. The validation process includes several steps: preparing a data set for validation; selecting a strategy for testing the neural network with validation data; and determining actions to take in accordance with validation results. More detailed discussion on validation is given in later sections.

Before leaving this section, we would like to briefly introduce another important and commonly used type of neural network: Radial Basis Function (RBF) neural networks. RBF network is a special type of nonlinear feed-forward neural network.[11] Like MLP, RBF network is also capable of approximating any nonlinear functions. But it has several distinct features that are different from MLP. First, an RBF network has only one hidden layer. The transfer functions of hidden neurons in a RBF network are radially symmetric about a set of "centers," $x_i, 0 \leq i \leq N$, that is, transfer function (i.e., basis function) is in the form: $\varphi(\cdot) = \varphi(\|x - x_i\|)$, where the net internal activity $\|x - x_i\|$ (the argument to transfer function) is computed as the Euclidean distance between an input vector x and a local center x_i, i.e., $\|x - x_i\| = (x - x_i)'(x - x_i)$. Gaussian functions are commonly used as basis functions for RBF networks:

$$\varphi(\|x - x_i\|) = \exp\left(-\frac{\|x - x_i\|}{\beta}\right) \qquad \text{(II.1.6)}$$

Note that β is simply a constant scale factor. Each hidden neuron usually represents a local center $x_i(0 \leq i \leq N)$. The output neuron of a RBF network is linear, i.e., it is a weighted sum of outputs from all hidden neurons:

$$y_o = \sum_{k=1}^{N} W_k y_{hk} = \sum_{k=1}^{N} W_k \varphi(\|x - x_k\|) \qquad \text{(II.1.7)}$$

where $y_{hk} = $ output from kth hidden neuron. The set of weights $\{w_1, \ldots, w_k\}$ can be trained or solved directly in simple cases.[6,11,21] This feature has usually made the design of RBF networks more efficient than standard MLPs. Intuitively, when an input x is close to a local center x_i, the distance $\|x - x_i\|$ will be close to 0 (zero), and that neuron's output $y_{hi} = \varphi(\|x - x_i\|)$ will be close to 1. This means that the trained weight w_i (e.g., representing a class type i) will have a dominant effect in predicting the network's output Y_o.

1.2.3 Basic Issues in the Design of ANN

There are two types of important and practical issues in the design and implementation of ANNs. One is related to the **improvement of learning efficiency**, and the other is concerned with **appropriate design of network architecture**. We discuss these issues in the following two subsections.

1.2.3.1 Improvement of ANN's Learning Efficiency

For many practical problems, simple backpropagation training takes a very long time due to the nature of gradient descent.[6,10,11] First, the algorithm is insensitive to the local shape of performance function (i.e., error surface) due to a *fixed* learning rate. A large learning rate often causes the learning steps bouncing between the opposite sides of a deep "valley" instead of following the contour to reach the "bottom" (a local minimum). On the other hand, a small learning rate results in a very slow convergence on a relatively flat surface (i.e., causing unreasonably long period for training). Second, a nonlinear network usually has many local minima on its error surface. Pure gradient descent search is easily trapped by these local minima. The convergence to a global minimum is not guaranteed. To address these issues and improve simple backpropagation algorithm, several techniques have been widely used and are presented as follows.

1. **Adaptive learning rate.**[6,11,20] has been adopted to improve network training efficiency. Initially, a large learning rate is used to reduce the error quickly. When the error increases to a certain degree, the learning rate is automatically reduced to allow smaller-step search for better solutions. By changing the learning rate adaptively (i.e., increasing the rate when error is decreasing and reducing the rate when error is increasing), this technique makes learning steps more responsive to the local shape of (performance) error surface and speeds up the network training process without increasing instability.

2. **Momentum models.**[6,11] A momentum factor is used to help network training to overcome local minima. Through a scalar factor $m (0 \leq m \leq 1)$, the weight change at a stage k (e.g., kth iteration), ΔW^k, is determined based on a trade-off between the weight change at previous stage and local gradient descent G (see formula II.1.8). The larger the value of momentum factor m, the stronger the impact of previous weight change on the current weight adjustment. This helps the network to get out of a local minimum and not be trapped by a local gradient during training iterations.

$$\Delta W^k = m \Delta W^{k-1} + (1 - m)G \qquad (\text{II.1.8})$$

It should be pointed out that adaptive learning rate and momentum techniques are often combined to provide better network training results.

3. Backpropagation can also be improved through **optimization-based methods**. One such technique is Newton's method.[6,11] Instead of using only first-order information (i.e., gradient vector), Newton's method also uses second-order quantities (i.e., Hessian matrix) to identify the optimum weight change (one that minimizes the error function). However, Newton's method requires a lot of computational resource (memory, storage space) to calculate and store Hessian matrix. Other more efficient methods have been developed to overcome this drawback. One example is conjugate-gradient method.[11] This method avoids the computation and storage of Hessian matrix. During each iteration, it first determines an optimal step size to move along the current search direction, then identifies a new search direction by combining the new gradient descent direction with previous search direction. Another significant improvement was attempted by incorporating Levenberg-Marquardt (LM) approximation scheme.[6,10] It is a numerical optimization-based technique and has shown to be very effective for improving backpropagation. Essentially, LM learning rule utilizes a scalar μ to control the shift between Gauss-Newton method and gradient descent scheme during training steps (see formula II.1.9). When μ is large, the learning process approximates gradient descent. When μ is small, the learning process essentially follows the Gauss-Newton rule. The value of μ increases if the error increases after an iteration step, and decreases when a step reduces the error. Note that J is the Jacobian matrix defined on error vector e.[6,10]

$$\Delta W = (J^T J + \mu I)^{-1} J^T e \qquad (\text{II.1.9})$$

1.2.3.2 Appropriate Design of ANN's Architecture

Determining the number of neurons in each hidden layer and the number of hidden layers is a critical decision in the design of neural networks. As mentioned earlier, an ANN is essentially a nonlinear mapping function $f(x, w)$ with w a parameter set. Increasing the number of hidden neurons enhances its ability to approximate (i.e., "fit") input/output data patterns, but also increases the number of free parameters (e.g., trainable weights w). This increases model complexity. In fact, a basic issue in designing ANN is the proper balance of model complexity and approximation ability. Too many hidden neurons result in too many trainable weights, which cause a neural network to become erratic

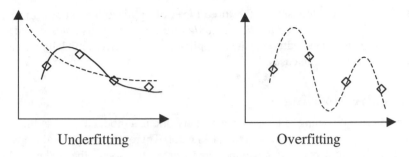

Underfitting   Overfitting

FIGURE II.1.4   Concept of underfitting and overfitting.

and unreliable.[11,19,21] This problem is usually referred to as "overfitting," i.e., too much flexibility makes the network behave erratically in between training data points (Figure II.1.4, right, dashed curve) and consequently results in poor generalization. On the other hand, too few hidden neurons restrict the learning capability of a neural network and deteriorate its approximation performance, i.e., causing "underfitting" (Figure II.1.4, left, dashed curve).[11,19,21] A basic principle in network design is to properly balance the requirement of generalization and approximation (note that generalization ability is directly affected by model complexity). In Figure II.1.4 (left), the solid curve represents a good fit.

Unfortunately there is no clear guideline for determining the number of neurons in hidden layer(s). It has been shown, however, that enough neurons must be provided to enable the network to perform the required mapping function satisfactorily.[11,14,21] Empirical rules have also been suggested to avoid overfitting. For instance, Winston[21] recommended that the number of trainable weights should not exceed the number of training samples (i.e., the number of input/output data sets used for training a network). For example, given the information that there are three input nodes and one output node, and a total of 41 data sets are used for training, ten hidden neurons may be used with a single hidden layer. Others[11,19] have also proposed heuristic rules for characterizing the relationship between the number of structural parameters of a neural network, the size of training/validation data sets, and desired error goal. There are several practical approaches to help designers determine appropriate network architecture for their applications:

- *Network pruning:*[17] In network pruning, the "right" size of a neural network is determined by examining the sensitivity of neurons' output within and between layers. For example, consider a neural network that predicts the class of incoming signals. If the change of the output of a hidden neuron has no effect on the classification results, then this neuron can be removed without affecting the network performance. In general, a network structure that produces satisfactory approximation is initially created. Redundant neurons are then identified through sensitivity analysis and eliminated in accordance with certain pruning rules.
- *Resampling approach:*[11,19] With the resampling approach, several networks of different complexity (e.g., different number of hidden neurons and/or different number of hidden layers) are created, and trained with the same set of data points. The approximation errors are collected. These networks are then tested with a different set of data through cross-validation to obtain generalization error (i.e., validation error). The *best* network architecture is determined by comparing and making trade-off between the approximation error and the validation error.

1.3   Typical Engineering Applications of ANN

As mentioned in Section 1.1, ANN has gained popular applications in many fields. To help engineers develop a better understanding of this technique relevant to their problem-solving practice, our introduction in this section focuses on the three types of applications that are most popular and provide good insight for new users: *predictive modeling, system/process control*, and *optimization*. The discussion emphasizes the nature of these applications and related modeling methods that can be generalized to many specific cases.

1.3.1 Predictive Modeling

The task of predictive modeling is to use a neural network as a model to predict the characteristics (e.g., attribute or magnitude) of some output of interest. There are two general steps in predictive modeling. First a neural network is developed and trained to capture the (unknown) dependency relationship between input and output variables. The trained neural network is then applied on new incoming data to predict system/process output(s). MLP and RBF networks are commonly used for predictive modeling.

There are two types of predictions that are usually made by neural networks: *classification* and *regression*.[19] In classification, the interest is to identify the underlying class to which a set of input patterns belongs. Examples include pattern recognition and image processing. For instance, an ANN can be trained to capture the distinct characteristics embedded in a set of target images, and then it can be applied to recognize incoming images that contain certain noises (i.e., to identify a target class to which an incoming image belongs). In the case of regression, the task is to predict the value of a continuous output variable given a set of input values. For example, in food extrusion process planning, a neural network can be applied to learn the process knowledge (i.e., the complex relationship between a set of system/process variables and a set of output variables), and use this knowledge to predict the texture properties of extruded food products.[25,27] In another case, a neural network was developed and trained with a set of experimental failure data, and then used to predict the damage rate in distribution packaging operations.[26]

1.3.2 System/Process Control

Neural networks have been used in the development of intelligent controllers.[9,11,18] In such applications, a neural network is first developed to learn the dynamics of the process/system under control, and then uses this knowledge to map the relationship between control signals and system response/output. In comparison with other control paradigms, neural controllers have certain advantages: they are able to learn (in real time) and generalize to unseen situations; and they are able to represent almost any nonlinear relationship between control variables and system output.

There are generally two types of neural control systems: supervised control vs. unsupervised control. In supervised control, a teacher (e.g., a human or an automatic controller) is present. The system output is fed back to both a teacher and a neural network. The teacher supplies desired control signals that are compared with signals produced by the neural network. The errors captured are used for training the weights of the neural network. After successful training, the network replaces the teacher in controlling the system. A straightforward application of supervised control is training an ANN to model an existing controller (Figure II.1.5).

Figure II.1.6 shows an example of unsupervised neural control. In this example, a neural network is trained to capture the inverse dynamics of the system under control, i.e., it represents an inverse model of the system. The network predicts process input u that corresponds to the desired output Y_d. The error signals, $Y - Y_d$, are captured and used in the learning process of the network. The controller learns continuously and is able to control the system with time-varying characteristics.

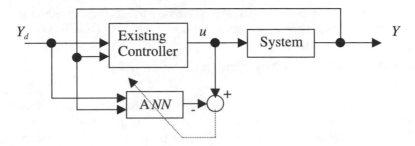

FIGURE II.1.5   Supervised control: modeling existing controller with neural network.

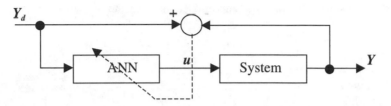

FIGURE II.1.6   Unsupervised control: direct inverse neural control.

1.3.3   Optimization

Recurrent neural networks (RNN) have been used in solving many kinds of optimization problems (e.g., linear programming, nonlinear programming, combinatorial optimization).[5] This type of neural network is based on an important model of artificial neurons called a "Hopfield model."[5,11] In a Hopfield network (Figure II.1.3), a neuron's dynamic is represented through a system of differential equations such as:

$$\tau \frac{du}{dt} = -au + w\varphi(u) + \theta \tag{II.1.10}$$

where u can be an input vector; w the set of trainable weights; and $\varphi(u)$ the transfer function. Given an initialization with a set of inputs, this dynamic system tries to converge to a set of stable states (equilibrium states), or equivalently to minimize a computational energy function $E(\cdot)$ (i.e., Liapunov function) by simultaneously adjusting its synaptic weights w. The stable states of this dynamic system correspond to the local minima of energy function $E(\cdot)$ associated with the network. Therefore, the process of converging to equilibrium states (of a Hopfield network) is equivalent to a minimization process of computational energy function $E(x)$. The Hopfield network can be easily implemented through analog modeling (e.g., through VLSI technology) in application. When developing such networks for most optimization problems (e.g., linear/nonlinear programming, combinatorial optimization), an important step is to construct an appropriate energy function $E(x)$. One common strategy is to use a penalty approach, i.e., relaxing an original constrained problem (II.1.11) into an unconstrained one (II.1.12) by including penalty terms in the objective function.

$$\text{Minimize } f(x) \text{ subject to } a_i(x) = 0 \text{ for } i = 1, \ldots, m \tag{II.1.11}$$

$$\text{Minimize } E(x) = f(x) + \sum_{i=1}^{m} \lambda_i P_i(a_i(x)) \tag{II.1.12}$$

The penalty terms $P_i(a_i(x))$ generate high cost if any constraint is violated. The computational process (i.e., the converging process of the neural network) tries to maintain a proper balance between the minimization of cost function and maximization of the number of constraints that are satisfied.

Among the difficult combinatorial optimization problems to which the ANN method has been applied include the quadratic assignment problem, graph partition problem, and traveling salesman problem.[5]

1.4   Basic Procedure: A Constructive Description

In this section, we describe a general constructive procedure for initialization, design, and implementation of a neural network application. Although the domain area of engineering application can be different from case to case, there are certain generic steps in planning and implementing an ANN application project. These steps are generic since they have been used by numerous successful applications. Although these steps are described in an "ordered" manner below, some steps are or can be implemented simultaneously. In general, there are nine steps in an ANN application:

- Define the problem and objectives
- Collect and preprocess data
- Design and construct network(s)
- Select learning strategies
- Train network(s)
- Verify and validate trained networks
- Select the best network architecture
- Apply the network to application
- Update/maintain the network

We discuss each of these steps in detail as follows.

1.4.1   Define the Problem and Objectives

In defining a problem, the following three questions should be answered in sequence:

What is the problem? To answer this question, an investigator has to start with understanding the nature of an application (e.g., a process or a system), obtaining relevant domain knowledge and information, and using them to identify the need and interest that lead to the definition of a problem.

Is the problem suitable for ANN? Once the need/interest for solving an application problem is identified, the investigator must determine whether it is suitable to apply ANN technique to solve the problem. This requires not only the understanding of problem domain and solution requirements, but also the knowledge and experience about different solution methods. A cost-benefit analysis is usually necessary.

What is the definition of the problem? A good problem definition includes not only a set of clearly stated objectives (e.g., expected solutions), but also a set of identified input/output variables, a definition of "environment" (i.e., "system" or "process"), and a set of constraints (e.g., conditions/boundaries of system or process variables). It is a challenging task to identify the set of input variables. These should be the system or process variables that have significant impact on the response or output variable(s). Unimportant variables should not be included in a neural network model. Expert opinions are often used to identify influential factors. But a more scientific approach such as *design and analysis of experiments* (DAE)[15] is recommended when the number of factors is large and their interactions present. By analyzing the variance contributions from the data collected through controlled experiments, DAE can efficiently and accurately identify the set of influential factors and their interactions, and therefore provide a right set of input variables for neural network modeling.

1.4.2 Collect and Preprocess Data

The decision to be made in this stage is "what data to collect and how?" There are two types of data that can be collected for an ANN study: *production data* and *experiment data*. The production data are observations collected in an uncontrolled environment, e.g., sensor readings from one shift of a production line. The experiment data, on the other hand, are collected through experimental design procedures (controlled data collection).[15] It is often necessary to preprocess the raw data collected through experiment or production runs. This includes tasks such as handling missing values and outliers. Scaling of raw input data is usually necessary to reduce the bias caused by different measuring units of original input variables. The most common type of scaling is normalization, i.e., each input value is scaled by its mean and standard deviation.[19]

1.4.3 Design and Construct ANN

The first task in design and implementation of an ANN is to select an appropriate software tool. Two options are (1) using a lower-level programming language or (2) using a higher-level neural network "shell." Using a lower-level language (e.g., C or C ++) has the advantage of flexibility, customization, and (possibly) execution efficiency. But it requires deeper knowledge and skill in computer programming. Commercially available ANN shells have become popular during the last decade as computer technology (especially software package) made significant progress. These shells provide highly modular commands and user-friendly interface that have turned the work of developing and training neural networks from a formidable job into an "easy" one. Tremendous gain on development efficiency has been achieved through the use of these shell languages.

The second step in design of an ANN is to select the type of network (perceptrons? MLPs? RBFs? Recurrent networks?). It is usually determined based on the type of the task to be performed by the proposed neural network. For instance, MLPs and RBFs are commonly used for prediction (including classification and regression), while RNNs have been preferred for optimization. The important decisions on the design of the most popular MLPs are given below:

- Design of input layer/nodes: Usually the number of input nodes is set equal to the number of input variables.
- Design of hidden layer(s)/nodes: The number of hidden layers and the number of neurons in each hidden layer are usually determined by a trial-and-error approach. The best size is usually determined through resampling or network pruning. Select the type of transfer functions for all hidden neurons.
- Design of output layer/node(s): Usually the number of output nodes is set equal to the number of output variables. Select the type of transfer functions for the output neurons.

Essentially, two things must be determined: the number of nodes in each of these layers and the type of transfer functions of neurons in hidden and output layers.

1.4.4 Select Learning Strategies

Learning strategy affects the efficiency and quality of network training. It is also a difficult decision for novel users of ANN. Fortunately most software shells usually provide a number of implemented learning strategies for users to choose. These independent modules can implement different learning methods as specified by a user. As an example, MATLAB© Neural Network ToolBox[6] provides users with eleven options of learning models, including simple backpropagation (gradient descent), adaptive learning rate, gradient descent with momentum, conjugate gradient algorithm, quasi-Newton algorithm, etc. There is, however, a trade-off between faster algorithms and cost of resources. Faster algorithms in general require more computing resource (e.g., memory/storage space, CPU time). It is often recommended for a user to try several learning methods through pilot runs, and then select the best one for application.

1.4.5 Train Neural Networks

There are several issues in training an ANN. First a decision needs to be made regarding an appropriate partition of input/output data sets prepared for a study. Total data sets (or "data points") are usually partitioned into two parts: a training set T and a validation set V. The training set T contains data points that are used to train the network, while the validation set V contains data points that are used to "validate" the performance of trained networks. The issue is how many data points should be allocated for training and how many for validation. Again, there is no hard rule here to follow, but there are many empirical rules of thumb.[11,21] For instance, many applications have simply used a fixed ratio rule, such as 70–30 (70% for training and 30% for validation) or 60–40. The partition also depends on the method of validation (see next subsection). In general, given the structure of a neural network, there must be enough training data to adapt all free parameters (link weights) so that the network can mimic well enough the behavior of the underlying distribution from which the set of training/validation data was taken.

1.4.6 Validate Trained ANN

After an ANN is properly trained, it needs to be validated on its performance of generalization, i.e., using the knowledge learned from the training to perform the required task with "unseen" data inputs. For instance, we may take a trained MLP and feed it with data points that are not used during the training process to see whether it can predict well at these "unseen" data points. The error captured by testing the neural network with the unseen data points is the error of generalization. A neural network that generalizes poorly (i.e., with large validation error) usually must be re-designed in terms of network architecture. As mentioned earlier, there are two common approaches for validation: simple validation and cross-validation.

Simple validation involves two steps: feeding a trained network with the data points from validation set V, and calculating the validation error V_e, which is usually defined as an average error over all validation points, e.g.:

$$V_e = \sum_{i=1}^{K} \frac{V_{ei}}{K} \tag{II.1.13}$$

where K = size of validation set, and $V_{ei} = |O_{pi} - O_{ai}|$ for $i = 1, \ldots, K$. Quantity O_{pi} is a predicted value, while O_{ai} is the actual value from set V. Cross-validation[11,19] is an effective method when the size of collected input/output data sets is very limited by the nature of an application. Let $m = |T|$ = size of training set, and $n = |V|$ = size of validation set. With cross-validation, a neural network is repeatedly trained with m (out of total $m + n$) data sets, and tested with the rest n (out of total $m + n$) data sets. The validation error is calculated as an average across all $(m+n)!/(n!)(m!)$ test runs. (Note that symbol $m!$ denotes a mathematical operation: the factorial of m, i.e., $1 \times 2 \times \cdots \times m$.) For example, suppose that there are only nine data points in total. We make a partition of 8-to-1, i.e., using eight points for training and one for validation each time. There are a total of $9!/(1!)(8!) = 9$ different 8-to-1 partitions (i.e., we need to run nine times). This special case is also referred to as the "leave-one-out" approach.[19]

1.4.7 Select the Best ANN

The best neural network depends on application-specific criteria used to define what is meant by "best." Usually it is the network that produces the minimum validation error (i.e., one that achieves the best generalization results), and can be implemented efficiently with available computing resource.

1.4.8 Apply the ANN

Once a neural network is trained and validated, it can then be applied to solving the intended application problem. In many cases, this means to generalize the knowledge of the neural network to unseen problem instances. For example, in one application, engineers used a neural network, trained with a set of field data, to predict the roll force in a cold mill production system.[4] An important thing to always keep in mind is that an ANN is developed and trained with a *finite set* of data (e.g., many applications used only a dozen sets of input/output data points) due to the cost and difficulty in data collection. This finite set of data provides only limited information about the underlying distribution that governs the true dependency relationship between system inputs and outputs.[19] This limitation inherently restricts the generalization ability of an ANN. It is not uncommon to see some "strange" (unexpected) value occurred during the application of ANN in practice. Our advice is to conduct an analysis of causes, or validate the results from ANN with a designed experiment (if affordable) or other techniques.

1.4.9 Update/Maintain ANN

Although neural networks have been commonly referred to as "universal approximators," they are inherently limited by their structures and the finite data used in training and validation. As sensor technology is making significant progress (e.g., on-line measuring becomes more and more affordable and popular), more data will become available. This is good news for the improvement of a developed ANN, since some of the new data may come from a region (of the underlying data distribution) that was not represented (or poorly represented) by the original training/validation data. Apparently the new knowledge embedded in such new data should be captured and added to the neural network to improve its performance (generalization) and reusability. Updating ANN can be done either on-line or off-line.

1.5 Recent Advancement: Neuro-Fuzzy and Neuro-Genetic Algorithm Applications

The recent advancement in the research and application of ANN has shown a trend of combining this technique with other artificial intelligent or computing techniques for better knowledge representation and computation. This section provides a very brief discussion on the "fusion" of neural network, fuzzy set, and fuzzy logic (FSFL), and genetic algorithm (GA). The intention is to show that, in many applications, by combining these techniques one can gain much more benefit than by just using a single technique alone. It is impossible in this chapter to give even a "fairly brief" introduction on FSFL or GA, due to the space limitation. Readers who are interested in knowing more on FSFL and/or GA are encouraged to read some introductory materials on these subjects.[7,8,13,22,23]

Although ANNs, FSFL, and GA have been successfully used in solving difficult engineering problems and gained wide recognition, each of these techniques has certain limitations or disadvantages. Neural networks can effectively learn the input-output relationship of a complex system, and generalize the learned relationship to process new signals. However, neural networks have been criticized for their "black box" nature, i.e., the knowledge representation of a neural network is not explicit to users.[1,3,25] The relationship captured by a neural network is embedded among its structural parameters (e.g., link weights) and is difficult for humans to understand.[1,3]

FSFL, on the other hand, can effectively model qualitative or subjective knowledge, and represent complex relationship (knowledge) in a logical form that can be easily understood by humans. Particularly, FSFL can capture the impreciseness and vagueness of qualitative knowledge by using linguistic variables (i.e., fuzzy sets) and a logic that fits naturally the human way of thinking. In

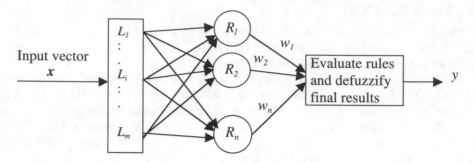

FIGURE II.1.7   A neural-fuzzy system.

fact, human logic is of fuzzy nature, not in black/white (B/W). For example, consider classifying an object's weight with 40 lbs as a threshold. "B/W" logic would simply classify any object as "heavy" if it weighs more than 40 lbs, and "not heavy" if less than 40 lbs. However, a human's belief on an object with 39.9 lbs and another with 5 lbs is certainly different! One is very close to "heavy" and the other is far from "heavy," although they would fall under the same category under B/W logic.

Fuzzy logic[13] captures such belief by modeling each attribute using a membership function, and associating each object with a membership function through a degree of belief (i.e., membership grade). Decision rules can be formed with fuzzified antecedents and consequents to represent logical relationship between input-output variables. For example, "IF (Temperature *is* TOO-HIGH) THEN (Fan_Speed *is* VERY-FAST)". Here TOO-HIGH and VERY-FAST are linguistic values (fuzzy sets) of fuzzified variable Temperature and Fan_Speed.

When system complexity increases, it becomes very difficult to specify the right membership functions and select the right rules to model and predict system behavior. Combining neural network with fuzzy logic maximizes the advantages of both techniques while it minimizes their drawbacks, and improves learning and performance of intelligent systems. Therefore it opens a much wider area for engineering application. There are several approaches of combining fuzzy logic and neural network,[12] such as (1) using fuzzy logic for designing neural networks and (2) using neural network for designing fuzzy logic systems.

One popular type of neural-fuzzy system for control application is shown in Figure II.1.7. When an input vector x comes in, it is fuzzified (i.e., converted into fuzzy variables) to match with the closest linguistic descriptor (i.e., a fuzzy set, say, Li = TOO-HIGH). The fuzzified signals are then propagated to a hidden layer that computes the activation degree for each decision rule $R_i (1 \leq i \leq n)$. The last layer of the network evaluates the conclusion of each rule based on its activation degree and weight, and produces and "defuzzifies" a final output y. During training process, the error signals between inferred output and measured output (teaching value) are used to adjust the weights $w_i (1 \leq i \leq n)$ in accordance with a gradient descent approach.

Genetic algorithm (GA) is a stochastic optimization technique that performs global search and does not require an analytical modeling structure. It imitates the natural evolution of biological systems. By allowing worse solutions to compete with good ones for survival, generation by generation, GA continuously improves solutions and eventually identifies the best ones over entire solution space. In GA, a solution is coded as a "chromosome." The algorithm starts with a set of randomly initiated solutions called a population. The chromosomes evolve through successive generations. During each generation, the chromosomes are evaluated by an objective function commonly referred to as fitness function.[8] New chromosomes, called offspring, are formed by either crossover operation or mutation operation. A new generation is formed by a selection operation that chooses candidates from present generation (parents) and offspring. Fitter chromosomes (i.e., ones with higher fitness values) have higher probabilities of being selected. The evolution continues until either a desired solution is obtained or a predetermined generation size is reached.[7,8]

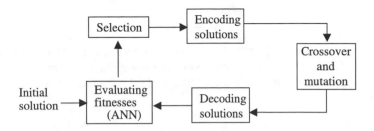

FIGURE II.1.8 A NN-GA-based system for optimizing food extrusion process.

Combining ANN with GA can provide an alternative way for solving design and optimization problems.[12] A trained ANN can be used as a function estimator to evaluate the fitness (e.g., cost) of chromosomes (solutions). The process involves two phases. First, a neural network is developed and trained to capture and map the nonlinear relationship between input and output variables. The trained network is then used as a fitness function to evaluate solutions that evolve generation by generation to approach specified optimal conditions. One application that combines neural network and genetic algorithm was in the optimization of process conditions for food extrusion processes.[27] The system used a trained neural network as a fitness function to predict process output value for each set of input values (Figure II.1.8). An initial set of solutions (first generation) is randomly generated and fed to the system. A neural network (ANN) then computes the fitness values of each solution in the generation. To create a new generation, the selection block selects "good" solutions from the current generation according to the Darwinian principle, i.e., the solutions with larger fitness values have a better chance of being chosen. Newly selected solutions are then encoded in "genotype" form (i.e., chromosomes). The encoded solutions are randomly selected for crossover and mutation operations. The new generation is then decoded in "phenotype" form (original solution form) and sent to the fitness function block for evaluation. The process is repeated until either a maximum iteration limit is met or an acceptable optimal solution is found.

1.6   Conclusions

As presented in earlier sections, artificial neural network has gained popular applications in science and engineering, and achieved remarkable success. However, it is important to realize that, like any other technique, ANN has certain limitations, e.g., its black box nature in knowledge representation. It is also important for novel practitioners to know that using only ANN alone in solving many engineering problems is subject to a higher risk. It is always recommended to use, if possible, more than one tool to validate the results. For instance, it is better to use statistical method/analysis to validate the performance of neural networks (e.g., in classification and regression). Another important suggestion is to use the "right" neural network model. This requires not only the knowledge about neural network theory and practice, but also a good understanding of application domain. In practice, much more time is spent in developing a good conceptual model that really represents the system/process being studied through sound communications with domain experts (e.g., process engineers). Finally, the materials presented in this chapter are intended to provide readers with a very brief introduction of the subject. For anyone who has no experience with neural networks and who is serious about using the technique it is highly recommended to consult with experts and/or study more advanced material on the subject.

Acknowledgment

The author would like to thank the Kluwer Academic Publishers and in particular, Drs. S.G. Tzafestas and C.S. Tzafestas, for their kind permissions to use some original work from their publication. The author would also like to thank Dr. G. Cai, from IBM, for his kind help in reviewing and proofreading this manuscript.

References

1. Andrews, R., Diederich, J., and Tickle, A.B., Survey and critique of techniques for extracting rules from trained neural networks, *Knowledge Based Systems*, 8(2), 374, 1995.
2. Brown, M. and Harris, C., *Neurofuzzy Adaptive Modeling and Control*, Prentice Hall, New York, 1994.
3. Browne, A., Ed., *Neural Network Analysis, Architecture and Applications*, IOP Publishing Ltd., London, UK, 1997.
4. Cho, S., Cho, Y., and Yoon, S., Reliable roll force prediction in cold mill using multiple neural network, *IEEE Trans. Neural Networks*, 8(4), 874, 1997.
5. Cichocki, A. and Unbehauen, R., *Neural Networks for Optimization and Signal Processing*, John Wiley & Sons, Inc., New York, 1993.
6. Demuth, H. and Beale, M., *Neural Network Toolbox, For Use with MATLAB*, The Math Works, Inc., Massachusetts, 1998.
7. Gen, M. and Cheng, R., *Genetic Algorithms and Engineering Design*, John Wiley & Sons, Inc., New York, 1997.
8. Goldberg, D.E., *Genetic Algorithms in Search, Optimization, and Machine Learning*, Addison-Wesley, Reading, Massachusetts, 1989.
9. Gulpta, M.M. and Sinha, N.K. Ed., *Intelligent Control Systems, Theory and Applications*, IEEE Press, New York, 1996.
10. Hagan, M.T. and Menhaj, M.B., Training feedforward networks with Marquardt Algorithm, *IEEE Trans. Neural Networks*, 5(6), 989, 1994.
11. Haykin, S., *Neural Networks, A Comprehensive Foundation*, Macmillan College Publishing Company, New York, 1994.
12. Jain, L.C. and Martin, N.M., *Fusion of Neural Networks, Fuzzy Sets, and Genetic Algorithms*, CRC Press LLC, New York, 1999.
13. Klir, G.J. and Yuan, B., *Fuzzy Sets and Fuzzy Logic: Theory and Applications*, Prentice Hall PTR, Upper Saddle River, New Jersey, 1995.
14. Mars, P., Chen, J.R., and Nambiar, R., *Learning Algorithm, Theory and Applications in Signal Processing, Control and Communications*, CRC Press, Inc., Boca Raton, Florida, 1994.
15. Montgomery, D.C., *Design and Analysis of Experiments*, 4th ed., John Wiley & Sons, Inc., New York, 1997.
16. Petri, K.L., Billo, R.E., and Bidanda, B., A neural network process model for abrasive flow machining operations, *J. Manufact. Syst.*, 17(1), 52, 1998.
17. Sietsma, J. and Dow, R.J.F., Neural net pruning – why and how, in *Proc. IEEE Int. Conf. Neural Networks*, 1998.
18. Tzafestas, S.G. Ed., *Methods and Applications of Intelligent Control*, Kluwer Academic Publishers, Dordrecht, Netherlands, 1997.
19. Vladimir, C. and Mulier, F., *Learning From Data: Concepts, Theory, and Methods*, John Wiley & Sons, Inc., New York, 1998.

20. Vogl, T.P. et al., Accelerating the convergence of the back-propagation method, *Biol. Cybernetics*, 59, 257, 1988.

21. Winston, P.H., *Artificial Intelligence*, Addison-Wesley Publ. Co., New York, 1992.

22. Zadeh, L.A., Fuzzy sets, *Information and Control*, 12(2), 94, 1965.

23. Zadeh, L.A., Outline of a new approach to the analysis of complex systems and decision processes, *IEEE Trans. Systems, Man, and Cybernetics*, 1(1), 28, 1973.

24. Zbigniew, M., *Genetic Algorithms + Data Structures = Evolution Programs*, Springer-Verlag, New York, 1992.

25. Zhou, M. and Paik, J., Integrating neural network and symbolic inference for predictions in food extrusion process, in *Proc. 13th Int. Conf. Indust. Engineer. Appl. Artificial Intelligence & Expert Systems*, New Orleans, Louisiana, 567, 2000.

26. Zhou, M. and Paik, J., Damage prediction using neural networks, *Int. J. Indust. Engineer.*, 7(2), 140, 2000.

27. Zhou, M., A neural network and genetic algorithm based approach for optimization of food extrusion process parameters, in *Proc. 10th Indust. Engineer. Res. Conf.*, Dallas, Texas, 2001.

2

Recurrent Neural Fuzzy Network Techniques and Applications

Chia-Feng Juang
National Chung Hsing University

2.1 Introduction

Over the past few years there has been a considerable interest in applying neural or neural fuzzy networks to the solving of problems in every area. This interest has included the areas of control, communication, and pattern recognition, etc.[1] Recently, interest in using these networks for solving temporal problems has been increasing. These types of networks are usually categorized as recurrent networks. Unlike the feedforward neural network whose output is a function of its current inputs only and is limited to static mapping, recurrent neural networks perform dynamic mapping. For problems where at least one system state variable exists which cannot be observed, recurrent networks are needed. Some commonly seen network structures are the simple recurrent networks,[2,3] whose feedback structure comes from feeding the hidden layer output back to the network input; the sequential neural network,[4] whose feedback comes from both the network output and the input itself; and the fully recurrent neural network,[5] where all nodes are fully connected.

Other special structures include the second-order recurrent neural network,[6] the block-structured recurrent neural network,[7] the memory neural network,[8] and the recurrent radial basis function network.[8] Basically, these recurrent neural networks are obtained by adding trainable temporal elements to feedforward neural networks (like multilayer perceptron networks[10] and radial basis function networks[9,11]) to make the output history-sensitive. Like feedforward neural networks, these networks function as black boxes; we do not know the meaning of each weight and node in these networks. Recently, the concept of incorporating fuzzy logic into a recurrent network was proposed.[12–19] In Gorrini and Bersini,[12] the concept of recurrent fuzzy rules is proposed, while in Omlin,

Thomber, and Gilies,[18] the fuzzy finite machine is encoded into a recurrent network, and in Unal and Khan,[19] a neural fuzzy network is implemented as a fuzzy finite machine. Since the neural fuzzy networks have many advantages over the feedforward neural networks, it seems worth constructing a recurrent network based on a neural fuzzy network. In this chapter, we propose such a *recurrent neural fuzzy network*. The proposed network will possess the same advantages over the pure recurrent neural networks, and extend the application domain of the normal neural fuzzy networks to temporal problems.

The recurrent neural fuzzy network introduced in this chapter is called the recurrent self-organizing neural fuzzy inference network (RSONFIN).[20] The RSONFIN expands the basic ability of a neural fuzzy network to cope with temporal problems via the inclusion of some internal memories, called *context elements*. In the perspective of fuzzy logic, these context elements are expressed in the form of internal fuzzy reasoning. More clearly, with these context elements, the network performs the following reasoning:

$$\text{Rule } i : \text{IF } x_1(t) \text{ is } A_{i1} \text{ and } \ldots \text{ and } x_n(t) \text{ is } A_{in} \text{ and } h_i(t) \text{ is } G$$
$$\text{THEN } y_1(t+1) \text{ is } B_{i1} \text{ and } h_1(t+1) \text{ is } w_{1_i} \text{ and } \ldots \text{ and } h_m(t+1) \text{ is } w_{mi},$$

where x_i is the input, y_1 is the output, A_{i1}, A_{in}, G and B_{i1} are fuzzy sets, h_i is the internal variable, w_{i1} and w_{mi} are fuzzy singletons, and n and m are the numbers of input and internal variables, respectively. The dynamic reasoning implies that the inference output $y_1(t+1)$ is affected by the internal variable $h_i(t)$, and the current internal output $h_i(t+1)$ is a function of previous output value $h_i(t)$; i.e., the internal variable h_i itself forms the dynamic reasoning. To apply RSONFIN to different learning problems, the RSONFIN design with Supervised learning environment (RSONFIN-S) and RSONFIN designed by Genetic algorithm (RSONFIN-G) are proposed.

To reduce the network design effort, the automatic adaptation of the network topology to the learning task is a tendency.[21–25] In contrast to other neural fuzzy networks, where the network structure is fixed and the rules should be assigned in advance, there are no rules initially in the RSONFIN-S; all of them are constructed during on-line learning. Two learning phases, the structure as well as parameter learning phases, are used to accomplish this task. The structure learning phase is responsible for the generation of fuzzy if-then rules as well as the judgment of feedback configuration, and the parameter learning phase for the tuning of free parameters of each dynamic rule (like the shapes and positions of membership functions and the singleton values). In the structure learning phase, since the way the input space is partitioned strongly affects the number of rules generated, an efficient partition scheme to reduce the number of rules is required.

In RSONFIN-S, an aligned-clustering-based partition scheme is proposed. This scheme partitions the input space in a flexible way and a fuzzy measure scheme is performed on each input dimension to eliminate the unnecessary terms during on-line learning. For the consequent-part identification of structure learning, a clustering-based scheme is proposed. Based on this scheme, the context elements are on-line generated, and then the whole network is constructed. For parameter learning, a recursive learning algorithm is developed based on the ordered derivative scheme.[32] This algorithm can tune the free parameters in the precondition and consequent part of fuzzy rules, and weights of feedback connections simultaneously to minimize an output error function. All of these processes are done on-line, so the network can be used for normal operation at any time as learning proceeds.

RSONFIN-S is applied to problems where supervised learning data is available. For problems where supervised input-output training pairs are unavailable or expensive to obtain, genetic algorithm is applied to recurrent networks design. Genetic algorithms (GAs) are stochastic search algorithms based on the mechanics of natural selection and natural genetics. Because GAs do not require or use derivative information, one appropriate application for their use is the circumstance where gradient information is unavailable or costly to obtain. GAs also provide a means to optimize ill-defined irregular problems, and can be tailored to the needs of different situations. Owning robust and excellent search properties, GAs have been applied to the design of fuzzy networks.

Karr[33] applied GAs to the design of the membership functions of a fuzzy controller, with the fuzzy rule set assigned in advance. Since membership functions and rule sets are co-dependent, simultaneous design of these two approaches would be a more appropriate methodology. Based upon this concept, many researchers have applied GAs to optimize both the parameters of the membership functions and the rule sets. Differences between the approaches lie mainly in the type of coding and the way the membership functions are optimized. In Homarfar and McCormick,[34] the authors proposed the use of a GA to simultaneously design membership functions and rule sets in a fuzzy controller. In Lee and Takagi,[35] the authors proposed the use of GAs to simultaneously optimize the rule base of a TSK type fuzzy controller, including membership functions, the number of rules, and consequent parameters.

Since, in a supervised learning environment, the performance of a recurrent neural fuzzy network has shown to be superior to that of a recurrent neural network, for GA-based recurrent network design problem, the adoption of a recurrent fuzzy network GA seems to be the better choice. Based upon this motivation, RSONFIN-G is proposed. Unlike other GA-based fuzzy design approaches, where the input space is partitioned in a grid-type,[33,34] the precondition part of the spatial fuzzy rules in RSONFIN-G is partitioned in a flexible way, and is designed concurrently with the consequent parts of both the spatial and temporal rules. The supervisory of RSONFIN-G will be verified by comparing its performance to its recurrent neural network counterpart, whose feedback structure also comes from feeding the hidden layer output back to the input.

Overall, the advantages of the RSONFIN against other recurrent network models[10-19] are summarized as follows:

1. Unlike other recurrent network models where the network structure is a normal neural network and functions as a black box, the RSONFIN is a fuzzy inference network. Each node and weight has its own meaning and functions as an element in a fuzzy reasoning process.
2. For most recurrent network models, the user has to specify the network structure in advance. However, for the RSONFIN-S, no pre-assignment of the network structure is required, since the RSONFIN-S can on-line construct itself automatically.
3. RSONFIN-G improves the design accuracy when compared with other recurrent neural networks.
4. As will be shown in Section 1.5, the RSONFIN is characterized by small network size and fast learning speed.

This chapter is organized as follows. Section 1.2 describes the structure and functions of the RSONFIN. The on-line structure/parameter learning algorithm of the RSONFIN-S is presented in Section 1.3, which contains four parts: the input/output space partitioning, fuzzy rule construction, feedback structure identification, and parameter learning. In Section 1.4, the design of RSONFIN-G is presented. In Section 1.5, the RSONFIN is applied to solve several dynamic problems including the dynamic identification, and dynamic plant control. Comparisons with some existing recurrent neural networks are also made. Finally, conclusions are made in the last section.

2.2 Structure of the RSONFIN

In this section, the structure of the RSONFIN shown in Figure II.2.1 is introduced. The RSONFIN consists of nodes, each with some finite fan-in of connections represented by weight values from other nodes and some fan-out of connections to other nodes. Basically, it is a six-layered neural

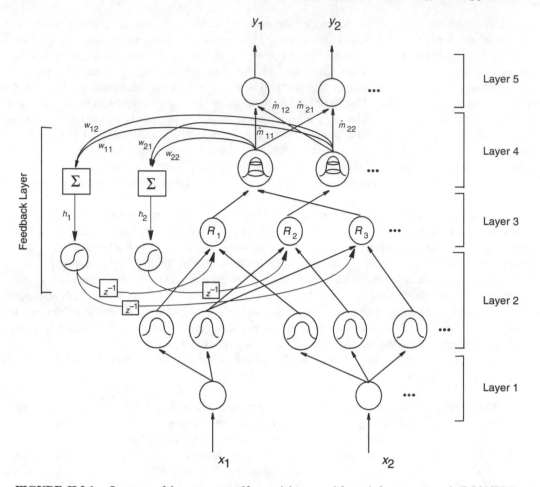

FIGURE II.2.1   Structure of the recurrent self-organizing neural fuzzy inference network (RSONFIN).

fuzzy network embedded with dynamic feedback connections (the feedback layer in Figure II.2.1) that brings the temporal processing ability into a feed-forward neural fuzzy network.

Nodes in layer 1 are input nodes and perform direct transmission of the input variables to the next layer. Nodes in layer 2 are called input term nodes and act as membership functions to express the input fuzzy linguistic variables. Inputs to layer 2 come from the external variables x. A local membership function, the Gaussian membership function is adopted, in which the mean value and standard variation are adjustable through the learning process. Here, a Gaussian membership function is employed because a multidimensional Gaussian membership function can be easily decomposed into the product of one-dimensional membership functions. Each node in layer 3 is called a rule node and represents a recurrent fuzzy rule. Unlike other clustering-based precondition part partition methods, the number of rule nodes in this layer is not necessarily equal to the number of fuzzy sets corresponding to each external linguistic input variable. Links to each node represent the preconditions of it. Each rule node performs precondition matching of a fuzzy rule. The fan-in of a node comes from two sources: one from layer 2 and the other from the feedback layer. The former represents the rule's spatial firing degree, and the latter the rule's temporal firing degree. Layer 4 is called the consequent layer, and the nodes in this layer are called output term nodes. Each output term node represents a multidimensional fuzzy set (described by a multidimensional Gaussian function) obtained during the clustering operation in the structure learning phase. Each node in layer 5 is called an output linguistic node and corresponds to one output linguistic variable. This layer performs the defuzzification operation. The nodes in this layer together with the links attached to them accomplish this task. For the feedback layer, it calculates the value of the internal variable h and the firing strength of the internal variable to its corresponding membership function, where the firing strength contributes to the matching degree of a rule node in layer 3. As shown in Figure II.2.1, two types of nodes are used in this layer, the square node named as context node and the circle node named as feedback term node, where each context node is associated with a feedback term node. The number of context nodes (and thus the number of feedback term nodes) are the same as that of output term nodes in layer 4. Each context node and its associated feedback term node corresponds to one output term node. The inputs to a context node are from all the output term nodes, and the output of its associated feedback term node is fed to the rule nodes whose consequent is the output term node corresponded to this context node.

To give a clear understanding of the network, functions of RSONFIN are described layer by layer below. For notation convenience, the net input to the ith node in layer k is denoted by $u_i^{(k)}$ and the output value by $a^{(k)}$.

Layer 1: No computation is done in this layer. The node only transmits input values to the next layer directly. That is:

$$a^{(1)} = u_i^{(1)} \tag{II.2.1}$$

Layer 2: The operation performed in this layer is:

$$a^{(2)} = \exp\left\{ -\frac{(u_i^{(2)} - m_{ij})^2}{\sigma_{ij}^2} \right\} \tag{II.2.2}$$

where m_{ij} and σ_{ij} are, respectively, the center (or mean) and the width (or variance) of the Gaussian membership function of the jth term of the ith input variable x_i. Neural fuzzy networks commonly partition the input space using a clustering method where each input variable has the same number of fuzzy terms; the number of fuzzy terms need not be identical for each input variable of a RSONFIN.

Layer 3: The following AND operation on each rule node is used to integrate its fan-in values:

$$a^{(3)} = a^{(6)} \cdot \prod_i u_i^{(3)} = a^{(6)} \cdot e^{-[D_i(x_i - m_i)]^T [D_i(x - m_i)]} \tag{II.2.3}$$

where $D_i = diag(1/\sigma_{i1}, \ldots, 1/\sigma_{in})$, $m_i = (m_{i1}, \ldots, m_{in})$ and $a^{(6)}$ is the output of the feedback term node, which will be described in the feedback layer part in this section. Obviously, the output $a^{(3)}$ of a rule node represents the firing strength of its corresponding rule.

Layer 4: In this layer, only the center of each Gaussian membership function is delivered to the next layer, so the width is used for output clustering only. Different nodes in layer 3 may be connected to a same node in this layer, meaning that the same consequent is specified for different rules. The function of each output term node performs the following fuzzy OR operation:

$$a^{(4)} = \sum_i u_i^{(4)} \tag{II.2.4}$$

to integrate the fired rules which have the same consequent part. The above fuzzy OR operation is a modified bounded sum operation in fuzzy theory. Again, its use is for computational convenience. For the same reason, it was also used in other neural fuzzy networks.[36,37] Although the use of simple summation as the fuzzy OR operation in Equation II.2.4 would give values larger than one, which are strictly speaking not fuzzy by definition, succeeding normalization in Equation II.2.5 makes the summation contributing to each output term smaller than or equal to one.

Layer 5: The function performed in this layer is:

$$y_j = a^{(5)} = \frac{\sum_i u_i^{(5)} \hat{m}_{ji}}{\sum_i u_i^{(5)}} \tag{II.2.5}$$

where $u_i^{(5)} = a_i^{(4)}$ and $\hat{m}_{ji}$, the link weight, is the center of the membership function of the jth term of the ith output linguistic variable.

Feedback Layer: The context node functions as a defuzzifier:

$$h_j = \sum_i a_i^{(4)} w_{ji} \tag{II.2.6}$$

where the internal variable h_j is interpreted as the inference result of the hidden (internal) rule, and w_{ji} is the link weight from the jth node in layer 4 to the ith internal variable. The link weight (w_{ji}) represents a fuzzy singleton in the consequent part of a rule, and also a fuzzy term of the internal variable h_j. For an internal variable, fuzzy singleton instead of fuzzy membership function is used as its fuzzy term; a fuzzy membership function on an internal variable does not make much sense in the network due to the use of LMOM defuzzification operation, where only the center of the Gaussian membership function is used. This is different from the situation for the input and output linguistic variables, where the widths of fuzzy membership functions are used for clustering the input and desired output training data. In Equation II.2.6, the simple weighted sum is calculated.[38,39] Instead of using the weighted sum of each rule's outputs as the inference result, the conventional average weighted sum, $h_j = \sum_i a_i^{(4)} w_{ji} / \sum_i a_i^{(4)}$, can also be used.[39,40]

As to the feedback term node, unlike the case in the space domain where a local membership function is used, a global membership function is adopted on the universe of discourse of the internal variable to simplify network structure and meet the global property of the temporal history. Here, the global property means that for a cluster in the space domain, its history path (memorized by the internal variables) can be anywhere in the space at different times, so a global membership function, which covers the universe of discourse of the internal variable, is used to rank the influence degree each internal variable contributes to a rule. In this paper, the membership function $f(u) = 1/(1+e^{-u})$ is used for each internal variable. With this choice, the feedback term node evaluates the output by:

$$a^{(6)} = \frac{1}{1 + e^{-h_i}} \tag{II.2.7}$$

This output is connected to the rule nodes in layer 3, which connect to the same output term node in layer 4. The outputs of feedback term nodes memorize the firing history of the fuzzy rules.

With the aforementioned node functions in each layer, the RSONFIN realizes the following dynamic fuzzy reasoning:

> Rule i : IF $x_1(t)$ is A_{i1} and $\ldots$ and $x_n(t)$ is A_{in} and $h_i(t)$ is G
>
> THEN $y_1(t+1)$ is B_{i1} and $y_2(t+1)$ is B_{i2}
>
> and $h_1(t+1)$ is w_{1i} and$\ldots$ and $h_m(t+1)$ is w_{mi},

where x_i is the input variable, y_i is the output variable, A_{i1}, A_{in}, G, B_{i1}, and B_{i2} are fuzzy sets, h_i is the internal variable, w_{1i} and w_{mi} are fuzzy singletons, and n and m are the number of input and internal variables, respectively.

From the above fuzzy rule, we can see the spatial mapping relationship between the input variable $x(t)$ and output variable $y(t)$, the influence of $h_i(t)$ on this spatial mapping, and the influence degree of each rule on other ones as explained as follows. Owing to the monotonic increasing property of membership function G_i, a higher value of h_i means a higher firing strength (or influence). From the value of $h_i(t + 1)$ in the consequent part of each rule, we may see the influence degree of the rule on other ones, or the influence of a spatial cluster on other clusters in the input region. For cases where *a priori* knowledge of the spatial mapping is clear and known, we may put the *a priori* knowledge in the dynamic rule with w in each rule setting to zero, meaning the influence degree of h_i to every rule is of the same value, and there is no temporal influence in the beginning. Further temporal relationship, the tuning of w, may be learned by succeeding parameter learning.

2.3 Supervised Learning Algorithms for the RSONFIN (RSONFIN-S)

Two types of learning, structure and parameter learning, are used concurrently for constructing the RSONFIN. The structure learning includes the precondition, consequent, and feedback structure identification of a dynamic fuzzy if-then rule. Here the precondition structure identification corresponds to the input space partitioning and can be formulated as a combinatorial optimization problem with the following two objectives: to minimize the number of rules generated and to minimize the number of fuzzy sets on the universe of discourse of each input variable. The consequent structure identification is to decide when to generate a new membership function for the output variable based upon clustering. As to the feedback structure identification, the main task is to decide the number of internal variables with its corresponding feedback fuzzy terms and the connection of these terms to each rule. For the parameter learning, based upon supervised learning, an ordered derivative learning algorithm is derived to update the free parameters in the RSONFIN-S. The RSONFIN-S can be used for normal operation at any time during the learning process without repeated training on the input-output patterns when on-line operation is required. There are no rules (i.e., no nodes in the network except the input/output linguistic nodes) in the RSONFIN initially. They are created dynamically as learning proceeds upon receiving on-line incoming training data by performing the following learning processes simultaneously:

1. Spatial Rule Construction.
2. Feedback Structure Identification.
3. Parameter Identification.

In the above, processes 1 and 2 belong to the structure learning phase and process 3 belongs to the parameter learning phase. The details of these learning processes are described in the rest of this section.

2.3.1 Spatial Rule Construction

The way the input space is partitioned determines the number of rules. Even though the precondition part of a rule in the RSONFIN includes the external inputs which represent the spatial information and the internal variable values which represent the temporal information, only the spatial information is used for clustering, due to its local mapping property.

Geometrically, a rule corresponds to a cluster in the input space with m_i and D_i representing the center and variance of that cluster. For each incoming pattern x, the strength a rule is fired can be interpreted as the degree the incoming pattern belongs to the corresponding cluster. For computational efficiency, we can use the spatial firing strength derived in Equation II.2.3 directly as this degree measure:

$$F^i(x) = \prod_i u_i^{(3)} = e^{-[D_i(x-m_i)]^T [D_i(x-m_i)]} \tag{II.2.8}$$

where $F^i(x) \in [0, 1]$. In the above equation, the term $[D_i(x - m_i)]^T[D_i(x - m_i)]$ is, in fact, the distance between x and the center of cluster i. Using this measure, we can obtain the following criterion for the generation of a new fuzzy rule. Let $x(t)$ be the newly incoming pattern. Find:

$$J = \arg \max_{1 \le j \le c(t)} F^j(x) \tag{II.2.9}$$

where $c(t)$ is the number of existing rules at time t. If $F^J \le \overline{F}_{in}(t)$, then a new rule is generated, where $\overline{F}_{in}(t) \in (0, 1)$ is a set threshold that decays during the learning process. Once a new rule is generated, the next step is to assign initial centers and widths of the corresponding membership functions. Since our goal is to minimize an objective function, and the centers and widths are all adjustable later in the parameter learning phase, it is of little sense to spend much time on the assignment of centers and widths for finding a perfect cluster. Hence, we can simply set:

$$m_{(c(t)+1)} = x \tag{II.2.10}$$

$$D_{(c(t)+1)} = \frac{-1}{\beta} \cdot diag(1/\ln(F^J) \ \dots \ 1/\ln(F^J)) \tag{II.2.11}$$

where $\beta \ge 0$ decides the overlap degree between two clusters. After a rule is generated, the next step is to decompose the multidimensional membership function formed in Equation II.2.10 to the corresponding one-dimensional membership functions for each input variable. For the Gaussian membership function used in the RSONFIN-S, the task can be easily done as:

$$e^{-[D_i(x-m_i)]^T[D_i(x-m_i)]} = \prod_j \exp\left(-\frac{(x_j - m_{ij})^2}{\sigma_{ij}^2}\right) \tag{II.2.12}$$

where m_{ij} and σ_{ij} are, respectively, the projected center and width of the membership function in each dimension. To reduce the number of fuzzy sets of each input variable and to avoid the existence of redundant fuzzy sets, we should check the similarities between them in each input dimension. Since bell-shaped membership functions are used in the RSONFIN, we use the formula of the similarity measure of two fuzzy sets with bell-shaped membership functions derived previously in Lin and Lee.[43] Suppose the fuzzy sets to be measured are fuzzy sets A and B with membership function $\mu_A(x) = \exp\{-(x - m_1)^2/\sigma_1^2\}$ and $\mu_B(x) = \exp\{-(x - m_2)^2/\sigma_2^2\}$, respectively. Assume $m_1 \ge m_2$, we can compute $|A \cap B|$ by:

$$|A \cap B| = \frac{1}{2} \frac{h^2[m_2 - m_1 + \sqrt{\pi}(\sigma_1 + \sigma_2)]}{\sqrt{\pi}(\sigma_1 + \sigma_2)} + \frac{1}{2} \frac{h^2[m_2 - m_1 + \sqrt{\pi}(\sigma_1 - \sigma_2)]}{\sqrt{\pi}(\sigma_2 - \sigma_1)}$$

$$+ \frac{1}{2} \frac{h^2[m_2 - m_1 - \sqrt{\pi}(\sigma_1 + \sigma_2)]}{\sqrt{\pi}(\sigma_1 - \sigma_2)}$$

where $h(x) = \max\{0, x\}$. So the approximate similarity measure is:

$$E(A, B) = \frac{|A \cap B|}{|A \cup B|} = \frac{|A \cap B|}{\sigma_1\sqrt{\pi} + \sigma_2\sqrt{\pi} - |A \cap B|}$$

where we use the fact that $|A| + |B| = |A \cap B| + |A \cup B|$. The detailed algorithm is as follows.

Let $\mu(m_i, \sigma_i)$ represent the Gaussian membership function with center m_i and width σ_i. The whole algorithm for the generation of new fuzzy rules as well as fuzzy sets for each input variable is as follows. Suppose no rules are existent initially.

IF x is the first incoming pattern THEN do

    PART 1. { Generate a new rule,
        with center $m_1 = x$, width $D_1 = diag(1/\sigma_{init}, \ldots, 1/\sigma_{init})$
        where σ_{init} is a prespecified constant.
        After decomposition, we have n one-dimensional membership functions,
        with $m_{1i} = x_i$ and $\sigma_{1i} = \sigma_{init}, i = 1 \ldots n$
        }
    ELSE for each newly incoming pattern x, do

    PART 2. { find $J = \arg\max_{1 \le j \le c(t)} F^j(x)$,
        IF $F^J \ge \overline{F}_{in}(t)$
        { after a period of time
        perform fuzzy measure and eliminate unnecessary membership functions }
        ELSE
        { $c(t + 1) = c(t) + 1$
        generate a new fuzzy rule, with:

$$m_{c(t+1)} = x, \quad D_{c(t+1)} = (-1/\beta) \cdot diag(1/\ln(F^J) \ldots 1/\ln(F^J))$$

        After decomposition, we have:

$$m_{new-i} = x_i, \sigma_{new-i} = -\beta \cdot \ln(F^J), i = 1 \ldots n,$$

        Do the following fuzzy measure for each input variable:

$$\{\deg ree(i, t) = \max_{1 \le j \le k_i} E[\mu(m_{new-i}, \sigma_{new-i}), \mu(m_{ji}, \sigma_{ji})],$$

        where k_i is the number of partitions of the ith input variable.
        IF $\deg ree(i, t) \le \rho$,
        THEN adopt this new membership function and set $k_i = k_i + 1$,
        ELSE set the projected membership function as the closest one.}
        }
    }

In the above algorithm, ρ is a scalar similarity criterion; higher similarity between two fuzzy sets is allowed for larger ρ.

    The generation of a new input cluster corresponds to the generation of a new fuzzy rule. At the same time, we have to decide the consequent part of the generated rule. For the output space partitioning, the same measure in Equation II.2.9 is used. The criterion for the generation of a new output cluster is related to the construction of a rule. Suppose a new input cluster is formed after the presentation of the current input-output training pair (x, d). The consequent part is constructed by the following algorithm:

    IF there are no output clusters,
    do { PART 1 in the above algorithm, with x replaced by d }
    ELSE
    do { find, $J = \arg\max_j F^j(d)$,
    IF $F^J \ge \overline{F}_{out}(t)$
    connect input cluster $c(t + 1)$ to the existing output cluster J
    ELSE
    generate a new output cluster,
    connect input cluster $c(t + 1)$ to the newly generated output cluster.
    }.

The algorithm is based on the fact that different preconditions of different rules may be mapped to the same consequent fuzzy set. Since only the center of each output membership function is used for defuzzification, the consequent part of each rule may simply be regarded as a singleton. Compared to the general fuzzy rule-based models with singleton output,[44] where each rule has its own individual singleton value, fewer parameters are needed in the consequent part of the RSONFIN, especially for the case with a large number of rules.

2.3.2 Feedback Structure Identification

In learning process 2, the number of generated clusters in the consequent part is affected by the problem to be solved. The number of output clusters is large for complex problems and is small for simple ones. Naturally, in the feedback layer, more internal variables are required for more complex problems. Knowing this relationship (i.e., the increment of internal variables as well as output clusters for solving a more complex problem), for simplicity, we simply set the number of internal variables equal to the number of output clusters in the consequent part (i.e., the number of output term nodes in layer 4). With this setting, for each output cluster, the corresponding internal variable is used to record the temporal history that should participate in the precondition part of that output cluster. Hence, during the on-line learning, an internal variable (and thus a context node) is created once an output cluster is created. The fan-in of the context node comes from all the nodes in layer 4, with the link weight assigned with a small random value [−1, 1] initially. After an internal variable is generated (meaning a context node is created), the next step is to decide its effect on each rule node. As mentioned in Section 2.2, only a global membership function is assigned to each internal variable and acts as the feedback term node of the corresponding context node. In general, each rule has its own corresponding internal variable, which is to memorize the history of the rule. But for the rules that have the same consequent part (i.e., connect to the same output term node), the same internal variable is assigned to these rules. Thus, we can effectively reduce the parameter number in the feedback layer.

2.3.3 Parameter Identification

After the network structure is adjusted according to the current training pattern, the network then enters the parameter identification phase to adjust the parameters of the network optimally based on the same training pattern. Notice that the following parameter learning is performed on the whole network after structure learning, no matter whether the nodes (links) are newly added or are existent originally. Since the RSONFIN-S is a dynamic system with feedback connections, the learning algorithm used in the feedforward radial basis function networks[45] or adaptive fuzzy systems[46] cannot be applied to it directly. Also, due to the on-line learning property of the RSONFIN-S, the off-line learning algorithms for the recurrent neural networks, like backpropagation through time and time-dependent recurrent backpropagation,[5] cannot be applied here. Instead, the ordered derivative,[32] which is a partial derivative whose constant and varying terms are defined using an ordered set of equations, is used to derive our learning algorithm. The ordered set of equations are described in Section 2.2 in each layer and are summarized in Equations II.2.13–II.2.16. Considering the single-output case for clarity, our goal is to minimize the error function:

$$E(t+1) = \frac{1}{2}(y_j(t+1) - y_j^d(t+1))^2 \qquad \text{(II.2.13)}$$

where $y_j^d(t+1)$ is the desired output and $y_j(t+1)$ is the current output. For each training pattern, starting at the input nodes, a forward pass is used to compute the activity levels of all the nodes in the network to obtain the current output $y_j(t+1)$. In the following, dependency on time t will be omitted unless emphasis on temporal relationships is required.

Summarizing the node functions defined in Section 2.2, the function performed by the network is:

$$y_j(t+1) = \frac{\sum_i u_i^{(5)} \hat{m}_{ji}}{\sum_i u_i^{(5)}} \tag{II.2.14}$$

$$u_i^{(5)} = \sum_k a_k^{(6)} u_{ik}^{(4)} = \sum_k a_k^{(6)} \prod_i u_{ki}^{(3)} = \sum_k a_k^{(6)} \exp\left\{ -\sum_i \left(\frac{x_i(t) - m_{ik}}{\sigma_{ik}} \right)^2 \right\} \tag{II.2.15}$$

where:

$$a_k^{(6)} = \frac{1}{1 + e^{-h_k}} \tag{II.2.16}$$

and:

$$h_k(t) = \sum_\ell w_{k\ell} a_\ell^{(4)}(t-1) = \sum_\ell w_{k\ell} a_\ell^{(6)}(t-1) \cdot \exp\left\{ -\sum_j \left(\frac{x_j(t-1) - m_{j\ell}}{\sigma_{j\ell}} \right)^2 \right\} \tag{II.2.17}$$

With the above formula and the error function defined in Equation II.2.13, we can derive the update rules for the free parameters in the RSONFIN as follows. The detailed algorithm may refer to Juang and Lin.[20]

- Update rule of $\hat{m}_{ji}$ (the center of the output membership function)
  The update rule of $\hat{m}_{ji}$ is:

$$\hat{m}_{ji}(t+1) = \hat{m}_{ji}(t) - \eta \frac{\partial^+ E}{\partial \hat{m}_{ji}}(t+1) = \hat{m}_{ji}(t) - (y_j(t+1) - y_j^d(t+1)) \frac{u_i^{(5)}}{\sum_i u_i^{(5)}} \tag{II.2.18}$$

- Update rule of m_{pq} (the center of the membership function in the precondition part)
  The update rule of m_{pq} is:

$$m_{pq}(t+1) = m_{pq}(t) - \eta \frac{\partial^+ E}{\partial m_{pq}}(t+1) \tag{II.2.19}$$

where:

$$\frac{\partial^+ E}{\partial m_{pq}}(t+1) = (y_j(t+1) - y_j^d(t+1)) \sum_k \frac{\partial y_j(t+1)}{\partial a_k^{(3)}(t)} \frac{\partial^+ a_k^{(3)}}{\partial m_{pq}}(t), \tag{II.2.20}$$

- Update rule of σ_{pq} (the width of the membership function in the precondition part)
  The update rule of σ_{pq} is:

$$\sigma_{pq}(t+1) = \sigma_{pq}(t) - \eta \frac{\partial E}{\partial \sigma_{pq}}(t+1) \tag{II.2.21}$$

where:

$$\frac{\partial E}{\partial \sigma_{pq}}(t+1) = (y_j(t+1) - y_j^d(t+1)) \sum_k \frac{\partial y_j(t+1)}{\partial a_k^{(3)}(t)} \frac{\partial^+ a_k^{(3)}}{\partial \sigma_{pq}}(t) \tag{II.2.22}$$

- Update rule of w_{pq} (the memory weight parameter in the feedback layer)
  The update rule of w_{pq} is:

$$w_{pq}(t+1) = w_{pq}(t) - \eta_w \frac{\partial E}{\partial w_{pq}}(t+1) \qquad \text{(II.2.23)}$$

where:

$$\frac{\partial E}{\partial w_{pq}}(t+1) = (y_j(t+1) - y_j^d(t+1)) \sum_k \frac{\partial y_j(t+1)}{\partial a_k^{(3)}} \frac{\partial^+ a_k^{(3)}}{\partial w_{pq}}(t) \qquad \text{(II.2.24)}$$

Note that two different learning constants are used in the above equations—η_w for the tuning of memory weight w, and η for the remaining parameters. Except the memory weight parameter w, which is assigned randomly initially, the other parameters all have good initial values assigned during the structure learning phase. Owing to this good initial assignment, the convergence of these parameters is usually faster than that of the weight parameter w. To increase the learning speed of temporal relationship (i.e., tuning of the weight parameter w), we may set the learning constant η_w several times larger than η, so the convergence speed of all parameters is about the same. The learning algorithm derived above is used in the following examples. Notice that according to the real-time recurrent learning (RTRL) scheme,[47] we can also obtain the same parameter learning rules for the RSONFIN-S. Of course, other existing on-line learning algorithms[48,49] for tuning the weights of recurrent neural networks can be used.

2.4   RSONFIN Design by Genetic Algorithm (RSONFIN-G)

Genetic algorithms (GAs) are stochastic search algorithms based on the mechanics of natural selection and natural genetics. GAs do not require or use derivative information; one appropriate application is the problem where gradient information is unavailable or costly to obtain. GAs also provide a means to optimize ill-defined irregular problems, and can be tailored to the needs of different situations. In GA, a candidate solution for a specific problem is called an individual or a chromosome and consists of a linear list of genes. Each individual represents a point in the search space and hence a possible solution to the problem. A population consists of a finite number of individuals. Each individual is decided by an evaluating mechanism to obtain its fitness value. Based on this fitness value and undergoing genetic operators, a new population is generated iteratively, with each successive population referred to as a generation. GAs use three basic operators (the reproduction, crossover, and mutation) to manipulate the genetic composition of a population.

Reproduction is a process by which the most highly rated individuals in the current generation are reproduced in the new generation. The crossover operator produces two offspring (new candidate solutions) by recombining the information from two parents. There are two processing steps in this operation. In the first step, a given number of crossing sites are selected along the parent individual uniformly at random. In the second step, two new individuals are formed by exchanging alternate pairs of selection between the selected sites. Mutation is a random alteration of some gene values in an individual. The alleles of each gene are candidates for mutation, and their function is determined by the mutation probability.

Detailed design of RSONFIN by GA is presented in this section. In RSONFIN-G, for simplicity, we assume that the number of nodes in layer 4 is equal to that in layer 3. The algorithm consists of three major operators: reproduction, crossover, and mutation. Before going into the details of these three genetic operators, the issues of coding and initialization are discussed. Coding concerns the way membership functions and spatial/temporal consequent part parameters are represented

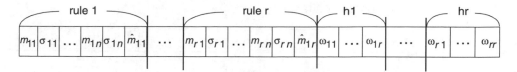

FIGURE II.2.2   Coding of RSONFIN with n input variables and r fuzzy rules into a chromosome.

as chromosomes. Initialization is the proper assignment of learning constants before entering the evolution process. The whole learning process is described step by step below.

- Coding. The coding of the RSONFIN is shown in Figure II.2.2. If there are r rules, n input variables, and single output, then the order of parameters coded into the chromosome is as follows. First, for rule 1, the mean m and standard variation σ of input variables 1 to n are coded, followed by the consequent part parameters $\hat{m}$. Following rule 1, rules 2 to r are coded in the same way one after another. Up to now, the parameters representing the spatial relation are coded. Next, we code the temporal relations (the link weight in feedback layer) into the chromosomes. First, weights w connected to the same hidden variable h_1 are coded in an order according to its source, i.e., rule 1 to rule r. The weights connecting to h_2 to h_r are coded in the same way. For each chromosome, a gene is represented by a floating point number. A flexible partition of the domain of a fuzzy rule is allowed, and no pre-partition of the input variables as in Karr[33] and Homaifar and McCormick[34] is required with the coding scheme. This flexible partition may increase the design accuracy of RSONFIN-G and reduce the number of dynamic fuzzy rules.
- Initialization. Before proceeding with the RSONFIN G design, the number of fuzzy rule r that constitutes the RSONFIN should be assigned. Furthermore, P_s individuals forming the population should be randomly generated and the mutation probability p_m should be assigned.
- Reproduction. Reproduction is a process in which individual strings are copied according to their fitness values. To perform the process, the population is first sorted according to the fitness value of each individual. Based on the elitist strategy, the top half of the best-performing individuals in the population, the elites, will advance to the next generation directly. The remaining half will be generated by performing crossover operations on individuals on the top half of the parent generation.
- Crossover. While reproduction directs the search toward the best of the existing individuals, it doesn't create any new individuals. This is performed mainly by crossover operations. In order to select the individuals for crossover, tournament selection[34] instead of a simple GA roulette wheel selection[58] is performed on the top half of the best-performing individuals. In the tournament selection, two or more individuals (two in this paper) in the top half of the population are selected at random, and their fitness values are compared. The individual with the highest fitness value is selected as one parent. The other parent is selected in the same way. Performing crossover on the selected parents creates the offspring. Here, two-point crossover is performed. One crossover site is located at the gene position coded for spatial rule relations, the other at the position for temporal rule consequences. This crossover operation is illustrated in Figure II.2.3. After operation, the individuals with poor performance will be replaced by the newly produced offspring.
- Mutation. Mutation is an operator whereby the allele of a gene is altered randomly. With mutation, new genetic materials can be introduced into the population. Mutation should be used sparingly because it is a random search operator; otherwise, with high mutation rates, the algorithm will become little more than a random search.

The flow chart of RSONFIN-G is shown in Figure II.2.4.

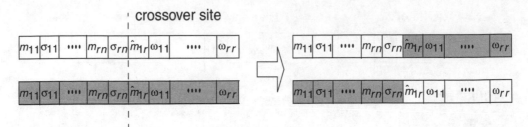

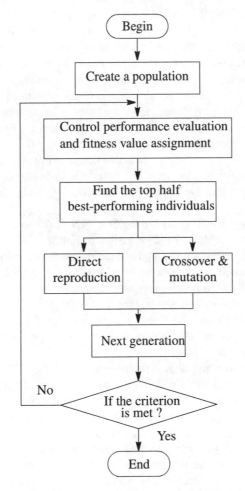

FIGURE II.2.3   Crossover operation on the chromosomes.

FIGURE II.2.4   The flow chart of RSONFIN-G.

2.5 Illustrative Examples

To verify the performance of the RSONFIN for temporal problems, several examples are presented in this section, and performance comparisons with some existing recurrent neural networks are also made. The examples illustrated here include the problem of dynamic plant identification and control. In the following simulations, the parameter β is set to 0.6 in the RSONFIN-S learning algorithm.

2.5.1   RSONFIN-S for Dynamic Plant Identification

The systems to be identified here are dynamic systems whose outputs are functions of past inputs and past outputs as well. This dynamic identification problem is more complicated than the static one because the identification model to be used (e.g., artificial neural network) should have some internal memory. Although we can model the system with a memoryless feedforward network by feeding all the necessary past inputs and outputs of the system as explicit inputs to the network, some drawbacks still exist with this method. One of them is that the exact order of past values for feeding into the network is unknown in practice. Other drawbacks and detailed discussions can be found in Sastry, Santharam, and Unnikrishnan[8] and Williams.[50]

2.5.1.1   SISO Identification

The plant to be identified in this example is guided by the difference equation:

$$y_p(k + 1) = 0.72y_p(k) + 0.025y_p(k - 1)u(k - 1) + 0.01u^2(k - 2) + 0.2u(k - 3)$$

Here the current output of the plant depends on two previous outputs and four previous inputs. To identify the system, a feedforward network with six input nodes for feeding the appropriate past values of y_p and u is required. Here, only two values, $y_p(k)$ and $u(k)$, are fed to the RSONFIN-S and the output $y_p(k + 1)$ is determined. In training the RSONFIN-S, we use only 15,000 time steps and, the input is an *iid* uniform sequence over $[-1.2, 1.2]$ for about one third of the training time, and a single sinusoid given by $0.5 \sin(\pi k/55) + 0.5 \sin(\pi k/20)$ for the remaining time. In applying the RSONFIN-S to this dynamic identification problem, the learning rate $\eta = \eta_w = 0.4$, $\rho = 0.8$, $\overline{F}_{in} = 0.06$, and $\overline{F}_{out} = 0.8$ are chosen. After training, six input clusters (rules) and two output clusters and internal variables are generated, and the number of fuzzy sets on $u(k)$ and $y_p(k)$ are six and six, respectively. The learned six rules are presented as follows:

Rule 1: IF $u(k)$ is $\mu(0.810, 0.456)$ and $y_p(k)$ is $\mu(-0.143, 0.373)$ and $h_1(k)$ is G
  THEN $y(k)$ is $\mu(-0.641, 0.200)$ and $h_1(k + 1)$ is -0.674 and $h_2(k + 1)$ is -0.031
Rule 2: IF $u(k)$ is $\mu(-0.026, 0.688)$ and $y_p(k)$ is $\mu(-0.193, 0.510)$ and $h_1(k)$ is G
  THEN $y(k)$ is $\mu(-0.641, 0.200)$ and $h_1(k + 1)$ is -0.674 and $h_2(k + 1)$ is -0.031
Rule 3: IF $u(k)$ is $\mu(-1.353, 0.522)$ and $y_p(k)$ is $\mu(-0.287, 0.615)$ and $h_1(k)$ is G
  THEN $y(k)$ is $\mu(-0.641, 0.200)$ and $h_1(k + 1)$ is -0.674 and $h_2(k + 1)$ is -0.031
Rule 4: IF $u(k)$ is $\mu(1.130, 0.317)$ and $y_p(k)$ is $\mu(-0.322, 0.247)$ and $h_1(k)$ is G
  THEN $y(k)$ is $\mu(-0.641, 0.200)$ and $h_1(k + 1)$ is -0.674 and $h_2(k + 1)$ is -0.031
Rule 5: IF $u(k)$ is $\mu(1.257, 0.188)$ and $y_p(k)$ is $\mu(0.053, 2.081)$ and $h_1(k)$ is G
  THEN $y(k)$ is $\mu(-0.641, 0.200)$ and $h_1(k + 1)$ is -0.6740 and $h_2(k + 1)$ is -0.031
Rule 6: IF $u(k)$ is $\mu(0.596, 1.320)$ and $y_p(k)$ is $\mu(0.163, 0.490)$ and $h_2(k)$ is G
  THEN $y(k)$ is $\mu(0.729, 0.103)$ and $h_1(k + 1)$ is -0.478 and $h_2(k + 1)$ is -0.634

To see the identified result, the following input is adopted for test:

$$u(k) = \begin{cases} 0.5 \sin(\pi k/55) + 0.5 \sin(\pi k/20) & 0 \leq k \leq 450 \\ 0.45 \sin(\pi k/25) + 0.15 \sin(\pi k/32) + 0.9 \sin(\pi k/10) & 451 < k \leq 800 \end{cases}$$

Figure II.2.5 shows the outputs of the plant (denoted as a solid curve) and the RSONFIN-S (denoted as a dotted curve) for the test input. The Root Mean Square (RMS) error for the test data is 0.0609 and is listed in Table II.2.1. For performance comparison, Elman's Recurrent Neural Network (ERNN),[2] whose feedback connections also come from feeding the hidden states back to the input nodes as TRFN does, is simulated. The number of hidden nodes in ERNN is five, and there are 40 free parameters in total. With the same training time step and data, the resulting RMSE of ERNN for the

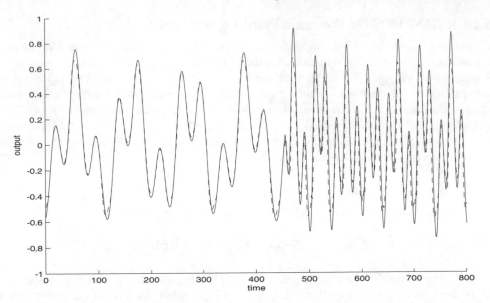

FIGURE II.2.5   Outputs of the SISO plant (solid curve) and model RSONFIN-S (dotted curve).

test data is shown in Table II.2.1. The superiority of RSONFIN-S over ERNN is verified from this comparison.

2.5.1.2 MIMO Identification

The MIMO plant to be identified in this example is the same as that used in Sastry, Santharam, and Unnikrishnan[8] and Narendra and Parthasarathy.[51] This plant has two inputs and two outputs, so there are four input nodes and two output nodes in the RSONFIN-S. The plant is specified by:

$$y_{p1}(k+1) = 0.5 \left[\frac{y_{p1}(k)}{1 + y_{p2}^2(k)} + u_1(k-1) \right],$$

$$y_{p1}(k+1) = 0.5 \left[\frac{y_{p1}(k)y_{p2}(k)}{1 + y_{p2}^2(k)} + u_2(k-1) \right].$$

In applying the RSONFIN-S to this dynamic identification problem, the learning rate $\eta = \eta_w = 0.65$, $\rho = 0.85$, $\overline{F}_{in} = 0.22$ and $\overline{F}_{out} = 0.17$ are chosen. During the training phase, 11,000 time steps are used and the input sequence is similar to that used in SISO identification; for about half of the training time the input is an *iid* uniform sequence over $[-2, 2]$ and for the remaining time, the input is a single sinusoid given by $\sin(\pi k/45)$. After training, eight input clusters (rules), three output clusters, and internal variables are generated. The number of fuzzy sets on the $u_1(k)$, $y_1(k)$, $u_2(k)$ and $y_2(k)$ are 6, 7, 7, 8, respectively. To see the identified result, the following input is adopted for test:

$$u_1(k) = u_2(k) = \begin{cases} 0.2\sin(\pi k/50) + 0.3\sin(\pi k/64) + 0.6\sin(\pi k/20) & 0 \le k < 500 \\ -0.8 & 500 \le k < 600 \\ 0.7 & 600 \le k < 700 \\ -0.5 & 700 \le k < 800 \\ \sin(\pi k/75) & 800 \le k < 1000 \end{cases}$$

TABLE II.2.1   Comparisons of the RSONFIN-S with Existing Recurrent Neural Networks

Network Structure	SISO Identification		MIMO Identification	
	RSONFIN-S	ERNN	RSONFIN-S	Memory Neural Network (6:1)[8]
Network parameter	40	30	71	131
Training time	15000 time steps	15000 time steps	11000 time steps	77000 time steps
RMS error	0.0609	0.0729	$y_{p1} = 0.044$ $y_{p2} = 0.050$	$y_{p1} = 0.136$ $y_{p2} = 0.181$

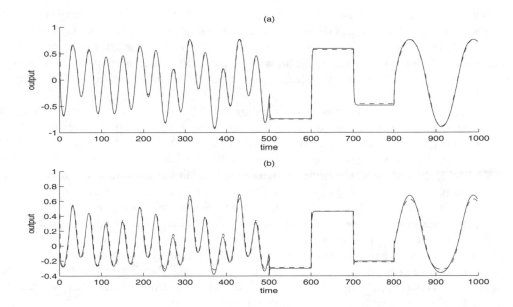

FIGURE II.2.6   Outputs of the MIMO plant (solid curve) and model RSONFIN-S (dotted curve). (a) The first output $y_{p1}(k+1)$. (b) The second output $y_{p2}(k+1)$.

The identified results are shown Figure II.2.6. Performance comparison of the RSONFIN-S and the memory neural network for the same identification task is made in Table II.2.1. Through the comparison, we find that the RSONFIN-S needs fewer training time steps and network parameters, and achieves higher accuracy than the memory neural network.

2.5.2   RSONFIN-G for Dynamic Plant Control

In dynamic system identification, where the precise input-output pattern is available, RSONFIN-S may handle the situation. However, for other problems, such as dynamic system control, where precise control input-output training patterns are unavailable or expensive to collect, a new learning algorithm or design configuration is required. As to time-delayed plant control, one generally adopted controller design approach is the Generalized Predictive Control (GPC).[52] GPC is presented based, as originally, upon a linear model, so it is not suitable for nonlinear plant control. To cope with this problem, some nonlinear controller model designs based on GPC are proposed.[53–55] Most of

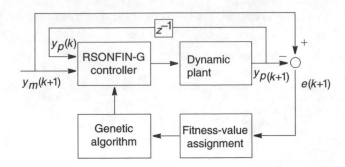

FIGURE II.2.7   The controller design configuration by RSONFIN-G.

these belong to fuzzy-model-based predictive control. In this model, a fuzzy controller with the consequents of linear GPC form is designed. Parameter design algorithm in linear GPC is applied to this model. The drawback of this model is that we should know in advance the order of input and output terms of the linear GPC model in the fuzzy consequent. Other controller design approaches for dynamic systems based upon supervised learning are the direct inverse, direct, and indirect adaptive control.[51]

For direct inverse control, the control configuration fails when the inverse of the controlled plant is nonexistent. This is true for most dynamic plants. For direct adaptive control, we should know the form of the controlled plant. For an unknown plant, this approach can't be applied. For indirect adaptive control, the controlled plant should first be identified and then a controller is designed based on this identification network. Controller design based upon this configuration is complex, and a good control performance is achieved only if a high precision identification model is obtained.

Although some fuzzy neural networks have been proposed and applied to dynamic system control, there are still disadvantages in these network structures and the controller design configurations are mainly based on the above-mentioned methods. In Zhang and Morris,[56] a recurrent neuro-fuzzy model is put forward as a way to build prediction model for nonlinear process, and based on this model, a predictive controller is designed by GPC. In Lee and Tens,[57] a recurrent fuzzy neural network (RFNN) is applied to dynamic plant control, and the controller is designed by direct and indirect adaptive control methods mentioned above.

In Juang and Lin,[20] RSONFIN is also applied to a dynamic plant control based on direct inverse control which works only when the inverse of the plant exists. In contrast to the above supervised learning-based controller design, RSONFIN-G is applied to a dynamic plant control. The proposed RSONFIN-G is a model-free approach, and we do not have to collect the input-output training data in advance. The controller design configuration by RSONFIN-G is shown in Figure II.2.7.

The dynamic plant to be controlled is given by:

$$y_p(k+1) = \frac{y_p(k)y_p(k-1) + 0.5u(k)}{1 + y_p^2(k) + y_p^2(k-1)}$$

Since there is no direct availability of precise input-output training data, GA is used for the RSONFIN controller design. The tracking output is given by:

$$y_m(k+1) = 0.3\sin\left(\frac{2\pi k}{45}\right) + 0.3\sin\left(\frac{\pi k}{20}\right), \qquad 0 \le k < 110$$

There are 110 pieces of data used for training. The RSONFIN controller is constituted by 4 rules, with 36 free parameters in total. Since a dynamic controller, the RSONFIN controller, is used, the input

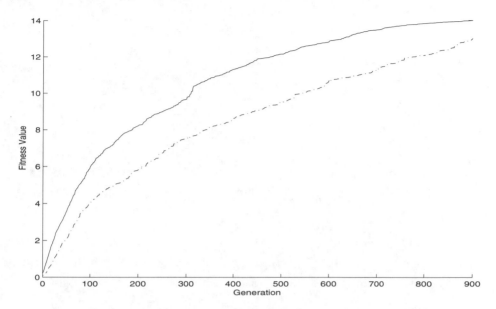

FIGURE II.2.8 The averaged best-so-far values on each generation for RSONFIN-G (solid curve) and ERNN (dotted curve) designed by the same GA.

to the controller consists of only current state $y_p(k)$ and reference output $y_m(k + 1)$. The controller output is $u(k)$. For GA, a population size $P_S = 50$ and mutation probability $P_m = 0.01$ are chosen. The fitness value of each individual is assigned as inverse of the squared control error sum values over 110 training time steps. The evolution is processed for 900 generations and is repeated for 50 runs. The averaged best-so-far fitness value over 50 runs for each generation is shown in Figure II.2.8. One successful control result is shown in Figure II.2.9. The corresponding four rules in RSONFIN controller are as follows:

Rule 1: IF $y_m(k + 1)$ is $\mu(-0.98, 1.46)$ and $y_p(k)$ is $\mu(-1.52, 1.48)$ and h_1 is G
  THEN u is 0.0125 and h_1 is -1.32 and h_2 is -1.40 and h_3 is 2.96 and h_4 is -0.11
Rule 2: IF $y_m(k + 1)$ is $\mu(-1.6, 0.86)$ and $y_p(k)$ is $\mu(-0.80, 1.03)$ and h_2 is G
  THEN u is 2.47 and h_1 is -2.93 and h_2 is 2.96 and h_3 is 2.86 and h_4 is 3.00
Rule 3: IF $y_m(k + 1)$ is $\mu(1.78, 0.95)$ and $y_p(k)$ is $\mu(-0.82, 1.32)$ and h_3 is G
  THEN u is 1.80 and h_1 is -2.56 and h_2 is 2.90 and h_3 is 2.44 and h_4 is -2.98
Rule 4: IF $y_m(k + 1)$ is $\mu(-1.04, 0.45)$ and $y_p(k)$ is $\mu(0.48, 0.59)$ and h_4 is G
  THEN u is -3.33 and h_1 is -2.97 and h_2 is 2.95 and h_3 is -2.97 and h_4 is 2.97

For comparison, Elman's Recurrent Neural Network (ERNN) designed by the same GA is simulated. The input and output of this recurrent network are the same as those of RSONFIN. The number of hidden nodes is 5, resulting in a total number of 40 free parameters. The averaged best-so-far fitness values over 50 runs for each generation are shown in Figure II.2.8. From these results, we see that an obviously better control result is achieved for RSONFIN than that achieved by ERNN when the same GA design approach is applied.

2.6 Conclusion

A recurrent neural fuzzy inference network, RSONFIN, with on-line self-organizing learning capability, is proposed in this chapter. Basically, this network is constructed by expanding the

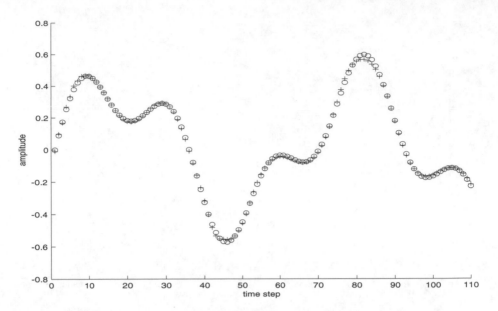

FIGURE II.2.9   The reference output (denoted as "o") and the RSONFIN-G controlled output (denoted as "+").

powerful ability of a neural fuzzy network to deal with temporal problems. To construct RSONFIN undeer different learning environments, RSONFIN-S and RSONFIN-G are proposed. The RSONFIN-S itself realizes dynamic fuzzy reasoning by creating recursive fuzzy rules, which are generated automatically and optimally during on-line operation via concurrent structure and parameter learning. The structure identification process proposed in this paper can effectively reduce the rule number and network size, and the derived order-derivative-based parameter learning algorithm can optimally tune the parameters on both the feedforward and feedback connections.

The RSONFIN-S can be used for normal operation at any time as learning proceeds without any assignment of fuzzy rules in advance. For problems, where the supervised learning data is unavailable or costly to obtain, then RSONFIN-G is applied. Simulations in many temporal problems have demonstrated the ability of the RSONFIN. As a contrast, the role played by the RSONFIN in the recurrent neural network domains is parallel to the role played by the feedforward neural fuzzy network in the feedforward neural network domain. The former networks are good at dynamic mapping, while the latter networks are for static mapping. Both the RSONFIN and the feedforward neural fuzzy networks own the same advantages (such as fast learning, small network size, easy to incorporate expert knowledge) over their pure neural network counterparts.

References

1. Lin, C.T. and Lee, C.S.G., *Neural Fuzzy Systems: A Neural-Fuzzy Synergism to Intelligent Systems*, New York, Prentice Hall, May, 1996.
2. Elman, J.L., Finding structure in time, *Cognitive Science*, 14, 179, 1990.
3. Kohonen, T., *Self-Organization and Associative Memory*, 3rd ed., Berlin, Springer-Verlag, 1989.
4. Jordan, M.I., Attractor dynamics and parallelism in a connectionist sequential machine, *Proc. 8th Ann. Conf. Cognitive Science Society*, Amherst, 531, 1986.
5. Pearlmutter, B.A., Learning state space trajectories in recurrent neural networks, *Int. J. Conf. Neural Networks*, Washington, 2, 365, 1989.

6. Giles, C.L., Kuhn, G.M., and Williams, R.J., Dynamic recurrent neural networks: theory and applications, *IEEE Trans. Neural Networks,* 5, 153, 1994.

7. Santini, S., Bimbo, A.D., and Jain, R., Block-structured recurrent neural networks, *Neural Networks*, 8, 135, 1995.

8. Sastry, P.S., Santharam, G., and Unnikrishnan, K.P., Memory neural networks for identification and control of dynamic systems, *IEEE Trans. Neural Networks*, 5, 306, 1994.

9. Billings, S.A. and Fung, C.F., Recurrent radial basis function networks for adaptive noise cancellation, *Neural Networks*, 8, 273, 1995.

10. Hertz, J., Krogh, A., and Palmer, R.G., *Introduction to the Theory of Neural Computation,* New York, Addison-Wesley, 1991, 7.

11. Miyoshi, T. et al., Learning chaotic dynamics in recurrent RBF network, *Proc. IEEE Int. Conf. Neural Networks,* 588.

12. Gorrini,V. and Bersini, H., Recurrent fuzzy systems, *Proc. IEEE Int. Conf. Fuzzy Systems,* 1, 193, 1994.

13. Grantner, J. and Patyra, M., Synthesis and analysis of fuzzy logic finite state machine models, *Proc. IEEE Int. Conf. Fuzzy Systems,* 1, 205, 1994.

14. Khan, E. and Unal, F., Recurrent fuzzy logic using neural networks, *Advances in fuzzy logic, neural network, and genetic algorithms* (T. Furuhashi, Ed.), Lecture Notes in Artifical Intelligence, Berlin, Springer Verlag, 1995.

15. Kosmatopoulos, E. and Christodoulou, M., Neural networks for identification of fuzzy dynamical systems: Approximation, convergence, and stability and an application to identification of vehicle highway systems, tech. rep., Department of Electronic and Computer Engineering, Technical University of Crete, 1995.

16. Kosmatopoulos, E. and Christodoulou, M., Structural properties of gradient recurrent high-order neural networks, *IEEE Trans. Circuits and Systems,* 1995.

17. Kosmatopoulos, E. et al., High-order neural networks for identification of dynamic systems, *IEEE Trans. Neural Networks,* 6, 422, 1995.

18. Omlin, C., Thornber, K., and Gilies, C., Representation of fuzzy finite-state automata in continuous recurrent neural networks, *IEEE Int. Conf. Neural Networks,* 1996.

19. Unal, F. and Khan, E., A fuzzy finite state machine implementation based on a neural fuzzy system, *Proc. IEEE Int. Conf. Fuzzy Systems*, 3, 1749, 1994.

20. Juang, C.F. and Lin, C.T., A recurrent self-organizing neural fuzzy inference network, *IEEE Trans. Neural Networks*, 10, 828, 1999.

21. Mézard, M. and Nadal, J.P., Learning in feedforward layered networks: The tiling algorithm, *J. Phys.,* 22, 2192, 1989.

22. Fahlman, S.E. and Lebiere, C., The cascade-correlation learning architecture. In D.S. Touretzky, Ed., *Advances in Neural Information Processing Systems II,* San Mateo, CA, Morgan Kaufmann, 542, 1990.

23. Lee, T.C., *Structure Level Adaptation for Artifical Neural Networks,* Boston, Kluwer Academic, 1991.

24. Martinelli, G., Mascioli, F.M., and Bei, G., Cascade neural network for binary mapping, *IEEE Trans. Neural Networks,* 4, 148, 1993.

25. Chen, D. et al., Constructive learning of recurrent neural networks, *Proc. IEEE Int. Conf. Neural Networks,* San Francisco, 3, 1196, 1993.

26. Jang, J.S., ANFIS: Adaptive-network-based fuzzy inference system, *IEEE Trans. Syst., Man, Cybern.,* 23, 665, 1993.

27. Tanaka, K., Sano, M., and Watanabe, H., Modeling and control of carbon monoxide concentration using a neuro-fuzzy technique, *IEEE Trans. Fuzzy Systems,* 3, 271, 1995.

28. Kosko, B., *Neural Networks and Fuzzy Systems*, Englewood Cliffs, NJ, Prentice-Hall, 1992.

29. Lin, C.J. and Lin, C.T., Reinforcement learning for ART-based fuzzy adaptive learning control networks, *IEEE Trans. Neural Networks*, 7, 709, 1996.

30. Wang, L. and Langari, R., Building Sugeno-type models using fuzzy discretization and orthogonal parameter estimation techniques, *IEEE Trans. Fuzzy Systems*, 3, 454, 1995.

31. Ruspini, E.H., Recent development in fuzzy clustering, *Fuzzy Set and Possibility Theory*, New York, North Holland, 113, 1982.

32. Werbos, P., Beyond regression: New tools for prediction and analysis in the behavior sciences, Ph.D. thesis, Harvard University, Cambridge, MA, August, 1974.

33. Karr, C.L., Design of an adaptive fuzzy logic controller using a genetic algorithm, *Proc. 4th Int. Conf. Genetic Algorithms*, 450, 1991.

34. Homaifar, A. and McCormick, E., Simultaneous design of membership functions and rule sets for fuzzy controllers using genetic algorithms, *IEEE Trans. Fuzzy Systems*, 3, 129–139, 1995.

35. Lee, M. and Takagi, H., Integrating design stages of fuzzy systems using genetic algorithms, *Proc. 2nd IEEE Int. Conf. Fuzzy Systems*, San Francisco, CA, 612–617, 1993.

36. Berenji, H.R. and Khedkar, P., Learning and tuning fuzzy logic controllers through reinforcements, *IEEE Trans. Neural Networks*, 3, 724, 1992.

37. Horikawa, S.S., Furuhasi, T., and Uchikawa, Y., On fuzzy modeling using fuzzy neural networks with the back-propagation algorithm, *IEEE Trans. Neural Networks*, 3, 801, 1992.

38. Takagi, H. and Hayashi, I., NN-driven fuzzy reasoning, *Int. J. Approximate Reasoning*, 5, 191, 1991.

39. Jang, J.S.R. and Sun, C.T., Functional equivalence between radial basis function networks and fuzzy inference system, *IEEE Trans. Neural Networks*, 4, 156, 1993.

40. Takagi, T. and Sugeno, M., Derivation of fuzzy control rules from human operator's control actions, *Proc. IFAC Symp. on Fuzzy Information, Knowledge Representation and Decision Analysis*, 55, July 1983.

41. Platt, J., A resource allocating network for function interpolation, *Neural Computation*, 3, 213, 1991.

42. Nie, J. and Linkens, D.A., Learning control using fuzzified self-organizing radial basis function network, *IEEE Trans. Fuzzy Systems*, 40, 280, Nov., 1993.

43. Lin, C.T. and Lee, C.S.G., Reinforcement structure/parameter learning for neural-network-based fuzzy logic control systems, *IEEE Trans. Fuzzy Systems*, 2, 46, Feb., 1994.

44. Wang, L.X. and Mendel, J.M., Fuzzy basis functions, universal approximation, and orthogonal least-squares learning, *IEEE Trans. Neural Networks*, 3, 807, 1992.

45. Moody, J. and Darken, C.J., Fast learning in networks of locally tuned processing units, *Neural Comput.*, 1, 281, 1989.

46. Wang, L.X., *Adaptive Fuzzy Systems and Control*, Englewood Cliffs, NJ, Prentice-Hall, 1994.

47. Williams, R.J. and Zipser, D., A learning algorithm for continually running recurrent neural networks, *Neural Computation*, 1, 270, 1989.

48. Pearlmutter, B.A., Gradient calculations for dynamic recurrent neural networks: a survey, *IEEE Trans. Neural Networks*, 6, 1212, 1995.

49. Piché, S.W., Steepest descent algorithms for neural network controllers and filters, *IEEE Trans. Neural Networks*, 5, 198, 1994.

50. Williams, R.J., Adaptive state representation and estimation using recurrent connectionist networks, *Neural Networks for Control*, W.T. Miller, R.S. Sutton, and P.J. Werbos, Eds., Cambridge, MA, MIT Press, 1990.

51. Narendra, K.S. and Parthasarathy, K., Identification and control of dynamical systems using neural networks, *IEEE Trans. Neural Network*, 1, 4, 1990.

52. Clarke, D.W., Mohtadi, C., and Tuffs, P.S., Generalized predictive control part I, the basic algorithm, *Automatica*, 23, 137, 1988.

53. Kim, J.H. et al. Generalized predictive control using fuzzy neural network model, *IEEE Int. Conf. Neural Networks,* 2596, 1994.
54. Cipriano, A. and Ramos, M., Fuzzy model based control for a mineral flotation plant, *IEEE Int. Conf. Ind. Electronics, Control and Instrumentation*, 1375, 1994.
55. Kim, J.H. et al., Fuzzy model based predictive control, *IEEE Int. Conf. Fuzzy Systems*, 405, 1998.
56. Zhang, J. and Morris, A.J., Recurrent neuro-fuzzy networks for nonlinear process modeling, *IEEE Trans. Neural Networks*, 10, 313, 1999.
57. Lee, C.H. and Teng, C.C., Identification and control of dynamic systems using recurrent fuzzy neural networks, *IEEE Trans. Fuzzy Systems*, 8, 349–366, Aug., 2000.
58. Goldberg, D.E., Genetic Algorithms in Search Optimization and Machine Learning, Reading, MA, Addison-Wesley, 1989.

3

Neural Network Systems Techniques and Applications in Signal Processing

Ying Tan
University of Science and Technology of China

Abstract. In this chapter, we discuss three kinds of neural networks and their applications in signal processing. In the first part, a second-order fast learning algorithm for multilayer feedforward neural network is proposed, in view of feedback control theory after analyzing the slow convergence property of traditional error back-propagation (BP) algorithm. The region of the learning factors for the proposed algorithm to achieve better performance is discussed in detail. Digital simulations on typical XOR problem support the theoretical analysis of the proposed algorithm, in which the training speed of our algorithm is about 50 and 20 times faster than the standard BP algorithm and the momentum BP algorithm, respectively. As a practical application, we apply this fast-learning algorithm to train a neural network for automatic detection system of fire alarm and do a number of experiments. It turns out that the system has not only high recognition accuracy but also fast training speed.

In the second part of this chapter, we discuss another topic on the matrix eigendecomposition by neural networks. After outlining the typical neural networks for the eigenstructure extraction problem, we propose a novel neural network for eigen-decomposition of a positive semi-definite matrix. The network is of globally Lyapunov stability and real-time eigen structure extraction ability. The network cannot only arrive at its stable state in the magnitude of the circuit time constant, but also their output voltage in steady-state is just the eigenvector corresponding to the minimum eigenvalue of its connection weight matrix, which is mapped by data covariance matrix without any pre-processing. Furthermore, by taking appropriate processing for the data covariance matrix, we can further extract all other eigenparameters one by one. Extensive simulations show the effectiveness of the proposed neural network and its capability.

In the third part, we propose an analog neural network and apply it to designing 2-D FIR filter with arbitrary response. According to the theory of eigenfilter, we realized the synthesis of 2-D FIR filter with arbitrary response by using analog neural networks, which is implemented by computing the eigenvector of a real symmetry positive-definite matrix. Following the famous eigenfilter method, we defined a cost function and mapped it as the energy function of the network. This network can give the coefficients of 2-D FIR filter in real time when implemented by hardware. Further, this method has been extended to the filter design of complex-value coefficients at last.

Key words: Multilayer neural networks, second-order fast learning algorithm, learning factors, automatic detection of fire alarm, recurrent neural networks, principal component analysis, matrix eigen-decomposition, real-time eigensolver, 2-D FIR filter with arbitrary response.

3.1   Introduction

The multilayer feedforward neural network is one of the most important artificial neural network models. Many problems of practical interest can be solved by three-layer neural networks, including the exclusive-or problem, symmetry determination problem, and T-C matching problem, et al. To train a multilayer neural network there exist a number of learning algorithms of which the backpropagation (BP) algorithm is the most popular.[1] The BP algorithm developed by Rumelhart et al.,[2] in 1985, is a first-order iterative gradient search algorithm designed to minimize the mean-square error (MSE) between the actual output of a multilayer feedforward network and the desired output. It is also well known that the BP algorithm shows good performance in learning but exhibits slow learning speed in many cases.

So far, many algorithms speeding up the training of a multilayer feedforward network have been proposed, such as BP algorithm adding momentum term, Becker's second-order quasi-Newton method,[3] Singhal's high-order extended Kalman filtering algorithm,[4] Yeh's projection method learning rule,[5] etc. All these algorithms are at the cost of computation complexity.

Since the BP algorithm is a first-order static search algorithm, there usually exist large oscillations in the case of complicated MSE surfaces, and the weights cannot be guaranteed to stay in the neighborhood of the minimal point of the MSE super-surface. So its convergence speed is usually very slow.

The convergence of the BP training algorithm with momentum term (so-called MBP) is sometimes faster than the standard BP algorithm (SBP). The introduced momentum term is equivalent to a recursive low-pass filter applied to the MSE surface gradient with weights. So the "high-frequency" oscillation through MSE's valley can be effectively attenuated.

From the control theory point of view, the damped term added in MBP algorithm is equivalent to introducing an inert section into the process of weight update. Therefore, although it can remove the weight oscillation and approach the neighborhood of its minimal point, the initial "descent time" of the weight will be increased. However, it is well known that moderate overshoot is good to shorten the initial "descent time" of weights, especially to multiple-values problems, which can be seen as the similar effect of "simulated annealing" and then the global minimal point may be reached easily.[6]

Motivated by this idea, a second-order learning algorithm for multilayer perceptrons is proposed in this chapter from the feedback control principle point of view. Furthermore, a better parameter's area for its learning factors, which is obtained through a moderate trade-off between damping and oscillating, is also given.

After Okayama[7] applied the feedforward neural network with standard BP training algorithm to the automatic detection of fire alarm (ADFA) and obtained satisfactory application results in 1991, the neural-network-based detection method for the ADFA problem has received extensive attention. Until now a number of neural-network-based models have been proposed,[8] among which an efficient experimental model was recently proposed in Deng.[8] Here we applied our new fast training algorithm to the experimental model for the ADFA and gave a number of experimental results.

The problem of matrix eigen-decomposition has received extensive research in both mathematics and engineering. Many researchers developed a variety of iterative algorithms such as power method, Jacobi method, Givens-Housholder method, and QR decomposition algorithm, etc. The eigen-decomposition of a matrix is not only very important to the theoretical research, but also has significance to many practical application domains such as modern automatic control, signal processing, radar, communication, etc. In particular, real-time matrix decomposition is very much a critical technique in some actual applications, such as satellite guidance and on-line signal detection, etc., which is a very difficult task for digital computers.[9]

Recently, the artificial neural network has been extensively researched and utilized in many practical engineering fields, due to its adaptiveness and self-organization, nonlinear processing, and massive collective computational ability.[10–12] By use of its massive parallel information processing ability, the aritificial neural networks have already played important roles in many applications such as combinatorial optimization, linear and nonlinear planning, association memory, pattern recognition, matrix inversion calculation, eigen-decomposition, etc.[11,12]

Since the 1980s, a number of neural network models have been proposed by many researchers to extract (or compute) the eigenparameter (such as eigenvalues and corresponding eigenvectors) of a given matrix or a set of sample data. Generally these models can be categorized as two classes: feedforward-type and feedback-type. The former has extensively researched and developed many efficient methods that mainly include: Oja's single linear neuron model for principal component analysis (PCA)[13,14] and the weighted subspace rule (WSR) of N-neuron linear model,[15] Mathew's G-N method,[16] Yang's PAST method,[17] Sanger's generalized hebbian algorithm (GHA) for multiple PC's extraction,[18] Chauvin's penalty function method for minor component extraction,[19] Kung and Diamantaras's adaptive principal component extraction (APEX) algorithm,[10,20] Xu's least mean square error reconstruction (LMSER) principle,[21] and Peper and Noda's dot product decorrelation (DPD) rule.[22] The latter contains Tan and He's main-subs transform network.[23–25]

Although they can extract the eigenparameters of data covariance matrix on condition of unknown statistics of input data, all of the feedforward-type models adopt linear neurons and must repeatedly modify the connection strength during the process of eigen decomposition. So it is impossible to be applied to many real-time signal processing realms. In contrast to these iterative algorithms of feedforward models, the feedback-type interconnection neural networks are capable of computing eigenparameters of matrix in real time by utilizing their inherent dynamic evolutionary mechanism and combining it with the matrix eigen decomposition process.

A novel feedback-type neural network is proposed in this chapter for solving the eigenstructure extraction of a symmetrical positive semi-definite matrix which is directly mapped as the network connection strength matrix. The proposed network can not only evaluate the smallest eigenvalue and its corresponding eigenvector of the data covariance (viz., connection strength) matrix in real time, but also extract all other eigenparameters by properly tackling the matrix.[26]

In the community of signal processing, it is well known that there exist two methods for designing 2-D FIR digital filter, i.e., weighting design method and optimization computational design method. The former is of high computational efficiency but only gives an approximate solution. While the latter is able to obtain better solution, it needs a large number of computational amount, since it is to minimize a proper cost function which is constructed according to given ideal response with respect to filter coefficients.

In 1987, Vaidyanathan and Nguyen[27] introduced a new method for the design of 1-D linear phase FIR digital filter, i.e., eigenfilter method. In essence, it is to compute the minimal eigenvalue of a real symmetrical and positive definite matrix. This method is of great flexibility and easily imposed time and frequency constraints over McClellan-Parks algorithm.[28] Pei and Shyu[29] has already extended this eigenfilter approach to the designs of FIR Hilbert transformer, differentiator, advanced differentiator, and symmetric 2-D FIR filter. Current methods can design a linear phase filter or 2-D quadrant symmetric filter only, and cannot be used to a design nonlinear phase filter such as constant group delay filter and all-pass filter. In these situations, the amplitude and phase of the response function are arbitrary. In order to design this kind of arbitrary response filter efficiently, we extend Nguyen's work and consider the complexity of 2-D FIR filter. We adopt the analog neural network to solve this task in real time.

It is well known that the collective computational ability and emergent behavior of neural networks make them highly efficient tools to deal with various complex nonlinear optimization problems.[11] Up to now, many researchers studied the application of dynamic optimization method of neural network to filter design; e.g., an improved TH network for adaptive FIR filter, improved Hopfield network for synthesis of FIR filter, and a very simple circuit to determine the filter coefficient in frequency domain.[30–33] Here we propose an approach to design 2-D FIR filter with arbitrary response by using analog neural network.[9,34] This method transforms the LS error objective function into an energy function of analog artificial neural networks, then converts the design problem of the filter into the evolutionary process of a dynamic network.[35] If implemented by VLSI hardware, real-time 2-D FIR filter design can be easily obtained by making use of our proposed method in this chapter.

3.2   Second-Order Error Back-Propagation Algorithm

3.2.1   Algorithm

It is well known that the standard BP learning algorithm for multilayer feedforward neural network is as follows:

$$\Delta w(n) = -\eta g(n) \qquad \text{(II.3.1)}$$

where $\Delta w(n) = w(n) - w(n-1)$ is the weight updating incremental of the network, $g(n) = \partial E / \partial w(n)$ is the gradient of MSE $E = \sum_p E_p$ (with respect to weight), η is learning rate, $n = 0, 1, 2, \ldots$ is the iteration index.

According to the symbol convention of above, the MBP algorithm can be given by:

$$\Delta w(n) = \alpha \Delta w(n-1) - \eta g(n) \qquad \text{(II.3.2)}$$

where $0 < \alpha < 1$.

To speed up the training process, we add an additional second preceding momentum term in Equation (II.3.2). So the weight update formula is seen as:

$$\Delta w(n) = \alpha \Delta w(n - 1) + \beta \Delta w(n - 2) - \eta g(n) \qquad \text{(II.3.3)}$$

where α and β are learning factors in this case, η has the same meaning as in Equation II.3.1.

Equation II.3.3 is our proposed second order BP algorithm (i.e., SOBP for short). Since we introduce two preceding momentum terms in the SBP algorithm, Equation II.3.3 becomes a second-order weights update method in which the moderate selection of learning factor values (presented in the next section) can get small overshoot, and at the same time shorten the initial "descent time" for weight update and then speed up the training progress of the net.

Equation II.3.3 can be seen as a second linear discrete feedback system when $g(n)$ is the input signal and $\Delta w(n)$ is the output signal. By taking Z-transform to both sides of Equation II.3.3, we have:

$$D(z) \bullet \Delta W(z) = -\eta G(z) \qquad \text{(II.3.4)}$$

where $\Delta W(z)$ and $G(z)$ are the Z-transforms of $\Delta w(n)$ and $g(n)$, respectively, and:

$$D(z) = 1 - \alpha z^{-1} - \beta z^{-2} \qquad \text{(II.3.5)}$$

The system transfer function $H(z)$ is written as:

$$H(z) = -\eta/D(z) \qquad \text{(II.3.6)}$$

In this approach, we can research the second-order linear discrete system in Z-domain instead of Equation II.3.3 in time domain. Of course, the "descent time" of weight can be shortened in terms of introducing two preceding momentum terms in the weight update equation of SBP algorithm and doing the reasonable tradeoff between damping and oscillating. The convergence performance of SOBP algorithm will be improved dramatically.

Figure II.3.1 shows the step-response curves of several discrete feedback control systems with different parameters. Curve 1 in Figure II.3.1 represents the response of a one-order inert system whose initial "raise time" (which corresponds to the initial "descent time" of weight of neural networks) is too long to settle the system down in a desirable time. The MBP algorithm is such a case. Curve 3 corresponds to a second-order damp system whose initial raise time is highly shortened, but the progress of oscillation process is too long to settle the system down in short time. This is in the case of a SOBP algorithm with inappropriate learning parameters. Curve 2 is a desirable case of our SOBP algorithm with proper learning parameters, which has not only short initial raise time but also very fast damped oscillation.

3.2.2 Region for Learning Factors

The choice of learning factors in Equation II.3.3 is very important to guarantee the convergence performance of the algorithm. There are two steps to determine the region of learning factors. The first step must ensure the stability of the system in Equation II.3.4, i.e., keeping the convergence of SOBP algorithm or the poles of system transfer function $H(z)$ of Equation II.3.6 located inside the unit circle in Z-plane. On the other hand, the second step of the choice of learning factors is, in order to obtain a better dynamic quality performance, how to do a tradeoff between damping and oscillating. First, let's discuss the necessary conditions of the choice of learning factors.

First of all, the poles of the system transfer function $H(z)$ are equal to the roots of the following equation:

$$D(z) = 0 \qquad \text{(II.3.7)}$$

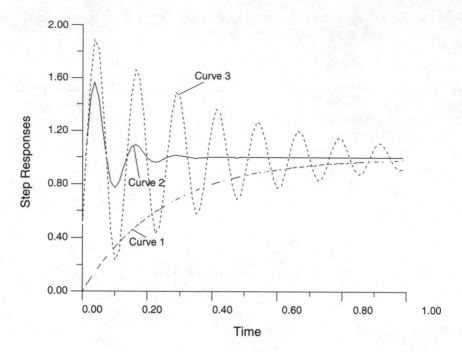

FIGURE II.3.1   Schematic step-response curves of feedback control systems with different parameters.

So the roots of Equation II.3.7, by considering Equation II.3.5, are as follows:

$$r_{1,2} = \frac{\alpha}{2} \pm \sqrt{\frac{\alpha^2}{4} + \beta} \tag{II.3.8}$$

To guarantee a system's stability, the roots of Equation II.3.8 must be within unit circle of Z-plane, i.e., satisfying:

$$|r_{1,2}| < 1 \tag{II.3.9}$$

The condition of Equation II.3.9 can be divided into the following two cases:

1. When the system in Equation II.3.5 has two real poles, the following conditions hold:

$$\begin{cases} \frac{\alpha^2}{4} + \beta \geq 0 \\ |r_{1,2}| < 1 \end{cases} \tag{II.3.10}$$

by solving for Equation II.3.10, we obtain:

$$\begin{cases} \beta \geq -\frac{\alpha^2}{4} \\ \beta < 1 - \alpha \\ \beta < 1 + \alpha \end{cases} \tag{II.3.11}$$

The region determined by Equation II.3.11 is plotted as area I in Figure II.3.2.

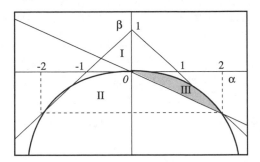

FIGURE II.3.2   The regions of learning factors.

2. When the system of Equation II.3.5 has a pair of conjugate complex (roots) poles, there also exist the following two conditions:

$$\begin{cases} \frac{\alpha^2}{4} + \beta < 0 \\ |r_{1,2}| < 1 \end{cases} \qquad (\text{II}.3.12)$$

By solving Equation II.3.12 and simplification, we have the conditions:

$$\begin{cases} \beta < -\frac{\alpha^2}{4} \\ \beta > -1 \end{cases} \qquad (\text{II}.3.13)$$

The region determined by Equation II.3.13 is shown as area II in Figure II.3.2.

In the second step, in order to obtain a better dynamic quality of the system, a small overshoot is permitted. So at first the choice of learning factors α and β must make system transfer function $H(z)$ have a pair of conjugate complex poles inside the unit circle of Z-plane, i.e., locating at area II of Figure II.3.2. On the other hand, if the overshoot of the system is big, it will oscillate and be harmful for the convergence or training progress. Therefore, the values of learning factors must be chosen in this way, in which a very small overshoot of the system is permitted and then damped rapidly.

Here we give an additional approximate condition, i.e., $\beta > -\alpha/2$, for the learning factors in Figure II.3.2, based on a number of experiments and trade-off between oscillating and damping. The final better area for α and β is shown as region III in Figure II.3.2.

3.3   Performance Analysis of SOBP Algorithm and Comparison

In order to verify our theoretical analysis in the above section, we simulate the SOBP algorithm on three-layer network and test it on the XOR problem. For comparison, we also simulate the SBP and the MBP algorithms on the XOR problem. The simulation software was programmed in terms of object-oriented programming language C++ under a WINDOW environment.

We started with a set of randomly assigned initial weights and biases, and applied the above three algorithms using the four exemplars repeatedly, i.e., each cycle of presentation of the learning samples consists of the successive presentation of the four patterns $\{00, 0\}, \{01, 1\}, \{10, 1\}, \{11, 0\}$. We use a three-layer feedforward neural network whose nodes (or neurons) of input layer, hidden layer, and output layers are 2, 2, and 1, respectively, and the bias of each node is adjustable. The nonlinear activation function of the neuron is chosen as sigmoid function.

It is shown by the experiment results that the learning speed of the SOBP algorithm is faster than that of SBP and MBP algorithms for XOR problem under every condition tested. The number of cycles of presentation needed for MSE to reach and stay less than 10^{-5}, averaged over 10 runs of experiments with different initial weights and biases, are listed in Table II.3.1. It turns out from

TABLE II.3.1 Average Number of Learning Cycles Required to Achieve a MSE of 10^{-5} on the XOR Problem for the Three Algorithms

Algorithm	SBP	MBP	SOBP
Parameter	$\eta = 0.8$	$\eta = 0.8, \alpha = 0.7$	$\eta = 0.8, \alpha = 0.7, \beta = -0.29$
Iterations	22173	11502	428

The numbers in the table are averaged results over 10 runs of experiments with different initial weights and biases.

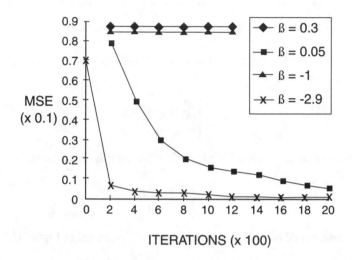

FIGURE II.3.3 Learning history of the SOBP algorithm with $\alpha = 0.7$ and several different values β.

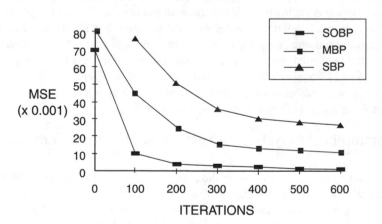

FIGURE II.3.4 Learning histories of SBP, MBP, and SOBP algorithms with the learning parameters of values shown in Table II.3.1.

Table II.3.1 that the iterations of SOBP are 50 and 20 times less than SBP and MBP algorithms, individually.

The MSE is computed after each training cycle to monitor the progress of the network's learning. Several different learning parameters were tested in Figures II.3.3 and II.3.4. In Figure II.3.3 is shown the learning history of the SOBP algorithm with several different learning parameters. When α is 0.7 and β is 0.3 or β is -1, the SOBP algorithm is dispersed, since values of the learning factors are outside the stable areas of Figure II.3.2. But when α is 0.7 and β is -0.29, -0.01, and 0.2, which are inside the convergent area of Figure II.3.2, the SOBP algorithm converges well and the convergence performance is best for the case of α is 0.7 and β is -0.29. All of the results agree with our theoretical analysis. Figure II.3.4 shows the learning history of the three algorithms with the same learning parameters, and it has been shown from the figure that the dynamic quality of the SOBP algorithm is the best one.

3.4 Application to Automatic Detection of Fire Alarm

3.4.1 Model of Automatic Detection of Fire Alarm

As a practical example, we apply our proposed algorithm to artificial neural-network-based automatic detection of fire alarm (ANN-ADFA). After Okayama applied the standard feedforward neural network to the ADFA in 1991, the neural network method for the ADFA system has received extensive attention. A number of neuro-models have been proposed.[8] An experimental model of the ANN-ADFA proposed in Deng[8] is shown in Figure II.3.5, which consists of three units that are sensor measurement unit, preprocessing unit, and network recognition unit.

$$x_i = T_i(s_i), \quad i = 1, 2, \ldots, n \tag{II.3.14}$$

$$\Psi : X \mapsto O \tag{II.3.15}$$

where s_i is the measurement datum of i-th sensor, x_i the output of i-th preprocessor, $X = (x_1, x_2, \ldots, x_n)^T$ and $O = (o_1, o_2, \ldots, o_m)^T$ denote the input and output vectors of the network, respectively, and superscript "T" represents vector transpose.

The sensor measurement unit mainly accomplishes the measurement job for various factors and materials causing fire, such as the densities of CO, O_2, H_2, and smoke, and their change rate as well as temperature. The preprocessor unit is used to properly transform and generalize the measurement data from the sensors so as to apply the processed data to the next unit. The output of the recognition network corresponds to the occurrence probability of the various stages of fire progress. Generally speaking, the fire progress can be divided into three stages, such as no flame stage, dim combustion stage, and on fire stage, whose occurrence probabilities are mapped as the output nodes of the recognition network in the experimental model.

3.4.2 Experimental Results

What follows is an actual experiment of the ANN-ADFA system under the condition of the experimental data from British Borehamwood Fire Institute. The recognition network is chosen as a three-layer neural network with five, six, and three neurons. The five neuron nodes in the first layer correspond to

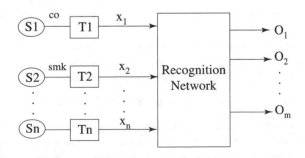

FIGURE II.3.5   An experimental model of the ANN-ADFA.

TABLE II.3.2 The Experimental and Simulated Results of the ANN-ADFA System

T	smk	CO	H_2	O_2	N-F (E)	N-F (S)	D-C (E)	D-C (S)	O-F (E)	O-F (S)
0.25	0.000	0.1	0.08	0.067	0.8351	0.8356	0.0000	0.0000	0.0000	0.0000
0.26	0.050	0.22	0.10	0.10	0.9744	0.9706	0.0000	0.0000	0.0000	0.0000
0.31	0.125	0.35	0.08	0.10	0.5727	0.5730	0.0183	0.0200	0.0000	0.0001
0.40	0.250	0.55	0.10	0.10	0.1695	0.1673	0.9741	0.9730	0.0561	0.0552
0.47	0.430	0.64	0.15	0.15	0.0713	0.0634	0.2133	0.0568	0.4868	0.5925
0.51	0.500	0.68	0.30	0.27	0.0253	0.0147	0.0045	0.0000	1.0000	0.9900
0.60	0.700	0.78	0.58	0.47	0.0075	0.0069	0.0000	0.0000	0.4868	0.6340
0.64	0.870	0.86	0.80	0.87	0.0019	0.0051	0.0000	0.0000	0.0561	0.0617

the densities of CO, H_2, and smoke, the density rate of O_2, and temperature, individually. The three neuron nodes in the third layer are the occurrence probabilities of the three stages of fire progress such as no flame (N-F), dim combustion (D-C), and on fire (O-F). The recognition network in the ANN-ADFA system has been trained by our proposed SOBP algorithm with learning parameter $\eta = 0.8, \alpha = 0.6, \beta = -0.1$. The total training error is less than 0.025 after 1000 iterations. The recognition results of the trained ANN-ADFA system for various fire alarm measurement data are listed in Table II.3.2, in which the column marked by letter 'E or S' denotes the experimental results or the simulation results, respectively.

It turns out that the ANN-ADFA system outperforms the traditional detection system of fire alarm on recognition accuracy, training speed, adaptivity, and robustness, etc. For the interested reader, the detailed relationships among the three stages of fire progress and the measurements of various sensors are found in the literature.[8,9]

3.5 Typical Feedforward Models for Eigenstructure Extraction

Principal component analysis (PCA) is used to extract main features of input signal and has wide applications in noise suppression and data compression. Feedforward-type neural network models provide an iterative way to perform eigenparameter extraction (such as major component, minor component, or subspace) if the data samples become available only one by one. They offer computational advantages in on-line applications like image compression and feature extraction.

Consisting of one layer of artificial neurons, PCA neural networks receive their inputs through Hebbian connections implemented as the weights. The output of the jth neuron is given by:

$$y_j = g(\mathbf{x}^T \mathbf{w}_j) \quad j = 1, \ldots, P \tag{II.3.16}$$

where $\mathbf{x} = (x_1, \ldots, x_N)^T$ denotes the N-dimensional input vector of the network ($N \geq P$), $\mathbf{w}_j = (w_{1j}, \ldots, w_{Nj})^T$ denotes the weight vector of the jth neuron, and g denotes the activation function of the neurons.

Function g of Equation II.3.16 determines whether the network is linear or nonlinear. The usual choice for linear neural networks is $g(y) = y$. Weight vector $\mathbf{w}_j$ is updated every time a new data sample $\mathbf{x}$ becomes available by a Hebbian learning rule of the form:

$$\mathbf{w}_j(t + 1) = \mathbf{w}_j(t) + \gamma(t)[\mathbf{x}(t)y_j(t) - \mathbf{d}_j(\mathbf{x}(t), \mathbf{y}(t), \mathbf{w}_1(t), \ldots, \mathbf{w}_P(t))] \tag{II.3.17}$$

where the learning rate γ determines the amount of update and has the properties of $\lim_{t \to \infty} \gamma(t) = 0$ and $\sum_{t=0}^{t=\infty} \gamma(t) = \infty$, the vector $\mathbf{y} = (y_1, \ldots, y_P)$ denotes the network output, and the vector function $\mathbf{d}_j$ (the decorrelation and decay term) decorrelates the weight vectors and keeps their lengths bounded.

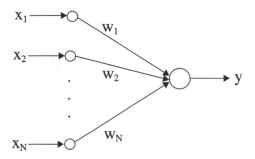

FIGURE II.3.6   Single linear neuron model.

Learning rules of Equation II.3.16 tend to bias the weight vectors toward the eigenvectors of the correlation matrix. Compared to the famous Hebbian learning rule, this form of learning algorithm, with an additional constrained term $\gamma(t)\mathbf{d}_j(\mathbf{x}(t), \mathbf{y}(t), \mathbf{w}_1(t), \ldots, \mathbf{w}_P(t))$, is referred to as constrained Hebbian learning rule. Existing feedforward-type neural networks for PCA are the variants of Equation II.3.16 through choices of different additional constrained terms. What follows is a brief description about several influential typical linear models for extracting eigenparameters. For convenient discussion hereafter, we assume $\mathbf{R} = E(\mathbf{xx}^T)$ denotes data covariance or correlation matrix whose eigenvalues and eigenvectors $\lambda_1 \geq \lambda_2 \geq \cdots \geq \lambda_N$ are $\underline{\mathbf{e}}_1, \underline{\mathbf{e}}_2, \ldots, \underline{\mathbf{e}}_N$, respectively.

3.5.1   Oja's PCA Learning Algorithm

Figure II.3.6 shows the single linear neuron model discussed by Oja for adaptively extracting principal components of data covariance matrix.

If the decorrelation and decay term $\mathbf{d}_j$ of Equation II.3.17 is chosen as:

$$\mathbf{d}(\mathbf{x}, y, \mathbf{w}) = y^2\mathbf{w} \tag{II.3.18}$$

then weight learning satisfies the following dynamic system equation according to Liung's statistical analysis theory[26] when γ is arbitrarily small and $\mathbf{x}$ is statistically independent to $\mathbf{w}$:

$$d\mathbf{w}(t)/dt = \mathbf{R}\mathbf{w}(t) - \mathbf{w}^T(t)\mathbf{R}\mathbf{w}(t)\mathbf{w}(t) \tag{II.3.19}$$

Theorem 1: Assume $\mathbf{w}_f$ to be the asymptotic stable point of Equation II.3.19, if there exists a compact set $A \subset D(\mathbf{w}_f)$ (the attraction domain of $\mathbf{w}_f$) and make Equation II.3.19 satisfying $P\{\mathbf{w}(t) \in A \text{ infinitely often}\} = 1$, then:

$$\lim_{t \to \infty} \mathbf{w}(t) = \mathbf{w}_f = \pm\underline{\mathbf{e}}_1 \tag{II.3.20}$$

We know that Oja's PCA learning algorithm can only compute the PC from an input signal. However, in many applications such as adaptive spectrum estimation, target recognition, and classfication, one needs the eigenvectors' corresponding largest P eigenvalues of covariance matrix. The following algorithms meet this requirement.

3.5.2   Adaptive Extraction of Multiple Principal Components

In order to extract multiple principal components, Oja generalized this one-neuron rule of Equation II.3.16 to an N-neuron rule of Equation II.3.20 and showed the N-neuron rule's ability to determine the so-called principal subspace, the space spanned by the eigenvectors corresponding to the largest P eigenvalues of the data covariance matrix.

$$d(\mathbf{x}, y, \mathbf{w}_1, \ldots, \mathbf{w}_P) = y_j \sum_{i=1}^{P} y_i \mathbf{w}_i \qquad \text{(II.3.21)}$$

After that, he gave a rule employing another type of asymmetry in the learning process, i.e., the weighted subspace rule (WSR). The decorrelation and bias term of weight updating in this case is the form of:

$$d(\mathbf{x}, y, \mathbf{w}_1, \ldots, \mathbf{w}_P) = \alpha_j y_j \sum_{i=1}^{P} y_i \mathbf{w}_i \qquad \text{(II.3.22)}$$

This rule resembles Oja's N-neuron rule, but differs in that to each weight vector a scalar parameter α_j is attached, each set to a different value.

In 1989, Sanger gave an asymmetric version of the N-neuron rule (Figure II.3.7), so-called the generalized Hebbian algorithm (GHA), whose weight iterative algorithm is also of the form of Equation II.3.16 with:

$$d(\mathbf{x}, y, \mathbf{w}_1, \ldots, \mathbf{w}_P) = y_j \sum_{i=1}^{j} y_i \mathbf{w}_i \qquad \text{(II.3.23)}$$

According to the weight update formula of Equation II.3.16 and II.3.23, we can prove weight learning satisfies the following dynamic system equations:

$$d\mathbf{W}(t)/dt = \mathbf{R}\mathbf{W}(t) - \mathbf{W}(t) \sup[\mathbf{W}^T(t)\mathbf{R}\mathbf{W}(t)] \qquad \text{(II.3.24)}$$

where operator sup[·] sets elements of lower diagonal of the matrix as zero:

$$\mathbf{W}(t) = [\mathbf{w}_1, \ldots, \mathbf{w}_P]$$

Theorem 2: Assume $\mathbf{W}_f$ is the asymptotic stable point of Equation II.3.24, if there exists a compact set $A \subset D(\mathbf{W}_f)$ (the attraction domain of $\mathbf{W}_f$) and make Equation II.3.24 satisfying $P\{\mathbf{W}(t) \in A$ infinitely often$\} = 1$, then:

$$\lim_{t \to \infty} \mathbf{W}(t) = \mathbf{W}_f = [\pm\underline{e}_1, \pm\underline{e}_2, \ldots, \pm\underline{e}_P] \qquad \text{(II.3.25)}$$

Using the GHA learning rule, we are able to extract largest P PCs in decreasing order of eigenvalue magnitude. Though it converges quickly and has good stability properties, GHA is an asymmetric learning algorithm in its network structure. This asymmetry is necessary to move GHA's weight vectors toward the correlation matrix's eigenvectors, but forces the weight vectors to learn sequentially rather than in parallel and requires an inhomogeneous connection structure.

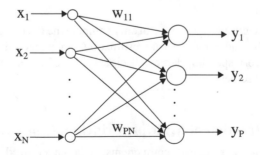

FIGURE II.3.7   Schematic diagram of GHA networks ($P < N$).

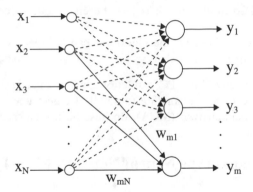

FIGURE II.3.8   APEX neural network.

3.5.3   Adaptive Principal Component Extraction

Kung et al. proposed an efficient adaptive principal component extraction (APEX) learning algorithm for multiple PCs extraction. Similar to lattice adaptive algorithm, APEX can recursively compute the connection weights of mth neuron by utilizing the weights of the former $(m-1)$ neurons. Figure II.3.8 shows the APEX neural network.

The input and output of the APEX neural network are related by:

$$\mathbf{y}(t) = \mathbf{W}(t)\mathbf{x}(t) \tag{II.3.26}$$

where $\mathbf{y}(t) = (y_1(t), y_2(t), \ldots, y_{m-1}(t))^T$, $\mathbf{W}(t) = \{w_{ij}\}(i = 1, \ldots, m-1; j = 1, \ldots, N)$ represents the connection strength matrix of the front $(m-1)$ neurons.

Kung's APEX algorithm can be written by:

$$\mathbf{w}_m(t+1) = \mathbf{w}_m(t) + \gamma(y_m(t)\mathbf{x}^T(t) - \mathbf{d}_m(\mathbf{x}, y_m, \mathbf{w}_m)) \tag{II.3.27}$$

$$\mathbf{d}_m(\mathbf{x}, y_m, \mathbf{w}_m) = y_m^2(t)\mathbf{w}_m(t) \tag{II.3.28}$$

$$\mathbf{s}(t+1) = \mathbf{s}(t) - \beta(y_m(t)\mathbf{y}^T(t) + y_m^2(t)\mathbf{s}(t)) \tag{II.3.29}$$

where $\mathbf{w}_m(t) = (\mathbf{w}_{m1}, \ldots, \mathbf{w}_{mN})^T$ denotes the connection weight vector of mth neuron and $\mathbf{s}(t) = (s_1, \ldots, s_{m-1})^T$ is the connection weight vector from the front $(m-1)$ neurons to mth neuron.

Obviously, Equations II.3.27 and II.3.28 are identical to Equation II.3.16 in Oja's PCA algorithm, and Equation II.3.29 is used to normalize the connection weights. For this network, we have the following theorem.

Theorem 3: If the connection weight matrix $\mathbf{W}$ of the front $(m-1)$ neurons of Figure II.3.8 has converged to the former eigenvector matrix of covariance matrix, then $\mathbf{w}_m(t)$ in Equation II.3.27 will converge to the mth eigenvector of covariance matrix.

When $m = 1$, APEX algorithm becomes the Oja's algorithm which can compute the major eigenvector of data covariance matrix. Therefore, from Theorem 4, APEX can automatically extract multiple PCs of input signal.

In summary, APEX has the following main features:

1. To compute a eigenvector, APEX only needs $O(N)$ times of multiplication but Foldiak's algorithm needs $O(PN)$.
2. For extracting mth eigenvector, Sanger's algorithm needs $(m+1)N$ computation amount but APEX only needs $2(m+N-1)$ computation amount.

3. The output of APEX neural network is just the eigenvalue of covariance matrix $\mathbf{R}$ since $\lim_{t \to \infty} E(y_m^2) = \lambda_m$.
4. Once the eigenvalue is obtained, the algorithm can be speeded up by flexibly choosing attenuation stepsizes of γ and β such as $\gamma = \beta = (M\lambda_{m-1})^{-1}$ to extract the mth eigenvector.

The above discussions are only about PC extraction, but in many actual applications we also want to extract the eigenvector corresponding to smallest eigenvalue of covariance matrix, i.e., the so-called minor component analysis (MCA). Because of limited space here, interested readers please refer to related literature.

3.6   A Novel Feedback-Type Neural Network Model

3.6.1   Network Architecture

Figure II.3.9 shows the architecture of our proposed neural network that consists of two subnetworks, i.e., so-called main net (MN) and secondary net (SN). SN, shown by the dashed-block, is composed of three subnetworks that are labeled by SN1, SN2, and SN3, respectively.

In Figure II.3.9, $\mathbf{D} = \{d_{ij}\}$, $(i, j = 1, 2, \dots, N)$ denotes the connection strength matrix between SN1 and MN. The symbol $-K$ denotes the connection strength between MN and SN3. What follows is a presentation of interconnection relationships among the subnetworks.

3.6.2   Networks' Relation

The MN consists of N neurons with monotonically increasing continuous activation function denoted by $g_i(u)$. The $\underline{\mathbf{U}} = [u_1(t), \dots, u_N(t)]^T$ and $\underline{\mathbf{V}} = [v_1(t), \dots, v_N(t)]^T$ are input and output voltage vectors of MN. In SN1, there are also N neurons with linear activation function of constant gain $k1$. The $\underline{\mathbf{V}}^{(1)} = [v_1^{(1)}(t), \dots, v_N^{(1)}(t)]^T$ denotes the output voltage vector of SN1. The connection relation between MN and SN1 can be expressed as:

$$v_l^{(1)}(t) = k_1 \sum_{j=1}^{N} d_{ij} v_j(t) \tag{II.3.30}$$

The SN2 is designed to receive the output of MN and to perform the function:

$$\beta(t) = k_2 \left(\sum_{i=1}^{N} v_i^2(t) - 1 \right) \tag{II.3.31}$$

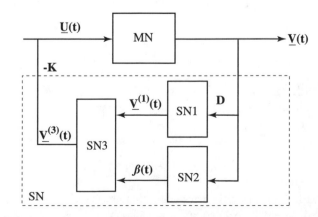

FIGURE II.3.9   Schematic diagram of our proposed neural networks.

The SN3 consists of two layers, whose first layer has N neurons to perform the multiplication of its two inputs, but whose second layer also has N neurons to perform the addition of its two inputs. Therefore, the output of SN3 can be represented as:

$$v_l^{(3)}(t) = v_l^{(1)}(t) + \beta(t)v_i(t) \tag{II.3.32}$$

Since the output $v_l^{(3)}(t)$ of SN3 multiplied by constant connection strength $-k$ is input to the ith neuron of MN, according to Kirhoff current law, we have:

$$C_i \frac{du_i(t)}{dt} = -kv_l^{(3)}(t) \quad (k > 0) \tag{II.3.33}$$

By substituting Equations II.3.30–II.3.32 into Equation II.3.33 and simplifying, we obtain:

$$C_i \frac{du_i(t)}{dt} = -k\left[\beta(t)v_i(t) + k_1 \sum_{j=1}^{N} d_{ij}v_j(t)\right] \quad (i = 1, 2, \ldots, N) \tag{II.3.34}$$

This set of differential equations defines the dynamics of our proposed feedback-type neural network.

3.6.3   Global Lyapunov Stability of the Proposed Neural Networks

Theorem 4: The neural network shown in Figure II.3.4 is globally Lyapunov stable if the energy function of the network is defined by:

$$E(t) = \sum_{i=1}^{N} v_i(t)v_l^{(1)}(t) + \beta^2(t) \tag{II.3.35}$$

The proof of this theorem can be found in Appendix A.

Corollary: the global minimizer of energy function of the network in Figure II.3.4 can be found by letting $\frac{du_i(t)}{dt} = 0$ or $\frac{dv_i(t)}{dt} = 0$ $(i = 1, 2, \ldots, N)$.

3.7   Eigenparameter Computation Ability of the Network

3.7.1   Smallest Eigenvalue and Its Corresponding Eigenvector

In order to find the minimum eigenvalue and its corresponding eigenvector of data covariance matrix $\mathbf{R}$, we map $\mathbf{R}$ as the connection strength matrix of our proposed network (i.e., $\mathbf{D} = \mathbf{R}$) and have the following theorem.

Theorem 5: If a positive semi-definite matrix $\mathbf{R}$ is directly used as the connection strength matrix $\mathbf{D}$ of our proposed neural network, the eigenvalues of $\mathbf{R}$ are $\{\lambda_i\}_{(\lambda_1 \geq \lambda_2 \geq \cdots \geq \lambda_N)}$, and $\{\underline{e}_i\}(i = 1, 2, \ldots, N)$ is an orthonormalized basis consisted of a set of eigenvectors, then:

  i. The network can reach its stable state from any initial state at magnitude of the circuit time constant.
 ii. $\mathbf{R}\underline{V}(\infty) = \lambda_N \underline{V}(\infty)$.
iii. $\beta(\infty) = -\lambda_N/\eta$.
 iv. $\|\underline{V}(\infty)\| = \sqrt{1 - \lambda_N/\eta}$, where η is a positive constant.

The proof of Theorem 5 can be found in Appendix B. Statement (i) of Theorem 1 means the net can arrive at its stable state in one multiplication operation (at magnitude of the circuit time constant) with the massive parallel information processing capability of the feedback neural network. From this perspective, we say our proposed neural network can perform matrix's eigendecomposition in real time. The second statement means that the output voltage vector of the MN is just the smallest eigenvector of connection strength matrix $\mathbf{R}$ that is mapped as the connection strength matrix without any preprocessing when the network arrives at its steady state. The third statement (iii) says the output of the SN2 is just the minimum eigenvalue of $\mathbf{R}$ multiplied by a constant. Statement (iv) means it is required to know the value of minimum eigenvalue by choosing η. Since λ_N will not be known *a priori*, we suggest the following practical lower bound:

$$\eta > \frac{Tr(\mathbf{R})}{N} \tag{II.3.36}$$

In many practical applications, if we choose $\eta \gg Tr(\mathbf{R})/N$, the output eigenvector of the main net is approximately a normalized one. We may add here that the proposed neural network can be extended to the case of a general symmetric matrix (which is not indefinite).

3.7.2 Scheme for Extracting Other Eigenparameters

From the preceding subsections discussed, we know that the output voltage vector of the main net is just the eigenvector with respect to the smallest eigenvalue of $\mathbf{R}$ after the transient process of the network. Therefore, the smallest eigenvalue and its corresponding eigenvector of the connection matrix can be determined by the proposed network. A natural question is how to find all other eigenparameters (that is, the other eigenvalues and their corresponding eigenvectors, respectively) of the $\mathbf{R}$? To solve the problem, we give a theorem as follows.

Theorem 6: If a new sequence of matrix $\mathbf{R}_j (j = 1, 2, \ldots, N)$ is constructed from matrix $\mathbf{R}$ in the following way:

$$\mathbf{R}_j = \mathbf{R} + \sum_{i=N-j+1}^{N} [Tr(\mathbf{R}) - \lambda_i]\underline{\mathbf{e}}_i \underline{\mathbf{e}}_i^T \tag{II.3.37}$$

and further assume $\lambda_i^j (i = 1, 2, \ldots, N, j = 1, 2, \ldots, N-1)$ are the eigenvalues of $\mathbf{R}_j$ in decreasing order of magnitude, we then have $\lambda_N^i = \lambda_j, (j = 1, 2, \ldots, N)$.

The proof of Theorem 6 is given in Appendix C.

To decrease the residual computations to construct the matrix $\mathbf{R}_j$, we can calculate it recursively:

$$\mathbf{R}_j = \mathbf{R}_{j-1} + \Delta\mathbf{R}_j \tag{II.3.38}$$

$$\Delta\mathbf{R}_j = [Tr(\mathbf{R}) - \lambda_{N-j+1}]\underline{\mathbf{e}}_{N-j+1}\underline{\mathbf{e}}_{N-j+1}^T \tag{II.3.39}$$

where $j = 1, 2, \ldots, N$ and $\mathbf{R}_0 = \mathbf{R}$.

Note that both the orthonormalized eigenvector basis $\underline{\mathbf{e}}_j$ and Equation II.3.39 can be computed by normalizing the output of the MN when $\mathbf{R}_j$ is used as its connection strength matrix or by choosing a very large η. In order to satisfy the condition of Equation II.3.36 during the process of eigen decomposition in any case, we are able to choose the parameter $\eta > Tr(\mathbf{R})$.

3.8 Simulation and Discussion

In order to verify our mathematical analyses and the real-time eigen-decomposition ability of the proposed neural network, first of all, we perform the computer simulation for the following data covariance matrix:

$$\mathbf{R} = \begin{bmatrix} 1.4178 & -0.1407 & 0.2416 & -0.1982 \\ -0.01407 & 0.3490 & -0.1199 & 0.0984 \\ 0.2416 & -0.1199 & 1.0297 & -0.1689 \\ -0.1982 & 0.0984 & -0.1689 & 0.6980 \end{bmatrix}$$

The minimum eigenvalue and its corresponding eigenvector of $\mathbf{R}$ calculated by the theoretical computation method are $\lambda_{min} = 0.75$, $\underline{\mathbf{V}}_{min} = (0.0722, 0.9778, 0.0987, -0.1699)^T$.

In our experiment, when we choose $\eta = 360$, $C = 50\,\mathrm{pF}$, $k_M = k_1 = 1$, the simulation results are listed in Table II.3.3.

The evolutionary trajectories of every component of MN's output $\underline{\mathbf{V}}_3^*$ corresponding to the second smallest eigenvalue $\lambda_3 = 0.6143$ are plotted in Figure II.3.10. It has been shown in the figure that each component of the output of the MN can approach its respective theoretical value in hundreds of nanoseconds regardless of its initial states. Table II.3.4 gives some values of the network's outputs corresponding to Figure II.3.5 in several initial steps.

TABLE II.3.3 The Obtained Eigenvalues and Corresponding Eigenvectors in Experiment 1

	Eigenvalues	Eigenvectors
1	$\lambda_1 = 1.6346$	$\underline{\mathbf{V}}_1^* = (-0.8406, 0.1540, -0.4422, 0.2723)^T$
2	$\lambda_2 = 0.9309$	$\underline{\mathbf{V}}_2^* = (-0.1523, -0.1173, -0.3222, -0.9269)^T$
3	$\lambda_3 = 0.6143$	$\underline{\mathbf{V}}_3^* = (0.5148, 0.0796, \quad 0.8312, 0.1942)^T$
4	$\lambda_4 = 0.3099$	$\underline{\mathbf{V}}_4^* = (0.0722, 0.9778, 0.0987, -0.1699)^T$

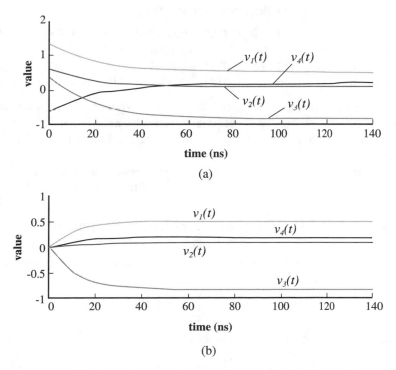

(a)

(b)

FIGURE II.3.10 The trajectories of the proposed network output corresponding to the second small eigenvalue with (a) random initial state and (b) fixed initial state.

TABLE II.3.4   The Calculated Values of the Initial Steps of the MN's Output Voltage Vector of the Proposed Network Corresponding to the Second Smallest Eigenvalue of the Data Covariance Matrix **R**

$t(nS)$	$v_1(t)$	$v_2(t)$	$v_3(t)$	$v_4(t)$
0	0.0000	0.0000	0.0000	0.0000
10	0.3055	0.0472	−0.4933	0.1152
20	0.4297	0.0664	−0.6938	0.1621
30	0.4802	0.0743	−0.7753	0.1811
—	—	—	—	—
100	0.5147	0.0796	−0.8311	0.1941

TABLE II.3.5   The Obtained Eigenvalues and Corresponding Eigenvectors in Experiment 2

	Eigenvalues	Eigenvectors
1	$\lambda_1 = 4.0868$	$\underline{\mathbf{V}}_1^* = (0.3291, 0.4268 + j0.2894, 0.6234 + j0.4497, 0.1593 + j0.0978)^T$
2	$\lambda_2 = 1.1094$	$\underline{\mathbf{V}}_2^* = (-0.1455 + j0.0987, -0.7696, 0.6092 + j0.0177, -0.0734 + j0.0033)^T$
3	$\lambda_3 = 0.5758$	$\underline{\mathbf{V}}_3^* = (0.7340 - j0.5295, -0.3610 + j0.0105, -0.1785, 0.1363 - j0.0102)^T$
4	$\lambda_4 = 0.2151$	$\underline{\mathbf{V}}_4^* = (-0.1741 + j0.1068, -0.1067, -j0.0048, -0.0766 - j0.0057, 0.9700)^T$

The averaged evolutionary time of the proposed NN from any initial state to its smallest energy state is about 120 ns. From this point of view, we say the proposed neural network can implement the extraction of the eigenstructure of data covariance matrix in real time.

Secondly, for the following complex covariance matrix R:

$$\begin{bmatrix} 0.9575 & 0.5465 - j0.3705 & 0.6694 - j0.4829 & 0.2509 - j0.1540 \\ 0.5465 + j0.3705 & 1.8214 & 1.1380 - j0.0331 & 0.4056 + j0.0184 \\ 0.6694 + j0.4829 & 1.1380 + j0.0331 & 2.8463 & 0.5061 + j0.0377 \\ 0.2509 + j0.1540 & 0.4056 - j0.0184 & 0.5061 - j0.0377 & 0.3620 \end{bmatrix}$$

we simulate our neural network again. In this case, we use two times the neuron number as in the real covariance matrix case above. The final simulation results are shown in Table II.3.5.

The trajectories of the real part of the output of MN corresponding to the smallest eigenvalue are plotted in Figure II.3.11.

It can be seen from above simulations that the experimental simulation results agree with the preceding theoretical analysis and show the effectiveness of the proposed network.

3.9   Design of 2-D FIR Filter with Arbitrary Response

The transfer function of a $2\text{-}D(N_1 \times N_2)$ causal nonrecusive filter can be expressed as:[9,31,32]

$$H(z_1, z_2) = \sum_{k_1=o}^{N_1-1} \sum_{k_2=0}^{N_2-1} h(k_1, k_2) z_1^{-k_1} z_2^{-k_2} \tag{II.3.40}$$

whose frequency domain response can be written as:

$$H(e^{i\omega_1}, e^{j\omega_2}) = \sum_{k_1=o}^{N_1-1} \sum_{k_2=0}^{N_2-1} h(k_1, k_2) e^{j(\omega_1 k_1 + \omega_2 k_2)} \tag{II.3.41}$$

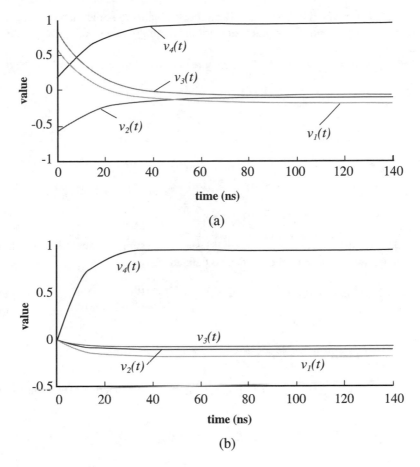

FIGURE II.3.11   The trajectories of each real part of the proposed network's output vector corresponding to the smallest eigenvalue with (a) random initial state and (b) zero initial state.

Let $D(\omega_1, \omega_2)$, $H(e^{j\omega_1}, e^{j\omega_2})$ and $W(\omega_1, \omega_2)$ be complex-valued ideal function, filter response function to be designed, and arbitrary weighting function, respectively. Here we suppose $h(m, n)$ to be a real function, then the objective function adopted by eigenfilter method is:[33]

$$E = \iint\limits_{P} |W(\omega_1, \omega_2)| |\alpha D(\omega_1, \omega_2) - \hat{H}(\omega_1, \omega_2)|^2 d\omega_1 d\omega_2$$

$$+ \iint\limits_{S} |W(\omega_1, \omega_2)| |\hat{H}(\omega_1, \omega_2)|^2 d\omega_1 d\omega_2 = E_P + E_S \qquad (\text{II.3.42})$$

where E_P represents weighted passband error, E_S represents weighted stopband error, P represents passband regions, S represents stopband regions.

$$\alpha = H(e^{j0}, e^{j0}) = \sum_{k_1=0}^{N_1-1} \sum_{k_2=0}^{N_2-1} h(k_1, k_2) \qquad (\text{II.3.43})$$

By properly choosing the proportional constants, Equation II.3.42 can be expressed into quadratic form as $E = \mathbf{a}^t \mathbf{Q} \mathbf{a}$, where t stands for transpose, $\mathbf{Q}$ is a real symmetrical and positive definite matrix, $\mathbf{a}$ is a real vector which consists of 2-D filter coefficients in some manner. According to the Rayleigh principle,[9,35] the minimal eigenvector corresponding to matrix $\mathbf{Q}$ always minimizes the total error E.

Through appropriate arrangement, the frequency response of Equation II.3.42 can be rewritten as:

$$H(e^{j\omega_1}, e^{j\omega_2}) = \mathbf{a}^t \mathbf{e}(e^{j\omega_1}, e^{j\omega_2}) \tag{II.3.44}$$

where

$$\mathbf{a} = [h(0,0), h(0,1), \ldots, h(0, N_2 - 1) \, h(1,0), h(1,1), \ldots, h(1, N_2 - 1) \ldots$$

$$h(N_1 - 1, 0), h(N_1, -1, 1), \ldots, h(N_1 - 1, N_2 - 1)]^t_{N_1 N_2 \times 1} \tag{II.3.45}$$

$$\mathbf{e}(e^{j\omega_1}, e^{j\omega_2}) = [1, e^{j\omega_2}, \ldots, e^{j(N_2 - 1)\omega_2} \ldots e^{j(N_1 - 1)\omega_1}, e^{j((N_1 - 1)\omega_1 + \omega_2)}, \ldots,$$

$$e^{j((N_1 - 1)\omega_1 + (N_2 - 1)\omega_2)}]^t_{(N_1 N_2 \times 1)} \tag{II.3.46}$$

For simplification, we denote $\mathbf{e}(e^{j\omega_1}, e^{j\omega_2})$ as $\mathbf{e}_\omega$. Usually, we suppose ideal filter is of infinite large attenuation in stopband. So stopband error can be defined as:

$$E_S = \iint_S |W(\omega_1, \omega_2)| \| \hat{H}(\omega_1, \omega_2)|^2 d\omega_1 d\omega_2 = \iint_S |W(\omega_1, \omega_2)| \mathbf{a}^t \mathbf{e}_\omega \mathbf{e}_\omega^+ \mathbf{a} d\omega_1 d\omega_2$$

$$= \mathbf{a}^t \left\{ \iint_S |W(\omega_1, \omega_2)| \mathbf{e}_\omega \mathbf{e}_\omega^+ d\omega_1 d\omega_2 \right\} \mathbf{a} = \mathbf{a}^t Q_S \mathbf{a} \tag{II.3.47}$$

where weighting function can choose the spectrum of any 2-D sequence. Then the (m, n)th entry can be expressed as:

$$[Q_S]_{m,n} = \iint_S e^{j((\lfloor \frac{m}{N_2} \rfloor - \lfloor \frac{n}{N_2} \rfloor)\omega_1 + (m-n)mod(N_2)\omega_2)} |W(\omega_1, \omega_2)| d\omega_1 d\omega_2$$

$$= \iint_S \cos\left(\left(\left\lfloor \frac{m}{N_2} \right\rfloor - \left\lfloor \frac{n}{N_2} \right\rfloor\right)\omega_1 + (m - n) \ \ mod(N_2)\omega_2\right) |W(\omega_1, \omega_2)| d\omega_1 d\omega_2$$

$$\tag{II.3.48}$$

On the other hand, the passband error of Equation II.3.42 can be expressed as:

$$E_P = \iint_P |W(\omega_1, \omega_2)| \mathbf{a}^t (\alpha D(\omega_1, \omega_2) - \mathbf{e}_\omega)(\alpha D(\omega_1, \omega_2) - \mathbf{e}_\omega)^+ \mathbf{a} d\omega_1 d\omega_2 = \mathbf{a}^t Q_P \mathbf{a}$$

$$\tag{II.3.49}$$

where $\alpha = H(\omega_1^*, \omega_2^*)/D(\omega_1^*, \omega_2^*)$, and $|D(\omega_1^*, \omega_2^*)| = \max_{\omega_1, \omega_2 \in P} |D(\omega_1, \omega_2)|$. Therefore, using Equation II.3.43, one can obtain $H(\omega_1^*, \omega_2^*) = \mathbf{a}^t \mathbf{e}_\omega^*$, where $\mathbf{e}_\omega^* = e_\omega|_{\omega_1^*, \omega_2^*}$. We call point (ω_1^*, ω_2^*) the reference frequency point in passband regions. Obviously, Q_P is also a symmetrical and positive definite matrix and its (m, n)th entry is:

$$[Q_P]_{m,n} = \iint_P |W(\omega_1, \omega_2)| \{Re[(\mathbf{e}_\omega^* D(\omega_1, \omega_2)/D(\omega_1^*, \omega_2^*) - \mathbf{e}_\omega)$$

$$(\mathbf{e}_\omega^* D(\omega_1, \omega_2)/D(\omega_1^*, \omega_2^*) - \mathbf{e}_\omega)^+]\}_{m,n} d\omega_1 d\omega_2 \tag{II.3.50}$$

It can be seen from the above discussion that, based on Equations II.3.47 and II.3.49, we can express the total error of Equation II.3.42 into a quadratic form of $\mathbf{a}$, i.e., $E = \mathbf{a}^t Q \mathbf{a}$, where $Q = Q_P + Q_S$ is a matrix with dimension $(N_1 N_2 \times N_1 N_2)$.

3.10 Neural Network Approach to Design FIR Filter

3.10.1 Design of 2-D Real-Valued Coefficient FIR Filter with Arbitrary Response

According to classic least square theory, to solve an LS problem one needs to compute the inverse of a matrix, which may be an ill-posed problem when the length of filter is long. Another efficient method to solve an LS problem is to adopt QR decomposition. However, at this moment we only need to compute the minimal eigenvector of a real symmetrical and positive definite Q. Vaidyanathan and Nguyen[27] and Nguyen[30] proposed a so-called eignfilter method, which is a very efficient new method to compute the minimal eigenvector of the matrix Q. Furthermore, for given matrix Q, there is a fast algorithm based on power method to design the eigenfilter. Even though there are so many efficient methods in literature, for the case of time-critical applications such as robot control and satellite guidance, real-time filter design is required. In that situation, all above methods cannot provide a satisfactory answer. To tackle this question, we make use of the collective emergent computation behavior and inherent parallel mechanism of artificial neural networks to retain the filter in real time.

The eigenvector corresponding to the minimal eigenvalue matrix Q (i.e., minimal eigenvector) amounts to the following constrained minimization optimization problem:[9,34,35]

$$\min_{\mathbf{a}} \mathbf{a}^t Q \mathbf{a}, \text{ s.t. } \mathbf{a}^t \mathbf{a} = 1 \tag{II.3.51}$$

where $\mathbf{a}$ is the filter coefficient column vector with dimension $N_1 N_2$, and the normalization constraint is to avoid a null solution.

Since the dimension of vector $\mathbf{a}$ is large and only minimal eigenvector is of interest, in order to efficiently and adaptively estimate minimal eigenvector, we propose a feedback neural network to solve the contrained optimization of Equation II.3.52 sucessfully on the basis of Tan and He.[34] Specifically, one constructs a resultant energy function with respect to Equation II.3.51:

$$E = \mathbf{a}^t Q \mathbf{a} + \mu P(\mathbf{a}) \tag{II.3.52}$$

where $P(\mathbf{a})$ is a penalty function to the unsatisfactory contraint. Thus the dynamic equation of the corresponding neural network can be expressed as:

$$\frac{d\mathbf{a}(t)}{dt} = 2Q\mathbf{a}(t) + \mu P_a(\mathbf{a}(t)) \tag{II.3.53}$$

According to Tan[9] and Tan and He,[34] by appropriately choosing the penal constant μ, the attraction region of global minimum of the network described by Equation II.3.53 is the whole space. Therefore, the network can evolve from any initial vector $\mathbf{a}_0$ freely and will settle down after the number of time constant of the circuit. At this moment, the output of the network just gives the minimal eigenvector of matrix Q. Thus, once the matrix Q is computed according to the design specifications, one can easily obtain its minimal eigenvector by using the collective calculation ability of the neural network. Then the corresponding filter impulse response is easily obtained from vector $\mathbf{a}$:

$$h(m, n) = \mathbf{a}(mN_2 + n) \quad 0 \leq m \leq N_1 - 1, 0 \leq n \leq N_2 - 1, \tag{II.3.54}$$

In summary, from the above discussion, we express the design problem of an LS 2-D FIR filter as a eigensolving problem. Given passband region P and stopband S, weighting function, and the ideal response over passband, choosing reference point via Equation II.3.43, compute matrix Q according to Equations II.3.48 and II.3.50 by using the digital integration approach. Then compute the eigenvector corresponding to the minimal eigenvalue of Q in real time. The optimal 2-D FIR filter coefficient h is just the minimal eigenvector of matrix Q. The main difference, compared to the existing methods in Pei and Shyu,[29] Nguyen,[30] Nashashibi and Charalambous,[31] Bhattacharya and Antoniou,[32] Martinelli and Perfetti,[33] Tan and He,[34] and Tan and Deng,[35] is that the proposed method here can be used to design 2-D filter with arbitrary frequency response and multiple passband regions.

3.10.2 Design of 2-D Complex-Valued Coefficient FIR Filter with Arbitrary Response

By using the similar derivative approach of the previous section, the design method for real-valued coefficient filter is easily extended to the case of complex-valued coefficient filters. The principal difference between them is that vector $\mathbf{a}$ consists of the real and imaginary parts of $h(m, n)$. Moreover, $\mathbf{e}_w$ is in some special formation.

Let $h_R(m, n)$ and $h_I(m, n)$ be the real and imaginary parts of complex-valued coefficient filter $h(m, n)$; then the frequency response is given by:

$$H(e^{j\omega_1}, e^{j\omega_2}) = \sum_{m=0}^{N_1-1} \sum_{n=0}^{N_2-1} [h_R(m, n) + jh_I(m, n)] e^{j(m\omega_1+n\omega_2)} = \mathbf{a}^t \mathbf{e}_\omega \qquad (\text{II.3.55})$$

where:

$$\begin{aligned}
\mathbf{a} &= [h_R(0, 0), h_R(0, 1), \ldots, h_R(0, N_2 - 1) \ldots h_R(N_1 - 1, 0), h_R(N_1 - 1, 1), \ldots, \\
&\quad h_R(N_1 - 1, N_2 - 1), h_I(0, 0), h_I(0, 1), \ldots, h_I(0, N_2 - 1) \ldots h_I(N_1 - 1, 0), \\
&\quad h_I(N_1 - 1, 1), \ldots, h_I(N_1 - 1, N_2 - 1)]^t_{(2N_1 N_2 \times 1)}
\end{aligned}$$

$$\begin{aligned}
\mathbf{e}_\omega &= [1, e^{j\omega_2}, \ldots, e^{j(N_2-1)\omega_2} \ldots e^{j(N_1-1)\omega_1}, e^{j((N_1-1)\omega_1+\omega_2)}, \ldots, e^{j((N_1-1)\omega_1+(N_2-1)\omega_2)}, \\
&\quad j, je^{j\omega_2}, \ldots, je^{j(N_2-1)\omega_2} \ldots je^{j(N_1-1)\omega_1}, je^{j((N_1-1)\omega_1+\omega_2)}, \ldots, \\
&\quad je^{j((N_1-1)\omega_1+(N_2-1)\omega_2)}]^t_{(N_1 N_2 \times 1)}
\end{aligned}$$

It turns out that the expression forms of the frequency response $H(e^{j\omega_1}, e^{j\omega_2})$ in Equations II.3.55 and II.3.44 are completely the same, but their distinction is that $\mathbf{a}$ and $\mathbf{e}$ are in different formations. It is clearly seen from Equation II.3.55 that the dimensions of $\mathbf{a}$ and $\mathbf{e}$ are twice in real case. As a result, given ideal function D and passband and stopband indexes, one can calculate the corresponding matrix Q. The goal vector is just the minimal eigenvector of the computed matrix Q. By rearranging the elements of vector $\mathbf{a}$, we can obtain the resulting designed filter coefficients $\mathbf{h}$ as follows:

$$h(m, n) - \mathbf{a}(mN_2 + n) + j\mathbf{a}((N_1 + m)N_2 + n) \quad 0 \leq m < N_1, 0 \leq n < N_2 \qquad (\text{II.3.56})$$

3.11   Concluding Remarks

From the perspective of the feedback control principle, this chapter proposed a SOBP training algorithm for the multilayer feedforward neural network and discussed the good region selection of the learning factors in detail from the adjustment control theory point of view. The computer experiments demonstrated, by simulation on the XOR problem, that the theoretical analysis of this chapter is correct and the proposed SOBP algorithm outperforms the well-known SBP algorithm and MBP algorithm. Therefore, the training progress for multilayer feedforward neural network can be dramatically speeded up by using this new training algorithm. As an actual application example, we apply this algorithm to train the recognition network of the ANN-ADFA system. It turns out by a number of experiments that these kinds of neural network methods have extensive applicable prospects in detecting the fire alarm.

After briefly presenting several typical feedforward-type linear neural networks for the eigenparameter extraction, we have proposed a new neural network solving the matrix eigen-decomposition problem. Using the massively parallel processing capability of neural networks, the proposed network can deal with the problem of determining minimum eigenvalue and its corresponding eigenvector of the connection strength matrix at the magnitude of circuit time constant. By taking proper processing for the connection strength matrix, we can further extract all other eigenparameters. This network is very suitable for the applications in which the processing time is critical. Both the mathematical analysis and experimental results verify the effectiveness of our neural network.

Finally, we discussed the topic on design of 2-D FIR filter with arbitrary response by using analog neural network, which can be conveniently implemented in hardware and has good real-time performance. In addition, the method given here is very easily extended to high-dimensional filter design. Therefore, the neural network, in particular the feedback interconnected artificial neural networks, will play very important roles in many future real-time applications.

In summary, neural networks have received extensive research and utilization in many engineering fields. In particular, their application in signal processing is one of the most important applications of neural networks, which pushed forward the research of signal processing greatly and played an important role in this field.

Acknowledgment

This work was supported by the Distinguished Youth Scientific and Technological Foundation of An Hui Province (2000–2003), Hefei, An Hui Province.

Appendix A: Proof of Theorem 4

By substituting Equations II.3.30–II.3.33 into Equation II.3.35 and using matrix notation, we obtain:

$$E(t) = k_1 \underline{V}^T D \underline{V} + k_2^2 (\underline{V}^T \underline{V} - 1)^2 \qquad \text{(II.3.A.1)}$$

Since we assume the connection strength matrix D to be nonnegative definite (i.e., $\underline{V}^T D \underline{V} \geq 0$) and $k_1 > 0, k_2 > 0$, the energy function of Equation II.3.A.1 therefore is nonnegative function (i.e., $E(t) > 0$).

After differentiating Equation II.3.A.1 with respect to time, we can reach:

$$\frac{dE(t)}{dt} = -2k \sum_i C_i [g_i^{-1}(v_i(t))] \bullet \left[\frac{dv_i(t)}{dt} \right]^2 \qquad \text{(II.3.A.2)}$$

Since C_i is a positive constant, and $g_i^{-1}(\bullet)$ is of monotonic increase, each term in Equation II.3.A.2 is nonnegative. So we conclude that the energy of (A.1) is a monotonically decreased function of time regardless of its initial state, i.e.:

$$\frac{dE(t)}{dt} \leq 0 \qquad \text{(II.3.A.3)}$$

According to Lyapunov's stability theorem, we can deduce that our proposed neural network is stable under the energy function definition of Equation II.3.35 and has an unique energy minimal point that is the global minimizer of $E(t)$.

Note that from Equations II.3.A.2–II.3.A.3 we have $\frac{dE(t)}{dt} = 0$ if and only if $\frac{dv_i(t)}{dt} = 0$, where $i = 1, 2, \ldots, N$.

Appendix B: Proof of Theorem 5

In matrix notation, Equation II.3.34 can be rewritten as:

$$\frac{C}{k} \frac{d\underline{U}(t)}{dt} = -k_1 \mathbf{R} \underline{V}(t) - \beta(t) \underline{V}(t) \tag{II.3.B.1}$$

where:

$$d\underline{U}(t)/dt = [du_1(t)/dt, du_2(t)/dt, \ldots, du_N(t)/dt]^T \text{ and } \underline{V}(t) = [v_1(t), v_2(t), \ldots, v_N(t)]^T$$

From the corollary, we have the condition $\frac{d\underline{U}(\infty)}{dt} = 0$ holds. Thus, Equation II.3.B.1 becomes:

$$\mathbf{R}\underline{V}^* = \lambda^* \underline{V}^* \tag{II.3.B.2}$$

where:

$$\underline{V}^* = \underline{V}(\infty), \lambda^* = -\beta(\infty)/k_1$$

If we choose $k_1 = 1$, then $\lambda^* = -\beta(\infty)$. This represents the output of SN2 as equal to the minimum eigenvalue with an opposite sign when the network settled down. By the definition, λ^* is an eigenvalue and $\underline{V}^*$ is the corresponding eigenvector of the $\mathbf{R}$. In the following, we will prove that λ^* and $\underline{V}^*$ are the minimum eigenvalue and its corresponding eigenvector of $\mathbf{R}$, respectively.

If we assume $\underline{e}_i (i = 1, 2, \ldots, N)$ is an orthonormalized basis consisting of a set of eigenvectors, then we have:

$$\underline{V}(t) = \sum_{i=1}^{N} \eta_i(t) \underline{e}_i \tag{II.3.B.3}$$

Substituting Equation II.3.B.3 into Equation II.3.B.1, we get:

$$\frac{d\eta_i(t)}{dt} = -k_C (k_1 \lambda_i + \beta(t)) \eta_i(t) \quad (i = 1, 2, \ldots, N) \tag{II.3.B.4}$$

By taking variable replacement of $\xi_i(t) = \frac{\eta_i(t)}{\eta_N(t)} (i < N)$ and considering Equation II.3.B.4, we obtain:

$$\frac{d\xi_i(t)}{dt} = k''(\lambda_N - \lambda_i) \xi_i(t)(i < N) \tag{II.3.B.5}$$

From Equation II.3.B.5, we solve for $\xi_i(t)$ and obtain:

$$\xi_i(t) = k_4 e^{k''(\lambda_N - \lambda_i)t} (i < N) \tag{II.3.B.6}$$

where $k_C = k_M k/C, k'' = k_C k_1$ and k_4 in Equations II.3.B.4–II.3.B.6 are constants.

Case 1: If λ_N is single-fold eigenvalue of $\mathbf{R}$, from Equation II.3.B.6, we have $\lim_{t \to \infty} \xi_i(t) = 0$ $(i = 1, 2, \ldots, N - 1)$ (i.e., $\lim_{t \to \infty} \eta_i(t) = 0$). Finally, Equation II.3.B.3 becomes:

$$\underline{V}^* = \eta_N(\infty) \mathbf{e}_N \tag{II.3.B.7}$$

Case 2: If λ_N is a M-fold eigenvalue of $\mathbf{R}$, we have $\lim_{t\to\infty} \xi_i(t) = 0$ $(i = 1, 2, \ldots, N - M)$ (i.e., $\lim_{t\to\infty} \eta_i(t) = 0$).

$$\underline{\mathbf{V}}^* = \sum_{i=M}^{N} \eta_i(\infty)\underline{\mathbf{e}}_i \tag{II.3.B.8}$$

By multiplying both sides of Equation II.3.B.8 with $\mathbf{R}$ and considering Equation II.3.B.8 again, we get:

$$\mathbf{R}\underline{\mathbf{V}}^* = \lambda_N \underline{\mathbf{V}}^* \tag{II.3.B.9}$$

Referring to Equation II.3.B.2, the obtained minimum eigenvalue is:

$$\lambda_N = -\beta(\infty)/k_1 = -\eta(\|\underline{\mathbf{V}}^*\|^2 - 1) \tag{II.3.B.10}$$

where $\eta = k_2/k_1$. The norm of the minimum eigenvector is given by:

$$\|\underline{\mathbf{V}}^*\| = \sqrt{1 - \lambda_N/\eta} \tag{II.3.B.11}$$

An important point to be noted from Equation II.3.B.11 is that the norm of the solution is predetermined by the value of η and λ_N; the higher the value of η, the closer this norm is to unity.

Appendix C: Proof of Theorem 6

Since $\mathbf{R} = \sum_{i=1}^{N} \lambda_i \underline{\mathbf{e}}_i \underline{\mathbf{e}}_i^T$ and $Tr(\mathbf{R}) = \sum_{i=1}^{N} \lambda_i$, substituting them into Equation II.3.B.2, we have:

$$\mathbf{R}_j = \sum_{i=1}^{N} \lambda_i \underline{\mathbf{e}}_i \underline{\mathbf{e}}_i^T + \sum_{i=N-j+1}^{N} [Tr(\mathbf{R}) - \lambda_i]\underline{\mathbf{e}}_i \underline{\mathbf{e}}_i^T = \sum_{i=1}^{N} \lambda_i^j \underline{\mathbf{e}}_i \underline{\mathbf{e}}_i^T \tag{II.3.C.1}$$

where:

$$\lambda_i^j = \begin{cases} Tr(\mathbf{R}) & i = N - j + 1, N - j + 2, \ldots, N, \\ \lambda_i & i = 1, 2, \ldots, N - j. \end{cases} \tag{II.3.C.2}$$

Therefore, from Equations II.3.C.1 and II.3.C.2, the smallest eigenvalue and corresponding eigenvector of the new matrix $\mathbf{R}_j$ are just the $(N - j)$-th eigenvalue and its corresponding eigenvector of $\mathbf{R}$ in ascending order, respectively.

References

1. Lippman, R.P., An introduction to computing with neural nets, *IEEE ASSP Magazine*, 20, 4, 1987.
2. Rumelhart, D.E., Hinton, G.E., and Williams, R.J., Chapter 8: Learning internal representations, in *PDP*, vol. 1, D.E. Rumelhart and J.L. McClelland, Eds., MIT Press, Cambridge, 1986.
3. Becker, S. et al., Improving the convergence of back-propagation learning with second-order methods. In: *Proc. of 1988 Connectionist Models Summer School*, Carnegie-Mellon University, 1988.
4. Singhal, S. et al. (Eds.), Training multilayer perceptrons with the extended kalman algorithm. *Adv. Neural Inform. Proc. Syst.*, 1989.
5. Yeh, S.J. and Stack, H., A fast learning algorithm for multilayer neural network based on projection methods, *Neural Network Theory and Applications*, Academic Press, San Diego, 323, 1991.

6. Tan, Y. and He, Z.Y., A second-order learning algorithm for multilayer neural networks. In *Proc. IEEE Int. Conf. Neural Network and Signal Processing*, 1, 188, 1995.

7. Okayama, Y., Approach to detection of fires in their early stage by odor sensors and neurons and neural net. In *Proc. 3rd Int. Sym. Fire Safety Science*, 372, 1991.

8. Deng, C., Research on feedforward neural network and its application in fire science. M.S. Thesis, University of Science and Technology of China, chapter 3, Hefei, China, June 1996.

9. Tan, Y., Neural networks optimization computation: Theory and its applications to signal processing, Southeast University Ph.D. Dissertation, Nanjing, China, July 1997.

10. Kung, S.Y., *Digital neural network*. PTR Prentice-Hall, Inc., 1993.

11. Hopfield, J.J. and Tank, D.W., Neural computations of decisions in optimization problems, *Biological Cybern.*, 52, 141, 1985.

12. Van Hulle, M.M., A goal programming network for LP, *Biological Cybern.*, 65, 243, 1991.

13. Oja, E., A simplified neuron model as a principal component analyzer, *J. Math. Biol.*, 15, 267, 1982.

14. Oja, E. and Karhunen, J., On stochastic approximation of the eigenvectors and eigenvalues of the expectation of a random matrix, *J. Math. Anal. and Appl.*, 106, 69, 1985.

15. Oja, E., Ogawa, H., and Wangwiwattana, J., Principal component analysis by homogeneous neural networks. Part I: The weighed subspace criterion; Part II: Analysis and extensions of the learning algorithm, *IEICE Trans. INF. & SYST.*, E-75D, 3, 366, 1992.

16. Mathew, G. et al., Gauss-Newton adaptive subspace estimation. In *Proc. IEEE Int. Conf. Acoust., Speech, Signal Processing*, 1201, 1992.

17. Yang, B., Subspace tracking based on projection approach and recursive least squares methods. In *Proc. IEEE Int. Conf. Acoust., Speech, Signal Processing*, 451, 1993.

18. Sanger, T.D., Optimal unsupervised learning in a single layer linear feedforward neural network, *Neural Networks*, 2, 3, 459, 1989.

19. Chauvin, Y., Principal component analysis by gradient descent on a constrained linear Hebbian cell. In *Proc. Int. Joint Conf. Neural Networks*, 1, 373, 1989.

20. Kung, S.Y. and Diamantaras, K.I., A neural network learning algorithm for adaptive principal component extraction (APEX). In *Proc. IEEE Int. Conf. Acoust., Speech, Signal Processing*, 2, 861, 1990.

21. Xu, L., Least mean square error reconstruction principle for self-organizing neural nets, *Neural Networks*, 6, 4, 627, 1993.

22. Peper, F. and Noda, H., A symmetric linear neural network that learns principal components and their variances, *IEEE Trans. Neural Networks*, 7, 4, 1042, 1996.

23. Tan, Y. and He, Z.Y., A neural network structure for determination of minimum eigenvalue and its corresponding eigenvector of a symmetrical matrix. In *Proc. IEEE Int. Conf. Neural Network and Signal Processing*, 1, 512, 1995.

24. Tan, Y. and He, Z.Y., Neural network approaches for the extraction of the eigenstructure. Neural Networks for Signal Processing VI—*Proc. 1996 IEEE Workshop*, Kyoto, Japan, 21, 1996.

25. Tan, Y. and He, Z.Y., A neural network method for matrix eigensolver in real time, *J. Southeast University (English Edition)*, 13, 1, 22, 1997.

26. Ljung, L., Analysis of recursive stochastic algorithm, *IEEE Trans. Automatic Control*, 22, 551, 1977.

27. Vaidyanathan, P.P. and Nguyen, T.Q., Eigenfilters: A new approach to least squares FIR filter design and applications including Nyquist filters, *IEEE Trans. Circuits Syst.*, 34, 11, 1987.

28. McClellan, J.H. and Parks, T., A unified approach to the design of optimum FIR linear-phase digital filters, *IEEE Trans. Circuit Theory*, 20, 697, 1973.

29. Pei, S.C. and Shyu, J.J., 2-D FIR eigenfilters: A least-squares approach. *IEEE Trans. on Circuit and System*, 37, 1, 24, 1990.

30. Nguyen, T.Q., The eigenfilter for the design of linear-phase filters with arbitrary magnitude response. In *Proc. IEEE Int. Symp. Acoust., Speech, Signal Processing*, Toronto, Canada, May 1991.

31. Nashashibi, A. and Charalambous, C., 2-D FIR eigenfilters. In *Proc. IEEE Int. Symp. Circuit and System*, Helsinki, Finland, 1037, June 1988.

32. Bhattacharya, D. and Antoniou, A., Design of 2-D FIR filters by feedback neural networks. In *Proc. IEEE Int. Symp. Circuit and System*, 1297, 1995.

33. Martinelli, G. and Perfetti, R., Neural network for real time synthesis of FIR filters, *Electron. Lett.*, 25, 17, 1199, 1989.

34. Tan, Y. and He, Z.Y., A neural network method for the matrix eigensolver in real time, *J. of Southeast University, (English Edition)*, 13, 1, 22, 1997.

35. Tan, Y. and Deng, C., Solving for a quadratic programming with a quadratic constraint based on a neural network frame, *Neurocomputing*, 30, 117, 2000.

4

Neural Network and Fuzzy Control Techniques for Robotic Systems

Kazuo Kiguchi
Saga University

Keigo Watanabe
Saga University

Toshio Fukuda
Nagoya University

Abstract. In this chapter, advanced neural network and fuzzy control techniques for robotic systems such as robot manipulators are introduced. Position/force control is one of the most important and fundamental operations of robot manipulators to perform sophisticated tasks. It is known that neural network and fuzzy control techniques are very effective for position/force control, especially in an unknown/uncertain environment. By applying these soft computing techniques, undesired overshooting and oscillation caused by the unknown/uncertain dynamics of a robot manipulator and an object/environment can be decreased efficiently and precise position/force control can be realized.

4.1 Introduction

Soft computing techniques such as neural networks, fuzzy reasoning, or genetic algorithms attract many researchers' attention in the field of robotics because of their significant abilities. It is known that learning and adaptation ability can be obtained by applying neural networks; human knowledge or experience can be incorporated into control law by applying fuzzy control; and global optimization can be performed by applying genetic algorithms. Adaptation/learning ability of neural networks and flexibility of fuzzy control is especially effective for robot systems working in an unknown/uncertain environment.

In this chapter, we introduce advanced neural network and fuzzy control techniques for robotic systems. Here, a robot manipulator is used as an example of robotic systems since a robot manipulator is one of the most popular and important robot systems. Position/force control[1-8] is one of the most important and fundamental operations of robot manipulators to perform sophisticated tasks. It is known that neural network and fuzzy control techniques are very effective for position/force control, especially in an unknown/uncertain environment. Many types of neural networks and fuzzy position/force controllers have been developed for control of robot manipulators up to the present.[7–21] In this chapter, the latest soft computing techniques (mainly neural network and fuzzy control techniques) are introduced to realize advanced robot position/force control.

The rest of this chapter is organized as follows. Section 4.2 presents position/force control methods of robot manipulators. Neuro control techniques for position/force control of robot manipulators are introduced in Section 4.3. Neural network and fuzzy control techniques for position/force control of robot manipulators are introduced in Section 4.4. Some conclusions are made in Section 4.5.

4.2   Position/Force Control of Robot Manipulators

A robot manipulator is one of the most popular and fundamental robots mainly used in industries. One of the most important and fundamental tasks of the robot manipulators is position/force control.[1–8] It is necessary to realize the appropriate amount of force to perform sophisticated tasks such as grinding, deburring, wiping, or assembling of objects with the robot.[22] Consequently, position/force control is necessary to perform sophisticated tasks by robot manipulators. In the case when dynamic properties of object/environment are unknown or uncertain, however, precise force control is difficult to be realized with the conventional controller since proper control gains cannot be defined. Therefore, adaptation/learning ability is required of the force controller in order to deal with an unknown or uncertain object/environment.

In order to perform sophisticated tasks, sometimes the position of the end-effector of the robot manipulator on the object/environment must be controlled during force control operation. Friction between the robot manipulator and the object/environment must be compensated for as much as possible to perform precise position control. However, amount of friction between the robot manipulator and the object/environment varies depending on amount of applied force, condition of the object/environment surface, condition of the end-effector, temperature, etc. Therefore, adaptation/learning ability is also required of the position controller in order to compensate for friction efficiently.

There are two major approaches, hybrid position/force control[1,3] and impedance control,[2] for robot force control. The basic concept of hybrid position/force is dividing the task space for force control direction (orthogonal to the constraint surface) and position control direction (moving along the constraint surface). Featherstone, Thiebaut, and Katib[4] extended this method for multi-contact compliant motions for assembly tasks. The basic concept of impedance control is controlling the relation between the robot motion and the reaction force from an environment, considering the robot manipulator as a mass-spring-damper system. In this chapter, we use hybrid position/force in order to control both position and force precisely.

In this chapter, a 3DOF planar robot manipulator (Figure II.4.1) is used as a robot manipulator model. The mathematical equation of the planar robot manipulator is written as:

$$M(q)\ddot{q} + h(q, \dot{q}) + F_{jc}\,\text{sgn}\,(\dot{q}) = \tau - J^T f \tag{II.4.1}$$

where M is inertia matrix, h denotes Coriolis and centrifugal components, F_{jc} is Coulomb friction of the robot manipulator joint, τ is output torque, J is Jacobian, f is the applied force to the environment, and q is an angular vector.

The acceleration of the end-effector of the robot manipulator in the Cartesian coordinate system is written as:

$$\ddot{x} = J\ddot{q} + \dot{J}\dot{q} \tag{II.4.2}$$

where x is a position vector in the Cartesian coordinate system.

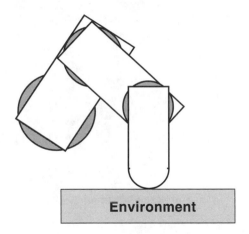

FIGURE II.4.1   3DOF planar robot manipulator.

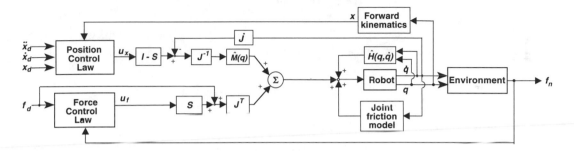

FIGURE II.4.2   Block-diagram of hybrid position/force control system.

Using Equations II.4.1 and II.4.2 and the selection matrix S, which selects the direction for force control that is normal to the constraint surface of the object/environment, and for position control that moves along the constraint surface in Cartesian coordinate system, the equation of hybrid position/force control is written as:

$$\tau = \hat{M}(q)J^{-1}[(1-S)u_x - \dot{J}\dot{q}] + \hat{h}(q,\dot{q})$$
$$+ \hat{F}_{jc}\,\text{sgn}\,(\dot{q}) + J^T f_d + J^T S u_f \tag{II.4.3}$$

where $\hat{M}$ is the estimated inertia matrix, $\hat{h}$ represents the estimated Coriolis and centrifugal components, $\hat{F}_{jc}$ denotes the estimated friction model of the robot manipulator joint, f_d represents the desired force, u_p is the position control commond, and u_f is the force control command. The resolved acceleration method,[23] in which command vectors are written as the following equations, is often used in the conventional hybrid position/force control:

$$u_p = \ddot{x}_d + K_{pd}(\dot{x}_d - \dot{x}) + K_{pp}(x_d - x) \tag{II.4.4}$$

$$u_f = K_{fp}(f_d - f) \tag{II.4.5}$$

where x_d denotes the desired position, K_{pd} is a velocity gain for position control, K_{pp} is a position gain for position control, and $\check{K}_{fp}$ is a gain for force control. The dynamic properties of the object/environment have to be known in order to control contact force using the conventional control

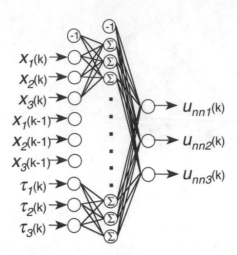

FIGURE II.4.3   The neuro position/force controller.

techniques, since the dynamics of the object/environment affects the dynamics of the whole system. The block-diagram of the hybrid position/force control system is depicted in Figure II.4.2.

4.3 Neuro Control Techniques for Robot Manipulators

4.3.1 Neuro Position/Force Control

Adaptation/learning ability of neural networks plays an important role in robotics. Unknown/ uncertain dynamics of the robot manipulator and the object/environment, modeling error, friction, and unexpected disturbances can be compensated for by the effect of adaptation/learning ability of neural networks. Several kinds of neuro position/force controllers have been proposed.[7–18]. Some of them applied adaptation ability of the neural networks,[9–14] others applied iteratively learning ability of neural networks,[15,16] or classification ability of neural networks[17,18] for position/force control in an unknown/uncertain object/environment. More detailed explanation about these studies is given in Katic[7] and Kicuchi and Fukuda.[8] In this chapter, we briefly explain the neuro position/force controller[11] depicted in Figure II.4.3 as an example of adaptive type neuro position/force controllers. In Figure II.4.3, Σ means sum of the inputs. In this neuro hybrid position/force controller, Equation II.4.3 is changed to:

$$\tau = \hat{M}(q)J^{-1}[(I - S)u_x - \dot{J}\dot{q}] + \hat{h}(q,\dot{q})$$
$$+ F_{jc}\,\mathrm{sgn}\,(\dot{q}) + J^T f_d + J^T S u_f + J^T u_{nn} \qquad (\text{II.4.6})$$

where u_{nn} represents the force command vector, i.e., the output from the neural network.

The neural network consists of three layers (input layer, hidden layer, and output layer). Input signals to the neural network are current and previous positions of the robot manipulator in Cartesian coordinate system, and the computed torque commands from the conventional hybrid position/force controller for each motor (i.e., the calculation result of Equation II.4.3). Output signals from the neural network are the force commands for the robot manipulator in Cartesian coordinate system. The block-diagram of the neuro position/force control system is depicted in Figure II.4.4.

Activation function in the neuron characterizes the nonlinear mapping ability of neural networks. Several kinds of nonlinear functions, such as sigmoid function, Gaussian function, etc., can be used for activation functions in the neuron. In this neural network, sigmoid function is used as an activation

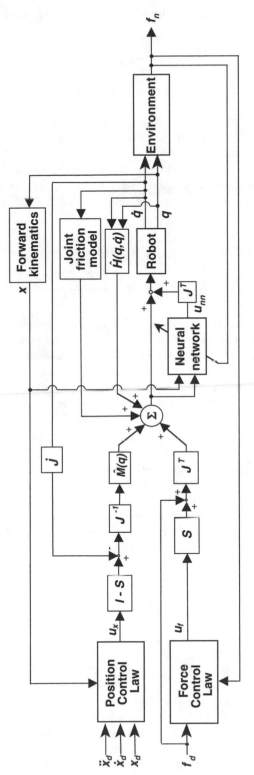

FIGURE II.4.4 Block-diagram of neuro position/force control system.

function of neurons in the hidden layer and the output layer. The equation of sigmoid function is written as:

$$f_s(u) = \frac{2}{1 + e^{-u}} - 1 \tag{II.4.7}$$

$$u = w_i x + w_o \tag{II.4.8}$$

where w_o is the threshold value and w_i is the weight. This neural network is used to compensate for unknown/uncertain dynamics of the robot manipulator and the object/environment, modeling error, friction, and unexpected disturbances.

Adaptation/learning of the neural network is carried out using the back-propagation learning algorithm[24] to minimize the evaluation function; every weight of the neural network is adjusted on-line to eliminate the output errors using the back-propagation learning algorithm. Usually, the evaluation function is given as:

$$E = \frac{1}{2}(x_d - x)^2 \qquad \text{for position control} \tag{II.4.9}$$

$$E = \frac{1}{2}(f_d - f)^2 \qquad \text{for force control.} \tag{II.4.10}$$

As far as using the squared error function written above as the evaluation function, however, the neural network does not learn unknown/uncertain dynamics of the robot manipulator and the object/environment, although it eliminates the control error. Therefore, another evaluation function[12,25,26] is recommended. In Kiguchi and Fukuda,[12,25] fuzzy controlled evaluation functions were proposed. The evaluation ratio of error term and velocity term is fuzzy controlled in Reference 12, and the evaluation value itself is fuzzy controlled in Reference 25. Recently, genetic programming was applied to generate effective evaluation functions automatically.[26]

In order to evaluate the effectiveness of the neuro controller, force control simulation in uncertain environment has been performed. The results are shown in Figure II.4.5. In this simulation, the desired force is given as a combination of step signals which change between 20 N and 10 N every 2 sec. The position of the robot manipulator is controlled to push the same point on the object/environment perpendicularly for force control. The property of the environment is supposed to be uncertain. The sampling time is set to 1 m/sec. Uncertain joint friction and modeling errors of the robot manipulator have been taken into account as in the real experiment. One can see that the desired force can be realized from the simulation results, although overshooting and a little oscillation occur when the desired force is changed. Overshooting and oscillation would decrease if the neuro controller is iteratively trained in the same condition.

4.3.2 Damping Neurons for Neural Networks

Although a typical artificial neuron (like a neuron in the neural network explained in the previous section) is modeled as a static system, a biological neuron is a dynamic system. In the human brain, transmission of electrochemical signals is dynamically performed to process information.[27] Consequently, process of signals in the biological neural networks is dynamic. On the other hand, almost all physical systems such as robot manipulators and objects/environments possess their own dynamics. Therefore, processing of signals in the artificial neural networks should also be dynamic in order to make the controller adapt itself to the object/environment dynamics efficiently.

In order to take into account the dynamics of neural networks, recurrent neural networks are sometimes used.[28,29] However, structure and adaptation/learning algorithm of the recurrent neural networks is complicated. Besides, recurrent neural networks themselves do not seem to be suitable for generating control signals, although they are suitable for identification of dynamic systems.

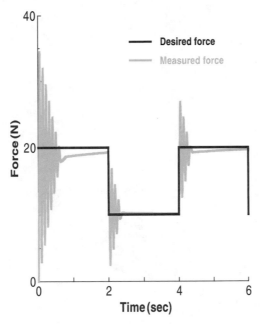

FIGURE II.4.5 Simulation result of neuro position/force control.

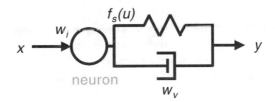

FIGURE II.4.6 Physical image of the damping neuron.

In this section, the damping neuron, in which the output signal from the neuron has damping effect, is introduced and applied to construct the neuro controller. The damping neuron in which the viscosity is given to the output signal is considered. Figure II.4.6 describes the physical image of the damping neuron. Regarding the output signal from the neuron as displacement as shown in Figure II.4.6, the total output signal (i.e., the input signal to the neurons in the next layer) is obtained by summing the amount of displacement and the weighted amount of change of displacement. By applying the damping neurons, the neuro controller is able to effectively learn and compensate for unknown/uncertain dynamics of the robot manipulator and the object/environment during the position/force control, since the output signal from the active neuron is accelerated. Furthermore, with viscous friction at each robot joint, and nonlinearities such as Coriolis and centrifugal of the robot manipulator in addition to the dynamics of the object/environment (including unknown or unmodeled velocity terms), it might be easier to learn such velocity terms if neurons themselves in the neuro controller contain viscosities.

The equation of the output from the damping neuron is written as:

$$y = f_s(u) + w_v \dot{f}_s(u) \tag{II.4.11}$$

where y is the total output from the damping neuron, w_v represents the weight for change of output.

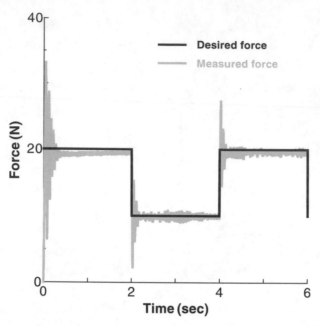

FIGURE II.4.7   Simulation result of neuro position/force control with damping neurons.

For the adaptation/learning of the damping neural network (i.e., the neural network that consists of damping neurons), the conventional back-propagation learning algorithm can be applied. The weights for the damping term (change of output) of the damping neuron can be adjusted on-line as well as the other weights by the back-propagation learning algorithm, considering that the output from the damping neuron expressed in Equation II.4.11 is an output signal from the neuron. The special learning algorithm is not required for the training of damping neurons. The simplicity of the structure and the learning is the advantage of the damping neural network over the recurrent neural network.

Force control simulation has been performed to evaluate the effectiveness of the damping neural network. The architecture of the damping neural network is the same as that of the neural network explained in Section 4.3.1. We have applied the damping neurons in the hidden layer and the output layer. The simulation condition is the same as that in Section 4.3.1. The results are shown in Figure II.4.7. One can see that the undesired oscillation becomes smaller compared with the result shown in Figure II.4.5 due to the effective adaptation ability of damping neurons. The effectiveness of damping neuron has been confirmed by experiment.[12]

4.4   Neural Network and Fuzzy Techniques for Robot Manipulators

4.4.1   Fuzzy-Neuro Position/Force Control

Neuro control is effective (especially with damping neurons) for adaptation of the force controller to unknown/uncertain dynamics of the object/environment. However, undesired response such as overshooting or oscillation might occur until the controller adapts to the unknown/uncertain dynamics of the object/environment. We have to avoid them, since unexpected overshooting or oscillation might cause some damage to the object/environment or the robot manipulator itself. Fuzzy-neuro control,[30] the combination of fuzzy control and adaptive type neuro control, is one of the best control methods to solve this problem. It is known that fuzzy control is able to deal with human knowledge and experience. Consequently, our knowledge and experience can be applied to avoid overshooting in unknown/uncertain object/environment by applying the fuzzy-neuro controller to position/force control. Thus, the fuzzy-neuro controller is the same as the trained neuro controller using human linguistic control rules as teaching data, although it does not require actual prelearning. Therefore, human-like adaptive control is realized by applying the fuzzy-neuro controller. In order to apply fuzzy-neuro control to hybrid position/force control, a fuzzy-neuro position controller and a fuzzy-neuro force controller have to be prepared, as depicted in Figure II.4.8.

Our knowledge and experience, which realize precise control without undesired overshooting and oscillation, are expressed as fuzzy IF-THEN control rules and then transferred to the structure of a neural network, in order to design the fuzzy-neuro controller. Figure II.4.9 shows the architecture of the fuzzy-neuro force controller. Here, Σ means sum of the inputs, Π means multiplication of the inputs. The fuzzy-neuro force controller consists of five layers (input, fuzzifier, rule, defuzzifier, and output layer). The fuzzifier layer, which generates membership functions, consists of 10 neurons, the rule layer consists of 19 neurons, and the defuzzifier layer consists of 2 neurons.

There are two input signals to the controller. One is the force error and the other is the momentum of the robot manipulator in the direction of the force control. Usually fuzzy control uses error and error change rate for input signals to the controller. In the case of force control, it is not easy to use error change rate, since the signals from a force sensor are noisy. The robot manipulator's pushing velocity against the object/environment surface might be used instead of force error change rate. As far as robot manipulators are concerned, however, the sense of the velocity depends on the amount of their inertia matrices. This means that the sense of velocity might be different even with the same robot manipulator if its configuration is different. Therefore, the membership functions of the velocity should be modified whenever the configuration of the robot manipulator is changed. This is not practical. Using the momentum of the robot manipulator instead of force error change rate is practical, since the amount of the inertia is taken into account. The equation of the momentum of the robot manipulator in the direction of the force control is written as:

$$M_o = M_x(q)v \tag{II.4.12}$$

$$M_x(q) = J^{-T}(q)M(q)J^{-1}(q) \tag{II.4.13}$$

where M_o represents the momentum of the manipulator in the direction of force control, v is the velocity of the robot manipulator in the direction of force control, and $M_x(q)$ denotes the inertia matrix in the Cartesian coordinate system when the angular position vector of the robot manipulator is q.

Five kinds of fuzzy linguistic variables (NB: Negative Big; NS: Negative Small; ZO: Zero; PS: Positive Small; and PB: Positive Big) are prepared for each input signal (error and robot manipulator momentum). The membership functions are defined in the fuzzifier layer using Gausian function, which is written as the Equation II.4.14, and sigmoid function, which is written as Equation II.4.16:

$$f_G(u_G) = e^{-u_G^2} \tag{II.4.14}$$

$$u_G = \frac{w_o + x}{w_i} \tag{II.4.15}$$

$$f_S(u_S) = \frac{1}{1 + e^{-u_S}} \tag{II.4.16}$$

$$u_S = w_i x + w_o \tag{II.4.17}$$

Calculated membership functions in the fuzzifier layer are sent to the rule layer and multiplied in the neurons in the rule layer according to fuzzy IF-THEN rules. There are two outputs from each neuron in the rule layer. One of them is multiplied by the weight and summed in the next layer, and

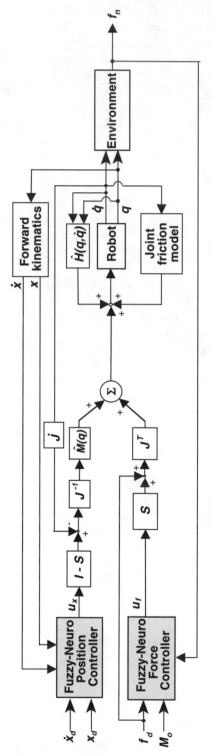

FIGURE II.4.8   Block-diagram of fuzzy-neuro position/force control system.

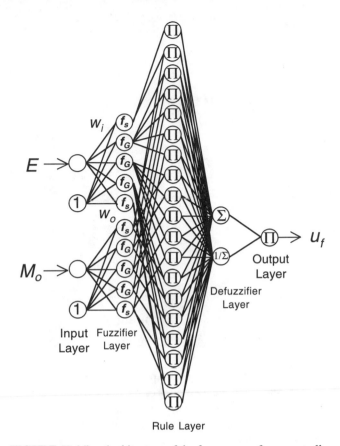

FIGURE II.4.9   Architecture of the fuzzy-neuro force controller.

the other is summed and inverted in the defuzzifier layer. Multiplied value of these outputs from the defuzzifier layer will be the output of the fuzzy-neuro force controller (i.e., the force command in the Cartesian coordinate system).

Force control experiment has been carried out to evaluate the effectiveness of the fuzzy-neuro force controller. In the experiment, the desired force is given as a combination of step signals which change between 20 N and 10 N every 2 sec. The position of the robot manipulator is controlled to push the same point on the object/environment perpendicularly for force control. Steel is used as an unknown/uncertain environment in this experiment. The sampling time is set to 2 m/sec. The experiment result is shown in Figure II.4.10. One can see that precise force control can be performed without undesired overshooting and oscillation.

The fuzzy-neuro position controller is the same as the fuzzy-neuro force controller except for input signals. The position error and its change rate are used as input signals to the fuzzy-neuro position controller. The effectiveness of the fuzzy-neuro position controller is evaluated in Section 4.4.3 since it is related to friction compensation.

4.4.2   Application of Multiple Controllers

The fuzzy-neuro controller is effective for force control in unknown/uncertain object/environment as explained in the previous section. However, if the dynamic properties of the unknown object/environment are extremely different from the initially estimated ones, unexpected overdamped or underdamped responses might occur until the controller adapts to the object/environment. In order

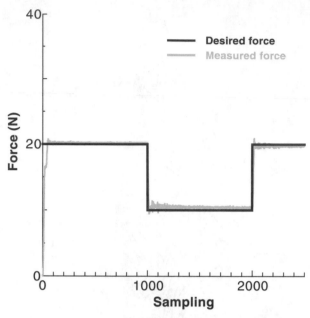

FIGURE II.4.10   Experimental result of fuzzy-neuro position/force control.

to cope with this problem, the on-line controller adjustment methods, which adjust the controller inputs immediately in accordance with the dynamics of the object/environment, have been proposed using off-line trained neural network[31] and a fuzzy reasoning.[22] These adjustment methods work well, although they simply adjust the antecedent part of control rules.

In this section, we present another fuzzy-neuro robot force control method in which suitable fuzzy-neuro force controllers are selected immediately from initially prepared multiple fuzzy-neuro force controllers designed for various kinds of objects/environments, and harmonized with a proper ratio using fuzzy reasoning according to the dynamic properties of the unknown/uncertain object/environment. In this control method, on-line classification of the unknown/uncertain object/environment is carried out with an off-line trained neural network; then fuzzy reasoning for controller selection and harmonization is performed based on the classification information (i.e., the output from the classification neural network). Consequently, an exactly suitable fuzzy-neuro controller is selected when the object/environment is the same as an initially expected one, and some suitable fuzzy-neuro controllers are selected and harmonized with a proper ratio when the object/environment is an unexpected one. A suitable fuzzy-neuro force controller is supposed to be designed for each object/environment which is expected to be used for the tasks. Therefore, the number of fuzzy-neuro force controllers is the same as the number of the expected object/environments to be used. The block-diagram of this control system is depicted in Figure II.4.11.

4.4.2.1 Object/Environment Classification Using a Neural Network

An off-line trained multi-layer neural network is used to classify an unknown/uncertain object/environment in an on-line manner. The output from the neural network (i.e., classification result) is used to select the suitable fuzzy-neuro force controllers. The inputs to the neural network are deformed displacement, deforming velocity, and deforming acceleration of the object/environment surface and reaction force from the object/environment. The off-line training data for the neural network regarding the initially expected environments to be used are supposed to be obtained by pre-experiment. Using these training data, the neural network is off-line trained with the back-propagation learning algorithm to output a larger number for harder objects/environments.

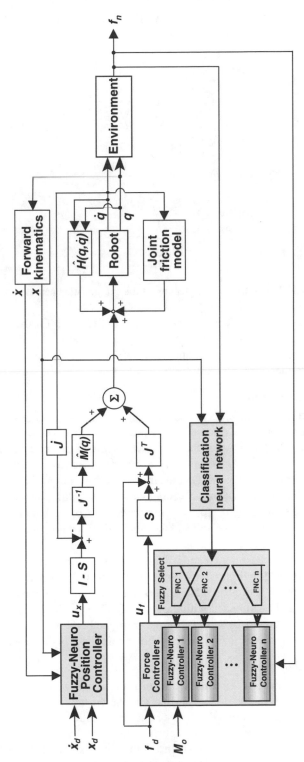

FIGURE II.4.11 Block-diagram of multiple fuzzy-neuro position/force control system.

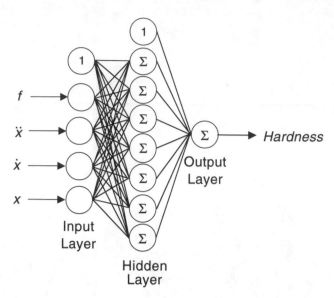

FIGURE II.4.12   Architecture of the classification neural network.

The training patterns of every object/environment are used in turn for the effective training of the neural network. The architecture of the classification neural network is depicted in Figure II.4.12. In general, sample objects/environments and their training patterns should be prepared as much as possible to generalize the neural network.

4.4.2.2   Selection of Controllers

Suitable fuzzy-neuro force controllers are selected using fuzzy reasoning based on the output from the classification neural network explained in the previous subsection. If the output from the neural network indicates that the hardness of the unknown/uncertain object/environment is the same as, or very similar to, that of one of the initially expected objects/environments, the most suitable fuzzy-neuro force controller is selected. On the other hand, if the output from the neural network does not indicate that the unknown/uncertain object/environment is different from the initially expected objects/environments, some suitable fuzzy-neuro force controllers which have been designed for similar objects/environments are selected and harmonized with a proper ratio using fuzzy reasoning according to the degree of similarity.

For example, if the fuzzy reasoning results in the degrees of similarity of the unknown environment against the initially expected environment A, B, and others are 0.2, 0.8, and 0.0, respectively, as shown in Figure II.4.13, the fuzzy-neuro force controllers designed for the environment A (FNC A) and B (FNC B) are selected and harmonized with the ratio of 0.2 : 0.8. Consequently, the output from the controller is the combination of 20% of the output from the FNC A and 80% of the output from the FNC B. This will be the force command for the robot manipulator in the Cartesian coordinate system.

4.4.2.3   On-Line Adaptation

The desired force response can be generated even if the initially designed fuzzy-neuro force controllers are not perfect, because of the on-line adaptation ability of the fuzzy-neuro controller. The squared error function is adopted as the evaluation function of the back-propagation learning algorithm. In the case when only one fuzzy-neuro force controller is selected, the learning algorithm is exactly the same as that of basic fuzzy-neuro controllers explained in Section 4.4.1. In the case when several fuzzy-neuro force controllers are selected, the same learning algorithm can also be applied since the whole structure of the controller is regarded as a neural network in which one

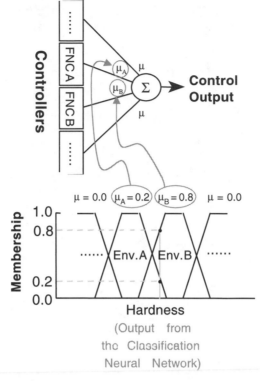

FIGURE II.4.13   Example of controller selection.

more layer (considering the ratio of the selected controllers to be the weights) is added to the basic fuzzy-neuro force controllers.

4.4.2.4   Experiment

Force control experiment has been performed to evaluate the effectiveness of the control method. In the experiment, the desired force is given as a combination of step signals which change between 20 N and 10 N every 2 sec. The position of the robot manipulator is controlled to push the same point on the object/environment perpendicularly for force control. Four kinds of material (steel, rubber, styrofoam, and sponge) have been prepared as the unknown/uncertain environments. The sampling time is set to 2 msec. The off-line training of the classification neural network has been carried out to output 2.0, 1.2, 0.6, and 0.0 for the environment of steel, rubber, styrofoam, and sponge, respectively. The fuzzy-neuro force controllers for the environment of steel, rubber, and sponge have been prepared, while that for the styrofoam environment has not been prepared, assuming that the environment of steel, rubber, and sponge are the initially expected environments to be used.

In order to avoid unexpected sudden change of the output from the classification neural network caused by noise, external disturbance, etc., classification has been started 10 sampling times after the beginning and finished 50 sampling times after the beginning. Figure II.4.14 shows examples of output from the classification neural network for these environments during the force control experiment.

Figures II.4.15 and II.4.16 show the experimental results of force control with the styrofoam environment, for which the fuzzy-neuro force controller has not been designed, using the fuzzy-neuro force controller designed for the rubber environment and the fuzzy-neuro force controller designed for the sponge environment, respectively. The classification neural network has not been used

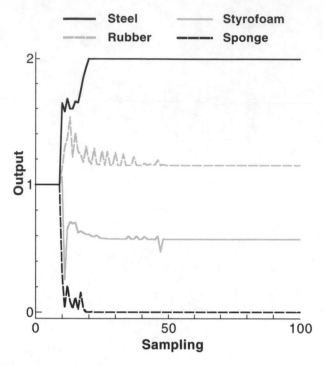

FIGURE II.4.14   Examples of output from the classification neural network.

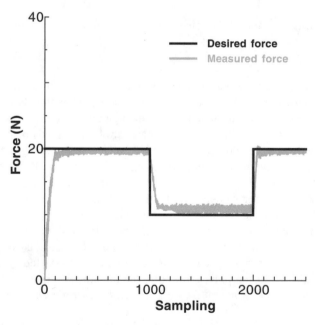

FIGURE II.4.15   Experimental result with the styrofoam environment using the fuzzy-neuro force controller designed for the rubber environment.

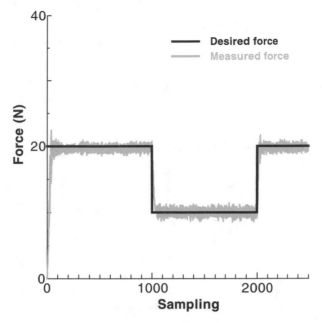

FIGURE II.4.16   Experimental result with the styrofoam environment using the fuzzy-neuro force controller designed for the sponge environment.

in these experiments. It can be said that overdamped response occurs with the controller for the rubber environment, which is harder than the styrofoam environment, and a little overshooting and oscillation occurs with the controller for the sponge environment, which is softer than the styrofoam environment.

Figure II.4.17 shows the experimental results of force control with the environment of steel, rubber, and sponge, using the proposed force control method. These results show that the desired force response can be precisely realized with every environment because of the fuzzy selection of the proper fuzzy-neuro force controllers based on the output from the environment classification neural network. Figure II.4.18 shows the experimental result of force control with the styrofoam environment using the proposed force control method. The desired force response was realized with the styrofoam environment, although the fuzzy-neuro force controller for the styrofoam has not been prepared. One can see that the result is better (faster response and less overshooting and oscillation) than the results in Figures II.4.15 and II.4.16. This means the proper fuzzy-neuro force controllers have been selected and harmonized with the proper ratio for an unknown environment by the proposed method.

4.4.3 Adaptive Friction Compensation

In the case of contact tasks, friction between the robot manipulator end-effector and the object/ environment is one of the biggest problems for realizing precise positioning. It is difficult, however, to prepare a precise friction model for feedforward compensation because of the complexity of static and dynamic characteristics of friction such as the Stribeck effect, the Dahl effect, stick-slip motion, etc. In this subsection, adaptive fuzzy friction models are introduced to effectively compensate for friction between a robot manipulator end-effector and the environment in position control. Figure II.4.19 depicts the friction model. The friction models consist of negative viscous (including static friction), Coulomb, and viscous friction. A mathematical model which is similar to Tustin's model[32] has

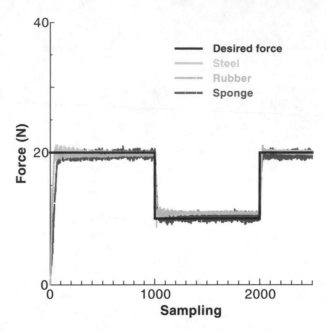

FIGURE II.4.17   Experimental result with the environment of steel, rubber, and sponge using the multiple fuzzy-neuro force controller.

been applied to express these friction models. The equations of these fuzzy friction models were written as:

$$F_{env} = F_s e^{-(\frac{v}{k})^2} \qquad\qquad \text{IF } v \text{ is ZERO} \qquad\qquad (II.4.18)$$

$$F_{env} = F_k \text{sgn}(v) + F_v v \qquad \text{IF } v \text{ is NOT ZERO} \qquad (II.4.19)$$

where F_s is the static friction, k is a constant defining the width of the function with units of velocity, F_k is the kinetic friction, and F_v is the viscous friction parameter. ZERO and NOT ZERO are fuzzy linguistic variables that indicate velocity of the end-effector moving on the object/environment. The membership functions of these fuzzy linguistic variables are depicted in Figure II.4.19. Equations II.4.18 and II.4.19 are smoothly switched based on the amount of velocity of the robot manipulator end-effector by applying these fuzzy linguistic variables in these friction models. These friction models are adjusted on-line by adjusting the coefficients of the equations using the back-propagation learning algorithm to minimize the following evaluation function:

$$E = \frac{1}{2}(F_{tan} - F_{env})^2 \qquad\qquad (II.4.20)$$

where F_{tan} is friction force in the direction of position control measured by a force sensor, and F_{env} is the output of the fuzzy friction models. Consequently, the explicit friction model can be obtained after on-line learning. The output of this fuzzy friction model is used for friction compensation. The block-diagram of the control system is shown in Figure II.4.20. The fuzzy friction model is applied in addition to the fuzzy-neuro controllers.

In order to evaluate the effectiveness of the adaptive fuzzy friction model, position/force control simulation in uncertain environment has been performed. In this simulation, the desired force is

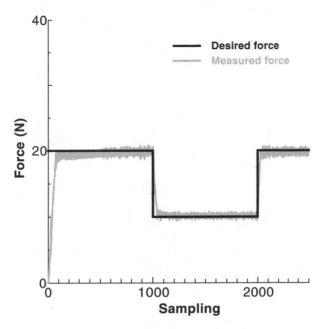

FIGURE II.4.18   Experimental result with the styrofoam environment using the multiple fuzzy-neuro force controller.

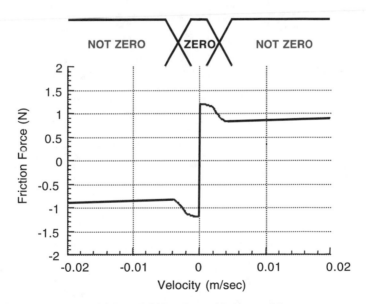

FIGURE II.4.19   Fuzzy friction model.

given as a combination of step signals which change between 20 N and 10 N every 2 sec. The desired trajectory on the environment surface is $0.02 \sin(0.2t)$ m. The angle of the robot manipulator end-effector is controlled to be perpendicular to the environment surface. The sampling time is set to 1 msec. In this simulation, the fuzzy-neuro force controller and the fuzzy-neuro position controller are used for hybrid position/force control (Figure II.4.20). The results are shown in Figure II.4.21. The position control error of the fuzzy-neuro position control method with the adaptive fuzzy friction

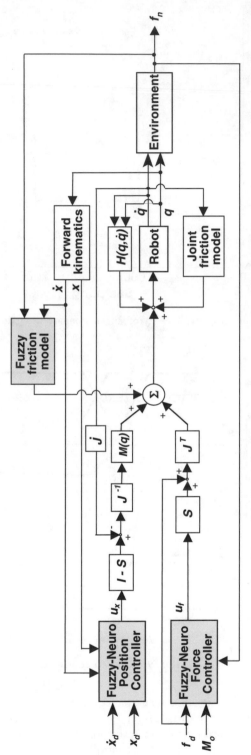

FIGURE II.4.20 Block-diagram of the control system with the fuzzy friction model.

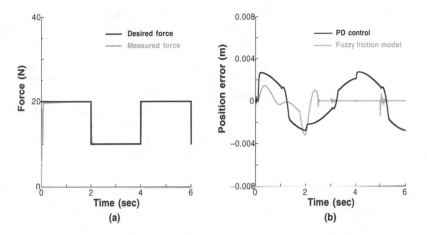

FIGURE II.4.21   Simulation results with the fuzzy friction model. (a) Force control result; (b) position error.

model has been compared with that of PD control method. One can verify the effectiveness of the fuzzy-neuro position controller with the adaptive fuzzy friction model. The adaptive friction model explained in this subsection can also be used for compensation of the robot manipulator joints' friction.[33]

4.5   Conclusions

Advanced neural network and fuzzy control techniques for robotic systems are introduced in this chapter. By applying these techniques, precise position/force control can be realized even in an unknown/uncertain environment. The simulation and experimental results show the effectiveness of these techniques. The neural network and fuzzy control techniques explained in this chapter are applicable not only to position/force control of robot manipulators in an unknown/uncertain environment, but also to control of other robotic systems, i.e., mobile robots, biorobots, etc.

References

1. Raibert, M.H. and Craig, J.J., Hybrid position/force control of manipulators, *Trans. ASME J. Dynamic Systems, Measurement, and Control*, vol. 102, pp. 126–133, 1981.
2. Hogan, N., Impedance control: An approach to manipulator: Parts I, II, III, *Trans. ASME J. Dynamic Systems, Measurement, and Control*, vol. 107, pp. 1–24, 1985.
3. Yoshikawa, T., Dynamic hybrid position/force control of robot manipulators—description of hand constraints and calculation of joint driving force, *IEEE J Robotics Automation*, vol. 3, no. 5, pp. 386–392, 1987.
4. Featherstone, R., Thiebaut, S.S., and Katib, O., A general contact model for dynamically-decoupled force/motion control, *Proc. 1999 IEEE Int. Conf. Robotics Automation*, pp. 3281–3286, 1999.
5. Whitney, D.E., Historical perspective and state of art in robot force control, *Int. J. Robotics Res.*, vol. 6, no. 1, pp. 3–14, 1987.
6. Zweng, G. and Hemami, A., An Overview of Robot Force Control, *Robotica*, vol. 15, no. 5, pp. 473–482, 1997.
7. Katic, D., Some Recent Issues in Connectionist Robot Control, *Proc. 3rd ECPD Int. Conf. Adv. Robotics, Intelligent Automation and Active Systems*, pp. 79–92, 1997.

8. Kiguchi, K. and Fukuda, T., A Survey of Force Control of Robot Manipulators Using Softcomputing Techniques, *Proc. IEEE Int. Conf. Systems, Man, and Cybernetics*, pp. II-764–II-769, 1999.

9. Tokita, M., Mitsuoka, T., Fukuda, T., and Kurihara, T., Force Control of Robot Manipulator by Neural Network Model, *J. Robotics Mechatronics*, vol. 2, no. 4, pp. 273–281, 1990.

10. Yamada, T. and Yabuta, T., Neural Network Controller Using Autotuning Method for Nonlinear Functions, *IEEE Trans. Neural Networks*, vol. 3, no. 4, pp. 595–601, 1992.

11. Kiguchi, K. and Necsulescu, D.S., Control of Multi-DOF Robots Using Neural Networks, *Proc. Knowledge-Based Systems Robotics Workshop*, pp. 747–754, 1993.

12. Kiguchi, K. and Fukuda, T., Neural Network Controllers for Robot Manipulators Application of Damping Neurons, *Adv. Robotics*, vol. 12, no. 3, pp. 191–208, 1998.

13. Jung, S. and Hsia, T.C., Neural Network Impedance Force Control of Robot Manipulator, *IEEE Trans. Ind. Electronics*, vol. 45, no. 3, pp. 451–461, 1998.

14. Lin, S.T. and Tzeng, S.J., Neural Network Force Control for Industrial Robots, *J. Intell. Robotic Systems*, vol. 24, pp. 253–268, 1999.

15. Cohen, M. and Flash, T., Learning Impedance Parameters for Robot Control Using Associative Search Network, *IEEE Trans. Robotics Automation*, vol. 7, no. 3, pp. 382–390, 1991.

16. Venkataraman, S.T., Gulati, S., and Toomarian, N., A Neural-Network-Based Identification of Environments Models for Compliant Control of Space Robots, *IEEE Trans. Robotics Automation*, vol. 9, no. 5, pp. 685–697, 1993.

17. Tsuji, T., Ito, K., and Morasso, P.G., Neural Network Learning of Robot Arm Impedance in Operational Space, *IEEE Trans. Systems, Man, and Cybernetics—Part B: Cybernetics*, vol. 26, no. 2, pp. 290–298, 1996.

18. Katic, D. and Vukobratovic, M., A Neural-Network-Based Classification of Environment Dynamics Models for Compliant Control of Manipulation, *IEEE Trans. Systems, Man, and Cybernetics*, vol. 28, no. 1, pp. 58–69, 1998.

19. Hsu, F.Y. and Fu, L.C., New Design of Adaptive Fuzzy Hybrid Force/Position Control for Robot Manipulators, *Proc. 1995 IEEE Int. Conf. Robotics Automation*, pp. 863–868, 1995.

20. Tarokh, M. and Bailey, S., Adaptive Fuzzy Force Control of Manipulators with Unknown Environment Parameters, *J. Robotic Systems*, vol. 14, no. 5, pp. 341–353, 1997.

21. Nagata, F., Watanabe, K., Kiguchi, K., Tsuda, K., Kawaguchi, S., Noda, Y., and Komino, M., Joystick teaching system for polishing robots using fuzzy compliance control, *Proc. IEEE Int. Symp. Comput. Intell. Robotics Automation*, pp. 362–367, 2001.

22. Kiguchi, K. and Fukuda, T., Intelligent position/force controller for industrial robot manipulators application of fuzzy neural networks, *IEEE Trans. Ind. Electronics*, vol. 44, no. 6, pp. 753–761, 1997.

23. Luh, J.Y.S., Walker, W.M., and Paul, R.P.C., Resolved acceleration control of manipulators, *IEEE Trans. Automatic Control*, vol. AC-25, no. 3, pp. 468–474, 1980.

24. Rumelhart, D.E., Hinton, G.E., and Williams, R.J., *Learning Internal Representations by Error Propagation*, Parallel Distributed Processing, MIT Press, Cambridge, MA, pp. 45–76, 1986.

25. Kiguchi, K. and Fukuda, T., Fuzzy evaluation for supervised learning of neural network robot controllers, *Proc. 3rd Int. Conf. Adv. Mechatronics*, pp. 802–807, 1998.

26. Kiguchi, K., Watanabe, K., Izumi, K., and Fukuda, T., Generation of evaluation function for robot force control using genetic programming, *Proc. J. IFSA 9th World Congress and 20th NAPFIPS Int. Conf.*, pp. 2767–2771, 2001.

27. Sejnowsky, T.J., Synapses get smarter, *Nature*, vol. 382, pp. 759–760, 1996.

28. Karakasoglu, A., Sudharsanan, S.I., and Sundareshan, M.K., Identification and decentralized adaptive control using dynamic neural networks with application to robotic manipulators, *IEEE Trans. Neural Networks*, vol. 4, no. 6, pp. 919–930, 1993.

29. Jin, L., Nikiforuk, P.N., and Gupta, M.M., Dynamic recurrent neural networks for control of unknown nonlinear systems, *Trans. ASME J. Dynamic Systems, Measurement and Control*, vol. 116, pp. 567–576, 1994.

30. Kiguchi, K. and Fukuda, T., Fuzzy neural controller for robot manipulator force control, *Proc. J. Conf. 4th IEEE Int. Conf. Fuzzy Systems and the 2nd Int. Fuzzy Engineer. Symp. (FUZZ-IEEE/IFES '95)*, vol. 2, pp. 869–874, 1995.

31. Kiguchi, K. and Fukuda, T., Fuzzy neural hybrid position/force control for robot manipulators, *J. Appl. Math. Comp. Sci.*, vol. 6, no. 3, pp. 101–121, 1996.

32. Tustin, A., The effects of backlash and of speed-dependent friction on the stability of closed-cycle control systems, *IEE J.*, vol. 94, part 2A, pp. 143–151, 1947.

33. Kiguchi, K., Watanabe, K., Izumi, K., and Fukuda, T., Two-stage adaptive robot position/force control using fuzzy reasoning and neural networks, *Adv. Robotics*, vol. 14, no. 3, pp. 153–168, 2000.

5

Processing of Linguistic Symbols in a Multi-Level Neural Network

Jong Joon Park
Seokyeong University

Abraham Kandel
University of South Florida

Abstract. This chapter proposes a procedure for embedding linguistic processing in a neural network and investigates a method of representing and coding linguistic terms for processing by a neural network scheme. We also propose the multi-level grading rule (MLGR) for the conversion of linguistic symbols into numeric values and for the quantization of numeric values into linguistic symbols. The MLGR method has been applied to three examples: modulo 10, next integer, and double the inputs problems, and their results are given with analysis and interpretations of the outputs.

From the results of the three examples, we find that the system produces better outputs when there is a smaller effective range for linguistic terms and a larger percentage of training data. Among the produced outputs, which were not given the sample training data, most of them have extremely reasonable meanings. The system's output results show that the system gives a high rate of correctness (greater than 60%, 80%

in the modulo 10, and in the next integer problem each) even though it trained with a small set of sample data (20%).*

Key words: Linguistic symbols, neural networks, symbolic processing, fuzzy logic.

5.1   Introduction

Fuzzy logic provides a means of linking symbolic processing and numeric computations. Symbolic processing usually manipulates linguistic symbols representing concepts and qualitative relationships; numeric computation is generally used with precise and detailed algorithms which manipulate quantitative information.[22-25] There are a variety of fuzzy implication operators, membership functions, and relation matrices used to control systems efficiently, [1,7,8] but the applications for which these apply are limited. Since a neural network is very flexible, it also can be trained to control systems in many different ways. Neural networks operate most effectively with numeric values, because their basic process is a summation of weighted values at each node. Thus, to process linguistic symbols in a neural network, it must be determined how to code linguistic symbols numerically in a robust manner so that small errors in the input cause only corresponding small changes in the output.

In this chapter we propose a procedure for embedding linguistic processing in a neural network and the multi-level grading rule (MLGR) for the conversion of a linguistic symbol into numerical values. To this end, we consider the issues of representing and coding linguistic terms: how do we decide the specific linguistic terms which correctly represent the characteristics of a problem, e.g., "extremely," "very," "little," "slightly," "tall," "middle," or "short"? What is the effective (decision) range of such linguistic terms resulting from the numerical values provided by the neural network? How can the influence of the input error be reduced on the output? How correct is the system's output when it is based (trained) on part of the data, or on the rest of untrained input data?

5.2   Neural Network Processing of Linguistic Symbols

5.2.1   Linguistic Symbols

The concepts of linguistic variables were first introduced in Zadeh.[22-25] The inputs and outputs in a neural network are represented in three ways: numeric, pictorial, and symbolic. They can also be combined with fuzzy sets.[2] The numeric representation is used to find statistical correlations and predictions. It is also used to process, control, or analyze inputs from sensors or instruments. The pictorial representation refers to the input or output being depicted in some pictorial way, such as a binary matrix representing a character, or as input from a scanner. The symbolic representation uses symbols to describe neuron firing patterns or to give values that represent the membership of the symbol. However, a symbolic representation must first be converted into a numeric one before it can be given to a neural network for processing.

To process a linguistic symbol, the system can treat its value as a feature name or as an attribute name. For example, some attributes of an apple are color, taste, size, etc. It can be classified as the features in each attribute as follows:

class : golden-apple, red-apple
feature color : red, green, yellow
feature size : large, medium, small

---

*Acknowledgment to Dr. Lois Hawkes for her helpful comments on an earlier draft of this chapter.

Fuzzy logic is the logic of approximate reasoning.[19,21] So, in fuzzy rules of inference, both premises and conclusions are allowed to be fuzzy propositions and the final conclusion of some fuzzy inferences is an approximate, rather than an exact, consequence.[4,5] The truth values of fuzzy logic are given in linguistic form.

Using fuzzy logic, the measurement of a linguistic variable is given by the membership function, which is defined over a domain of numeric values.[10,11] This measurement provides a measure of the compatibility of the numeric descriptions with a qualitative entity. The measurement may also give a measure of the possibility that the concept in a linguistic symbol might correspond to its set of numeric data.[26]

5.2.2 Neural Network Processing

Using backpropagation,[6,12,15,17] the network learns by having error signals (the differences between the outputs and the target data) propagate backward through the network, and modifying the interconnection strengths to reduce the errors as rapidly as possible.[3]

In the network of Figure II.5.1, the network input to the kth node (sum_k) is the weighted linear sum of all the outputs from the previous layer:

$$sum_k = \sum_{j=1}^{M} \omega_{kj} o_j, \tag{II.5.1}$$

where M is the number of nodes of layer i, ω_{kj} is the weight value of the interconnection from the node j of layer i to node k of layer $i + 1$, and o_j is the output of node j of layer i.

The average system error, i.e., the total square error of the outputs from the target, is given by:[12,15]

$$Err = \frac{1}{2P} \sum_{i=1}^{P} \sum_{j=1}^{K} (t_{ij} - o_{ij})^2, \tag{II.5.2}$$

where P is the number of patterns and K is the number of outputs, t_{ij} is the target or desired output value, and o_{ij} is the output of node j in the ith sample. For each pattern p and output attribute k, the error is:

$$Err_{pk} = |t_{pk} - o_{pk}|. \tag{II.5.3}$$

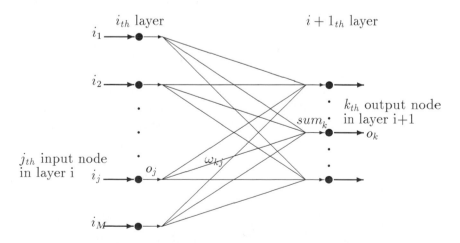

FIGURE II.5.1   The typical architecture of a feedforward neural network.

For backpropagation of error, the rate of change of error with respect to input node j is:[12,15]

$$\delta_j = f_j'(sum_j) \sum_{i=1}^{M} \delta_i \omega_{ij}. \tag{II.5.4}$$

where $f'(sum_j)$ is the first derivative of the sigmoidal activation function of $f(sum_j)$ and:

$$f(sum_j) = \frac{1}{1 + e^{-(sum_j + \theta_j)/\theta_0}}. \tag{II.5.5}$$

The weight difference $\Delta\omega_{ji}$ is:

$$\Delta\omega_{ji}(n+1) = \eta\delta_j o_i + \alpha\Delta\omega_{ji}(n), \text{ or}$$

$$\omega_{ji}(n+1) = \omega_{ji}(n) + \eta\delta_j o_i + \alpha[\omega_{ji}(n) - \omega_{ji}(n-1)], \tag{II.5.6}$$

where η $(0 < \eta)$ is a learning rate,[9,20] and α is a proportionality constant referred to as a momentum term $(0 \le \alpha \le 1)$. The momentum term α gives some inertia in the rate of change of weight values.

When training is completed, the trained network can accept any inputs which are similar to its training inputs and produce correct results. The ability of the network to generalize its learning to new situations and environments depends on the proper selection of the network architecture and training set.

In Section 5.2.3, a conversion method for linguistic symbols is proposed and is called the multilevel grading rule (MLGR).

5.2.3 The Multilevel Grading Rule (MLGR)

As stated in earlier sections, symbolic values should be converted into numeric values so that a linguistic symbol can be processed in a neural network system.

In order to have an equal range for each linguistic value, a multi-level grading rule (MLGR) for the conversion of linguistic symbols is proposed.[13] The value of the l_{th} level of a symbol which has L levels in total is represented as:

$$\frac{2(l-1)+1}{2L}, \tag{II.5.7}$$

and the numeric values in each level are shown in Table II.5.1, where L is the number of levels in a linguistic expression. Therefore, the maximum effective (decision) range is $\pm\frac{1}{2L}$, i.e., the length of $1/L$. In other words, the linguistic effective range must be less than $1/L$. Here the effective range means the range for which a linguistic symbol is in effect, i.e., the range of the symbol's value as shown in Figure II.5.2. The domain of this range is $[0, 1]$, because the input values of the neural network and the output of the sigmoidal activation function, Equation 5, lie between 0 and 1.

TABLE II.5.1 The Conversion Table of Symbolic Values into Numeric Values in Each Level

Level Number	1	2	$\cdots$	l	$\cdots$	$L-1$	L
Converted Numeric Value	$\frac{1}{2L}$	$\frac{3}{2L}$	$\cdots$	$\frac{2(l-1)+1}{2L}$	$\cdots$	$\frac{2L-3}{2L}$	$\frac{2L-1}{2L}$

TABLE II.5.2   Numerical Representation of Linguistic Terms and Their Conversion in Seven Levels

Level Number	Linguistic Expression	Numeric Value
1	extremely (very very) small	1/14
2	very small	3/14
3	small	5/14
4	medium	7/14
5	large	9/14
6	very large	11/14
7	extremely (very very) large	13/14

FIGURE II.5.2   Numerical representation of linguistic terms and their conversion in seven levels.

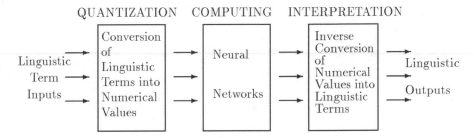

FIGURE II.5.3   Processing of linguistic terms in a neural network.

A simple example of the conversion of a linguistic-valued pattern into a numeric-valued one, defined with seven levels, is shown in Table II.5.2.[14] The correspondence between level numbers and numeric values is also represented in Figure II.5.2.

To process linguistic data in a neural network, we need an encoding scheme to convert the linguistic symbols into numerical values in a form suitable for presenting them to the input layer of the neural network. To obtain meaningful outputs from the neural network, we also need a decoding scheme which converts the numerical values — produced by the output layer of the neural network—back into linguistic terms. This process is depicted schematically in Figure II.5.3.

5.2.4   Example 1: Modulo 10 Problem

Now, the MLGR is applied to a system that has more levels in its inputs and outputs. The analysis of the results concerning the rate of correctness of their outputs are shown in Section 5.5.3.

<u>Modulo 10</u>: There are two inputs and one output in this system. The output is the remainder of the sum of two inputs divided by 10. The total number of levels is 10 in each input or output. Each level is named "zero," "one," ... , "nine" as follows:

Symbolic value	Numeric value	Quantum range (100% of Effective range)
zero	0.05	.00 ~ .09
one	0.15	.10 ~ .19
two	0.25	.20 ~ .29
three	0.35	.30 ~ .39
four	0.45	.40 ~ .49
five	0.55	.50 ~ .59
six	0.65	.60 ~ .69
seven	0.75	.70 ~ .79
eight	0.85	.80 ~ .89
nine	0.95	.90 ~ .99

The value of intervals between the levels is 0.05, using the MLGR given in Section 5.2.3. All the combinations of inputs and their outputs are as follows:

Number	Input 1	Input 2	Output
0	zero	zero	zero
1	zero	one	one
2	zero	two	two
3	zero	three	three
4	zero	four	four
5	zero	five	five
6	zero	six	six
7	zero	seven	seven
8	zero	eight	eight
9	zero	nine	nine
10	one	zero	one
11	one	one	two
12	one	...	...
⋮	⋮	⋮	⋮
98	...	eight	seven
99	nine	nine	eight

We return to this example in Section 5.5 for discussion.

This example has the characteristic of input independence; exchanging the inputs preserves the same output values. With this rule, the system can be trained more effectively since it can be done well with a smaller number of training data.

5.2.5   Example 2: Next Integer Problem

In this model, there are two inputs and two outputs with ten levels each. The output is the next integer number of the input value. Each level is named as "zero," "one," ... , "nine," as before. Table II.5.3 shows the converted symbols. We assume that the next integer of 99 is 0.

TABLE II.5.3   Converted Linguistic Symbols in Example 2

Numbers	Symbolic Representation	Meaning	Numeric Value	Output
0	zero zero	zero	0.05 0.05	zero one
1	zero one	one	0.05 0.15	zero two
2	zero two	two	0.05 0.25	zero three
3	zero three	three	0.05 0.35	zero four
⋮	⋮	⋮	⋮	⋮
10	one zero	ten	0.15 0.05	one one
11	one one	eleven	0.15 0.15	one two
⋮	⋮	⋮	⋮	⋮
98	nine eight	ninety eight	0.95 0.85	nine nine
99	nine nine	ninety nine	0.95 0.95	zero zero

TABLE II.5.4   Converted Linguistic Symbols in Example 3

	Inputs			Outputs	
Numbers	Symbolic Representation	Meaning	Value	Numeric Symbol	Numeric Value
0	zero zero	zero	.05 .05	zero zero zero	.25 .05 .05
1	zero one	one	.05 .15	zero zero two	.25 .05 .25
2	zero two	two	.05 .25	zero zero four	.25 .05 .45
3	zero three	three	.05 .35	zero zero six	.25 .05 .65
⋮	⋮	⋮	⋮	⋮	
10	one zero	ten	.15 .05	zero two zero	.25 .25 .05
11	one one	eleven	.15 .15	zero two two	.25 .25 .25
		. . .			
50	five one	fifty	.55 .05	one zero zero	.75 .05 .05
		. . .			
⋮	⋮	⋮	⋮	⋮	⋮
98	nine eight	ninety eight	.95 .85	one nine six	.75 .95 .65
99	nine nine	ninety nine	.95 .95	one nine eight	.75 .95 .85

5.2.6   Example 3: Double the Inputs Problem

This model has two inputs and three outputs with ten levels each except the first output; it has two levels. The output is the doubled integer to the input value. This is depicted in Table II.5.4.

The above two examples have the characteristic of input dependence; exchanging the input changes the output value.

In these examples, we assume that the membership value is either one or zero, and also assume that the majority outputs are the correct ones. We can treat the problem in another way: e.g., a system may have eight inputs and seven outputs with corresponding membership values, such as input 1 is "short," input 2 is "middle," and input 3 is "tall" for the feature of height (see Table II.5.5 and Figure II.5.4); inputs 4, 5, and 6 are "dark," "red," and "blond," respectively, for the feature of hair;

TABLE II.5.5   A Representation of Linguistic Term
"Height" in Fuzzy Sets

Linguistic Term "Height"	Input 1	Input 2	Input 3
short	5/6	3/6	1/6
middle	3/6	5/6	3/6
tall	1/6	3/6	5/6

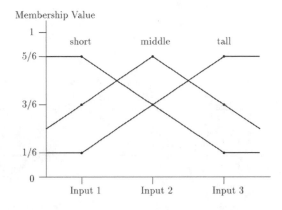

FIGURE II.5.4   A representation of linguistic term "height" in fuzzy sets.

inputs 7 and 8 are "blue" and "brown," respectively, for the feature of eyes, and give the membership values for each feature the corresponding converted values by MLGR for each attribute.

In the next section, we investigate how the system errors work in this linguistic neural network processing and what the relationship is between the linguistic effective range (of the output results) and the system errors.

5.3   The Relationship between the System Errors and the Linguistic Effective Range on Outputs

5.3.1   The Admissible System Error and the Linguistic Effective Range

Let K be the number of outputs, L_i the number of levels in the ith output, and P the number of sample patterns. Admissible system error, E_{adm}, is defined as the average system error which is sufficiently small and admissible in the system to have correct outputs. Δ_y is defined as half of the effective range of the linguistic term y in the l_{th} level and $1 \leq l \leq L$ (Figure II.5.5). Then, referring to Equation II.5.2, the average system error should be less than, or equal to, the admissible system error (E_{adm}) and is:

$$Err = \frac{1}{2P} \sum_{i=1}^{P} \sum_{j=1}^{K} (t_{ij} - o_{ij})^2 \leq E_{adm}, \tag{II.5.8}$$

$$E_{adm} = \frac{1}{2P} \sum_{i=1}^{P} \sum_{j=1}^{K} \Delta_{min}^2$$

$$= \frac{K}{2} \Delta_{min}^2,$$

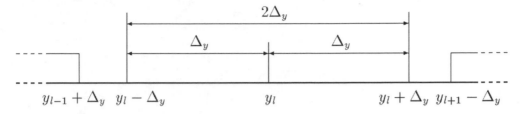

FIGURE II.5.5   The effective range Δ_y of a linguistic term y in the l_{th} level.

such that $\Delta_{min} = min\{\Delta_{11}\Delta_{12}, \ldots, \Delta_{ij}, \ldots, \Delta_{PK}\}$ for all $i = 1, 2, \ldots, P$ and $j = 1, 2, \ldots, K$, and Δ_{ij} is an admissible effective error range for each *ith* sample data and *jth* output term (set by a user). Here Δ_{min} is the minimum effective range in P patterns and K linguistic output attributes.

Referring to Equation II.5.2, the admissible individual sample (or system) error (E_{padm}) for a pattern p is:

$$E_p = \frac{1}{2} \sum_{j=1}^{K} Err_{pj}^2 \le E_{padm},$$

$$E_{padm} = \frac{1}{2} \sum_{j=1}^{K} \Delta_j^2,$$

such that $\Delta_j = min\{\Delta_{p1}, \Delta_{p2}, \ldots, \Delta_{pK}\}$ for all $j = 1, 2, \ldots, K$. So:

$$E_{padm} = \frac{K}{2} \Delta_p^2$$

$$= \frac{K}{2} \Delta_{min}^2$$

$$= E_{adm}, \tag{II.5.9}$$

where $\Delta_p = min\{\Delta_{p1}, \Delta_{p2}, \ldots, \Delta_{pK}\}$ and $\Delta_p = \Delta_{min}$, because $\Delta_{pk} = \Delta_{ik}$, for all sample data $i = 1, 2, \ldots, P$, i.e., the *kth* error, Δ_{ik}, is independent from the input sample data. Thus, for all samples (P) the individual sample error (E_p) should be less than or equal to E_{padm}.

Similarly, the admissible individual output error (E_{kadm}) for attribute k is:

$$E_k = \frac{1}{2} \sum_{i=1}^{P} (t_{ik} - o_{ik})^2 \le E_{kadm},$$

$$E_{kadm} = \frac{1}{2} \sum_{i=1}^{P} \Delta_i^2,$$

such that $\Delta_i = min\{\Delta_{1k}, \Delta_{2k}, \ldots, \Delta_{Pk}\}$ for all $i = 1, 2, \ldots, P$. Therefore,

$$E_{kadm} = \frac{P}{2} \Delta_k^2. \tag{II.5.10}$$

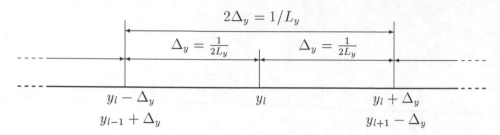

FIGURE II.5.6   The maximum effective range Δ_y of a linguistic term y_l in the l_{th} level.

5.3.2   The Maximum Admissible System Error

With L levels, the maximum error range Δ_y is $1/2L$, when $y_l - \Delta_y = y_{l-1} + \Delta_y$ and $y_l + \Delta_y = y_{l+1} - \Delta_y$, and $1 \leq l \leq L$, as shown in Figure II.5.6. So the maximum admissible system error (E_{max}) is:

$$E_{max} = \frac{K}{2}(min\{\Delta_1, \Delta_2, \ldots, \Delta_K\})^2$$

$$= \frac{K}{8(max\{L_1, L_2, \ldots, L_K\})^2} \tag{II.5.11}$$

and the average system error *Err* should be less than or equal to E_{max}, namely:

$$Err \leq E_{max}$$

$$= \frac{K}{8L^2_{max}}, \tag{II.5.12}$$

where $L_{max} = max\{L_1, L_2, \ldots, L_k\}$. The admissible individual maximum error (E_{pmax}) is:

$$E_{pmax} = \frac{1}{2}\sum_{j=1}^{K}\Delta^2_{min}$$

$$= \frac{K}{8L^2_{max}}. \tag{II.5.13}$$

Therefore, the error should be $Err \leq E_{max}$ or, for all the sample patterns $(p)E_p \leq E_{max}$. In the same way:

$$E_{kmax} = \frac{1}{2}\sum_{i=1}^{P}\Delta^2_k$$

$$= \frac{P}{8L^2_{max}}. \tag{II.5.14}$$

5.3.3   Error Checking Methods

The admissible system errors mentioned earlier are used to check the output error, and to terminate the system run (learning) in what is called a weak error checking method. The weak error checking method saves the system run-time but, in some cases, the outputs are faulty. Therefore, strong error checking is recommended.

For strong error checking—used for checking the error on every sample data (pattern) and on every output attribute—the individual output error, E_k (k_{th} attribute), should be:

$$E_k \leq \frac{1}{2L_k},$$
(II.5.15)

and is applied to all patterns and to all output attributes.

For example, when $K = 2$, $L_1 = 4$, $L_2 = 3$, and $P = 8$;

$$E_{max} = E_{pmax}$$
$$= 0.015625,$$
$$E_{kmax} = .0625,$$

and for strong error checking, individual output error $E_1 \leq 0.125$ and $E_2 \leq 0.166667$. Thus, in the weak error checking method, the average system error Err should be less than 0.015625 or, for all K attributes, the individual output error E_k should be less than 0.0625. However, in strong error checking method, the individual output errors E_1 should be less than 0.125, and E_2 should be less than 0.166667 for all patterns in P.

In the next section, the measure of a problem's simplicity and the rate of correctness are explained.

5.4   The Degree of Simplicity of a Problem and the Probability of Its Correct Outputs

In this section, we are going to examine a way to measure a problem's simplicity and examine the probabilistic view of the system's outputs and suggest a theorem and a following corollary.

5.4.1   The Degree of Complexity

A Boolean function is a function f with domain $\{F, T\}^n$ and range $\{F, T\}^m$, for some integer n and m as follows:

$$f : \{F, T\}^n \longrightarrow \{F, T\}^m,$$
(II.5.16)

where $\{F, T\}^n$ denotes the n-fold cartesian product of the set $\{F, T\}$ with itself. In a Boolean function, when a problem has n input and m output variables, there are 2^{m2^n} kinds of functions (rules).

Basically, "the degree of complexity of a problem or rule" is related to the characteristics of the problem: the rate of changing their bits (digits for multilevel variables). But in general, it is related to the number of inputs and outputs because when the problem has a larger number of inputs, it has a larger variety in its solutions and the system's rules become more complicated. In order to simplify the discussion, let us define the degree D of simplicity (or complexity) of a problem as the multiplication of the number of all inputs and their levels. Here, we ignore the redundant or dummy inputs.

Let the set of instances (total combination of input data) be **I**. Then:

$$\mathbf{I} = \mathbf{L}_{I1} \times \mathbf{L}_{I2} \times \cdots \times \mathbf{L}_{In},$$
(II.5.17)

$$I = |\mathbf{I}| = L_{I1} \times L_{I2} \cdots \times L_{In},$$
(II.5.18)

$$L_{Ii} = |\mathbf{L}_{Ii}|, \quad 1 \leq i \leq n,$$

where $\mathbf{L}_{Ii}$ is a set of input levels, L_{Ii} is the number of levels for the i_{th} input $(1 \leq i \leq n)$, and n is the number of inputs, the set of outputs $\mathbf{O}$ is:

$$\mathbf{O} = \mathbf{L}_{O1} \times \mathbf{L}_{O2} \times \cdots \times \mathbf{L}_{Om}, \tag{II.5.19}$$

$$O = |\mathbf{O}| = L_{O1} \times L_{O2} \ldots \times L_{Om}, \tag{II.5.20}$$

$$L_{Oj} = |\mathbf{L}_{Oj}|, \quad 1 \leq j \leq m,$$

where $\mathbf{L}_{Oj}$ is a set of output levels, L_{Oj} is the number of levels for the j_{th} output, $1 \leq j \leq m$, and m is the number of outputs. Then the set of functions with this input (source), $\mathbf{I}$, and output (target), $\mathbf{O}$, is:

$$\mathbf{F} = \{f | f : \mathbf{I} \longrightarrow \mathbf{O}\} \tag{II.5.21}$$

$$F = |\mathbf{F}| = O^I \tag{II.5.22}$$

$$= (L_{O1}L_{O2} \cdots L_{Om})^{(L_{I1}L_{I2}\cdots L_{In})}.$$

For example, if $L_{Ii} = 2$ and $L_{Oj} = 2$ for $1 \leq i \leq n$ and $1 \leq j \leq m$, then $F = 2^{m2^n}$. The possibility of a neural network system that could be one of these functions is $1/F$.

The set of total answers of the problem is:

$$\mathbf{A} = \{(x, y) | x \in \mathbf{I}, y \in \mathbf{O}\}, \tag{II.5.23}$$

and the set of correct answers $\mathbf{C}$ (or solutions) is a subset of $\mathbf{A}$:

$$\mathbf{C} \subseteq \mathbf{A}. \tag{II.5.24}$$

Usually, the length of $\mathbf{C}$ is less than the length of $\mathbf{A}$ because many contradictions and/or incorrect answers exist in $\mathbf{A}$. Let $\mathbf{T}$ be a set of training data, then:

$$\mathbf{T} \subseteq \mathbf{C}, \tag{II.5.25}$$

because the system has at least $\mathbf{T}$ correct outputs if the system has been successfully trained within an allowed system error. Here it is assumed that all the trained data are correct ones.

For simplicity, let $|\mathbf{I}|$ be the degree D of a problem. D can be reduced by giving some implied rules, such as "independence of the location of inputs" in the modulo 10 problems. Generally, the number of all the combinations of data inputs is:

$$D \equiv |\mathbf{I}| = L_{I1} \times L_{I2} \times \cdots \times L_{In},$$

where L_{Ii} is the number of levels for the i_{th} input.

For example, a binary logical OR function consists of two inputs and one output, and the number of levels is 2. Thus, the degree of the OR function is 4 ($= 2 \times 2$), and the total number of combinations of input data (the number of all possible data) is 4: there are also 4 rules as follows:

In1	In2	Out	Rule
F	F	F	F + F = F
F	T	T	F + T = T
T	F	T	T + F = T
T	T	T	T + T = T

Thus, to train an OR function neural network, the user needs four items of data. If trained well with these four items, this system will output all four correct answers.

In the examples, it is 100 (10 levels of the first input × 10 levels of the second input).

Therefore, we can treat the concepts of "the degree of a problem," "the degree of a rule," and "the degree of a function," as having the same meaning in terms of measurement of the simplicity of a problem. The output of a system will have a higher rate of correctness when the problem is simpler. In other words, if it has simpler rules, then it needs a smaller number of training data. For simplicity, let D be the degree of a problem, measuring the problem's simplicity, and be given as the total number of possible inputs.

The following section examines the probabilistic view of the system's result.

5.4.2 The Probabilistic Analysis of the Rate of Correctness

In this section, we review probability as it relates to the system's output results and its rate of correctness.

5.4.2.1 The Probability of the Rate of Correctness

Let $\mathbf{T}$ be a set of training data, then from Equations II.5.18 through Equation II.5.25:

$$\mathbf{T} \subseteq \mathbf{I} \times \mathbf{O}$$
$$\subseteq \{(x_1, x_2, \ldots, x_n, y_1, \ldots, y_m) | x_i \in \mathbf{L}_{Ii}, \quad 1 \leq i \leq n,$$
$$y_j \in \mathbf{L}_{Oj}, \text{ and } 1 \leq j \leq m\}, \tag{II.5.26}$$

$$T = |\mathbf{T}|, \tag{II.5.27}$$

$$N = |\mathbf{I}|$$
$$= |\mathbf{L}_{I1}| \times |\mathbf{L}_{I2}| \times \cdots \times |\mathbf{L}_{In}|$$
$$= D. \tag{II.5.28}$$

Let the probability of correctness for each of the system's outputs be a constant ($p_j = \frac{1}{L_{Oj}}$) from trial to trial where L_{Oj} is the number of levels of the j_{th} output. Then in a series of n independent trials, the total number of correct outputs (Y) for the target values is a binomial distribution, because whether the output is correct has one of two possible outcomes: "YES" or "NO" (or "TRUE" or "FALSE"), and each output result is independent from trial to trial. The probability of k correct outputs from n trials ($P(Y = k)$) is:

$$P(Y = k) = \binom{n}{k} p_j^k (1 - p_j)^{n-k}, \tag{II.5.29}$$

where $k = 0, 1, \ldots, n$ and n is the total number of independent trials of output correctness from the target values.

5.4.2.2 The Correctness Ratio in MLGR

Assume the number of inputs is n and the i_{th} input level is L_{Ii}, then the total number of trial outputs N is $L_I 1 \times L_{I2} \ldots \times L_{In}$, where $L_{Ii} = |\mathbf{L}_{Ii}|, 1 \leq i \leq n$. If the system has trained with T items of data $(0 \leq T \leq N)$, then the system has at least T correct outputs. The expected value of correctness for the j_{th} output is:

$$
\begin{aligned}
E_j(T) &= T + \sum_{k=1}^{N-T} k \times P(Y = k) \\
&= T + \sum_{k=1}^{N-T} k \times \binom{N-T}{k} \times p_j^k \times (1 - p_j)^{N-T-k} \\
&= T + \sum_{k=1}^{N-T} k \times \binom{N-T}{k} \times \left(\frac{1}{L_{Oj}}\right)^k \times \left(1 - \frac{1}{L_{Oj}}\right)^{N-T-k} \\
&= T + (N - T) \times \frac{1}{L_{Oj}}.
\end{aligned}
$$
(II.5.30)

There are two cases on the interpretation of the system's output: disjunctive outputs and conjunctive outputs. Case [I] (disjunctive outputs): each output in **O** is an OR condition, i.e., the correctness of one output in **O** is independent of the correctnesses of the other outputs in **O**; case [II] (conjunctive outputs): each output in **O** is an AND condition, i.e., to have one correct output, all m outputs in **O** should be correct. It depends on the user's design which type of the system's output is needed: disjunctive or conjunctive.

Case [I] (disjunctive outputs): the output correctnesses in the OR condition.

For the m outputs, the total expected value $E(T)$ is the sum of each expected value of outputs as follows:

$$
\begin{aligned}
E(T) &= \sum_{j=1}^{m} E_j(T) \\
&= mT + (N - T) \sum_{j=1}^{m} \frac{1}{L_{Oj}}
\end{aligned}
$$
(II.5.31)

Let us define the correctness rate as follows:

$$
\begin{aligned}
C(T) &\equiv \frac{E(T)}{\text{the total number of possible outputs}} \\
&= \frac{E(T)}{mN} \\
&= \frac{T}{N} + \frac{(N - T)}{mN} \sum_{j=1}^{m} \frac{1}{L_{Oj}}
\end{aligned}
$$
(II.5.32)

If $L_O = L_{O1} = L_{O2} = \cdots = L_{Om}$, then:

$$
E(T) = mT + m(N - T)\frac{1}{L_O}, \text{ and}
$$
(II.5.33)

$$
C(T) = \frac{T}{N} + \frac{(N - T)}{NL_O}.
$$
(II.5.34)

Case [II] (conjunctive outputs): the output correctnesses in the AND condition.

For the m outputs, the joint probability p is the multiplication of each probability p_j:

$$p = \prod_{j=1}^{m} p_j$$

$$= \prod_{j=1}^{m} \frac{1}{L_{Oj}},$$

where p_j is the probability of the correctness of the j_{th} output. The probability of k correct outputs $P(Y = k)$ in n independent trials is:

$$P(Y = k) = \binom{n}{k} p^k (1 - p)^n$$

$$= \binom{n}{k} \prod_{j=1}^{m} \left(\frac{1}{L_{Oj}}\right)^k \times \left(1 - \frac{1}{L_{Oj}}\right)^{N-T-k}$$

$$= (N - T) \prod_{j=1}^{m} \frac{1}{L_{Oj}}. \tag{II.5.35}$$

Thus, the total expected value $E(T)$ and the rate of correctness C(T) are:

$$E(T) = T + \sum_{j=1}^{N-T} k \times P(Y = k)$$

$$= T + (N - T) \prod_{j=1}^{m} \frac{1}{L_{Oj}}, \tag{II.5.36}$$

$$C(T) = \frac{T}{N} + \frac{(N - T)}{N} \prod_{j=1}^{m} \frac{1}{L_{Oj}}. \tag{II.5.37}$$

If $L_O = L_{O1} = L_{O2} = \cdots = L_{Om}$, then:

$$E(T) = T + (N - T)\frac{1}{L_O^m}, \text{ and} \tag{II.5.38}$$

$$C(T) = \frac{T}{N} + \frac{(N - T)}{NL_O^m}. \tag{II.5.39}$$

To examine the advantage of MLGR, let us convert the multi-level system into a binary-level system. To convert the j_{th} output level L_{Oj} into a binary level, the number of converted outputs is increased. Let the number of converted outputs of the j_{th} output be s_j as follows:

$$2 \leq L_{Oj} \leq 2^{s_j}$$

and:

$$s_j = \lceil \log_2 L_{Oj} \rceil \text{(the smallest integer greater than or equal to } \log_2 L_{Oj}).$$

The total number of new outputs m' is:

$$m' = \sum_{j=1}^{m} s_j$$

$$= \sum_{j=1}^{m} \lceil \log_2 L_{Oj} \rceil$$

In this converted binary-level system, $L'_O = L'_{O1} = L'_{O2} = \cdots = L'_{Om} = 2$.
From Equations II.5.31–II.5.34 and II.5.36–II.5.39,

$$E'(T) = T + (N - T) \sum_{j=1}^{m} \prod_{k=1}^{s_j} \frac{1}{L'_O}$$

$$= T + (N - T) \sum_{j=1}^{m} \frac{1}{2^{s_j}} \tag{II.5.40}$$

$$C'(T) = \frac{T}{N} + \frac{N - T}{m'N} \sum_{j=1}^{m} \prod_{k=1}^{s_j} \frac{1}{L'_O}$$

$$= \frac{T}{N} + \frac{N - T}{m'N} \sum_{j=1}^{m} \frac{1}{2^{s_j}} \tag{II.5.41}$$

from Equation II.5.32 (case [I]):

$$C'(T) = \frac{T}{N} + \frac{N - T}{m'N} \sum_{j=1}^{m} \frac{1}{2^{s_j}}$$

$$\leq \frac{T}{N} + \frac{N - T}{mN} \sum_{j=1}^{m} \frac{1}{L_{Oj}} = C(T),$$

and from Equation II.5.37 (case [II]):

$$C'(T) = \frac{T}{N} + \frac{N - T}{m'N} \sum_{j=1}^{m} \frac{1}{2^{s_j}}$$

$$\leq \frac{T}{N} + \frac{N - T}{N} \prod_{j=1}^{m} \frac{1}{L_{Oj}} = C(T).$$

Therefore, in both cases, the rate of correctness in MLGR, $C(T)$, is greater than the one in the binary leveled system, $C'(T)$.

5.5   Simulation Results

The following sections show the simulation results for the three examples of Section 5.2.

5.5.1   The Execution of the Neural Network System

The examples were run with five different effective ranges, full range (100%), 70, 50, 30, and 10% of maximum range (1/2L), which are used in determining linguistic values. In each effective range, it has tested with 10 different percentages of training data: 10, 20, 30, 40, 50, 60, 70, 80, 90, and 100%. To find the effect of the random numbers given to the initial weight values, 10 sequential executions were run in each range and training data. Thus, the total number of executions was 500. We set the learning rate as 0.5 and the momentum rate as 0.3 in example 1, and 0.9 for the learning rate and 0.7 for the momentum rate in example 2 and example 3. In all executions, a strong error checking is applied rather than a weak one, for deciding when to terminate the maximum iteration of 200,000 in example 1 and 50,000 in example 2 and example 3. Some outputs were significantly different from the corresponding targets, even though the average system error Err was less than the maximum admissible average system error E_{max}, and the system terminated successfully when the weak error checking method was used. Thus, many mismatches necessitated this strong error check.

Sections 5.5.2, 5.5.3, 5.5.4 show the simulation results of the example in Section 5.2.5.

5.5.2 Simulation Results of Example 1

With the example in Section 5.2.5, the system was executed with various parameters: five different modification factors (1, 1.5, 2, 2.5, and 3); 5 different effective ranges 100, 70, 50, 30, and 10%; 8 different numbers of nodes in their hidden layers (3, 5, 7, 9, 11, 13, 15, and 17); and 9 different percentages of training data (10, 20, 30, ... , 90%). The above system was run 10 times with the maximum iteration of 200,000 on each of the above parameters with different random sample (training) data (random sample numbered 1, 2, ... , 10).

5.5.2.1 Varying the Number of Nodes

As shown in Table II.5.6, the case of 3 nodes has only 12.7% of convergence while the case of 17 nodes has 92.67%. A node in a hidden layer represents a hyperplane which separates a cluster of data from others by the characteristics of the given data. If a system does not have a sufficient number of nodes in its hidden layer, it cannot classify the data well because of the small number of hyperplanes used to differentiate the clusters of data. In this example, the 3 nodes—3 hyperplanes—are not sufficient to classify the given data, while the 17 nodes are. This is confirmed by the average values (of 10 executions with random sampling) in the second column in Table II.5.6; the 10% of training data (that has less training data) has a high rate of convergence (79.6%) even though it has only 3 nodes, and the rates are significantly decreased by increasing the percentage of the sampling data. From the table, the more nodes in the hidden layer, the greater the percentage of convergence in training.

This neural network with MLGR is dependent on the number of nodes in the hidden layer, on the effective range of the training, and on the characteristics of the random number used. When the system has a greater percentage of convergence, it has a greater number of correct outputs because the weights of the trained neural network system are assigned well. If the system reached the maximum allowable iteration, the weight values would not be assigned so as to have a sufficiently small system error (Err of Equation II.5.2), and the system's outputs would not be correct; thus, the system would have a smaller rate of correctness. In other words, the character of the training data has not imbedded well into the weight values of the system. In general, for a larger number of levels of inputs and outputs, the system needs more precise control of errors and needs more nodes in its hidden layer.

5.5.2.2 Varying the Percentage of Training Data, Modification Factor, and the Percentage of Effective Range

Table II.5.6 shows that the case of 10% training data has an extraordinarily better convergence rate than the others. This is due to the system error becoming small enough quickly. This follows since it uses a small amount of training data to determine the differences between the targets and the system outputs. However, as shown in the following section, the rate of correctness is much lower in this case

TABLE II.5.6   The Average Percentage of System Convergence in Training by the Number of Nodes in the Hidden Layer and by the Percentage of the Training Data

% of Training Data	Number of Nodes								
	3	5	7	9	11	13	15	17	Avg.
10	79.6	99.2	98.4	98.0	98.8	98.0	99.2	98.8	96.25
20	16.0	46.8	64.0	75.6	82.0	73.6	84.8	89.2	66.50
30	10.0	29.6	52.8	70.8	79.2	78.0	84.0	82.0	60.80
40	4.9	47.2	71.6	84.0	85.6	81.6	86.4	89.6	68.85
50	3.6	48.0	68.8	82.8	85.6	91.2	94.8	92.4	70.90
60	0.4	38.4	73.6	88.4	90.0	94.4	92.8	95.6	71.70
70	0.0	67.6	91.6	98.4	97.2	96.0	97.6	95.2	80.45
80	0.0	76.0	91.2	98.8	98.0	98.0	94.4	95.2	81.45
90	0.0	73.6	95.6	98.4	95.6	98.4	97.6	96.0	81.90
Average	12.71	58.49	78.62	88.36	90.22	89.91	92.40	92.67	75.42

TABLE II.5.7   The Rate of Convergence Relative to the Percentage of Effective Range

Effective range (%)	10	30	50	70	100	Average
Rate of conv. (%)	51.06	78.06	80.78	82.39	85.25	75.51

(30.2%) than the other cases of different percentages of training data [83.5%, 99.8%]. In general, when the system has a larger amount of training data, it converges easily.

Table II.5.7 shows that the larger the effective range, the higher the rate of system convergence. This is because as the percentage of effective range increases, the system allows a larger range of average system error (Err).

The following section analyzes the system's outputs in conjunction with their rates of correctness.

5.5.3   The Rate of Correctness of the System's Outputs

In the case using 10% of the training data, the rate of possibility of correctness is 19%: the sum of 10 correctly trained data and 9 possibilities out of the remaining 90 sample data ($90 \times 1/10$). From Table II.5.8, the rate of correctness for the output of the trained system is 30.2%, much higher than the possibility of 19%. In the case using 20% of the training data, the output of the system is 60.3%, while the possibility is 28% ($20\% + 1/10 \times 80$) shown in Table II.5.8. In this table, the third column (the rate of correctness of all outputs) is the system's average output in 10 executions with random sampling of training data. The fourth column contains the average values excluding the outputs of maximally reached iterations from the third column results. This table shows that the values in the fourth column have better rates than those in the third column. So the system has a higher rate of correctness when the executions converge (terminate training successfully), i.e., the system's errors are sufficiently small. This table also shows that the example trained with the method of MLGR has a much higher rate of correctness than the simple possibility in all the above nine cases. Figure II.5.7 shows that the outputs of converged executions have a better rate of correctness than the ones in all outputs.

TABLE II.5.8   The Rate of Correctness

Percentage of Training Data	Rate of Corr. in Probability (%)	Rate of Corr. in all Outputs (%)	Rate of Corr. of only Conv. Outputs (%)
10	19	30.4	30.2
20	28	55.8	60.3
30	37	74.6	83.5
40	46	92.0	96.9
50	55	89.5	98.1
60	64	93.5	99.4
70	73	98.9	99.5
80	82	99.3	99.8
90	91	99.2	99.5
average	5	81.47	85.24

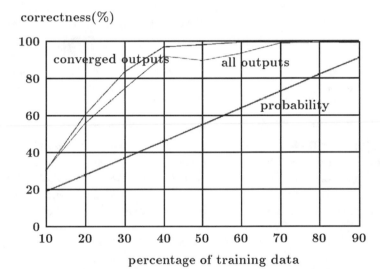

FIGURE II.5.7   The percentage of correctness vs. the training data.

5.5.3.1   Varying the Number of Nodes

To have some degree of significance, e.g., more than 90% correctness, the system needs to use a minimum of 40% of the training data; in probability theory, the system needs at least 90%. By applying a specific rule in training, the rate of correctness can be increased through changing the order of training data or through selecting some particular data from the sample data. However, the modulo 10 problem has the characteristic of independence on the input location. Thus, by exchanging the two items of input data, the training data can be doubled (but the percentage of training data is the same).

Considering the number of nodes in the hidden layer of the system, as shown in Table II.5.9, the case of 3 nodes has the worst result: 65.59% of correctness on its outputs. The reason for the lower rate in the case of 3 nodes is that the implicit rule of training data is not fully implanted into the weight values of the system because of the smaller number of nodes used to control the system error precisely. This table shows that the number of nodes are not important in the rate of correctness of the system's output when it has a sufficient number.

TABLE II.5.9 The Average Rate of Correctness vs. the Number of Nodes

Number of nodes	3	5	7	9	11	13	15	17
Rate of correctness (%)	65.59	85.36	85.88	84.89	85.40	85.42	84.83	84.75

TABLE II.5.10 The Average Rate of Correctness vs. Effective Range and the Percentage of Training Data

% of training	Effective Range (%)					
	10	30	50	70	100	Average
10	31.45	31.57	30.58	29.34	28.28	30.24
20	63.46	63.15	60.72	59.58	56.88	60.75
30	87.08	87.32	84.70	82.56	78.62	84.06
40	99.33	98.78	98.16	96.84	93.23	97.27
50	99.34	99.02	98.85	98.56	95.36	98.23
60	99.84	99.77	99.62	99.48	98.55	99.45
70	100	99.82	99.77	99.78	98.44	99.56
80	100	99.93	99.87	99.91	99.31	99.80
90	100	100	99.96	99.96	97.72	99.53
average	86.72	86.60	85.80	85.11	82.93	85.43

5.5.3.2 Varying the Effective Range and the Percentage of the Training Data

Table II.5.10 shows the results of the average values run in eight cases of nodes; five effective ranges; five modification factors; nine kinds of training data; and ten random sampling data, i.e., the results of 18,000 executions with the maximum number of iteration as 200,000 in each training.

This table shows that narrowing the effective range increases the rate of correctness in the system. This means that the narrower effective range has the more accurate answers. So, narrowing the effective range of the system gives better learning of the rules (characteristics) from the given training data. As shown in Figure II.5.7, the system using our MLGR method has a much higher rate of correctness than the plain probability.

The system results showed that the variation of modification factors from one through three by 0.5 steps did not affect the rate of correctness.

5.5.4 Simulation Results of Examples 2 and 3

Table II.5.11 shows the results of example 2, average number of correct outputs in 100 samples and 10 executions with 10 sets of random data. The numbers in parentheses indicate the number of successful terminations.

The results of average correctness in example 3 are shown in Table II.5.12.

In the above two examples (problems 2 and 3), the system has two hidden layers with seven units each, and the maximum iteration is 50,000. It has two inputs and two or three outputs, and has run 10 times with different sets of random data, and 10 different percentages of data in each run. If the system reaches into the maximum iteration, then the outputs may have errors that differ from the target values, since the weight values between the nodes are not tuned up well by the abrupt termination. Therefore, if the number of successful termination is small, the correctness numbers may not correct well but, in most cases, they have already tuned well and are correct, even though the termination was not good enough to be successfully completed.

TABLE II.5.11　The Average Rate of Correctness and the Number of Successful Termination vs. Percentage of Training Data in Example 2

% of Training	Effective Range (%)					Average
	10	30	50	70	100	
10	39.9(8)	38.8(10)	35.9(10)	35.3(10)	35.1(10)	37.0(9.6)
20	80.0(4)	78.3(10)	73.7(10)	68.8(10)	60.9(10)	72.3(8.8)
30	91.4(3)	90.5(8)	88.9(9)	86.7(9)	78.4(10)	87.2(7.8)
40	92.2(4)	92.3(7)	89.6(8)	87.4(8)	85.2(9)	89.3(7.2)
50	97.6(1)	97.4(9)	96.8(9)	96.4(9)	93.2(10)	96.3(7.6)
60	98.3(1)	98.3(8)	98.0(10)	97.2(10)	93.4(10)	97.0(7.8)
70	99.2(1)	99.2(8)	98.9(9)	98.7(9)	97.7(10)	98.7(7.4)
80	99.6(0)	99.6(8)	99.6(8)	99.4(8)	97.4(10)	99.1(6.8)
90	97.5(2)	97.5(7)	97.5(9)	97.5(9)	96.7(9)	97.3(7.2)
100	100.(0)	100.(8)	100.(9)	100.(9)	100.(10)	100.(7.2)
average	89.6(2.4)	89.2(8.3)	87.9(9.1)	86.7(9.1)	83.8(9.8)	87.44(7.74)

TABLE II.5.12　The Average Rate of Correctness and the Number of Successful Termination vs. Percentage of Training Data of Example 3

% of Training	Effective Range (%)					Average
	10	30	50	70	100	
10	13.3(8)	13.4(9)	13.2(9)	13.2(9)	13.6(9)	13.3(8.8)
20	29.7(7)	29.8(10)	30.0(10)	28.3(10)	25.6(10)	28.7(9.4)
30	46.0(0)	45.6(8)	45.5(8)	45.2(8)	42.8(9)	45.0(6.6)
40	64.9(0)	64.4(2)	64.1(7)	58.7(7)	52.0(8)	60.8(4.8)
50	82.2(0)	82.0(2)	81.5(6)	81.4(7)	75.6(8)	80.5(4.6)
60	99.2(0)	99.2(2)	98.9(8)	96.7(10)	90.4(10)	96.9(6.0)
70	99.6(0)	99.6(3)	99.4(8)	97.8(10)	92.4(10)	97.8(6.2)
80	99.9(0)	99.9(1)	99.8(8)	99.7(10)	97.4(10)	99.3(5.8)
90	99.9(0)	99.9(0)	99.9(5)	99.9(9)	99.0(10)	99.7(4.8)
100	98.1(0)	98.1(3)	98.1(4)	97.8(9)	98.1(10)	98.0(5.2)
average	73.3(1.5)	73.2(4.0)	73.0(7.3)	71.9(8.9)	68.7(9.4)	72.01(6.22)

The second problem (next integer) is simpler in its reasoning the outputs than the other two problems (modulo 10 and double the inputs). Thus, the correctness of the second problem is higher than the one of the first and third problems when it is compared by the same amount of training data. The most complex problem in the above three examples is the double the input problem.

The next section shows the comparison of the results of MLGR with those of a conventional method in example 2.

5.5.5　Comparison of MLGR Results with Conventional Ones

Example 2 in Section 5.2.5 has been run with the conventional method of seven input nodes and seven output nodes which have binary levels in each node, and two hidden layers (seven units in each layer) model.

TABLE II.5.13   The Average Rate of Correctness of Example 2 in a Conventional Method

| | Effective Range (%) | | | | | |
% of Training	10	30	50	70	100	Average
10	13.4	13.4	13.1	13.3	13.9	13.4
20	25.1	25.1	26.1	25.9	26.0	25.6
30	34.6	34.6	38.1	38.2	37.2	36.5
40	43.4	43.4	45.7	45.8	45.9	44.8
50	44.5	44.5	45.0	45.5	45.3	45.0
60	40.4	40.4	40.4	40.4	40.4	40.4
70	53.4	53.4	53.4	53.4	53.4	53.4
80	53.0	53.0	53.2	53.2	53.2	53.1
90	59.9	59.9	59.9	59.9	59.9	59.9
100	56.8	56.8	56.8	56.8	56.8	56.8
average	42.4	42.4	43.2	43.2	43.2	42.90

The results of the conventional method with example 2 is shown in Table II.5.13.

Comparing the results in Table II.5.11 to those in Table II.5.13, the MLGR method has a much higher correctness than the conventional one. It shows the improvement of 44.54% in average.

In the next section, conclusions are given and recommendations for future work are suggested.

5.6   Conclusion

5.6.1   The Result of Example 1

From Tables II.5.6 and II.5.7, in Section 5.5.2, it is shown that if the system has a larger number of inputs and outputs, then it needs more nodes in its hidden layer to control the system error precisely. It is also shown that the system converges easily when it has a bigger effective range.

As shown in Table II.5.8 and Figure II.5.7 in Section 5.5.3, MLGR has a much higher rate of correctness for the expected values than the one using simple probability. This means that a neural network trained with the MLGR method has good characteristics with respect to the training data. Also, the simpler the rule, the higher the rate of correctness. In this neural network system, the imbedded rules of the sample data are implanted into different forms than the formulas in a mathematical representation.

To have a higher rate of correctness in the outputs, as shown in Table II.5.9, the system needs a sufficient number of nodes in the hidden layer of the neural network. Table II.5.10 shows that the system can acquire a higher rate of correctness on the outputs by narrowing the effective range of the linguistic term.

5.6.2   The Results of Examples 2 and 3

Example 2 is the simplest problem among the three given examples and has the highest correctness ratio. It means that the more complex problem needs more data to be trained well, and is harder to converge.

As we see from the three examples explained in the previous sections, we are able to successfully apply a sigmoidal activation function and the grading rule (MLGR) to the linguistic processing.

In this chapter, the grading rule (MLGR) ($[2(l - 1) + 1]/2L$) has been suggested to convert a linguistic symbol into a numerical value of the l_{th} level ($1 \leq l \leq L$) for the linguistic processing in neural networks (quantization). This grading rule has the maximum effective range, Δ_{max}, of $1/2L$. After a large number of executions, it was decided that the weak error check was inadequate for linguistic symbol processing. Therefore, the strong error checking method is highly recommended for linguistic processing with neural networks.

Section 5.4.2 shows that the neural network system trained with the MLGR method has a better rate of correctness in its outputs than the non-MLGR system in some domain of problems. The great advantage of this MLGR method is in the higher rate of correctness on the untrained data. In other words, the characteristics of the sample data are imbedded very well, in the form of specific weight values, in the links of the neural network system by using MLGR. It also minimizes the effect of input errors on the outputs.

5.6.3   Further Research

We suggest that the MLGR can be successfully applied to fuzzy control. The fuzzy control system can be designed by changing the levels of the input variables into membership values in fuzzy set.

The advantages of such a neural network system are as follows:

1. The system will give good results even when it does not have enough information for training, as shown in Tables II.5.8, II.5.10, II.5.11, and II.5.12. In other words, the trained neural network can control a system well even though it encounters unknown situations. A conventional fuzzy control system cannot generate any output if it is given unknown input data, while the above neural network system with MLGR produces an output with a high probability of correctness (about 87% on average in the next integer problem).
2. The neural network system with linguistic processing can be adjusted to the changes in environmental factors.
3. Such a system can be analyzed for a variety of controls and possibilities, and this helps make it easy to design the system.

For further study, we suggest that more extensive statistical analysis is needed for various problems, so they could be classified by whether or not they are suitable for the MLGR method. In addition, we need to suggest how the system can be fine-tuned by choosing the appropriate parameters. Also, additional research is needed regarding the proper sampling of data and the determination of optimal training samples. Further interesting areas for investigation include determining better grading rules and the way of interpreting results relating to the complexity of the problem and its measurement.

References

1. Bouslama, F. and Ichikawa, A., Application of neural networks to fuzzy control, *Neural Networks*, vol. 6, pp. 791–799, 1993.
2. Bulsari, A., Training artificial neural networks for fuzzy logic, *Complex Systems* 6, pp. 443–457, 1992.
3. Burrascano, P., A norm selection criterion for the generalized delta rule, *IEEE Trans. Neural Networks*, vol. 2, no. 1, pp. 125–130, January 1991.
4. Cao, Z., Kandel, A., and Li, L., A new model of fuzzy reasoning, *Fuzzy Sets and Systems*, 36, pp. 311–325, 1990.
5. Lin, C.- T. and Lu, Y.-C., A neural fuzzy system with linguistic teaching signals, *IEEE Trans. Fuzzy Systems*, vol. 3, no. 2, May 1995.

6. Hirose, Y., Yamashita, K., and Hijiya, S., Back-propagation algorithm which varies the number of hidden units, *Neural Networks*, vol. 4, pp. 61–66, 1991.

7. Kandel, A., Li, L., and Cao, Z., Fuzzy inference and its applicability to control systems, Department of Computer Science and the SUS Center for A.I., FSU, Tallahassee, FL.

8. Kandel, A., On the control and evaluation of uncertain process, *Proc. JACC*, San Francisco, August 1980.

9. Kuan, C.- M. and Hornik, K., Convergence of learning algorithms with constant learning rates, *IEEE Trans. Neural Networks*, vol. 2, no. 5, pp. 484–489, September 1991.

10. Mamdani, E.H., An experiment in linguistic synthesis with a fuzzy controller, *Int. J. Man-Machine Studies*, 7, pp. 1–13, 1975.

11. Mamdani, E.H., Advances in the linguistic synthesis of fuzzy controllers, *Int. J. Man-Machine Studies*, 8, pp. 669–678, 1976.

12. Pao, Y.-H., *Adaptive Pattern Recognition and Neural Networks*, pp. 83–85, Addison-Wesley, Reading, MA., 1989.

13. Park, J.J., Neural Network Processing of Linguistic Symbols Using Multi-Level Grading Rule, Ph.D. dissertation, Florida State University, Tallahassee, FL 1993.

14. Park, J.J., Kandel, A., Langholz G., and Hawkes, L., Neural network processing of linguistic symbols, pp. 265–284, *Fuzzy Sets, Neural Networks, and Soft Computing*, R.R. Yager and L.A. Zadeh, Eds., Van Nostrand Reinhold, 1994.

15. Rumelhart, D.E., Hinton, G.E., and Williams, R.J., Learning internal representations by error propagation. In D.E. Rumelhart and J.L. McClelland (Eds.), *Parallel Distributed Processing: Explorations in the Microstructures of Cognition. vol. 1: Foundations*, pp. 318–362, MIT Press, Cambridge, MA, 1986.

16. Sankar, K.P. and Mitra, S., Multilayer perceptron, fuzzy sets, and classification, *IEEE Trans. Neural Networks*, vol. 3, no. 5, pp. 683–697, September 1992.

17. Shoemaker, P.A., Carlin, M.J., and Shimabukuro, R.L., Back-Propagation Learning With Trinary Quantization of Weight Updates, *Neural Networks*, vol. 4, pp. 231–141, 1991.

18. Horikawa, S., Furuhashi, T., and Uchikawa, Y., On fuzzy modeling using fuzzy neural networks with the back-propagation algorithm, *IEEE Trans. Neural Networks*, vol. 3, no. 5, pp. 801–806, September 1992.

19. Takagi, Hideyuki, Suzuki, Noriyuki, Koda, Toshiyuki, and Kojima, Yoshihiro, Neural networks designed on approximate reasoning architecture and their applications, *IEEE Trans. Neural Networks*, vol. 3, no. 5, September 1992.

20. Weir, M.K., A method for self-determination of adaptive learning rates in back propagation, *Neural Networks*, vol. 4, pp. 371–379, 1991.

21. Zadeh, L.A., *A Theory of Approximate Reasoning, Machine Intelligence 9*, J.E. Hayes, D. Michie, and L.I. Mikulich (Eds.), John Wiley & Sons, pp. 149–194, 1979.

22. Zadeh, L.A., Outline of a new approach to the analysis of complex systems and decision processes, *IEEE Trans. Systems, Man and Cybernetics*, vol. 3, pp. 28–44, 1973.

23. Zadeh, L.A., The concept of a linguistic variable and its application to approximate reasoning-I, *Information Sciences*, 8, pp. 199–249, 1975.

24. Zadeh, L.A., The concept of a linguistic variable and its application to approximate reasoning-II, *Information Sciences*, 8, pp. 301–357, 1975.

25. Zadeh, L.A., The concept of a linguistic variable and its application to approximate reasoning-III, *Information Sciences*, 9, pp. 43–80, 1975.

26. Zadeh, L.A., Fuzzy logic—Computing with words, *IEEE Trans. Fuzzy Systems*, vol. 4, no. 2, May 1996.

6

Applied Fuzzy Intelligence System: Fuzzy Batching of Bulk Material

Zimic Nikolaj
University of Ljubljana

Mraz Miha
University of Ljubljana

Virant Jernej
University of Ljubljana

Ficzko Jelena
University of Ljubljana

Abstract.   This chapter represents a design procedure of the fuzzy logic controller for batching different kinds of bulk material. The fuzzy logic controller was implemented in software and is an extension of existing electronic scale. The controller for batching is based on Mamdani's type of decision. After some initial adaptations of the fuzzy logic controller, the optimal time of batching was obtained and the batching error became smaller (it did not exceed 0.1% of desired mass). The design procedure of control consists of modeling, simulating, and optimizing the fuzzy logic controller behavior, which is described in the second part of the chapter.

Key words: Batching scale, fuzzy logic controller, MATLAB, modeling, simulation.

6.1 Introduction

This chapter concerns the design of a control system for batching bulk materials. In the set of specifications, two are especially important. First, an inexpensive solution is desirable, and second, the batching scale should be able to adapt to different types of material (sugar, potato chips, sweets, etc.). For these two reasons it was decided to implement a batching scale with a fuzzy inference system as this does not require a powerful processing unit. Although the solution is inexpensive, the design of the batching scale, regarding its electronic and mechanical parts, is in accordance with OIML (Organisation Internationale de Metrologie Legale) Recommendation R6, Automatic gravimetric filling instruments.[8]

The fuzzy inference system (FIS) under consideration here is implemented in software and represents an extension of an existing electronically controlled scale. This system is based on the Mamdani's type of controller. Particular attention was devoted to the time complexity of the procedure for obtaining the FIS output. After some initial adaptations, the optimal time of batching was obtained and in this case the batching error was smaller than a scale division, i.e., it did not exceed 0.1% of the desired mass.

The design of a fuzzy batching scale requires a simulation procedure with suitable support to describe scale working. For this type of computation and visualization we have used programming environments MATLAB, SIMULINK, and Fuzzy Logic Toolbox.

6.2 Description of the Electronic Batching Scale

6.2.1 Building Parts and Working Concept of the Batching Scale

The objective of our work was to produce an electronic batching scale for dosing the desired mass at the point of a load receptor. The current batching scale is designed for batching bulk material in the range from 0.5–10 kg.[1] The main scheme of the batching scale is presented in Figure II.6.1. The batching cycle starts with the material located in a special container **Store**. It runs through a feeding device **Channel** leading to a **LoadReceptor**. The **Channel** is connected to a **Vibrator** which controls the flow of material by means of vibration, increased power of vibration meaning increased speed of flow. The flow can be completely stopped by a **Trap-Door** attached to the end of the **Channel**. Under the **Channel** there is a **LoadReceptor**. The amount of mass in the **LoadReceptor** is measured with a sensor of mass. When the desired amount of material is transferred to the **LoadReceptor**, the **TrapDoor** is closed and the system pauses. In that time interval, the remaining material which is in the air after closing the **Trap-Door** falls into the **LoadReceptor**. The latter is then turned over and its contents are fed to the packing part of the device. A new batching cycle can start after the **LoadReceptor** has been emptied.

The main requirements for the start of the batching process are that there is a stock of material in the **Store** location and that the **LoadReceptor** is empty and stationary. The process of batching then goes through the following phases:

- The dosing of material
- The termination of dosing after the chosen mass is received in the **LoadReceptor**
- The time interval of pause (time used for the falling material in the air which has to reach the **LoadReceptor**)
- The contents of the **LoadReceptor** are fed to the packing part of appliance

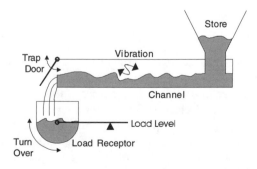

FIGURE II.6.1   Filling process in the observed batching scale.

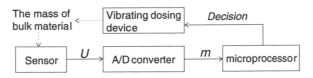

FIGURE II.6.2   Basic scheme of batching scale operation.

The batching scale is built from the following parts:

- A sensor located at the **LoadReceptor**, which transduces mass to an electrical voltage
- An A/D converter which transforms the voltage to a digital value
- A microprocessing unit which performs the processing of FIS
- A memory unit which memorizes the calibration parameters
- A keyboard
- A display unit

The control cycle for the working process is presented in Figure II.6.2. At the first stage the sensor produces an electric voltage U which depends on the mass in the **LoadReceptor**. At the second stage the voltage U is passed to the A/D converter, which converts it to a digital value m. This value is then entered into the microprocessor where the decision about the dosing speed and shut-off is made.

The mass value m is presented on the display as the number of unit objects. The unit of mass (or a particle) represents the smallest value which could be recognized, measured, and presented on the display of the batching scale. The resolution of the A/D converter should be higher than the resolution of the unit objects presented on display. In this way, the mass measured is also correct when the **LoadReceptor** is overdosed.

A special case of batching is when the measured mass m is close to zero. The main European standard for batching procedures[8] determines that the zero area is lower than the mass of a unit ($\pm 1/4$ of a scale unit).

The batching precision is an important specification. Batching should be sufficiently precise to enable material to be batched and packed within precise declared limits. The batching scale must measure the mass to the required accuracy and report whenever a discrepancy between the actual and the desired mass occurs.

Another important specification when designing a batching device is the batching time. High batching precision could easily be achieved by increasing the batching time, i.e., by decreasing the speed of the flow of material. However, in practice, it is desirable to have as short a batching cycle as possible, while meeting the required batching precision.

6.2.2   Calibration of Batching Scale

The batching scale is built from mechanical and electronic parts. Even if they are bought in the same series from the same manufacturer, the parts will vary in their characteristic within stated tolerances. Due to this, every batching scale has to be calibrated, first after its completion and at regular intervals afterward.

The calibration process determines the value which will be presented on the display in response to the electric voltage produced by the sensor. The calibration procedure is relatively simple, due to the linear relation between the measured mass and the digital value from the A/D converter. Two points have to be determined, and these are normally the point of empty **LoadReceptor**, and the point when the **LoadReceptor** is loaded with a mass greater than the dosing mass. The case

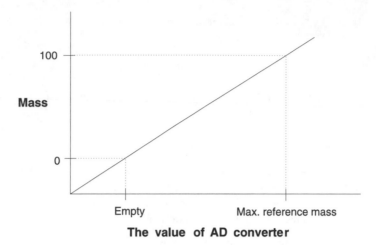

FIGURE II.6.3   The case of measured value transformation to mass.

of linear dependency is presented in Figure II.6.3. The procedure of calibration is done in the following steps:

- The ≫ zero ≪ point is determined when the **LoadReceptor** is empty
- The **LoadReceptor** is loaded with reference mass (the reference point is obtained)
- The acquired data are saved to eeprom memory

Electronic batching scales are usually controlled by PID controllers. Their setup procedure is relatively long, and responding to every change in the control strategy is almost impossible. The fuzzy approach presented in the following sections enables very fast and very cheap prototyping of a relatively good control process.

6.3   Fuzzy Control of Batching Scale

6.3.1   Introduction to Fuzzy Control

L. Zadeh presented the theoretical approach to fuzzy sets in 1965.[2] The fuzzy approach enables the solution of various problems with unclear, uncertain, ambiguous or not explicitly expressed or stated information about the observed system. The theory of fuzzy sets introduces procedures which deal with fuzzy numbers, fuzzy sets, fuzzy variables, fuzzy propositions, fuzzy predicates, fuzzy operations, fuzzy functions, etc. Another important principle introduced from fuzzy logic is the extension principle, which enables the generalization of the major part of the basics of mathematics. It took the relatively long period of 10 years to involve fuzzy logic in practical control applications.[3]

The procedures of decision making can be built on the basis of ordinary (crisp) logic. The most simple of these procedures is realized with the set of logical rules, which use the operation of implication. A lot of fuzzy implication types are also well known in fuzzy logic, so we could also build sets (or lists) of fuzzy logic rules. The reasoning carried out on the basis of such a set is called approximate reasoning. The procedure of approximate reasoning consists of *fuzzification*, *fuzzy inference*, and *defuzzification*.

A fuzzy logic controller (FLC) is a control device that is described by a knowledge-based system consisting of IF-THEN rules, with uncertain predicates and a fuzzy inference mechanism.[4]

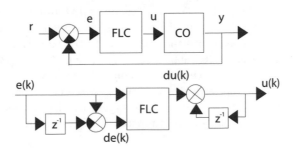

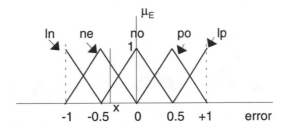

FIGURE II.6.4 Basic feedback fuzzy control system (up) and structure around the FLC (down).

FIGURE II.6.5 Membership function of fuzzy variable e.

Figure II.6.4 shows FLC in the feedback control of an object control (**CO**) (up) and control structure on input and output of FLC (down). From Figure II.6.4 Equations II.6.1 follow:

$$du(k) = F(e(k), de(k))$$
$$u(k) = u(k - 1) + du(k) \tag{II.6.1}$$
$$de(k) = e(k) - e(k - 1).$$

Here k is the time-step, u the output, and e the input to FLC. The difference $e = y - r$ is the control error where r is the reference value and y the output from the controlled object CO. Error e, error change de, and output change du are linguistic (fuzzy) variables; values of these variables are linguistic values (terms). An example of the definition of variable e is shown in Figure II.6.5. The variable has five terms: **large negative (ln)**, **negative (ne)**, **normal (no)**, **positive (po)**, and **large positive (lp)**. Each of these is a fuzzy set (fuzzy number) with a fuzzy spread of 0.5. If variable e has the space $X = [-1, +1]$, $x \in X$, where x represents **error**, we have the possibility to use the propositions 'e is **term**' where **term** is one of five given linguistic values. Let us take $x = -0.4$. Its impact on the reasoning process is expressed as follows:

$$\mu_{ln}(-0.4) = 0, \ \mu_{ne}(-0.4) = 0.8, \ \mu_{no}(-0.4) = 0.2, \ \mu_{po}(-0.4) = 0, \mu_{lp}(-0.4) = 0. \quad (II.6.2)$$

In Equation II.6.2 only the memberships $\mu_{ne}(-0.4)$ and $\mu_{no}(-0.4)$ differ from 0. This means that in the case of $x = -0.4$, only those two memberships have the influence to fuzzy inference. In the given case the list of fuzzy rules can be set as follows:

IF e is **ln** AND de is **sm** THEN U is **pz**

IF e is **ne** AND de is **sm** THEN U is **bi**

IF e is **no** AND de is **la** THEN U is **ze**

...

$$(II.6.3)$$

Terms **sm, la, bi, pz**, and **ze** are the terms of fuzzy variables *de* (error change) and *U* FLC (output). The value min $(\mu_{\mathbf{no}}, \mu_{-0.4}) = 0.2$ is called a firing degree α_e, which cuts an output term **ze**. The concrete result of approximate reasoning in the case of the third rule is the fuzzy set $r_3 = \min (\mathbf{ze}, 0.2)$.

All fuzzy rules together make, after approximate reasoning, output fuzzy set $R = \cup r_i$. The list of rules for the above-mentioned typical fuzzy controller with two input variables (*e, de*) and one output variable may also be presented with a table. Such a table could be called an algorithm of fuzzy controller. If the object of control CO in Figure II.6.4 is a nonfuzzy system, it cannot receive a fuzzy resultative set *R* on its input. In this case we have to involve defuzzification in the control process.

A lot of defuzzification methods are known. The most popular of them is the *center of gravity* method (COG). In this case, the resultant set *R* is treated as an area in output space *V*, and we would like to find its gravity point. In this way the crisp value is obtained, which is used as an output from FLC using Equation II.6.4.

$$du = \frac{\sum_{i=1}^{m} u_i w_i}{\sum_{i=1}^{m} w_i}. \tag{II.6.4}$$

The u_i is a value from output space *U* and w_i a value of resultant output membership $\mu_R(u_i)$ in an observed point $(u_i \in U)$.

6.3.2 Fuzzy Controller of Batching Scale

The introduction of fuzzy control into controlling a batching process is based on Figure II.6.2, where the measured mass *m* enters into the decision process in time *t*. On the basis of entered mass $m(t)$ and previously memorized mass $m(t-1)$, we can calculate $\Delta m(t)$ using relation $\Delta m(t) = m(t) - m(t-1)$. The last value could be called the *speed of flow* of the material. When $m(t)$ is compared to the desired mass, two outputs are processed. The first is the intensity of vibration (the speed of flow of material) *vibro(t)* and the second is a crisp variable *Trap(t)*, which has two possible values (the **TrapDoor** is opened or closed). The basic scheme of the controller is shown in Figure II.6.6.

The batching cycle is controlled with FLC until the measured value $m(t)$ falls into the interval of *Acceptable mass area* presented in Figure II.6.7. After that, the FLC closes the TrapDoor and stops vibrating. The fuzzy part of the batching cycle has ended and the "post-processing phase"

FIGURE II.6.6   Scheme of the fuzzy controller of batching scale.

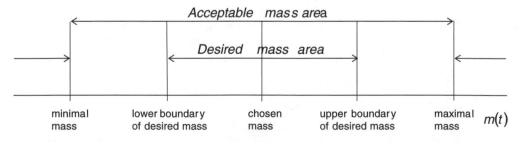

FIGURE II.6.7   Intervals used after FLC stops the process of dosing the material.

begins. The delay time used for falling material in the air is located between the FLC processing and the post-processing phase. Due to this extra added mass of material, we introduce the *overflow correction factor* described later.

The interval border points are:

- *Minimal mass* (below this mass the measured $m(t)$ is too low)
- *Lower boundary of desired mass*
- *Chosen mass* (chosen value for batching procedure)
- *Higher boundary of desired mass*
- *Maximal mass* (above this mass the measured $m(t)$ is too high)
- *Acceptable mass area* (interval [minimal mass, maximal mass])
- *Desired mass area* (interval [lower boundary, higher boundary of desired mass])

All those intervals are crisp intervals, but the post-processing phase could also be done with fuzzified borders. The post-processing is done in two steps:

- If the final $m(t)$ is in the *acceptable mass area,* the mass corresponds to batching criteria, but parameters of the FLC are changed with the goal of more precise dosed mass compared to chosen mass in next batching cycle; in the same time, the speed of dosing could be increased (description of this procedure is presented in the section on optimizing the FLC)
- If the final $m(t)$ is higher than the *maximal mass,* then an error message is displayed

Due to extra mass in the air after closing the **TrapDoor**, the chosen mass is less than the expected mass which will fall into the **LoadReceptor**. Because of this the *overflow correction factor* (*Foc*) is introduced. It represents the mass of the material that is still falling into the batching container after closing the **TrapDoor**. To control the overflow mass we use Equation II.6.5:

$$Foc(n + 1) = Foc(n) - \frac{m_{desired} - m(t)}{2}, \ Foc(0) = 0. \tag{II.6.5}$$

At the beginning, the *Foc* is set to 0 so that dosing occurs without correction. *Foc* is changed depending on the difference between the desired mass $m_{desired}$ and the temporary measured mass $m(t)$. Half the difference (see Equation II.6.5) is taken with the goal of decreasing sensitivity of the system to momentary error in time. The chosen mass from Figure II.6.7 is decreased.

6.3.3 Fuzzy Controller Characteristics

The set of rules are constructed on the basis of two inputs: *mass* ($m(t)$) and *flow* of mass ($\Delta m(t)$) (Table II.6.1). The values of *mass* and *flow* input terms are **very small, small, medium, almost precise, precise**, and **very small, small, medium, large, very large**. The meanings of *vibro* output terms are **small, medium** and **large**.

TABLE II.6.1　FLC Control Matrix

FLC		Flow				
		VS	S	M	L	VL
	VS	L	L	L	L	L
	S	L	L	L	M	M
Mass	M	L	L	M	M	S
	AP	M	M	S	S	S
	P	S	S	S	S	S

6.4   Simulation of Dosing Batching Scale

We will try to model the FLC as a *fuzzy inference system* (FIS) using a specialized Fuzzy Logic Toolbox in MATLAB & SIMULINK environments.[5–7] Let us suppose the following facts about the FIS:

- We will use the normalized spaces of variables *mass, flow,* and *vibro,* as shown in Figure II.6.6
- The simulation is carried out for a chosen mass of 3 Kg
- The results are presented with **Auto-Scale Graph** block in MATLAB environment

6.4.1   The Model of Batching Scale Used for Simulation

The complete simulation scheme is presented in Figure II.6.8. On the top left side of the figure are block models of:

- *Dosing*: blocks **DosageValue, Vib, Gain, Power of vibration, Sum1, Disturb** (it generates disturbances in the flow of material in **Channel**); **NextBatch** triggers the next batching cycle after the successfully finished batching of 3 Kg of the mass
- *Trap door of channel*: blocks **Dela, NOT, Trap**
- *Accumulating of mass in load receptor*: blocks **Sum, Clear, Load, Store, Read**

On the top right side of Figure II.6.8 we have block models of:

- *FIS system with loopback for dosing procedure*: blocks **Fact1, Mux1, FIS, Fact2, Delay3, Gain** (power of vibration), **Diff** for *Flow*

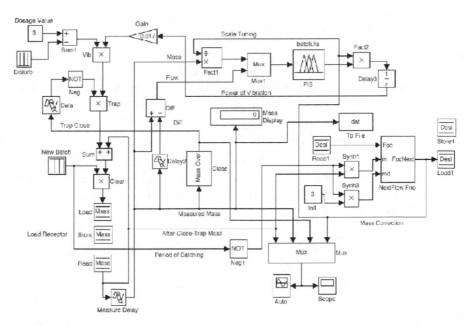

FIGURE II.6.8   Simulation of fuzzy logic batching, model BATCH.MDL with fuzzy inference system BATCH.FIS.

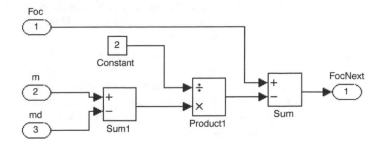

FIGURE II.6.9   Subsystem for the calculation of overflow factor *Foc*.

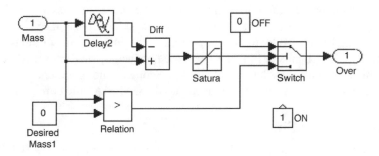

FIGURE II.6.10   Subsystem **Close**.

On the right side of Figure II.6.8 we have modeling blocks of:

- *Processing* of **NextFlowFnc:** this gives us *MassCorrection* factor for mass which is falling down after the door at the end of the channel has closed; this material is registered with sensor *MeasureMass*; **Mass** and **Desi** are both memorizing variables; blocks **Store** and **Store1** are declaring addresses used for memorizing
- *The observation of simulation flow*: the mass is numerically measured with block **Mass Display; Auto (Auto-Scale Graph)** or **Scope** is used for signal observation during the simulation

Let us look at two subsystems used in the observed model BATCH.MDL from Figure II.6.8. Subsystem **NextFlowFnc** is shown in Figure II.6.9. It calculates the *Foc* overflow correction factor. The summary of the system is equivalent to Equation II.6.5. The evaluation of this block takes place only once per batching cycle, so we need synchronization with the clock of batching cycle **NewBatch**. The second subsystem from Figure II.6.8 called **Close** is presented in Figure II.6.10. It produces the output **Over**, which becomes 1, when the difference between the measured mass $m(t)$ and chosen mass $m_{desired}$ falls into the interval [0, 0.25]. Block **Relation** is used to eliminate the situation where the value 1 in **Over** could appear at the beginning of dosing. The most important part of the model in Figure II.6.8 is FIS block, which is processing the sequence of fuzzification, fuzzy inference, and defuzzification.

6.4.2   Simulation

Our intention is to run a simulation based on the file BATCH.MDL in SIMULINK and MATLAB environment. Due to dependence of the model on FIS block, we first have to load the FIS model file. The load is triggered with the command below:

```
% Initialization for program BATCH.MDL
aa=readfis('batch').
```

During the simulation we observe the time from two different viewpoints. The most important is the simulation time which is measured in simulation sec (simulation steps). It runs through the whole simulation. In our case we have a time of 300 simulation sec. In case of the occurrence of disturbances, we extend this time to 800 simulation sec. The second time is the time of batching, called the ≫period of batching≪, and initialized with a time of 50 simulation sec. This time could be even shorter. The reality demand is that as many measurement cycles as possible are done in a unit of time. This must be without the loss of accuracy of batching. The maximal error could be 0.1%. The period of batching is given by the sequential vector (II.6.6), where the numerator is a value of control and the denominator is a simulation step (time).

$$\textbf{NewBatch } \textit{Value/Time} = (1/0\,,\,\ldots,\,1/48\,,\,0/49,\,1/50). \qquad \text{(II.6.6)}$$

The process of batching is observed in 50 time points. The measured mass $m(t)$ in **LoadReceptor** is lower than the *After-Close_Trap Mass* in the phase of transitional phenomena. After it both masses are equalized using FIS controller to $m_{desired}$ intialized by **Init**. The flow of mass is obtained with the difference **Diff**. Two inputs to FIS subsystem are obtained by multiplexor **Mux1**.

If the graphical presentations done from **Auto** or **Scope** are not good enough, we could also do some analyses on the pure data which are saved into the DAT.MAT file during the simulation.

6.4.3   The Results of Simulation

In Figure II.6.11 four different simulation results obtained by running the BATCH.MDL model file are shown. The meanings of the data are:

$1 = After\_Close\text{-}Trap\_Mass$
$2 = Measured\ Mass\ m(t))$
$3 = Over$ (end of observed batching)
$4 = Mass\ Correction$

We can see from Figure II.6.11 that the chosen mass 3 Kg is decreased until the sum of chosen mass and the falling mass (mass in the air after the closed door) is the same as the chosen mass at the

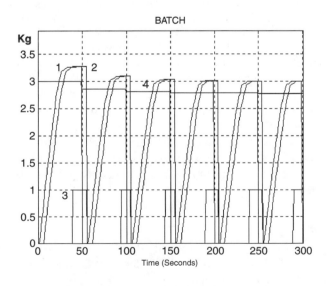

FIGURE II.6.11   First six batchings of mass 3 Kg.

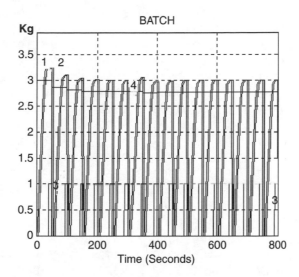

FIGURE II.6.12   First 16 batchings with a disturbance at 300th simulation second.

beginning of batching. The measured mass reaches the chosen mass in three to four batching cycles. Those batchings are rejected. All following batchings of 3 Kg mass are precise with the error less than 0.1%. Signal *Over* = 1 means that the new batching cycle can be started.

Figure II.6.12 is similar to Figure II.6.11. The simulation time is extended because of the disturbance introduced to dosing at the 300th simulation sec. The normal dosing procedure is constant with the value of 3 Kg with the vibration gain **Gain** = 0.017. Actually, the product of these two values should be constant. Due to this, we could take a higher mass of material (vibrations should be higher) and the **Gain** should be smaller. When the disturbance occurs, we have reduced the input of material to 60% of the previous one. The power of vibration has not been changed. The duration of the disturbance is presented with a sequential vector:

Disturb *Value/Time*

$$= (0/0 , \ldots , 0/299, \ 1/300, \ldots , 1/319, \ 0/320 , \ldots , 0/600). \qquad \text{(II.6.7)}$$

Figure II.6.12 shows how fast the system reacts to a disturbance and how fast the previous stationary regular measurement of 3 Kg mass is achieved. Due to the disturbance, the seventh cycle shows the greatest difference from the chosen value, so it is rejected.

All simulation results are saved to DAT.MAT file for *MeasuredMass* variable values. Let us present some chosen data from simulation intervals from 310 to 407 and from 753 to 801 simulation time steps. In the first case the disturbance is present and in the second it is not (the system eliminates it).

```
310.0000 0            359.0000 0            753.0000 2.9999
311.0000 0            360.0000 0            754.0000 2.9999
312.0000 0            361.0000 0            755.0000 2.9999
312.0000 0.0875       362.0000 0            755.0000 0
313.0000 0.1821       362.0000 0.1298       756.0000 0
314.0000 0.2766       363.0000 0.2701       757.0000 0
315.0000 0.3712       364.0000 0.4104       757.0000 0
316.0000 0.4658       365.0000 0.5506       758.0000 0
```

317.0000	0.5604	366.0000	0.6909	759.0000	0
318.0000	0.6549	367.0000	0.8312	760.0000	0
319.0000	0.7415	368.0000	0.9715	761.0000	0
320.0000	0.8276	369.0000	1.0950	762.0000	0
321.0000	0.9131	370.0000	1.2205	762.0000	0.1318
322.0000	0.9968	371.0000	1.3461	763.0000	0.2736
323.0000	1.0829	372.0000	1.4717	764.0000	0.4154
...		373.0000	1.5972	765.0000	0.5572
354.9842	3.0617	...		766.0000	0.6989
354.9842	3.0617	394.0000	2.9651	...	
354.9999	3.0617	395.0000	2.9665	793.0000	2.9957
354.9999	3.0617	396.0000	2.9672	794.0000	2.9978
354.9999	3.0617	397.0000	2.9672	795.0000	2.9992
355.0000	3.0617	398.0000	2.9672	796.0000	2.9999
355.0000	3.0617	399.0000	2.9672	797.0000	2.9999
355.0000	3.0617	400.0000	2.9672	798.0000	2.9999
355.0000	3.0617	401.0000	2.9672	798.0000	2.9999
355.0000	3.0617	402.0000	2.9672	799.0000	2.9999
355.0000	3.0617	403.0000	2.9672	800.0000	2.9999
355.0000	3.0617	404.0000	2.9672	801.0000	2.9999
355.0000	3.0617	405.0000	2.9672		
355.0000	3.0617	406.0000	0		
356.0000	0	407.0000	0		
357.0000	0				
357.0000	0				
358.0000	0				

It is evident from the data that, in the case of disturbance, there are two consecutive rejected measurements with $error(i+1) = +6.17\%$ and $error(i+2) = -3.28\%$. When the system eliminates the disturbance, we have an approximate error around 0.01%. The end of the sequence of rejected measurements is at 355 and 405, and the end of the sequence of successful measurements is at 755 and 801.

6.4.4   Subsystem FIS of Batching Scale Fuzzy Controller

Let us now determine the characteristics of the FLC for a batching scale. The input variables are *mass* and *flow*, and output variables *vibro* and *trap* as presented in the third section of this chapter. We will be interested mainly in forming a fuzzy output variable *vibro* (power of vibration).

All observed spaces are normalized to interval [0, 1]. For example, in a real batching process we can take the input space [0, 255]. It can be divided into four regions [0, 63], [64, 127], [128, 191], and [192, 255] with the possibility to set five terms. If we use such an input space, all incoming values have to be divided by 255. The value $x/255$ cannot be higher than 1. Equally distributed terms for input and output variable are shown in Figure II.6.13.

The term set of the input variable *mass* includes five linguistic values: **very small (VS), small (S), medium (M), almost precise (AP),** and **precise (P)**. In Figure II.6.13, all of them have added the character 'M' for *mass* or character 'V' for *Vibra*. Membership functions of *mass*, as will be seen later, are subject to change during the optimization phase. Therefore, its linguistic values will be dynamic. The term set of the input variable *flow* also has five linguistic values (therefore it is not seen in Figure II.6.13). They are named as: **very small (VS), small (S), medium (M), large (L),** and **very large (VL)**. The character 'F' is added for flow meaning. Membership functions of the **flow** are static in time. The term set of output variable **vibro** includes three terms: **small (S), middle (M),** and **large (L)**. Output space is also normalized to [0, 1].

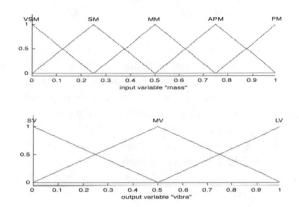

FIGURE II.6.13   Membership functions for *mass* and *vibro* variables.

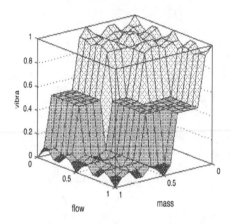

FIGURE II.6.14   Crisp value of variable *Out** in the case of COG defuzzification.

In order to obtain crisp output from the **FIS** that controls the vibration of the batching channel and thereby the flow of material, the following sequence is employed:

1. Computing the level of firing of each rule R_i, $i = 1, 2, \ldots, 25$ (see Table II.6.1)
2. Approximative reasoning
3. Output rule aggregation which gives output fuzzy set *Out*
4. Defuzzification calculated on the basis of COG method as it follows from Equation II.6.4 or better from Equation II.6.8

$$Out^* = \frac{\sum_j \mu_{Out}(j)\, Out_j}{\sum \mu_{Out}(j)}. \tag{II.6.8}$$

*Out** is a crisp output value, which represents the temporary value of variable *vibro*. In Figure II.6.14 is a 3D plot, which is constructed through the FIS-editor of Fuzzy Logic Toolbox (View option). COG cannot give the value *vibro* = 1, since it cannot reach the edges of the output variable space. Such a problem could be eliminated if we use a MOM method (Mean Of Maximum), which could be simply chosen in Fuzzy Logic Toolbox environment instead of COG.

All the data used in FIS model (BATCH.FIS file) described above are presented in the data list labeled with 'aa' label from index 1 to 81.

6.5 Optimizing of Dosing Time

The dosing time is treated as a time which is used to finish one batching cycle. The main goal of the application is to do as many possible measurements in a given time period with the greatest possible accuracy. In addition, optimization should also be done on the time used for processing the FLC decision. Since a fast response system is desirable, this time is important. A response is composed of the time of batching and the time of FLC parameter calculation.

The simulation has shown us that the dosing time can be changed by changing membership functions, which describe the mass. During the simulation, eight different groups of input membership functions have been declared. The calculation of input value membership to fuzzy sets determined in advance is a relatively ≫heavy≪ procedure for an 8-bit processor. The first step to simplify this procedure is to limit it to triangular membership functions. The results of simulation with such a limitation are still satisfactory.

```
showfis(aa)
1.  Name batchp                        61.    [1 5]
2.  Type mamdani                       62.    [2 1]
3.  Inputs/Outputs [2 1]               63.    [2 2]
4.  NumInputMFs [5 5]                  64.    [2 3]
5.  NumOutputMFs 3                     65.    [2 4]
6.  NumRules 25                        66.    [2 5]
7.  AndMethod min                      67.    [3 1]
8.  OrMethod max                       68.    [3 2]
9.  ImpMethod min                      69.    [3 3]
10. AggMethod max                      70.    [3 4]
11. DefuzzMethod centroid             71.    [3 5]
12. InLabels mass                      72.    [4 1]
13. flow                               73.    [4 2]
14. OutLabels vibra                    74.    [4 3]
15. InRange [0 1]                      75.    [4 4]
16.         [0 1]                      76.    [4 5]
17. OutRange [0 1]                     77.    [5 1]
18. InMFLabels VSM                     78.    [5 2]
19.            SM                       79.    [5 3]
20.            MM                       80.    [5 4]
21.            APM                      81.    [5 5]
22.            PM                       57.    Rule Consequent 3
23.            VSF                      58.    3
24.            SF                       59.    3
25.            MF                       60.    3
26.            LF                       61.    3
27.            VLF                      62.    3
28. OutMFLabels SV                     63.    3
29.            MV                       64.    3
30.            LV                       65.    2
31. InMFTypes trimf                    66.    2
40.           trimf                    67.    3
41. OutMFTypes trimf                   68.    3
42.           trimf                    69.    2
```

```
43.                trimf            70.    2
44. InMFParams  [-10 0 0.25 0]     71.    1
45.             [0 0.25 0.5 0]     72.    2
46.             [0.25 0.5 0.75 0]  73.    2
47.             [0.5 0.75 1 0]     75.    1
49.             [-10 0 0.25 0]     76.    1
50.             [0 0.25 0.5 0]     77.    1
51.             [0.25 0.5 0.75 0]  78.    1
52.             [0.5 0.75 1 0]     79.    1
53.             [0.75 1 10 0]      80.    1
54. OutMFParams [-1 0 0.5 0]       81.    1
55.             [0 0.5 1 0]        57.-81. Rule Weight 1
56.             [0.5 1 2 0]        57.-81. Rule Connection1
57. Rule Antecedent [1 1]
58.                 [1 2]
59.                 [1 3]
60.                 [1 4]
```

Dosing batching scale uses one of the eight groups of membership functions. The selected level determines the speed of dosing. The next chosen set of membership functions depends mainly on the results of temporary dosing. If the results are favorable, then a set of membership functions is chosen which enables the higher speed of dosing. In the case of poorer results, a set of functions is chosen which enables slower and more precise dosing. In the case of average results, the functions are not changed. Membership functions are quite frequently changed during the working of the batching scale.

In Figure II.6.13 are presented terms for input variable *mass*. The equally distributed terms in space of input variable represent the membership functions for the layer of mass m. The terms of this presentation are not equally spaced. This can be seen from Figure II.6.15, where terms for the eighth layer are presented. The normalized space of variable *mass* should be observed with the measure of percentages: $m = 1$ means 100% and $m = 0$ means 0%. The terms of layers, which are not linearly distributed, are translated to membership functions of layer 1 using the function:

$$m^* = f_i(m), \quad i = 1, \ldots, 8. \tag{II.6.9}$$

If m^* and suitable membership is calculated (e.g., for the eighth layer) using the input value and the membership functions for layer 1, it is the same as the calculation of membership of mass m using the membership functions of the eighth layer.

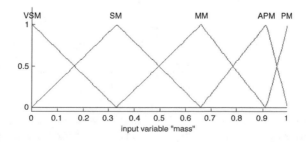

FIGURE II.6.15   Membership functions of input variable *mass*, layer 8.

TABLE II.6.2   Parameters Used for Translating f_i

Level	$f_i(0)$	$50 = f_i(k_{1i})$	$75 = f_i(k_{2i})$	$100 = f_i(k_{3i})$
1	0	50	75	100
2	0	62	82	100
3	0	70	85	100
4	0	76	88	100
5	0	80	90	100
6	0	82	93	100
7	0	84	94	100
8	0	87	95	100

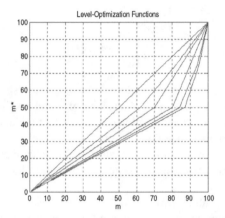

FIGURE II.6.16   Functions used for evaluating m^*.

For the first layer the mass m^* is the same as input mass m. In other cases, mass m^* is evaluated using the function f_i, shown in Figure II.6.16. Functions f_i are obtained from three linear sections. The first is from 0–50%, the second between 50 and 75%, and the third from 75–100% of full scale of the appliance. The evaluation of f_i is then translated into the evaluation of one of three linear functions. Parameters used for translating f_i are presented in Table II.6.2.

The mass m^* can be evaluated using the following expressions:

$$m^* = m\,\frac{50}{k_{1i}}, \qquad m \le k_{1i}$$

$$m^* = 50 + (m - k_{1i})\,\frac{75 - 50}{k_{2i} - k_{1i}}, \qquad k_{1i} \le m \le k_{2i}$$

$$m^* = 75 + (m - k_{2i})\,\frac{100 - 75}{k_{3i} - k_{2i}}, \qquad k_{2i} \le m \le k_{3i}. \qquad (\text{II.6.10})$$

The membership to input values can be evaluated in two ways as shown in Figure II.6.17. The first approach uses the membership functions which may be built in advance. The second approach uses the calculated mass m^* and linearly distributed membership functions.

The measured mass is presented on the scale's display with the value from interval [0, 10000]. In this way the accuracy of scale is obtained. The information about mass is stored in the microprocessor as an integer type number. Due to the need for a robust control where such a resolution of data on input is not needed, the mass is normalized into the interval [0, 255]

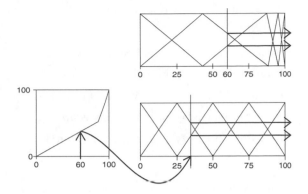

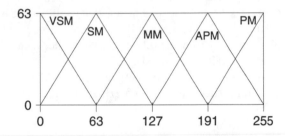

FIGURE II.6.17 Evaluation of influence of membership functions.

FIGURE II.6.18 The approach of normalizing membership in the space of *mass*.

which requires 1 byte used for memorizing the value. The membership, which is ordinarily presented with interval [0,1], is normalized with [0, 63] interval too, for ease of computing. The selection of interval depends on the distribution of five membership functions. If the angle of triangles is around 45°, then the function space is from interval [0, 63] when the normalized input interval is [0, 255].

From the above description it can be seen that membership evaluation is done only on integer values, which is very important for the speed of response of FLC realized in an 8-bit microprocessor. Such processing units usually have no floating point arithmetic unit. Normalized membership functions are presented in Figure II.6.18. We can see that input variable is normalized into the interval from 0–255 and the membership is normalized into the interval from 0–63. The membership is evaluated using Equation II.6.11:

$$\mu_{VSM}(x) = 63 - x; \quad \mu_{SM}(x) = x; \qquad 0 \le x \le 63$$
$$\mu_{SM}(x) = 127 - x; \quad \mu_{MM}(x) = x - 64; \qquad 64 \le x \le 127$$
$$\mu_{MM}(x) = 191 - x; \quad \mu_{APM}(x) = x - 128; \qquad 128 \le x \le 191$$
$$\mu_{APM}(x) = 255 - x; \quad \mu_{PM}(x) = x - 192; \qquad 192 \le x \le 255. \qquad \text{(II.6.11)}$$

All other memberships, not included in the equations, are zero valued in the observed interval. From the above equations, it can be seen that evaluation of memberships is translated to a decrementation process, which is not treated as a "hard" processing problem. Presented membership evaluations relate to the input variable *mass*. The same approach could be used for evaluating the membership function of input variable *flow*.

From the viewpoint of evaluation, the fuzzy rules of calculus can also be optimized. The set of rules consists of the following formats:

IF *mass* is **VS** AND *flow* is **VS** THEN *vibro* is **L**

IF *mass* is **VS** AND *flow* is **S** THEN *vibro* is **L**

IF *mass* is **VS** AND *flow* is **M** THEN *vibro* is **L**

IF *mass* is **VS** AND *flow* is **L** THEN *vibro* is **L**

IF *mass* is **VS** AND *flow* is **VL** THEN *vibro* is **L**

. . .

$$(\text{II}.6.12)$$

The maximum number of rules with regard to number of input terms of variables *mass* (II.6.6) and *flow* (II.6.6) is 25. They have previously been listed in the list of BATCH.FIS program. Matrix representation of all fuzzy rules is shown in Table II.6.1 in the third section of this chapter.

Using optional input values of mass, variable influence factors differ from zero at only two terms. If the normalized value of mass is around 130, that means that the influence is different only at terms MM and APM (see Figure II.6.18). We have a similar situation at the input *flow* variable's values. The output function is influenced only from the rules whose influence factors are different from zero. That means that only 4 rules are evaluated instead of 25.

The result of rule evaluation is the resulting output fuzzy set *Out*. Due to the crisp value required from the appliance, the *Out* fuzzy value has to be defuzzified. We use the COG method presented in Equation II.6.8.

Similarly to the input membership functions, the output functions are also normalized for more effective evaluation. The output value's space is determined within the interval [0, 63]. Those values are selected due to the use of integer arithmetics in the microprocessor. The main condition of evaluation is that even intermediate results cannot exceed the maximal value, which can be presented with 16 bits (32768). The evaluated output of FLC is used to control the vibrator appliance.

As can be seen from Figure II.6.15, the optimization procedure which used f_i functions has led us from equally (regularly) to nonequally distributed terms of variable *mass*. Due to this, we have to change the terms of variable *mass* in starting FLC modeled in file BATCH.FIS. A simulation done in such an environment is presented in Figure II.6.19. The meaning of items 1–4 is the same as on

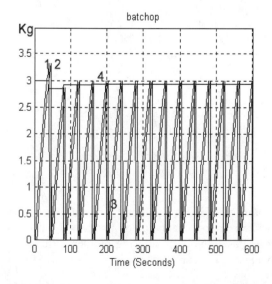

FIGURE II.6.19   Batching done in ideal conditions.

Figures II.6.11 or II.6.12. The simulation takes 600 simulation sec, which gives 1.5 batching done per min compared to 1.2 batching per min presented in the case from Figure II.6.11.

The accuracy achieved in simulation from steps 524 to 567 is presented in a partial list of simulation results from DAT.MAT file. In the list there are two finished batchings. The first is at step 525, where we have the error -0.1%, and the second is at step 565, where the error is zero. The results presented are obtained with evaluation on four places after the floating point. In intermediate steps we can see how the **LoadReceptor** is filling with the chosen mass of 3 Kg.

```
524.0000 2.9999          . . .
525.0000 2.9999          557.0000 2.5525
525.0000 0               558.0000 2.6255
526.0000 0               559.0000 2.6966
526.0000 0.0547          560.0000 2.7647
527.0000 0.0547          561.0000 2.8291
527.0000 0.1101          562.0000 2.8887
528.0000 0.1661          563.0000 2.9444
529.0000 0.2137          564.0000 3.0000
530.0000 0.2392          565.0000 3.0000
531.0000 0.2648          565.0000 0
532.0000 0.3883          566.0000 0
533.0000 0.3883          566.0000 0.0547
   . . .                 567.0000 0.0547
```

Considering the optimized distribution of terms of fuzzy input variable *mass* the surface from Figure II.6.14 is no more valid. It is better to say that it is valid only in the case when layer = 1 and we have linear distribution of terms. Considering the optimization procedure, we have a new relation between the input and output of the observed FIS. For this case a new FIS system BATCOPT.FIS model is prepared, which gives us the surface view presented in Figure II.6.20. We can see that transitions between functional values are more intensive and at the same time smoother, compared to the case in Figure II.6.14.

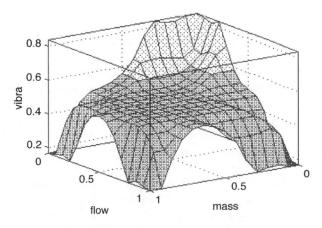

FIGURE II.6.20 Surface view for optimized FIS BACHOPT.FIS.

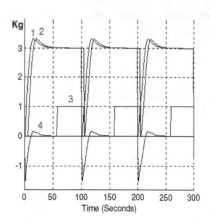

FIGURE II.6.21   Batching with another type of overflow correction.

6.6   Another Approach of Fuzzy Control

The mass which has passed the **TrapDoor** and has not yet fallen to **LoadReceptor** after closing the **TrapDoor** could be also treated with a decreased mass in the phase of measurement as we are doing in hand batching. Let us consider this kind of approach. If we want to eliminate the influence of excess mass with FIS in one observed measurement, we should decrease variable *MeasuredMass* with *MassCorrection*. It should be done in **Load, Store**, and **Read** blocks which are presented on the left side of Figure II.6.8. We put the difference of *mass* and *MassCorrection* in front of operation **Clear** (see blocks **Hold, Gain1** in **Dif**). The results of such a simulation are shown in Figure II.6.21, where we have courses from 1–4. Different items of mass are presented in this figure. At **Gain1** = 0.2 we have a correction of mass 4, which increases from a negative value to 0. It means that mass became equal to *MeasuredMass*.

6.7   Conclusion

Let us conclude with a basic description of simulation and logical behavior of fuzzy control. Batching with a fuzzy inference system was implemented using a Dallas DS80C320, an 8-bit microprocessor. This microprocessor has also performed the measurements of mass in the **LoadReceptor** so that the measuring and batching were implemented as two modules of one program. Due to the rather low processing capability of the microprocessor, different optimization procedures were introduced to reduce the batching time. For the same reason, integer arithmetic was used. When the program was optimized in terms of computation time, it provided 25 control cycles per sec. Each control cycle includes measurement of mass, computation of fuzzy inferences, and crisp FIS output.

The real experiment measures sugar with a chosen mass of 1 Kg. After the initial adaptations (about 20 batching cycles were performed at a lower speed), the optimal time of batching (4 sec), i.e., a filling rate of 15 batching cycles per min, was obtained. Approximately half the time was spent on filling and half on emptying the **LoadReceptor**. The speed of simulation presented in this chapter is only about 1–2 batchings per min. However, a simulation sec is not equal to a sec in real time. Constants with a big influence on the speed of batching are **DosageValue, Gain, NewBatch**, and **Measure Delay**. We take the following parameters for our simulation:

BATCH.MDL:
DosageValue = 3 Kg
Gain = 0.017
NewBatch = 50 sec period
Measure Delay = 5 sec

If the above data are exchanged with the following data:

MTCHspeed.MDL:
DosageValue = 10 Kg
Gain = 0.025
NewBatch = 5 sec period
Measure Delay = 1 sec

we obtain the following list of simulation data:

```
. . .
350.9862    3.0000      361.0000    3.0000
350.9862    3.0000      361.0000         0
351.0000    3.0000      362.0000         0
351.0000         0      362.0000    0.0233
352.0000         0      363.0000    0.0233
352.0000    0.0233      363.0000    1.5544
353.0000    0.0233      364.0000    3.0000
353.0000    1.5544      365.0000    3.0000
354.0000    3.0000      365.4466    3.0000
355.0000    3.0000      365.4466    3.0000
355.4466    3.0000      365.9862    3.0000
355.4466    3.0000      365.9862    3.0000
355.9862    3.0000      366.0000    3.0000
355.9862    3.0000      366.0000         0
356.0000    3.0000      367.0000         0
356.0000         0      367.0000    0.0233
357.0000         0      368.0000    0.0233
357.0000    0.0233      368.0000    1.5544
358.0000    0.0233      368.0000    3.0000
358.0000    1.5544      370.0000    3.0000
359.0000    3.0000      370.4466    3.0000
360.0000    3.0000      370.4466    3.0000
360.4466    3.0000      370.9862    3.0000
360.4466    3.0000      370.9862    3.0000
360.9862    3.0000      371.0000    3.0000
360.9862    3.0000      . . .
```

We can see that the batching error is very small and that between two measured values of 3.0000 Kg, 5 sec take place. We have 7 sec to take away the measured mass from the batching scale. The whole process of the batching cycle takes 12 sec. This means that the speed of batching is 5 batchings per min. If the speed of the pre- and post-processes, which are hidden in the row of results 0000 3.000 3.0000 3.0000 3.0000 3.0000, is high enough, then we could reach 5 simulation sec as the time used for one batching cycle. The fastest period 0 0.0233 0.0233 1.5544 3.000 enables 12 batching cycles per min.

When the parameters of the scale are correctly set, the batching error is smaller than a scale division, which means that the batching error is $\pm$ 3g or 0.1% of the desired mass. This precision level and batching time were obtained after some initial cycles, provided that the flow of material was constant during the time in which the fuzzy system was adapted; i.e., it found the optimal membership functions.

The fuzzy batching scale was put into practice for batching sugar and the results obtained were very favorable (see Zimic et al.[1]). The same batching scale was used later with a spiral screw instead of a vibrator as a filling device. In this case the batching mass was 50 Kg and the batching precision was also 0.1%.

References

1. Zimic, N. et al., *Batching with a Robust Fuzzy Logic Controller*, IFSICC '96 Maui, HI.
2. Zadeh, L.A., Fuzzy sets, *Information and Control*, 8, 1965, pp. 338–353.
3. Mamdani, E.H. and Assilian, S., An experiment in linguistic synthesis with a fuzzy logic controller, *Int. J. Man-Machine Studies*, 7, 1975, pp. 1–13.
4. Yager, R.R. and Filev, D.P., *Essentials of Fuzzy Modeling and Control*, John Wiley, New York, 1994.
5. MATLAB, Using MATLAB, The MathWorks Inc., Natick, MA, 1997.
6. SIMULINK, Using SIMULINK, The MathWorks Inc., Natick, MA, 1997.
7. Fuzzy Logic Toolbox, User's Guide, The MathWorks Inc., Natick, MA, 1998.
8. OIML: Recommendation R6, Automatic gravimetric filling instruments, Organisation Internationale de Metrologie Legale, 1994.

7

Techniques for Designing and Refining Linguistic Fuzzy Models to Improve Their Accuracy[*]

Rafael Alcalá
University of Granada

Jorge Casillas
University of Granada

Oscar Cordón
University of Granada

Francisco Herrera
University of Granada

7.1 Introduction

Fuzzy rule-based systems (FRBSs) constitute an extension of classical rule-based systems, because they deal with IF-THEN rules where antecedents and/or consequents are composed of fuzzy logic statements, instead of classical logic rules. This consideration presents two essential advantages: the key features of knowledge captured by fuzzy sets involve handling uncertainty, and inference methods become more robust and flexible with approximate reasoning methods of fuzzy logic.

---

[*]This research has been supported by Spanish CICYT, project PB98-1319

The most usual application of FRBSs is *system modeling*,[4,77] which in this field may be considered as an approach used to model a system making use of a descriptive language based on fuzzy logic with fuzzy predicates.[91] Fuzzy modeling (FM) (i.e., system modeling with FRBSs) usually comes with two contradictory requirements to the obtained model: the *interpretability* (capability to express the behavior of the real system in an understandable way) and the *accuracy* (capability to faithfully represent the real system).

While linguistic FM (LFM)—mainly developed by linguistic (Mamdani-type) FRBSs[69,70]—is focused on the interpretability, precise FM (PFM)—mainly developed by Takagi-Sugeno-Kang FRBSs[90,94]—is focused on the accuracy. Since both criteria are of vital importance in system modeling, the balance between them has started to receive attention in the fuzzy community in the last few years.[21]

Roughly speaking, the balance is usually attained from two different perspectives: either the LFM is extended to obtain more accurate models or the PFM is improved to obtain more interpretable models. The former approach is usually developed by automatically defining the membership functions,[32,48,59] improving the fuzzy rule set derivation,[43] or extending the model structure.[25,75,76] The latter approach is usually developed by reducing the fuzzy rule set,[73,101,102] reducing the number of fuzzy sets (with the subsequent merging of rules),[83,85] or exploiting the local description of the rules.[6,103]

This chapter is focused on the LFM side approach to find the balance between interpretability and accuracy. Thus, different mechanisms to improve the accuracy of LFM will be reviewed and proposed. The contribution supports the opinion that good accuracy degrees are obtained by combining a preliminary simple design of the linguistic model with an *a posteriori* refinement.

The chapter is organized as follows. Section 7.2 reviews different improvements to the accuracy of LFM and makes an experimental study comparing representative methods of them. Section 7.3 proposes a learning method to quickly obtain simple and accurate linguistic models. Section 7.4 introduces a method to adjust initial linguistic models and analyzes its combination with the previous learning method. Section 7.5 points out some conclusions. Finally, two appendices that, respectively, introduce the considered applications and one of the metaheuristics considered are also included.

7.2   Linguistic Fuzzy Modeling

System modeling is the action and effect of approach to a model, i.e., to a theoretical scheme that simplifies a real system or complex reality with the aim of easing its understanding. Thanks to this model, the real system can be explained, controlled, simulated, predicted, and even improved. The development of *reliable* and *comprehensible* models is the main objective in system modeling. If not, the model loses its usefulness.

There are at least three different paradigms in system modeling. The most traditional approach is the *white box modeling*, which assumes that a thorough knowledge of the system's nature and a suitable mathematical scheme to represent it are available. As opposed to that, the *black box modeling*[89] is performed entirely from data using no additional *a priori* knowledge and considering a sufficiently general structure. Whereas the white box modeling has serious difficulties when complex and poorly understood systems are considered, the black box modeling deals with structures and associated parameters that usually do not have any physical significance.[2] Therefore, generally the former approach does not adequately obtain reliable models, while the latter one does not adequately obtain comprehensible models.

A third, intermediate approach arises as a combination of the said paradigms, the *gray box modeling*,[47] where certain known parts of the system are modeled considering the prior understood, and the unknown or less certain parts are identified with black box procedures. With this approach, the mentioned disadvantages are palliated, and a better balance between reliability and comprehensibility is attained.

Nowadays, one of the most successful tools to develop gray box models is *fuzzy modeling* (FM),[62] which is an approach used to model a system making use of a descriptive language based on fuzzy logic with fuzzy predicates.[91] FM usually considers model structures (fuzzy systems) in the form of fuzzy rule-based systems (FRBSs), and constructs them by means of different parametric system identification techniques. Fuzzy systems have demonstrated their ability for control,[38] modeling,[77] or classification[23] in a huge number of applications. The keys for their success and interest are the ability to incorporate human expert knowledge—which is the information mostly provided for many real-world systems and is described by vague and imprecise statements—and the facility to express the behavior of the system with a language easily interpretable by human beings.

As a system modeling discipline, FM is mainly characterized by two features that assess the quality of the obtained fuzzy models:

- *Interpretability*—It refers to the capability of the fuzzy model to express the behavior of the system in an understandable way. This is a subjective property that depends on several factors, mainly the model structure, the number of input variables, the number of fuzzy rules, the number of linguistic terms, and the shape of the fuzzy sets. With the term "interpretability" we encompass different criteria appearing in the literature, i.e., *compactness, completeness, consistency,* or *transparency.*
- *Accuracy*—It refers to the capability of the fuzzy model to faithfully represent the modeled system. The closer the model to the system, the higher its accuracy. As closeness, we understand the similarity between the responses of the real system and the fuzzy model. This is why the term "approximation" is also used to express the accuracy, being a fuzzy model of a fuzzy function approximation model.

As Zadeh stated in its *Principle of Incompatibility,*[107] "*as the complexity of a system increases, our ability to make precise and yet significant statements about its behavior diminishes until a threshold is reached beyond which precision and significance (or relevance) become almost mutually exclusive characteristics.*"

Therefore, to obtain high degrees of interpretability and accuracy is a contradictory purpose and, in practice, one of the two properties prevails over the other one. Depending on what requirement is mainly pursued, the FM field may be divided into two different areas:

- *Linguistic fuzzy modeling (LFM)*—The main objective is to obtain fuzzy models with a good interpretability.
- *Precise fuzzy modeling (PFM)*—The main objective is to obtain fuzzy models with a good accuracy.

Nevertheless, recent efforts to get a balance between the two requirements by either improving the accuracy of LFM or the interpretability of PFM are being addressed by the scientific community. This chapter is focused on the former approach. The following section shows some possibilities to perform it.

7.2.1 Improving the Accuracy in Linguistic Fuzzy Modeling

LFM is usually developed by linguistic FRBSs, also known as Mamdani-type FRBS.[69,70] A crucial reason why this approach is worth considering is that it may remain verbally interpretable, giving the concept of linguistic variable[108] a central role. Linguistic FRBSs are formed by linguistic rules with the following structure:

$$\textbf{IF } X_1 \text{ is } A_1 \text{ and } \ldots \text{ and } X_n \text{ is } A_n \textbf{ THEN } Y \text{ is } B,$$

with X_i (Y) being input (output) linguistic variables, and with A_i (B) being linguistic labels with fuzzy sets associated defining their meaning. These linguistic labels will be taken from a global *semantic* defining the set of possible fuzzy sets used for each variable. This structure provides a natural framework to include expert knowledge in the form of fuzzy rules.

In these systems, the knowledge base (KB)—the component of the FRBS that includes the knowledge about the problem being solved—is composed of the rule base (RB), constituted by the collection of linguistic rules themselves joined by means of the connective *also*, and the data base (DB), containing the term sets and the membership functions defining their semantics.

In spite of its interesting advantages, the linguistic FRBS has certain inflexibility due to the use of a global semantic that gives a meaning to the used fuzzy sets. Effectively, the use of linguistic variables imposes the following constraints:[5,14]

1. There is a lack of flexibility in the FRBS because of the rigid partitioning of the input and output spaces.
2. When the system input variables are dependent themselves, it is very hard to fuzzy partition the input spaces.
3. The homogeneous partitioning of the input and output spaces when the input-output mapping varies in complexity within the space is inefficient and does not scale to high-dimensional spaces.
4. The size of the KB directly depends on the number of variables and linguistic terms in the system. The derivation of an accurate linguistic FRBS requires a significant granularity amount, i.e., it needs the creation of new linguistic terms. This granularity increase causes the number of rules to rise significantly, which may make the system lose the capability of being interpretable by human beings. In the most of the cases, it would be possible to obtain an equivalent FRBS having a much lesser number of rules if there would not exist that input space rigid partitioning.

However, it is possible to make some considerations to face this drawback.[16] Basically, two ways of improving the accuracy in LFM can be considered by performing the improvement in:

- The *modeling process*, extending the model design to other components different from the RB, i.e., the DB, or considering more sophisticated derivations of it
- The *model structure*, slightly changing the rule structure to make it more flexible

In the following three subsections, some examples of the improvements appearing in the specialized literature are introduced.

7.2.1.1 Data Base Design

Basic LFM methods are exclusively focused on determining the set of fuzzy rules composing the RB of the model.[95,99,105] In these cases, the DB is usually obtained from expert information (if available) or by a normalization process and it remains fixed during the RB derivation process.

However, the automatic design of the DB has shown to be a very suitable mechanism to increase the approximation capability of the linguistic models. Generally speaking, the procedure involves either defining the most appropriate shapes for the membership functions that give meaning to the fuzzy sets associated to the considered linguistic terms, or determining the optimum number of linguistic terms used in the variable fuzzy partitions, i.e., the granularity.[33,39]

In the following, we show different approaches of the DB design regarding the way of integrating it in the derivation process and the effects caused to the membership functions. Some considerations on the interpretability preservation are also discussed.

7.2.1.1.1 *Integration of the DB Design in the Derivation Process*
The design of the DB may be integrated within the whole derivation process with different schemata:

- *Preliminary design*—It involves extracting the DB *a priori* by induction from the available data set. This process is usually performed by nonsupervised clustering techniques.[65,78]

- *Embedded design*—This approach derives the DB using an embedded basic learning method.[29,32,33,41,52] The technique involves having a simple learning method that designs, from a specific DB, other components of the fuzzy linguistic model (e.g., the RB). Following a meta-learning process, the method generates different DBs and samples its efficacy running the basic learning method.
- *Simultaneous design*—The process of designing the DB is jointly developed with the derivation of other components, i.e., the RB in a simultaneous procedure.[42,51,58,60,68,88,98,100]
- *A posteriori design*—This approach, usually called DB *tuning*, involves refining the DB from a previous definition once the remaining components have been obtained. It is one of the most common procedures. Usually, the tuning process changes the membership function shapes[10,26,44,48,55,59,64,92] and the main requirement is to improve the accuracy of the linguistic model. Nevertheless, as shown in the previous section, sometimes another kind of *a posteriori* DB design is made to improve the interpretability (e.g., merging membership functions[39]).

Of course, several of these approaches can also be jointly considered. For example, in Liska and Melsheimer[63] a two-stage DB design is made by first deriving simultaneously the DB and the RB and then performing an *a posteriori* tuning. In Jin,[57] an initial generation of the RB with a subsequent three-stage DB design (input variable selection, simultaneous DB tuning and RB reduction, and DB fine tuning) is developed.

Preliminary, embedded, and *a posteriori* design approaches are usually combined with other methods to perform the whole derivation process in several sequential stages. Instead, the simultaneous design of the DB together with other components constitutes a whole derivation process itself.

The sequential derivation has the main advantage of reducing the search space since confined spaces are tackled at each stage. On the other hand, in the simultaneous derivation, the strong dependency of the components is properly addressed. However, the process becomes significantly more complex in the latter case because the search space significantly grows, making the selection of an appropriate search technique crucial.

Recently, the cooperative coevolutionary paradigm[82] has shown an increasing interest thanks to its high ability to manage with huge search spaces and decomposable problems, and new simultaneous derivation methods are currently emerging using this technique.[22,80,81]

Finally, we should say that the DB design gives more flexibility to the modeling process, but it runs the risk of losing interpretability and overfitting the problem, wherefore this task must be carefully made. Some mechanisms to keep a good interpretability are discussed later on.

7.2.1.1.2 *Definition of the Membership Function Shapes*

Another interesting aspect to consider when designing the DB is the way of defining the membership function shapes. The most usual approaches are the following:

- *Defining the membership function parameters*—The most common way of defining the membership functions is to change their definition parameters.[10,26,42,48,51,55,57–60,63,80,92,98] For example, if the following triangular-shape membership function is considered:

$$\mu(x) = \begin{cases} \frac{x-a}{b-a}, & \text{if } a \leq x < b \\ \frac{c-x}{c-b}, & \text{if } b \leq x \leq c, \\ 0, & \text{otherwise} \end{cases}$$

  changing the basic parameters—a, b, and c—will vary the shape of the fuzzy set associated to the membership function (Figure II.7.1), thus influencing the FRBS performance. The same yields for other shapes of membership functions (trapezoidal, gaussian, sigmoid, etc.).
- *Using linguistic modifiers*—Another way to define the membership function shapes of the DB is to use linguistic modifiers.[25,64] Section 7.2.1.2 describes them in depth. They involve

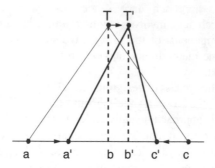

FIGURE II.7.1   Definition of the membership function shapes by their parameters.

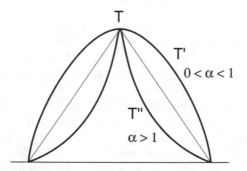

FIGURE II.7.2   Definition of the membership function shapes using linguistic modifiers.

considering more flexible alternative expressions for the membership functions to vary the compatibility degrees to the fuzzy sets. For example, a new membership function can be obtained raising the membership value to the power of α, i.e.:

$$\mu'(x) = (\mu(x))^{\alpha}, \ 0 < \alpha.$$

By changing the α value we may define different membership function shapes. Figure II.7.2 shows the effect of this approach.

- *Establishing the context*—A third possibility to change the membership functions is to define the context, i.e., their operation range. It is usually performed by *scaling functions* that map the input and output variables onto the universe of discourse over which the fuzzy sets are defined. From a linguistic point of view, the scaling function can be interpreted as a sort of context information.

We may distinguish between linear and nonlinear scaling functions, which are well known in classical control theory:

- *Linear context*[45]—A linear scaling function is of the form:

$$f(x) = \alpha \cdot x + \beta$$

The scaling factor α enlarges or reduces the operating range, which in turn decreases or increases the sensitivity of the controller with respect to that input variable, or the corresponding gain in case of an output variable. The parameter β shifts the operating range and plays the role of an offset to the corresponding variable.

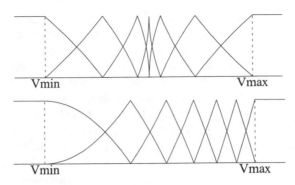

FIGURE II.7.3   Nonlinear scaling effects.

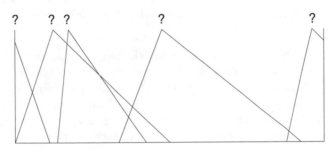

FIGURE II.7.4   Certain interpretability may be lost when designing the DB.

— *Nonlinear context*[46,67,79]—The main disadvantage of linear scaling is the fixed relative distribution of the membership functions. Nonlinear scaling provides a remedy to this problem as it modifies the relative distribution and changes the shape of the membership functions. A common nonlinear scaling function for a variable that is symmetric with respect to the origin is of the form:

$$f(x) = sign(x) \cdot |x|^{\alpha}.$$

Nonlinear scaling increases ($\alpha > 1$) or decreases ($\alpha < 1$) the relative sensitivity in the region around the origin and has the opposite effect at the boundaries of the operating range. Figure II.7.3 depicts some effects caused by the nonlinear scaling.

Derivation methods combining both kinds of context adaptations have also been proposed.[29]

Some other contributions propose methods to define not only the membership function shapes but also the membership function types (such as triangular, trapezoidal, gaussian, and sigmoid).[88]

7.2.1.1.3   *Interpretability Preservation when Designing the DB*
As said, the DB design may generate intricate semantics that could disturb the expert interpretation, thus losing some legibility degree. Figure II.7.4 illustrates an example where garbled membership functions may involve losing interpretability.

To decide if a specific DB is interpretable or not is a difficult and subjective task. Nevertheless, some researchers have become aware of this matter and several properties that ensure a good interpretability have been proposed.[96,97] Different constraints to the DB design process may be imposed to fulfill all or some of these properties. A selection of the most important properties and some possible solutions when designing the DB include:

- *Coverage property*—Every value of the universe of discourse should belong to, at least, a linguistic term. Alternately, a more strict criterion may be considered establishing a minimum level of coverage to be met for the entire universe of discourse.

  Possible solutions—This property may be satisfied, e.g., using variation intervals that ensure the overlap between two consecutive fuzzy sets[26] or employing strong fuzzy partitions.[10]

- *Normality property*—Every membership function should exhibit full matching with, at least, a value of the universe of discourse. That is, the fuzzy sets should be normal.

  Possible solutions—This property is easily kept, forcing the modal points of the extreme membership functions of the variable fuzzy partition to be contained in the universe of discourse.

- *Distinguishability property*—Each linguistic term should have a clear meaning and the associated fuzzy set should clearly define a range of the universe of discourse. In short, the membership functions should be distinct enough from each other.

  Possible solutions—This property may be satisfied, e.g., properly constraining the membership function parameters,[66] merging similar membership functions,[39] or establishing a semantic order among the linguistic terms.[7,8]

Of course, the constraints needed to ensure the semantic integrity make the derivation process less flexible but, besides easing the legibility, the risk of overfitting the problem is reduced.

When search techniques are used to design the DB, another possibility to ensure the integrity properties is to include interpretability measures in the objective function, thus guiding the trek to good solutions. Usually, measures of completeness, consistency,[58] compactness, or similarity[56] are considered. A more advanced criterion, called conciseness, is proposed in Suzuki et al.[93] by combining a fuzzy entropy measure (which distinguishes the shapes of the membership functions) with a deviation measure (which evaluates the discrepancy of a membership function from symmetry).

7.2.1.2   Extending the Model Structure

Another way to improve the LFM accuracy is to extend the usual linguistic model structure to be more flexible. Some specific possibilities are described in the following.

7.2.1.2.1   *To Use Linguistic Modifiers*

A possibility to relax the model structure allowing it to be more flexible is to include certain operators to slightly change the meaning of the linguistic labels involved in the system when necessary.[15,25,44] As Zadeh highlighted,[107] a way to do so without losing the description to a high degree is to use linguistic hedges or, in a wider sense, *linguistic modifiers*.

A linguistic modifier[11,12,106] is an operator that alters the membership functions for the fuzzy sets associated to the linguistic labels, giving a more or less precise definition as a result, depending on the case. Two of the most well-known linguistic modifiers are the concentration *very* and the dilation *more-or-less*. Their expressions are as follows:

$$\mu_T^{very}(x) = (\mu_T(x))^2, \quad \mu_T^{more\text{-}or\text{-}less}(x) = \sqrt{\mu_T(x)}.$$

Therefore, a new linguistic rule structure arises as follows:

$$\textbf{IF } X_1 \text{ is } lm_{X_1} A_1 \text{ and } \ldots \text{ and } X_n \text{ is } lm_{X_n} A_n \textbf{ THEN } Y \text{ is } lm_Y B,$$

with $lm_{X_i}(lm_Y)$ being the linguistic modifier to be used (including the identity operator) in the corresponding variable where the membership degree to the linguistic term is given by $\mu_{A_i}^{lm_{X_i}} (\mu_B^{lm_Y})$.

Section 7.4.1.1 details this kind of structure and introduces more linguistic modifier types.

7.2.1.2.2   *To Use Double-Consequent Rules*

This approach involves allowing the RB to present rules where each combination of antecedents may have two consequents associated when it is necessary to improve the model accuracy.[27,39,75] It is clear that this will improve the capability of the model to perform the interpolative reasoning and, thus, its performance. The rule structure obtained will be:

$$\textbf{IF } X_1 \text{ is } A_1 \text{ and } \ldots \text{ and } X_n \text{ is } A_n \textbf{ THEN } Y \text{ is } \{B_1, B_2\}.$$

Since each double-consequent fuzzy rule can be decomposed into two different rules with a single consequent, the usual plain fuzzy inference system can be applied. The only restriction imposed is that the defuzzification method must consider the matching degree of the rules fired, e.g., the *center of gravity weighted by the matching degree* defuzzification strategy[30] may be used.

When using two consequents per rule, the interpretation of the action performed by every rule may be confusing to some extent. However, we should note this fact does not constitute an inconsistency from the LFM point of view, but only a shift of the main labels making the final output of the rule lie in an intermediate zone between both consequents. Indeed, let us suppose that a specific combination of antecedents, "X_1 is A_1 and $\ldots$ and X_n is A_n," has two different consequents associated, B_1 and B_2. The resulting double-consequent rule may be interpreted as:[27]

$$\textbf{IF } X_1 \text{ is } A_1 \text{ and } \ldots \text{ and } X_n \text{ is } A_n \textbf{ THEN } Y \text{ is } between \ B_1 \text{ and } B_2.$$

7.2.1.2.3 *To Use Weighted Rules*
This approach involves using an additional parameter for each rule that indicates its importance degree in the inference process,[1,24,53,76,84,104] instead of considering all rules equally important as in the usual case. Thus, more flexibility to improve the interpolative reasoning and, therefore, the model performance, is achieved. The rule structure will be:

$$\textbf{IF } X_1 \text{ is } A_1 \text{ and } \ldots \text{ and } X_n \text{ is } A_n \textbf{ THEN } Y \text{ is } B \ with \ \lfloor w \rfloor,$$

with w being the real-valued rule weight. In this approach, some changes must be made to the classical inference system to consider the weighted action of each rule.

The operator *with*, which attaches a weight to a rule, may be defined in different ways. One of the most usual options is to multiply the matching degree of the antecedent by the corresponding weight before applying the implication operator, which relates antecedent and consequent. Another possibility is to change the conclusion derived from the implication operator according to the corresponding weight (e.g., changing the support of the obtained fuzzy set).

These weights are usually considered to handle inconsistencies with advanced inference methods[104] or neural networks.[24] Moreover, some proposals make use of them to improve the model accuracy with an automatic learning of weights using different techniques such as heuristic methods,[53,84] gradient descent processes,[76] or evolutionary algorithms.[1]

7.2.2 Experimental Comparative Study

In this section, an experimental comparison between LFM and improved LFM will be shown. To do so, nine learning methods with different characteristics will be regarded to develop these kinds of FM making use of linguistic FRBSs. Table II.7.1 shows the main characteristics of the methods considered.

The M-L-Tun-method and the LH-Tun-method will be considered as *a posteriori* tuning processes regardless of the learning method used to derive a previous KB. Therefore, in the experiments developed, these two tuning methods will be applied to the KBs previously generated by the two LFM methods (the WM-method and the T-method). The WR-method simply involves determining by means of a real-coded genetic algorithm, the weight (in [0, 1]) corresponding to each of the linguistic rules previously learned by the WM-method.

The behavior of these learning methods will be analyzed when solving two different real-world applications: the rice evaluation taste problem and the estimation of the electrical low-voltage line length problem. A brief introduction to them is presented in Appendix A.

TABLE II.7.1   Summary on the Analyzed LFM Methods and Their Main Characteristics

Ref.	Method	Modeling Type	Algorithm Type	IL	IS	SD	PD	MFP	LHDB	LHRB	DC	WR
99	WM	LFM	AHDD									
95	T	LFM	GA									
26	M-L-Tun	ILFM	GA	✓			✓	✓				
63	LMe	ILFM	GA	✓	✓	✓	✓					
26	M-L	ILFM	AHDD + GA	✓			✓	✓				
25	LH-Tun	ILFM	GA		✓					✓		
27	ALM	ILFM	AHDD + GA		✓						✓	
75	NIT	ILFM	AHDD	✓	✓				✓		✓	✓
—	WR	ILFM	GA		✓							✓

ILFM = improved LFM, **AHDD** = ad hoc data-driven method, **GA** = genetic algorithm-based method, **IL** = LFM improved in the learning, **IS** = LFM improved in the model structure, **MFP** = definition of membership function parameters, **LHDB** = use of a linguistic hedge for the whole DB, **SD** = simultaneous design of the DB and the RB, **PD** = *a posteriori* design of the DB, **LHRB** = rule structure with linguistic hedges, **DC** = double-consequent rules, **WR** = weighted rules.

7.2.2.1 Experiments

In the experiments developed, we consider the *mean square error* (MSE) to evaluate the quality of the results. We have defined MSE as:

$$MSE = \frac{1}{2 \cdot N} \sum_{l=1}^{N} (Y^l - y^l)^2,$$

with N being the data set size, Y being the output obtained from the FRBS, and y being the known desired output. The closer the measure to zero, the greater the model accuracy.

An initial DB constituted by a primary fuzzy partition for each variable will be considered when required. Every partition is formed by a specific number of labels with triangular-shaped equally distributed fuzzy sets giving meaning to them (as shown in Figure II.7.9), and the appropriate scaling factors to translate the generic universe of discourse into the one associated with each problem variable.

As regards the FRBS reasoning method used, we have selected the *minimum t-norm* playing the role of the implication and conjunctive operators, and the *center of gravity weighted by the matching* strategy acting as the defuzzification operator.[30]

Table II.7.2 collects the results obtained by the considered methods when solving the said applications. In that table, #R stands for the number of rules, and MSE_{tra} and MSE_{tst} for the values obtained over the training and test data sets, respectively. In the rice taste evaluation problem, the values shown in these columns have been computed as the average obtained by the ten models generated for each method (see Appendix A.7.1). In the case of the methods with double-consequent rules (the ALM-method and the NIT-method), the number of rules shown in the table corresponds to the rules obtained after decomposing them into simple ones. The best results obtained by each learning method group (LFM and improved LFM) for each problem are shown in boldface.

7.2.2.2 Analysis of the Obtained Results

From the obtained results, an analysis considering the behavior of the analyzed methods solving the different problems will be made from the accuracy, the interpretability, and the balance between both points of view:

TABLE II.7.2 Results Obtained by the Analyzed Methods in the Two Problems Considered

Method	#R	Rice Evaluation Taste		#R	Electrical Line Length	
		MSE_{tra}	MSE_{tst}		MSE_{tra}	MSE_{tst}
WM	**15**	0.013284	0.013119	**24**	222,654	239,962
T	16	**0.004949**	**0.005991**	49	**169,077**	175,739
WM + M-L-Tun	15	0.001106	0.002138	24	145,273	171,998
T + M-L-Tun	16	0.001153	0.002927	49	147,844	**164,993**
LMe	31	**0.000810**	0.002342	49	167,014	167,383
M-L	6	0.001075	0.002331	35	**137,905**	173,096
WM + LH-Tun	15	0.001758	**0.001731**	24	181,609	209,756
T + LH-Tun	16	0.000911	0.001902	49	139,013	188,945
ALM	**5**	0.003416	0.003984	**20**	155,866	178,601
NIT	64	0.002996	0.003520	64	173,230	190,808
WR	15	0.002588	0.003099	24	149,303	182,249

- *Accuracy*: We may note the great result shown by using different LFM improvements, generally obtaining more accurate linguistic models. Nevertheless, some improved linguistic models present certain overfitting of the problem losing generalization, as the case of the ones generated by the LMe-method, the T + M-L-Tun-method, and the T + LH-method in the rice problem, or the ALM-method, the NIT-method, and the WR-method in the electrical problem. This worse prediction ability is one of the risks when excessive flexibility is given to the model and shows the difficulty in overcoming LFM in some cases.

- *Interpretability*: In spite of obtaining good accuracy degrees, the improvements in LFM performed by some learning methods involve the use of a high number of rules in some cases (e.g., the LMe-method in the rice problem or the NIT-method in the two problems). Moreover, of the corresponding interpretability loss derived from such improvements (tuning of basic membership functions parameters, several consequents per rule, weighted rules, etc.), so many rules make the obtained model less understandable, which is an important requirement in LFM and should be taken into account. Opposite to these cases, some methods achieve significantly improving the behavior keeping a good interpretability. For example, Figure II.7.5 shows the RB generated by the T-method and the corresponding DB changed by the M-L-Tun-method in the electrical problem. As noted, a significantly more accurate model is obtained by the improved LFM method using the same RB and keeping a DB with good interpretability.

- *Balance*: Keeping in mind the said requirements in terms of accuracy (with good trade-off between approximation and generalization degrees) and interpretability, the following methods may be considered the ones that obtain the best solutions in each problem:

  - In the rice problem, the WM + LH-Tun-method (with a well-balanced accuracy) and the ALM-method (though it does not have a good accuracy, it has an excellent interpretability where the original semantic is kept and only six simple rules are used) may be the methods that obtain the best solutions. Their corresponding models are shown in Tables II.7.3(a) and (b), respectively.

  - The T + M-L-Tun-method generates a model with excellent approximation and prediction degrees in the electrical line length problem, having moreover a proper interpretability.

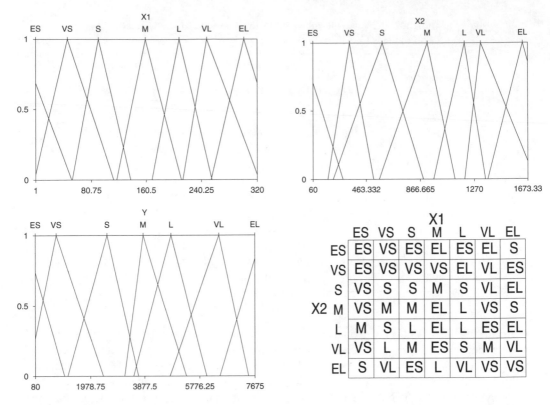

FIGURE II.7.5 RB (bottom right) and DB (top and bottom left) generated by the T + M-L-Tun-method.

As it can be seen, excellent models result from combining simple methods with a subsequent tuning process. This combination presents similar or better results than complex rule generation mechanisms, i.e., the LMe-method and the M-L-method, which also perform an *a posteriori* tuning, being significantly simpler and quicker.

This result encourages us to propose a simple method to obtain preliminary linguistic models and an *a posteriori* tuning process to refine the structures of such models. The following two sections introduce both proposals.

7.3 Accuracy Improvement when Designing Linguistic Fuzzy Models: Cooperation Induction

As it has been shown in the previous section, the difficulty presented by human experts expressing their knowledge in the form of fuzzy rules has made researchers develop automatic techniques to perform this task. In this sense, a large quantity of methods has been proposed to automatically generate fuzzy rules from numerical data (input-output pairs) that show the behavior of the real system. Usually, they combine different techniques such as neural networks[40,74] or genetic algorithms.[28,50]

However, the complexity of the rule generation mechanisms used by these methods may cause difficulty in design and a high-time-consuming learning process. Opposite to them, a family of efficient and simple methods guided by covering criteria of the data in the example set, called *ad hoc data-driven methods*, has been proposed in the literature.[27,54,75,99]

TABLE II.7.3 Some RBs Generated for a Specific Data Set Partition of the Rice Taste Evaluation Problem

(a) WM + LH-Tun-method's RB (MSE$_{tra/tst}$ = 0.002230/0.000901)

Rule	Flavor	Appearance	Taste	Stickiness	Toughness	Evaluation
R_1	m-or-l bad	m-or-l bad	very bad	not-sticky	m-or-l tender	low
R_2	m-or-l bad	very bad	very bad	very not-sticky	very tough	low
R_3	m-or-l bad	m-or-l good	very bad	not-sticky	very tender	very low
R_4	m-or-l bad	good	very good	very not-sticky	m-or-l tender	very low
R_5	very bad	good	good	very sticky	tender	low
R_6	m-or-l good	very bad	very bad	very not-sticky	m-or-l tender	m-or l low
R_7	m-or-l good	bad	very bad	m-or-l not-sticky	m-or-l tough	low
R_8	m-or-l good	bad	very good	very not-sticky	m-or-l tender	m-or-l low
R_9	very good	good	very bad	very not-sticky	very tough	low
R_{10}	very good	very good	very bad	very not-sticky	m-or-l tender	m-or-l high
R_{11}	very good	good	very bad	very sticky	tender	high
R_{12}	good	good	very good	m-or-l not-sticky	very tender	high
R_{13}	m-or-l good	m-or-l good	m-or-l good	m-or-l not-sticky	m-or-l tough	very high
R_{14}	very good	very good	good	sticky	m-or-l tender	m-or-l high
R_{15}	very good	good	very good	sticky	tough	m-or-l high

(b) ALM-method's RB (MSE$_{tra/tst}$ = 0.003838/0.002852)

Rule	Flavor	Appearance	Taste	Stickiness	Toughness	Evaluation
R_1	bad	bad	bad	not-sticky	tough	low
R_2	bad	good	bad	not-sticky	tender	low
R_3	good	bad	bad	not-sticky	tender	low
R_4	bad	good	good	not-sticky	tender	high
R_5	good	good	good	sticky	tender	high
R_6	good	good	good	sticky	tough	high

These methods present interesting advantages from the learning point of view:

- The algorithms are easily understandable and implementable.
- They are very quick at performing the learning process.
- Thanks to the two said advantages, they are very suitable for integration in more complex modeling procedures, which has made these methods gain a renewed interest. Indeed, we may find interesting contributions in the recent literature where these methods are used as a first approximation to understand the characteristics of the problem to be solved,[34] as a first stage of the modeling process to obtain a preliminary fuzzy model that is subsequently refined by other techniques,[26] or to be integrated as fitness criterion in meta-learning processes,[33] among others.

Nevertheless, in spite of these advantages, many times the obtained models are not as accurate as desired. It is due to the fact that these methods usually look for the rules with the best individual performance. It sometimes involves obtaining KBs with fuzzy rules that do not cooperate and, therefore, due to the interpolative reasoning developed by FRBSs, the results are not accurate enough.

This section faces the problem by proposing a new ad hoc data-driven methodology to improve the rule cooperation and thus the accuracy of the obtained models, the Cooperative Rules (COR) methodology. Once the rule antecedents (defining fuzzy subspaces) have been obtained, the operation mode will be composed of two stages: generation of a candidate consequent set for each subspace and search of the consequents with the best global performance.

To present the methodology, first the ad hoc data-driven methods are introduced and a taxonomy of them is proposed. Then the COR methodology and its use with a specific metaheuristic are described. Finally, some experimental results are shown.

7.3.1 Ad Hoc Data-Driven Linguistic Rule-Learning Methods

7.3.1.1 Characterization

Ad hoc data-driven linguistic rule learning methods[27,54,75,99] are characterized by four main features:

1. *They are based on working with an input-output data set*—$E = \{e_1, \ldots, e_l, \ldots, e_N\}$, with $e_l = (x_1^l, \ldots, x_n^l, y_1^l, \ldots, y_m^l)$, N being the data set size, n being the number of input variables, and m being the number of output variables—*representing the behavior of the problem being solved*. In this chapter, we will work with multiple-input single-output (MISO) systems, i.e., $m = 1$, $e_l = (x_1^l, \ldots, x_n^l, y^l)$.

2. *They consider a previous definition of the DB composed of the input and output primary fuzzy partitions*—It may be obtained from expert information (if available) or by a normalization process.

   The set of linguistic terms of the i-th input variable will be denoted by $\mathcal{A}_i$—with $i \in \{1, \ldots, n\}$—and the set of linguistic terms of the output variable will be denoted by $\mathcal{B}$, with $|\mathcal{A}_i|(|\mathcal{B}|)$ being the number of linguistic terms of the i-th input (output) variable.

3. *The learning of the linguistic rules is guided by covering criteria of the data in the example set* (hence the name *data-driven*).

4. *The learning mechanism is not based on any well-known optimization or search technique, but is specifically developed for this purpose* (hence the name *ad hoc*).

7.3.1.2 Taxonomy

We can distinguish between two different approaches to obtain the linguistic rules by ad hoc data-driven methods:

- *Methods guided by examples*—The linguistic rule generation process guided by the example data set obtains each rule from a specific example. In this way, the rule R^l is obtained from the example e_l. Actually, these rules belong to a candidate rule set, since after this stage a selection process is performed to generate the final RB. Figure II.7.6 graphically illustrates this rule generation process. One of the most well-known and widely used example-based methods is Wang and Mendel's method (WM-method).[99]

- *Methods guided by fuzzy grid*—Another possibility to generate the linguistic rules is to bracket the examples according to a fuzzy grid, and then obtain a rule for each group (subspace) taking into account all of them. The fuzzy grid is obtained by combining the input primary fuzzy partitions defined by the Cartesian product of the linguistic terms existing for each input variable, $\mathcal{A}_1 \times \cdots \times \mathcal{A}_n$.

Figure II.7.7 graphically shows this rule generation process, where $S_s = (\mathcal{A}_1^s, \ldots, \mathcal{A}_i^s, \ldots, \mathcal{A}_n^s)$—with $s \in \{1, \ldots, N_s\}$, $N_s = \prod_{i=1}^n |\mathcal{A}_i|$ being the number of multidimensional fuzzy input subspaces, and $\mathcal{A}_i^s \in \mathcal{A}_i$—denotes a particular fuzzy input subspace and R^s is the corresponding linguistic rule. An example of this second approach can be found in Cordón and Herrera,[27] where the authors adapt the fuzzy-grid-based learning method of Ishibuchi et al.[54] for simplified Takagi-Sugeno-type rules[94]

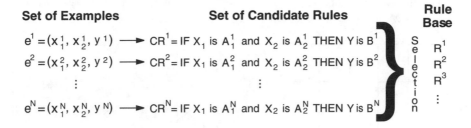

FIGURE II.7.6   Rule generation process followed by the methods guided by examples.

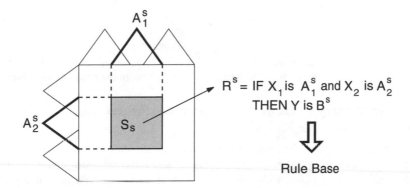

FIGURE II.7.7   Rule generation process followed by the methods guided by fuzzy grid.

(also known as singleton rules, i.e., rules with a single point instead a fuzzy set in the consequent) to allow it to generate linguistic rules (CH-method).

7.3.1.3   Relation between Example-Based and Fuzzy-Grid-Based Methods

While in an example-based method each example generates a single rule, the generation process performed by a fuzzy-grid-based method allows an example to contribute to the generation of several rules. Indeed, we may draw an analogy between both approaches considering the use of different kinds of grids (Figure II.7.8 graphically shows this fact):

- *Example-based method*—Since the antecedent of the rule generated from each example is obtained by taking the linguistic labels best covering the input values of this example, we see that an example-based method separates the example set according to a *crisp grid* bounded by the cross points between labels. Thus, an example may only belong to one subspace.
- *Fuzzy-grid-based method*—With the fuzzy grid used by this latter approach, an example may belong to more than one subspace. In Figure II.7.8(b), the examples lying in white zones have an influence on the generation of one rule, those lying in light gray zones influence two rules, and those lying in dark gray zones influence four rules.

It is not possible to determine which approach is the best. The example approach always obtains an equal number or fewer rules than the fuzzy grid one—as in the fuzzy grid approach the examples have an influence on a wider region, thus generating more rules. However, this fact may make the model obtained by an example-based method not as accurate as desired sometimes.

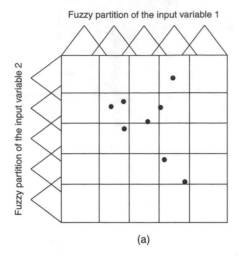

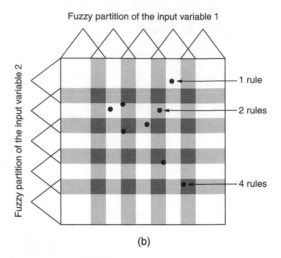

(a)   (b)

FIGURE II.7.8 (a) Example-based methods (crisp grid) — an example only contributes to the generation of one rule; (b) Fuzzy-grid-based methods — an example may contribute to the generation of several rules.

7.3.2 A Methodology to Improve Accuracy by Obtaining More Cooperative Rules

Having analyzed the behavior of ad hoc data-driven learning methods, this section proposes a new methodology to generate accurate and simple linguistic models by considering the global cooperation developed by the linguistic rules. Following this methodology, two learning methods guided by examples and fuzzy grid, respectively, will be introduced.

One of the most interesting features of an FRBS is the interpolative reasoning it develops. This characteristic plays a key role in the high performance of FRBSs and is a consequence of the cooperation among the linguistic rules composing the KB. It is a well-known fact that the output obtained from an FRBS is not usually due to a single linguistic rule, but to the cooperative action of several linguistic rules that have been fired because they match the system input to any degree.

However, the global interaction among the rules of the KB is not considered in the said ad hoc data-driven methods (since they select the rule with the best performance in each subspace). This causes the final RB obtained, in spite of presenting a good local behavior, to not cooperate suitably. Moreover, the fact of locally processing these rules makes these methods more sensitive to noise.

With the aim of addressing these drawbacks while keeping the interesting advantages of ad hoc data-driven methods, a new methodology to improve the accuracy obtaining better cooperation among the rules is proposed in Casillas, Cordón, and Herrera:[17,18] the COR methodology. Instead of selecting the consequent with the highest performance in each subspace like ad hoc data-driven methods usually do, the COR methodology considers the possibility of using another consequent, different from the best one, when it allows the FRBS to be more accurate thanks to having a KB with better cooperation.

In this way, its operation mode consists of two stages:

1. Obtain a set of candidate consequents for each subspace according to the examples contained in it. In addition to that, the *null consequent* ($\mathcal{N}$) is included in each set with the aim of giving the methodology the capability of removing rules with bad cooperation.[18] In this way, if such a term is selected for a specific rule, the corresponding rule will not take part in the RB finally learned.
2. Perform a *combinatorial search* among these sets looking for the combination of consequents with the best global accuracy.

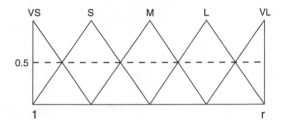

FIGURE II.7.9   Learning generic scheme followed by the COR methodology.

Since the methodology has the possibility of removing bad rules, in this contribution the fuzzy grid approach is followed to provide a richer search space and set of subspaces. A wider description of the COR-based rule generation process (for the fuzzy grid approach) is shown in Figure II.7.9, while an example of the operation mode for a simple problem with two input variables and three labels in the output fuzzy partition is graphically illustrated in Figure II.7.10.

Because the search space (Figure II.7.10(d)) tackled in step 2 of the algorithm is usually large, it is necessary to use approximate search techniques. Any combinatorial search approach may be used for such a purpose. In this chapter, the application of ant colony optimization (ACO) algorithms is analyzed. Other well-known techniques to develop the COR methodology were proposed in Casillas, Cordón, and Herrera.[20]

7.3.3   Learning with Ant Colony Optimization Algorithms

This section shows the application of ACO algorithms to the COR methodology. For an introduction to this kind of metaheuristic, refer to Appendix B. The proposed COR-based learning method has the following main aspects:[18]

- *Problem representation* — To use ACO algorithms in the COR methodology, it is convenient to see it as a combinatorial optimization problem with the capability of being represented on a graph. In this way, we can face the problem interpreting the COR methodology as the way of assigning consequents ($B_j \in \mathcal{B}$) — i.e., labels of the output fuzzy partition — to n-dimensional fuzzy input subspaces containing examples (S_s) with respect to an optimality criterion (MSE).

   Thus, the graph is constructed taking the steps described in Figure II.7.11. Following these steps, the graph corresponding to the example presented in Figure II.7.10 would be the one shown in Figure II.7.12.

- *Heuristic information* — The heuristic information on the potential preference of selecting a specific consequent (B_j), in each antecedent combination (subspace) is determined as follows:

$$\eta_{sj} = \max_{e_{ls} \in E'_s} \{CV(R^s_j, e_{ls})\} \cdot \frac{\sum_{e_{ls} \in E'_s} CV(R^S_j, e_{ls})}{|E'_s|}$$

   with $R^s_j = $ IF X_1 is A^s_1 and ... and X_n is A^s_n THEN Y is B_j.

   The heuristic information for the null consequent of each subspace will be the mean value of the remaining candidate consequents used in the corresponding subspace, i.e.:

$$\eta_{ic_i} = \frac{\sum_{j=1}^{c_i-1} \eta_{ij}}{c_i - 1}.$$

- *Pheromone initialization* — The initial pheromone value of each assignment is obtained as follows:

$$\tau_0 = \frac{\sum_{s=1}^{N_S} \max_{j=1}^{|\mathcal{B}|+1} \eta_{sj}}{N_S}.$$

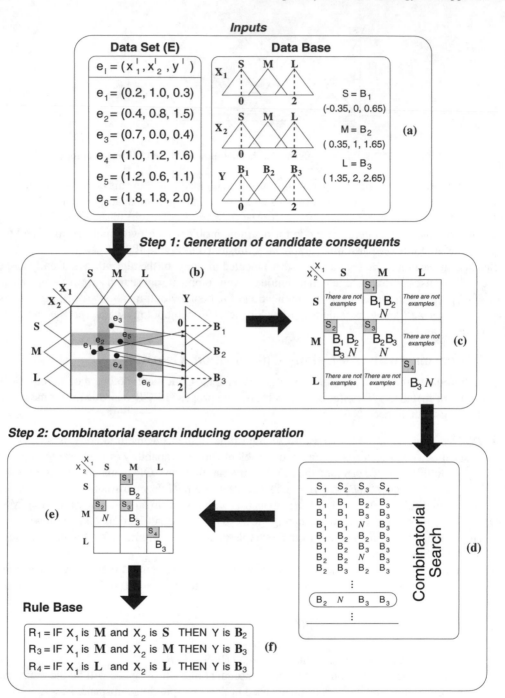

FIGURE II.7.10 COR-based learning process for a simple problem with two input variables ($n = 2$) and three labels in the output fuzzy partition ($|\mathcal{B}| = 3$): (a) data set (E) and DB previously defined; (b) the six examples are located in four ($N_S = 4$) different subspaces that determine the antecedent combinations and candidate consequents of the rules; (c) set of possible consequents for each subspace including the null consequent; (d) combinatorial search accomplished within the generated search space; (e) rule decision table for a specific combination; (f) RB generated from such a combination (the consideration of the null consequent for the second subspace involves removing the corresponding linguistic rule).

1. *Set a node for each subspace* — Use a node for each n-dimensional fuzzy input subspace containing examples (S_s), thus having a total of N_S subspace nodes.

2. *Link the subspaces to consequents* — The subspace S_s will be linked to the consequent $B_j \in \mathcal{B}$ if and only if it meets the following condition:

   $$\exists e_l \in E \text{ such that } \mu_{A_1^s}(x_1^l) \cdot \ldots \cdot \mu_{A_n^s}(x_n^l) \cdot \mu_{B_j}(y^l) \neq 0 \ .$$

   That is, if there is at least one example located in the fuzzy input subspace that is covered by such a consequent. Moreover, all the subspaces are linked to the null consequent $(\mathcal{N})$.

FIGURE II.7.11   Graph construction process.

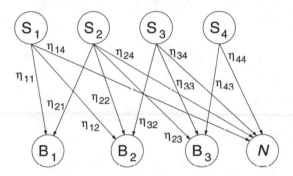

FIGURE II.7.12   ACO graph corresponding to the example of Figure II.7.10.

In this way, the initial pheromone will be the mean value of the path constructed taking the best consequent in each rule according to the heuristic information (greedy assignment).

- *Fitness function* — The said MSE function (defined in Section 7.2.2.1) is used.
- *ACO approach* — Once the previous components have been defined, an ACO algorithm has to be given to solve the problem. In this contribution, the well-known Ant Colony System (ACS) algorithm[36] is used. Its operation mode adapted to the COR methodology is the following:

- *Solution Construction*: The transition rule establishes a balance between biased exploration and exploitation of the available information. The node j (i.e., the consequent B_j) is selected for the rule R_i as follows:

$$j = \begin{cases} \arg \max_{u \in J_k(i)} \{(\tau_{iu})^\alpha \cdot (\eta_{iu})^\beta\}, & \text{if } q < q_0 \\ S, & \text{otherwise} \end{cases},$$

with $J_k(i) = \{j$ such that $\eta_{ij} \neq 0\}$ being the set of nodes attainable from S_i (i.e., the set of consequents that can be associated to it), τ_{ij} the pheromone of the trail (i, j), η_{ij} being the heuristic information, α and β parameters which determine the relative influence of the pheromone strength and the heuristic information, q a random variable uniformly distributed over $[0, 1]$, $q_0 \in [0, 1]$ a threshold defining the probability of selecting the more hopeful coupling (exploitation), and S a random node selected according to the following probability (biased exploration):

$$p_k(i, j) = \begin{cases} \frac{(\tau_{ij})^\alpha \cdot (\eta_{ij})^\beta}{\sum_{u \in J_k(i)} (\tau_{iu})^\alpha \cdot (\eta_{iu})^\beta}, & \text{if } j \in J_k(i) \\ 0, & \text{otherwise} \end{cases}$$

We should note that the transition rule becomes an assignment rule, but there is not a need for the ant to keep a tabu list with the previous assignments made, since the same consequent can be assigned to different rules.

- *Pheromone Trail Update Rule*: The pheromone trail update rule is performed in two stages, global and local:

  * *Global pheromone trail update rule*: The amount of pheromone released on a coupling is made by only one ant, that which generated the best solution (S_{best}) until now. The formula is the following:

  $$\tau_{ij} \leftarrow (1 - \rho) \cdot \tau_{ij} + \rho \cdot \Delta\tau_{ij},$$

  with

  $$\Delta\tau_{ij} = \begin{cases} \frac{1}{MSE(RB_{best})}, & \text{if } (i, j) \in S_{best} \\ 0, & \text{otherwise} \end{cases}$$

  with $\rho \in [0, 1]$ being the pheromone evaporation parameter.

  * *Local pheromone trail update rule*: Each time an ant covers a coupling, a local pheromone update is also performed as follows:

  $$\tau_{ij} \leftarrow (1 - \rho) \cdot \tau_{ij} + \rho \cdot \Delta\tau_{ij}.$$

  In this contribution, we consider $\Delta\tau_{ij} = \tau_0$.

- *Local Search*: This approach entails employing a local optimization technique to refine the solutions obtained after one or several iterations. The local search technique is usually applied to the solution generated by each ant. However, due to the time restrictions imposed and to keep the high speed of the proposed learning method, in our case *the local search process will only be applied to the best solution generated at each iteration*. After this process, the global pheromone trail is updated in the usual way. The proposed local search will consist of a simple hill-climbing algorithm.[18]

7.3.4 Experimental Study

In this section, some examples of application will be shown. With the aim of analyzing the behavior of the proposed learning method, we will compare them with some other fuzzy rule learning methods: the WM-method[99] guided by examples, the method proposed by Cordón and Herrera (CH-method)[27] guided by fuzzy grid, the NIT-method[75] guided by fuzzy grid, and the genetic-algorithm-based T-method.[95] As noted in Section 7.2.2, the WM-method and the T-method develop a basic LFM, while the NIT-method extends the LFM using weighted double-consequent rules. Table II.7.4 summarizes the characteristics of the methods compared.

The performance of these learning methods will be analyzed when solving the two real-world applications (see Appendix A) considered in the experimental study shown in Section 7.2.2. The same considerations related with the DB and inference engine employed are regarded. Table II.7.5 collects the results obtained by the analyzed methods. The best results are shown in boldface. In that table, EBS stands for the number of evaluations needed to obtain the best solution.

TABLE II.7.4   Methods Considered in This Experimental Study

Ref.	Method	Algorithm	Comments
99	WM	AHDD	Well-known learning method
27	CH	AHDD	Heuristic-information-based greedy algorithm
75	NIT	AHDD	Uses two weighted rules in each subspace
95	T	GA	Capability of learning the number of rules
—	COR	ACO	The COR-based method

AHDD = ad hoc data-driven, **GA** = genetic algorithm.

TABLE II.7.5   Results Obtained by the Analyzed Methods in the Two Problems Considered

Method	Rice evaluation taste				Electrical line length			
	#R	MSE_{tra}	MSE_{tst}	EBS	#R	MSE_{tra}	MSE_{tst}	EBS
WM	15	0.013284	0.013119	—	24	222,654	239,962	—
CH	32	0.009101	0.010123	—	32	239,393	275,953	—
NIT	64	0.008626	0.009851	—	64	185,395	170,489	—
T	16	**0.004949**	**0.005991**	10,021	49	**169,077**	175,739	26,706
COR	19	0.005715	0.006438	1,027	29	173,823	**160,753**	417

7.3.4.1 Analysis of the Results Obtained

The obtained results lead us to highlight the good behavior of the COR-method. The best results in terms of generalization (MSE_{tst}) degrees are always obtained by the COR-method, as well as using a reduced number of rules.

The COR-method significantly improves the accuracy of the models compared with noncooperative methods like the WM and CH ones. In Figure II.7.13 we show how, from the candidate consequent set generated in each subspace following the fuzzy grid approach (Figure II.7.13(a)), the CH-method performs the selection considering the best rule in each subspace (Figure II.7.13(b)); whereas the COR-method makes use of a cooperative criterion that takes into account the global behavior of the rules (Figure II.7.13(c)) with the capability of removing the rules with bad cooperation. Hence, the latter method selects a different consequent from the locally best one in some specific cells, making the accuracy of the obtained model notably better.

The main difference between the NIT-method and the COR-method is that the former makes use of the two best consequents in each subspace, giving a certainty factor to each, while the latter analyzes which rule provides higher global accuracy and only makes use of such a rule. Thus, the model obtained by the COR-method is more accurate and presents a significantly simpler KB than the NIT-method one, which is a very important aspect in LFM where the interpretability is the main requirement. Therefore, using the most globally cooperative rule in each subspace seems to be more suitable than taking the two best local rules.

Although the T-method and the COR-method generally obtain similar accuracy degrees, the latter performs a significantly quicker learning process than the former. This fact is because of two aspects. On the one hand, the search space reduction carried out by the proposed methodology accelerates the obtaining of good solutions. Figure II.7.14 illustrates the search space tackled by the T-method and the COR-method. The larger the search space, the more the difficulty to find good solutions. Thus, the reduction performed by the COR-method generally leads to either obtaining the best solutions or to needing less evaluations to find good solutions. On the other hand, the heuristic information considered by the COR-method properly guides the search.

X_1

	ES	VS	S	M	L	VL	EL
ES	{ES,VS}	{ES,VS,S}	{ES,VS,S}	{VS,S}			
VS	{ES,VS,S,M}	{ES,VS,S,M}	{ES,VS,S,M}	{ES,VS,S,M}	{S,M}		
S	{ES,VS,S,M,L,VL}	{ES,VS,S,M,L,VL}	{ES,VS,S,M,L,VL}	{ES,VS,S,M,L,VL}	{S,M}		
X_2 **M**	{ES,VS,S,M,L,VL}	{ES,VS,S,M,L,VL}	{VS,S,M,L,VL,EL}	{M,L,VL,EL}	{VL,EL}		{S,M}
L	{VS,S,M,L,VL}	{VS,S,M,L,VL}	{S,M,L,VL,EL}	{S,M,VL,EL}	{VL,EL}		{S,M}
VL	{VS,S,M,L}	{VS,S,L,M}	{S,M,L}	{S,M}			
EL	{S,M}	{S,M}					

(a) Candidate consequent sets in each subspace in the electrical problem

		ES	VS	S	M	L	VL	EL
				X1				
	ES	ES	ES	VS	VS			
	VS	VS	VS	VS	VS	S		
	S	VS	VS	M	S	M		
X 2	**M**	S	L	VL	VL	VL		S
	L	VS	M	S	EL	VL		S
	VL	VS	L	L	M			
	EL	M	M					

(b) CH-method's RB

		ES	VS	S	M	L	VL	EL
				X1				
	ES	ES	ES	VS	VS			
	VS	VS	VS	VS	S	S		
	S	VS	S	S	M	M		
X2	**M**	VS	M		EL	VL		S
	L	S	M	M	EL	VL		S
	VL	VS	L	M	M			
	EL							

(c) COR-method's RB

FIGURE II.7.13   CH-method vs. COR-method in the electrical problem.

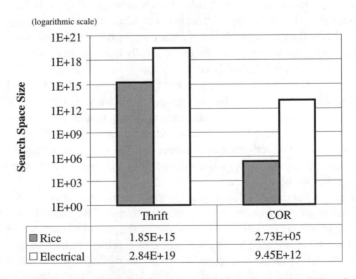

(logarithmic scale)

	Thrift	COR
■ Rice	1.85E+15	2.73E+05
□ Electrical	2.84E+19	9.45E+12

FIGURE II.7.14   Search space tackled by the two combinatorial search algorithms for each problem.

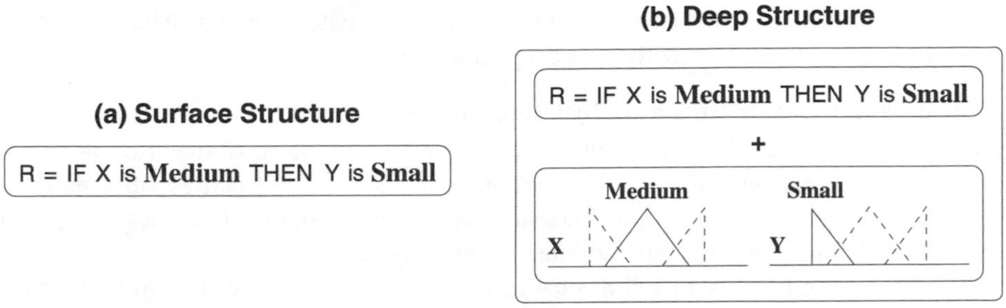

(b) Deep Structure

(a) Surface Structure

R = IF X is **Medium** THEN Y is **Small**

R = IF X is **Medium** THEN Y is **Small**

+

Medium   Small

X   Y

FIGURE II.7.15   Two different ways to define a fuzzy linguistic rule: (a) *surface structure* = symbolic representation; (b) *deep structure* = symbolic representation + membership function shapes.

7.4   Accuracy Improvement when Refining Linguistic Fuzzy Models: Deep Structure Tuning

Each of the fuzzy linguistic rules in a linguistic model may be represented at two different levels of description by defining two different structures:[109]

- *Surface structure*: It is a less specific description and involves defining the rule in its symbolic form as a relation between input and output linguistic variables. Figure II.7.15(a) shows an example of the surface structure of a fuzzy linguistic rule.
- *Deep structure*: It is a more specific description and consists of the surface structure together with the meaning of the membership functions associated with the linguistic terms of the variables. Figure II.7.15(b) illustrates the deep structure of a fuzzy linguistic rule.

In a linguistic FRBS, the surface structure of each fuzzy rule is encoded in the RB, while the deep structure is contained in the KB.

Thus, the derivation of a fuzzy model involves defining these surface and deep structures from numerical information. We can distinguish between two different derivation approaches:

- *Learning process*: It relates to the task of directly obtaining the fuzzy rule surface[95,99] or deep structures[87] from the available data.
- *Tuning process*: It assumes the existence of a previous definition for both structures—provided by learning processes or experts—and adjusts them with *slight* modifications to increase the system performance.[55,59]

Traditionally, the tuning process has been used to fit the deep structure by exclusively changing membership function shapes.[10,26,48,55,59,92] On the other hand, recent contributions perform a tuning of the surface structures adjusting the symbolic representations[25,44] with linguistic hedges.[108]

Nevertheless, no proposal combining these different tuning approaches has been considered until now. This section aims at introducing a method based on genetic algorithms that *fully* tunes the deep structures (including the surface ones) changing the linguistic term meanings together with integrating linguistic hedges in the fuzzy rules.

7.4.1   Tuning Surface and Deep Structures

Prior to proposing our tuning process in Section 7.4.2, this section introduces the way to adjust the FRBS by tuning its whole deep structures. To do that, the two following subsections, respectively, explain how to tune surface structures (symbolic representations) using linguistic hedges, and how to adapt the part of the deep structures that contains the membership function shapes. The combined action of both processes will then fully tune deep structures.

7.4.1.1 Tuning the Surface Structure by Using Linguistic Hedges

In Section 7.2.1.2, we see that the linguistic rule structure can be flexibilized by adding operators that qualify the meaning of each linguistic term involved in it (such as *very*, *more-or-less*, *slightly*, *extremely*, etc.). These operators make a clear semantic modification whose action is easily interpretable. To do that, the linguistic modifiers were proposed.

This extension to the LFM is a ideal mechanism to develop the surface structure tuning process, since the linguistic modifiers perform a slight modification of previously defined structures. Indeed, the nuances that the linguistic modifiers provided to the meaning of each label become actually a tuning of them. Moreover, we can say that such modifications only impinge on the rule surface structure, owing to the symbolic representation that is altered independently of the global meaning of the used linguistic terms. Because of this fact, the interpretability of the tuned rules is preserved to a high degree. For some proposals that perform this kind of tuning with linguistic hedges, the interested reader can refer to Cordón, del Jesus, and Herrera[25] and González and Pérez.[44]

For example, from the previously defined linguistic rule:

IF X_1 is high and X_2 is good **THEN** Y is small,

the following tuned rule would be obtained:

IF X_1 is *very* high and X_2 is good **THEN** Y is *more-or-less* small.

Actually, this tuning approach does not define a new meaning for the so-called *primary terms* — *high*, *good*, and *small* in our example — but it uses them as generators whose meaning is specified in the context. In other words, thanks to the *attributed-grammar semantics*[108] involved in linguistic variables, the final membership functions are computed from the knowledge of the primary term membership functions.

Of course, the fact of using linguistic hedges will have a significant influence in the FRBS performance, since the matching degree of the rule antecedents, as well as the shape of the output fuzzy set obtained from the inference process, will vary.

From a linguistic point of view, the linguistic modifiers are usually classified as those that reinforce the characterizations and those that weaken them. From a FM point of view, however, the most interesting effect caused by the modifiers is the alterations in the inference process (support, center of gravity, etc.) that they involve. From this perspective, we can distinguish among three different kinds of linguistic modifiers:

- *Powered modifiers*[106] — The membership degrees of the linguistic terms are modified with nonlinear scaling functions by rising them to the power of some factor (Figure II.7.16(a)):

$$\mu'(x) = (\mu(x))^\alpha, \ 0 < \alpha,$$

  For example, the operator "*very*" squares the membership degree of the linguistic term, i.e., $\mu_T^{very}(x) = (\mu_T(x))^2$.
  With $\alpha < 1$, the modifier dilates the fuzzy set, while with $\alpha > 1$ the modifier concentrates the fuzzy set. These kinds of linguistic modifiers are known as *linguistic hedges*.[106]

- *Expansive/reduced modifiers*[12,71]—These modifiers change the support and the core of the fuzzy sets enlarging or reducing them, but trying to keep a center of gravity similar to the original one (Figure II.7.16(b)). The effect over the fuzzy sets, though similar to the

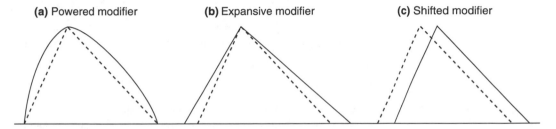

(a) Powered modifier **(b)** Expansive modifier **(c)** Shifted modifier

FIGURE II.7.16   Types of linguistic modifiers.

previous one (in terms of dilation and concentration), is rather more severe and is implemented with linear variations.

- *Shifted modifiers*[12,61]—These kinds of modifiers are defined by translations of the membership functions along their domains (Figure II.7.16(c)). The effect over the linguistic terms is more severe than the previous ones.

7.4.1.2   Tuning the Deep Structure with Linear and Nonlinear Alterations

Tuning the deep structure, more than adjusting the surface structure, involves fitting the characterization of the membership functions associated with the primary linguistic terms considered in the system. Thus, the meaning of the linguistic terms is changed from a previous definition, an initial DB in an FRBS.

To change the shapes of the membership functions, the parameters defining them must be altered. As shown in Section 7.2.1.1, we may change the individual membership functions with linear or nonlinear variations, which determine two different kinds of tuning approaches:

- On the one hand, we may perform *linear variations* by changing the parameters that define the membership functions.
- On the other hand, we may perform *nonlinear variations* by adding a nonlinear scaling factor to the membership functions and changing its value.

The latter approach (nonlinear tuning) has an important limitation with respect to the former one (linear tuning): support and core sets of the fuzzy set are not altered. Moreover, when a symmetrical fuzzy set is considered, its center of gravity is not changed. On the contrary, the membership degree of a value to the fuzzy set increases in a nonlinear way as it gets closer to the core.

These differences will have a significant importance in the inference process of the FRBS and, since each approach affects a different aspect of it, their actions are not contradictory but complementary.

7.4.2   A Genetic Tuning Process of Surface and Deep Structures

In this section, a tuning process based on genetic algorithms[72] will be introduced to jointly adjust the membership functions with linear membership function alterations and the rule surface structure using powered linguistic hedges. This combination will tune the model structure at two different levels of specification and will make linear and nonlinear variations of the fuzzy sets. The tuning process involves starting from a previous KB (RB + DB), either derived by any learning method or provided by experts.

Our proposal of FRBS genetic tuning for LFM has the following components:

- The objective (fitness function) is to minimize the MSE (defined in Section 7.2.2.1).
- A double coding scheme ($CS_P + CS_L$) for tuning both membership functions (CS_P)—i.e., the linguistic term meanings in the deep structure—and symbolic representations (CS_L)—i.e., surface structure—is used. CS_P will encode the membership function parameters and

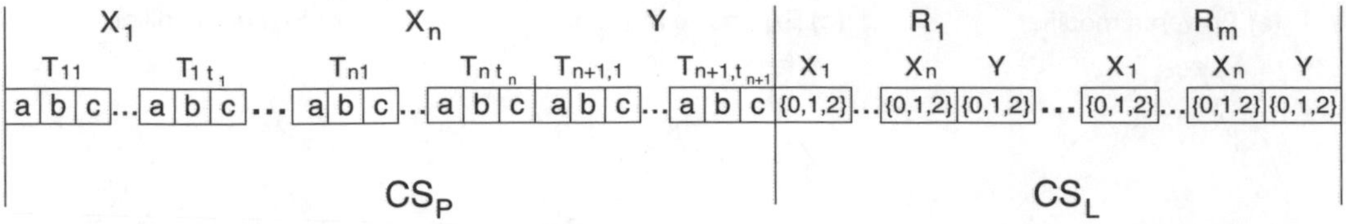

| X₁ | Xₙ | Y | R₁ | Rₘ |

| T_{11} | T_{1t_1} | T_{n1} | T_{nt_n} | $T_{n+1,1}$ | $T_{n+1,t_{n+1}}$ | X₁ | Xₙ | Y | X₁ | Xₙ | Y |
| a b c | ... | a b c | ... | a b c | ... | a b c | a b c | ... | a b c | {0,1,2} | ... {0,1,2} {0,1,2} | ... | {0,1,2} | ... {0,1,2} {0,1,2} |

$$CS_P \qquad\qquad\qquad\qquad CS_L$$

FIGURE II.7.17 Coding scheme for tuning deep structures with n being the number of input variables, T_{ij} the j-th linguistic term of the i-th variable (with $n+1$ being the output variable), t_i the number of linguistic terms of the i-th variable, and m the number of fuzzy linguistic rules.

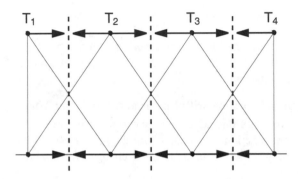

FIGURE II.7.18 Variation intervals for each membership function parameter to preserve meaningful fuzzy sets.

CS_L the linguistic hedges included in the different rules. Figure II.7.17 graphically shows such a scheme.

- For the CS_P part, a 3-tuple of real values for each triangular membership function is used, thus being the DB encoded into a real-coded chromosome built by joining the membership functions involved in each variable fuzzy partition. A variation interval is associated with every gene in the CS_P part to preserve meaningful fuzzy sets. Figure II.7.18 shows an example of the variation intervals considered for each membership function parameter according to the cross points between fuzzy sets.

- For the CS_L part, the coding scheme generates integer-coded strings of length $m \cdot (n+1)$ (with m being the number of rules and n being the number of input variables).

    Each gene can take any value in the set $\{0, 1, 2\}$ with the following correspondence to the linguistic hedge used:

$$c_{ij}^L = 0 \longleftrightarrow \text{ the "very" linguistic hedge is used,}$$

$$c_{ij}^L = 1 \longleftrightarrow \text{ no linguistic hedge is used,}$$

$$c_{ij}^L = 2 \longleftrightarrow \text{ the "more-or-less" linguistic hedge is used.}$$

with c_{ij}^L being the gene associated with the linguistic term used in the j-th variable of the i-th rule.

- When generating the initial population, some of the original information in the initial KB (automatically learned or provided by experts) will be mixed with random values.

    To introduce the original values in the CS_P part in the genetic population, the actual values will be directly considered. For the CS_L part, the modifiers used in the initial KB are encoded with said scheme. If no linguistic hedges were previously considered, alleles 1 will be used.

The following three steps are considered to initialize the population:

1. A chromosome that represents the initial DB and RB is included. Therefore, the genes in CS_P and CS_L parts will directly encode the values corresponding to the original KB.
2. A half of the population is generated with the CS_P part at random (within the variation interval for each gene) while the alleles in CS_L will encode the original values.
3. The remaining chromosomes are generated with the original values of the DB in the CS_P part, and alleles at random (within the set {0, 1, 2}) in the CS_L part.

- The crossover operator will depend on the chromosome part where it is applied:

  - In the CS_P part, the max-min-arithmetical crossover[49] is considered.

  - In the CS_L part, the standard two-point crossover is used.

  After recombining each part, the two best chromosomes among the eight (*four* different CS_P parts combined with *two* different CS_L parts) descendants obtained will be selected to replace their parents.
- The mutation operator will also depend on the chromosome part where it is applied:

  - In the CS_P part, the Michalewicz's nonuniform mutation operator[72] within the interval allowed for each gene is considered.

  - In the CS_L part, the mutation operator changes the gene to the allele 1 when a gene with alleles 0 or 2 must be mutated, and randomly to 0 or 2 when a gene with allele 1 must be mutated.

- A generational genetic algorithm with the Baker's stochastic universal sampling procedure[3] together with elitism is considered.

Figure II.7.19 illustrates the genetic tuning process proposed. Its performance will be analyzed in the following section.

7.4.3 Experimental Study

The experimental study will be focused on applying the proposed tuning process (the PL-tun-method) to a previously generated model. Two learning methods (the T-method and our COR-method) will be used to derive the initial RBs. Therefore, these methods will act as the learning module shown in Figure II.7.20, where the two-stage operation mode considered in this experimental study is graphically shown.

Table II.7.6 collects the best results obtained in the study of Section 7.2.2, the original values of the considered learning method before applying the tuning process, and the results obtained by the proposed two-stage learning methods in the two applications considered. The best results for both applications are shown in boldface.

First of all, in view of the obtained results, we notice that the proposed tuning process significantly improves the accuracy of the models derived by the T-method and the COR-method. From the obtained results, we can say that when a full tuning of deep structures (i.e., including the surface ones) is performed, more accurate fuzzy models are obtained. It is due to the PL-tun-method combining macroscopic and microscopic tuning effects[10] with two ways of changing the membership function shapes.

As regards the interpretability of the tuned models, Figure II.7.21 shows the tuned DB and RB obtained by the COR + PL-tun-method for a specific data set partition of the rice problem. As may be observed, the tuned DB remains a good interpretability, but significantly improves the accuracy.

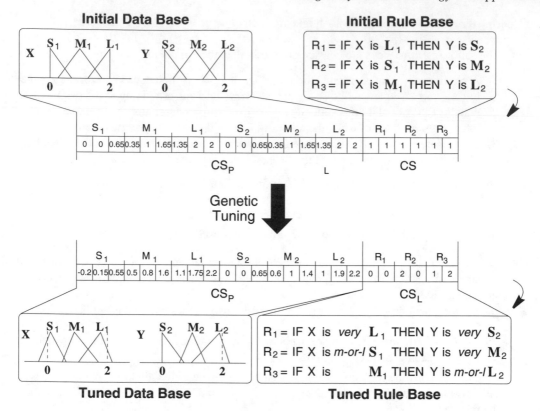

FIGURE II.7.19   An example of genetic tuning process for a single-input single-output FRBSs with three different linguistic terms for each variable and three fuzzy linguistic rules. *S* stands for *small*, *M* for *medium*, *L* for *large*, and *m-or-l* for "more-or-less."

TABLE II.7.6   Results Obtained in the Two Applications Considered

Tuning	\#R	Rice evaluation MSE_{tra}	MSE_{tst}	\#R	Electrical line length MSE_{tra}	MSE_{tst}
LMe	31	0.000810	0.002342	—	—	—
WM + LH-Tun	**15**	0.001758	**0.001731**	—	—	—
M-L	—	—	—	35	137,905	173,096
T + M-L-Tun	—	—	—	49	147,844	164,993
T	16	0.004949	0.005991	49	169,077	175,739
COR	19	0.005715	0.006438	29	173,823	**160,753**
T + PL	16	0.000660	0.002386	49	132,362	187,168
COR + PL	19	**0.000643**	0.002271	**29**	**130,922**	172,131

On the other hand, the use of linguistic hedges gives a good interpretation to the slight changes performed in the tuned RB.

When comparing the results obtained by the PL-tun-method with the best results obtained by the methods considered in the comparative study of Section 7.2.2, the good performance of our tuning

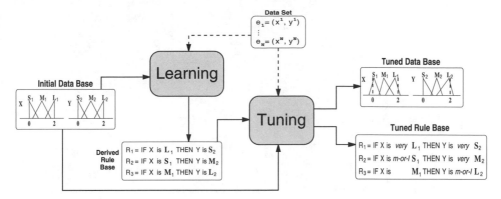

FIGURE II.7.20 The modeling process is performed in two stages: in the former one, a learning method is used to derive a RB from an initial DB; then, the tuning method adjusts the previously obtained RB and/or the initial DB.

method is corroborated. The best approximation degrees (MSE_{tra}) are obtained when the whole deep structures are tuned. Moreover, good generalization degrees (MSE_{tst}) are obtained.

With regard to the combination of the PL-tun-method with the preliminary learning method, better accuracy degrees—both in approximation and generalization—are obtained when the COR-method is used instead of the T-method, in spite of the latter one obtaining a more accurate preliminary model. In fact, the consideration of previous linguistic rules with a good cooperation among them, as the COR-method does, allows the tuning process to generate more accurate models. Moreover, the good cooperation of the previous model makes the process more robust against the overfitting problem (excessive approximation to the training data with a undesirable prediction).

To sum up, the obtained results lead us to think that the use of an appropriate tuning process significantly improves the accuracy without needing previous complex learning methods. Instead, previous models with an acceptable accuracy but good RB properties (i.e., cooperation, completeness, consistency, etc.) seems to be more suitable for being tuned.

7.5 Concluding Remarks

Currently, the most interesting tendency in FM is the consideration of mechanisms to find a good balance between interpretability and accuracy. One of the possibilities is to make the LFM more flexible to improve the accuracy of the linguistic models maintaining a good interpretability. To do so, two different kinds of linguistic improvements have been analyzed in this contribution:

- Improvements in the *learning process* by inducing cooperation among the linguistic rules and tuning their deep structures.
- Improvements in the *model structure* by using linguistic hedges.

An experimental study comparing different learning methods has shown that very interesting results are obtained combining simple linguistic rule learning methods, which do not make use of complex rule generation mechanisms, with a tuning process that refines the obtained models *a posteriori*. Inspired by this conclusion, the chapter has proposed a two-step mechanism to improve the accuracy: the former step quickly generates preliminary models and the latter step adjusts them.

To generate simple and accurate models, a learning methodology that considers the cooperation among the rules has been proposed. The method has shown good results combining accuracy and

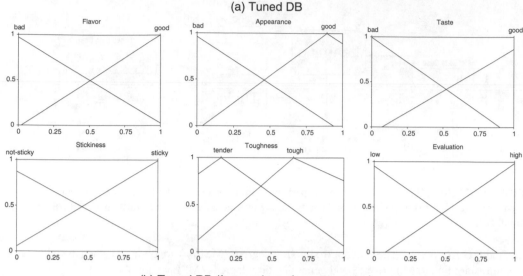

(a) Tuned DB

(b) Tuned RB (fuzzy rule surface structures)

Rule	Flavor		Appearance		Taste		Stickiness		Toughness		Evaluation	
R_1		bad	very	bad		bad	m-or-l	not sticky	m-or-l	tender	m-or-l	low
R_2	m-or-l	bad	very	bad	very	bad		not sticky		tough	m-or-l	low
R_3		bad	m-or-l	bad		good		not sticky	very	tough		low
R_4	m-or-l	bad		good		bad	m-or-l	not sticky		tender		low
R_5	m-or-l	bad	m-or-l	good	very	bad		not sticky		tough	very	low
R_6	very	bad	very	good		bad	m-or-l	sticky		tender	very	high
R_7		bad	m-or-l	good		bad		sticky	m-or-l	tough	m-or-l	low
R_8		bad	m-or-l	good		good	very	not sticky		tender	m-or-l	high
R_9	m-or-l	bad	m-or-l	good	very	good	m-or-l	sticky	very	tender	very	high
R_{10}	m-or-l	bad	very	good	very	good	very	sticky		tough		high
R_{11}	very	good	m-or-l	bad		bad	m-or-l	not sticky	m-or-l	tender	very	low
R_{12}		good	m-or-l	bad	m-or-l	bad	very	not sticky		tough	m-or-l	low
R_{13}	very	good	very	bad	very	bad	m-or-l	sticky	very	tender		high
R_{14}	very	good	very	bad	very	bad	m-or-l	sticky		tough	very	high
R_{15}	very	good	m-or-l	bad	m-or-l	good		sticky	m-or-l	tender		high
R_{16}	very	good	m-or-l	bad		good	very	sticky		tough	m-or-l	high
R_{17}		good	m-or-l	good	very	bad	m-or-l	not sticky	m-or-l	tender	m-or-l	low
R_{18}	m-or-l	good	very	good	very	bad	m-or-l	sticky		tough	m-or-l	high
R_{19}	very	good	very	good	m-or-l	good	very	not sticky	m-or-l	tender	m-or-l	high
R_{20}	very	good		good		good		sticky		tender	very	high
R_{21}	very	good		good		good	very	sticky		tough	very	high

FIGURE II.7.21   KB (fuzzy rule deep structures) generated by the COR + PL-tun-method for a specific data set partition of the rice problem. $MSE_{tra/tst}$ before tuning = 0.006296/0.003879, $MSE_{tra/tst}$ after tuning = 0.000651/0.002050.

interpretability. Moreover, the quick process performed makes the method very appropriate for a subsequent tuning process.

Tuning FRBSs involves adjusting some of the components composing the KB. To do that, as the fuzzy rule, symbolic representations as the shapes of the involved membership functions can be defined. This contribution has introduced a genetic tuning process for jointly refining these two components. To adjust the symbolic representations, the use of linguistic hedges to perform slight modifications keeping a good interpretability is proposed. To change the membership function shapes, linear variations are proposed. The application of our proposal to LFM has shown very

TABLE II.7.7   Summary on the Three Applications Considered and Their Main Characteristics

Application	#V	#Tra	#Tst	#LT	#P
Rice taste evaluation	5	75	30	2	10
Electrical line length	2	396	99	7	1

Note: **V** = number of input variables, **Tra** = training data set size, **Tst** = test data set size, **LT** = number of linguistic terms considered for each fuzzy partition, **P** = number of different partitions of the sample considered.

good results. This good performance mainly lies in the consideration of tuning at two different levels of significance and modifying the fuzzy set shapes with and without changes in their support sets.

Finally, we have shown that it is possible to generate linguistic models with good interpretability and accuracy degrees by combining a simple learning method that considers the cooperation among the rules with a subsequent process that refines the whole fuzzy rule deep structures.

Appendix A   Applications Considered in the Experimental Studies

Table II.7.7 collects the main characteristics of the two real-world applications considered for the different experiments developed in this chapter. They are briefly described in the following.

A.7.1   The Rice Taste Evaluation Problem

Subjective qualification of food taste is a very important but difficult problem. In the case of the rice taste qualification, it is usually put into effect by means of a subjective evaluation called the *sensory test*. In this test, a group of experts evaluates the rice according to a set of characteristics associated with it. These factors are: *flavor*, *appearance*, *taste*, *stickiness*, and *toughness*.[75] Because of the large quantity of relevant variables, the problem of rice taste analysis becomes very complex, thus leading to solution by means of modeling techniques capable of obtaining a model representing the nonlinear relationships existing in it. In order to do so, we are going to use the data set presented in Nozaki, Ishibuchi and Tanaka.[75] This set is composed of 105 data vectors collecting subjective evaluations of the six variables in question (the five mentioned and the overall evaluation of the kind of rice), made up by experts on this number of kinds of rice grown in Japan.

With the aim of not biasing the learning, we have randomly obtained ten different partitions of the mentioned set, composed by 75 pieces of data in the training set and 30 in the test set, to generate ten models in each experiment. Two labels, or uniform partitions with two linguistic terms, will be considered when required.

A.7.2   The Electrical Low-Voltage Line Length Problem

This problem involves finding a model that relates the total length of low-voltage line installed in Spanish rural towns.[31] This model will be used to estimate the total length of line being maintained by an electrical company. We were provided with a sample of 495 towns in which the length of line was actually measured, and the company used the model to extrapolate this length over more than 10,000 towns with these properties. We will limit ourselves to the estimation of the *total length of low-voltage line installed in a town*, given the inputs *number of inhabitants of the town* and *distance from the center of the town to the three furthest clients*. To develop the different experiments in this contribution, the sample has been randomly divided in two subsets, the training and test ones, with 80–20% of the original size, respectively. Thus, the training set contains 396 elements, while the test set is composed by 99 elements. These data sets are available on the web at http://decsai.ugr.es/~casillas/FMLib/. Seven labels, or the corresponding uniform partitions, will be considered when necessary.

Appendix B   Ant Colony Optimization Algorithms

A new family of bio-inspired algorithms has recently appeared, ACO algorithms.[9,35] Since the first proposal, the AS algorithm[37] (applied to the traveling salesman problem), numerous models have been developed to solve a wide set of optimization problems (refer to Bonabeau, Dorigo, and Theraulaz[9] and Dorigo and Di Caro[35] for a review of models and applications).

ACO algorithms model the behavior of real ant colonies. Particularly, they draw inspiration from the social behavior of these insects to provide food to the colony. In the food search process, consisting of the food find and the return to the nest, the ants deposit a substance called *pheromone*. The ants have the ability of sniffing the pheromone and the colony is guided by it during the search. When an ant is located in a branch, it decides to take the path according to the probability defined by the pheromone existing in each trail.

In this way, the depositions of pheromone terminate in constructing a track between the nest and the food that can be followed by new ants. The continued action of the colony members involving the length of the track is progressively reduced. The shortest paths are finally the more frequently visited ones and, therefore, the pheromone concentration is higher. On the contrary, the longest paths are less visited and the associated pheromone trail is evaporated.

Therefore, ACO algorithms are based on the cooperative action of multiple agents (ants) each of them generating a possible solution to the problem at each iteration of the algorithm. To do so, each ant travels a graph that represents the problem and makes use of two kinds of information, shared by the colony, that indicate the preference among the different edges between nodes:

- *Heuristic information*: It depends on the problem and is obtained before running the algorithm, keeping inalterable during the process. The heuristic value for the edge (i, j) is noted as η_{ij}.
- *Pheromone trail information*: It is modified through the algorithm running, depending on the paths taken by the ants and the goodness of the generated solutions. This information is represented by the pheromone trail. The value for the edge (i, j) is noted as τ_{ij}.

---

1. *Problem representation*: Interpret the problem to be solved as a graph or a similar structure easily traveled by ants.

2. *Heuristic information*: Define the way of assigning a heuristic preference to each choice that the ant has to take in each step to generate the solution.

3. *Pheromone initialization*: Establish an appropriate way of initializing the pheromone.

4. *Fitness function*: Define a fitness function to be optimized.

5. *ACO algorithm*: Select an ACO algorithm and apply it to the problem.

FIGURE II.7.22   Steps followed to apply ACO algorithms to a specific problem.

The basic operation mode is as follows:[37] at each iteration, a population of a specific number of ants progressively construct different tracks on the graph (i.e., solutions to the problem) according to a *probabilistic transition rule* that depends on the available information. After that, the pheromone trails are updated. This is done by first decreasing them by some constant factor (corresponding to the evaporation of the pheromone) and then reinforcing the solution attributes of the constructed solutions considering their quality. This task is developed by the *global pheromone trail update rule*.

Several extensions to this basic operation mode have been proposed. Their improvements mainly consist of using different transition and update rules, introducing new components, or adding a local search phase.[13,36,86]

To apply ACO algorithms to a specific problem, the five steps shown in Figure II.7.22 have to be performed.

References

1. Alcalá, R. et al., Cooperative coevolution for linguistic modeling with weighted double-consequent fuzzy rules. In *Proc. 2nd Int. Conf. Fuzzy Logic and Technology*, 237–240, Leicester, UK, 2001.
2. Babuška, R., *Fuzzy Modeling for Control*. Kluwer Academic, Norwell, MA, 1998.
3. Baker, J.E., Reducing bias and inefficiency in the selection algorithm. In *Proc. 2nd Int. Conf. Genetic Algorithms*, 14–21, Lawrence Erlbaum, Hillsdale, NJ, 1987.
4. Bárdossy, A. and Duckstein, L., *Fuzzy Rule-Based Modeling with Application to Geophysical, Biological and Engineering Systems*. CRC Press, Boca Raton, FL, 1995.
5. Bastian, A., How to handle the flexibility of linguistic variables with applications. *Int. J. Uncertainty, Fuzziness and Knowledge-Based Systems*, 2(4), 463–484, 1994.
6. Bikdash, M., A highly interpretable form of Sugeno inference systems. *IEEE Trans. on Fuzzy Systems*, 7(6), 686–696, 1999.
7. Bodenhofer, U., A similarity-based generalization of fuzzy orderings preserving the classical axioms. *Int. J. Uncertainty, Fuzziness and Knowledge-Based Systems*, 8(5), 593–610, 2000.
8. Bodenhofer, U. and Bauer, P., Towards an axiomatic approach to interpretability. In *Proc. 6th Int. Conf. Soft Computing*, Iizuka, Japan, 334–339, 2000.
9. Bonabeau, E., Dorigo, M., and Theraulaz, G., *Swarm intelligence. From natural to artificial systems*. Oxford University Press, Oxford, U.K., 1999.
10. Bonissone, P.P., Khedkar, P.S., and Chen, Y., Genetic algorithms for automated tuning of fuzzy controllers: a transportation application. In *Proc. 5th IEEE Int. Conf. Fuzzy Systems*, New Orleans, LA, 674–680, 1996.
11. Bouchon-Meunier, B., Linguistic variables in the knowledge base of an expert system. In *Cybernetics and systems: the way ahead*, J.Rose, Ed., 745–752. Thales Publications, Lytham St Annes, England, 1987.
12. Bouchon-Meunier, B. and Jia.Y., Linguistic modifiers and imprecise categories. *Int. J. Intelligent Systems*, 7, 25–36, 1992.
13. Bullnheimer, B., Hartl, R.F., and Strauss, C., A new rank-based version of the ant system: a computational study. *Central European Journal for Operations Research and Economics*, 7(1), 25–38, 1999.
14. Carse, B. Fogarty, T.C., and Munro, A., Evolving fuzzy-rule-based controllers using genetic algorithms. *Fuzzy Sets and Systems*, 80, 273–294, 1996.
15. Casillas, J. et al., Genetic tuning of fuzzy rule-based systems integrating linguistic hedges. In *Proc. 9th IFSA World Congress and 20th NAFIPS Int. Conf.*, 1570–1574, Vancouver, Canada, 2001.
16. Casillas, J., Cordón, O., and Herrera., F., Can linguistic modeling be as accurate as fuzzy modeling without losing its description to a high degree? Technical Report #DECSAI-00-01-20, Department

of Computer Science and Artificial Intelligence, University of Granada, Granada, Spain, 2000. Available at http://decsai.ugr.es/~casillas/.

17. Casillas, J., Cordón, O., and Herrera, F., Improving the Wang and Mendel's fuzzy rule learning method by inducing cooperation among rules. In *Proc. 8th Information Processing and Management of Uncertainty in Knowledge-Based Systems Conf.*, 1682–1688, Madrid, Spain, 2000.

18. Casillas, J., Cordón, O., and Herrera, F., Learning cooperative fuzzy linguistic rules using ant colony algorithms. Technical Report #DECSAI-00-01-19, Department of Computer Science and Artificial Intelligence, University of Granada, Granada, Spain, 2000. Available at http://decsai.ugr.es/~casillas/.

19. Casillas, J., Cordón, O., and Herrera, F., COR: A methodology to improve ad hoc data-driven linguistic rule learning methods by inducing cooperation among rules. *IEEE Trans. Systems, Man, and Cybernetics — Part B*: Cybernetics. To appear. Department of Computer Science and Artificial Intelligence, University of Granada, Granada, Spain, 2000. Available at http://decsai.ugr.es/~casillas/.

20. Casillas, J., Cordón, O., and Herrera, F., Different approaches to induce cooperation in fuzzy linguistic models under the COR methodology. In *Techniques for constructing intelligent systems*. B. Bouchon-Meunier, J. Gutiérrez-Ríos, L. Magdalena, and R.R. Yager, Eds., Springer-Verlag, Heidelberg, Germany, 2001. In press.

21. Casillas, J. et al., Eds., *Fuzzy Modeling and the Interpretability-Accuracy Trade-Off. Part II, Accuracy Improvements Preserving Interpretability.* Physica-Verlag, Heidelberg, Germany, 2001. In press.

22. Casillas, J. et al., Cooperative coevolution for learning fuzzy rule-based systems. In *Proc. 5th Int. Conf. Artificial Evolution*, 97–108, Le Creusot, France, 2001.

23. Chi, Z., Yan, H., and Pham, T., *Fuzzy algorithms with application to image processing and pattern recognition*. World Scientific, Singapore, 1996.

24. Cho, J.-S. and Park, D.-J., Novel fuzzy logic control based on weighting of partially inconsistent rules using neurtal network. *J. Intelligent and Fuzzy Systems*, 8, 99–110, 2000.

25. Cordón, O., del Jesus, M.J., and Herrera, F., Genetic learning of fuzzy rule-based classification systems cooperating with fuzzy reasoning methods. *Int. J. of Intelligent Systems*, 13, 1025–1053, 1998.

26. Cordón, O. and Herrera, F., A three-stage evolutionary process for learning descriptive and approximate fuzzy logic controller knowledge bases from examples. *Int. J. Approximate Reasoning*, 17(4), 369–407, 1997.

27. Cordón, O. and Herrera, F., A proposal for improving the accuracy of linguistic modeling. *IEEE Trans. on Fuzzy Systems*, 8(3), 335–344, 2000.

28. Cordón, O. et al., *Genetic fuzzy systems: evolutionary tuning and learning of fuzzy knowledge bases*. World Scientific, Singapore, 2001.

29. Cordón, O. et al., A genetic learning process for the scaling factors, granularity and contexts of the fuzzy rule-based system data base. *Information Sciences*, 136(1–4), 85–107, 2001.

30. Cordón, O., Herrera, F., and Peregrín. A., Applicability of the fuzzy operators in the design of fuzzy logic controllers. *Fuzzy Sets and Systems*, 86(1), 15–41, 1997.

31. Cordón, O., Herrera, F., and Sánchez. L., Solving electrical distribution problems using hybrid evolutionary data analysis techniques. *Applied Intelligence*, 10(1), 5–24, 1999.

32. Cordón, O., Herrera, F., and Villar, P., Generating the knowledge base of a fuzzy rule-based system by the genetic learning of the data base. *IEEE Trans. Fuzzy Systems*, 9(4), 667–674, 2001.

33. Cordón, O., Herrera, F., and Villar, P., Analysis and guidelines to obtain a good fuzzy partition granularity for fuzzy rule-based systems using simulated annealing. *Int. J. Approximate Reasoning*, 25(3), 187–215, 2000.

34. Cordón, O., Herrera, F., and Zwir, I., Linguistic modeling by hierarchical systems of linguistic rules. *IEEE Tran. Fuzzy Systems*, 10(1), 2–20, 2002.

35. Dorigo, M. and Di Caro, G., The ant colony optimization meta-heuristic. In *New ideas in optimization* D. Corne, M. Dorigo, and F. Glover, Eds., 11–32. McGraw-Hill, New York, 1999.

36. Dorigo, M. and Gambardella, L.M., Ant colony system: a cooperative learning approach to the traveling salesman problem. *IEEE Trans. Evolutionary Computation*, 1(1), 53–66, 1997.

37. Dorigo, M., Maniezzo, V., and Colorni. A., The ant system: optimization by a colony of cooperating agents. *IEEE Trans. Systems, Man, and Cybernetics—Part B: Cybernetics*, 26(1), 29–41, 1996.

38. Driankov, D., Hellendoorn, H., and Reinfrank, M., *An introduction to fuzzy control*. Springer-Verlag, Heidelberg, Germany, 1993.

39. Espinosa, J. and Vandewalle, J., Constructing fuzzy models with linguistic integrity from numerical data-AFRELI algorithm. *IEEE Trans. Fuzzy Systems*, 8(5), 591–600, 2000.

40. Fullér, R., *Introduction to neuro-fuzzy systems*. Springer-Verlag, Heidelberg, Germany, 2000.

41. Glorennec, P., Constrained optimization of FIS using an evolutionary method. In *Genetic algorithms and soft computing* F. Herrera and J.L. Verdegay, Eds., 349–368. Physica-Verlag, Heidelberg, Germany, 1996.

42. Gómez-Skarmeta, A.F. and Jiménez, F., Fuzzy modeling with hybrid systems. *Fuzzy Sets and Systems*, 104(2), 199–208, 1999.

43. González, A. and Pérez, R., A fuzzy theory refinement algorithm. *Int. J. Approximate Reasoning*, 19(3–4), 193–220, 1998.

44. González, A. and Pérez, R., A study about the inclusion of linguistic hedges in a fuzzy rule learning algorithm. *Int. J. Uncertainty, Fuzziness and Knowledge-Based Systems*, 7(3), 257–266, 1999.

45. Gudwin, R.R. and Gomide, F.A.C., Context adaptation in fuzzy processing. In *Proc. 1st Brazil-Japan J. Symp. Fuzzy Systems*, 15–20, Campinas, SP, Brazil, 1994.

46. Gudwin, R.R., Gomide, F.A.C., and Pedrycz, W., Context adaptation in fuzzy processing and genetic algorithms. *Int. J. Intelligent Systems*, 13(10–11), 929–948, 1998.

47. Hangos, K.M., Ed., *Special issue on grey box modelling*, volume 9(6) of *Int. J. Adaptive Control and Signal Processing*. John Wiley & Sons, New York, 1995.

48. Herrera, F., Lozano, M., and Verdegay, J.L., Tuning fuzzy controllers by genetic algorithms. *Int. J. Approximate Reasoning*, 12, 299–315, 1995.

49. Herrera, F., Lozano, M., and Verdegay, J.L., A learning process for fuzzy control rules using genetic algorithms. *Fuzzy Sets and Systems*, 100, 143–158, 1998.

50. Herrera, F. and Verdegay, J.L., Eds., *Genetic algorithms and soft computing*. Physica-Verlag, Heidelberg, Germany, 1996.

51. Homaifar, A. and McCormick, E., Simultaneous design of membership functions and rule sets for fuzzy controllers using genetic algoritms. *IEEE Trans. Fuzzy Systems*, 3(2), 129–139, 1995.

52. Ishibuchi, H. and Murata, T., A genetic-algorithm-based fuzzy partition method for pattern classification problems. In *Genetic algorithms and soft computing*, F. Herrera, and J.L. Verdegay, Eds., Physica-Verlag, Heidelberg, Germany, 555–578, 1996.

53. Ishibuchi, H. and Nakashima, T., Effect of rule weights in fuzzy rule-based classification systems. *IEEE Trans. Fuzzy Systems*, 9(4), 506–515, 2001.

54. Ishibuchi, H. et al., Empirical study on learning in fuzzy systems by rice taste analysis. *Fuzzy Sets and Systems*, 64, 129–144, 1994.

55. Jang, J.-S.R., ANFIS: adaptive-network-based fuzzy inference system. *IEEE Trans. Systems, Man, and Cybernetics*, 23(3), 665–685, 1993.

56. Jiménez, F. et al., A multi-objective evolutionary algorithm for fuzzy modeling. In *Proc. 9th IFSA World Congress 20th NAFIPS Int. Conf.*, 1222–1228, Vancouver, Canada, 2001.

57. Jin, Y., Fuzzy modeling of high-dimensional systems: complexity reduction and interpretability improvement. *IEEE Trans. Fuzzy Systems*, 8(2), 212–221, 2000.

58. Jin, Y., von Seelen, W., and Sendhoff, B., On generating FC^3 fuzzy rule systems from data using evolution strategies. *IEEE Trans. Systems, Man, and Cybernetics—Part B: Cybernetics*, 29(4), 829–845, 1999.

59. Karr, C.L., Genetic algorithms for fuzzy controllers. *AI Expert*, 6(2), 26–33, 1991.

60. KrishnaKumar, K. and Satyadas, A., GA-optimized fuzzy controller for spacecraft attitude control. In *Genetic algorithms in engineering and computer science*, J. Periaux, G. Winter, M. Galán, and P. Cuesta, Eds., 305–320. John Wiley & Sons, New York, 1995.

61. Lakoff, G., Hedges: a study in meaning criteria and the logic of fuzzy concepts. *J. Philosofical Logic*, 2, 458–508, 1973.

62. Lindskog, P., Fuzzy identification from a grey box modeling point of view. In *Fuzzy model identification*, H. Hellendoorn and D. Driankov, Eds., Springer-Verlag, Heidelberg, Germany, 3–50, 1997.

63. Liska, J. and Melsheimer, S.S., Complete design of fuzzy logic systems using genetic algorithms. In *Proc. 3rd IEEE Int. Conf. Fuzzy Systems*, Orlando, FL, 1377–1382, 1994.

64. Liu, B.-D., Chen, C.-Y., and Tsao., J.-Y., Design of adaptive fuzzy logic controller based on linguistic-hedge concepts and genetic algorithms. *IEEE Trans. Systems, Man, and Cybernetics— Part B: Cybernetics*, 31(1), 32–53, 2001.

65. López, S., Magdalena, L., and Velasco, J.R., Genetic fuzzy c-means algorithm for the automatic generation of fuzzy partitions. In *Information, Uncertainty, Fusion*, B. Bouchon-Meunier, R.R. Yager, and L.A. Zadeh, Eds., 407–418. Kluwer Scientific, Norwell, MA, 1999.

66. Lotfi, A., Andersen, H.C., and Tsoi, A.C., Interpretation preservation of adaptive fuzzy inference systems. *Int. J. Approximate Reasoning*, 15, 379–394, 1996.

67. Magdalena, L., Adapting the gain of an FLC with genetic algorithms. *Int. J. Approximate Reasoning*, 17(4), 327–349, 1997.

68. Magdalena, L. and Monasterio-Huelin, F., A fuzzy logic controller with learning through the evolution of its knowledge base. *Int. J. Approximate Reasoning*, 16(3), 335–358, 1997.

69. Mamdani, E.H., Applications of fuzzy algorithms for control a simple dynamic plant. *Proc. IEE 121*, 12, 1585–1588, 1974.

70. Mamdani, E.H. and Assilian, S., An experiment in linguistic systhesis with fuzzy logic controller. *Int. J. Man-Machine Studies*, 7, 1–13, 1975.

71. Marín-Blázquez, J.G., Shen, Q., and Gómez-Skarmeta, A.F., From approximative to descriptive models. In *Proc. 9th IEEE Int. Conf. Fuzzy Systems*, 829–834, San Antonio, TX, 2000.

72. Michalewicz, Z., *Genetic algorithms + data structures = evolution programs*. Springer-Verlag, Heidelberg, Germany, 3rd edition, 1996.

73. Mouzouris, G.C. and Mendel, J.M., A singular-value-QR decomposition-based method for training fuzzy logic systems in uncertain environments. *J. Intelligent and Fuzzy Systems*, 5, 367–374, 1997.

74. Nauck, D., Klawonn, F., and Kruse, R., *Foundations of neuro-fuzzy systems*. John Wiley & Sons, New York, 1997.

75. Nozaki, K., Ishibuchi, H., and Tanaka, H., A simple but powerful heuristic method for generating fuzzy rules from numerical data. *Fuzzy Sets and Systems*, 86(3), 251–270, 1997.

76. Pal, N.R. and Pal, K., Handling of inconsistent rules with an extended model of fuzzy reasoning. *J. Intelligent and Fuzzy Systems*, 7, 55–73, 1999.

77. Pedrycz, W., Ed., *Fuzzy modelling: paradigms and practice*. Kluwer Academic, Norwell, MA, 1996.

78. Pedrycz, W., Fuzzy equalization in the construction of fuzzy sets. *Fuzzy Sets and Systems*, 119(2), 329–335, 2001.

79. Pedrycz, W., Gudwin, R.R., and Gomide, F.A.C., Nonlinear context adaptation in the calibration of fuzzy sets. *Fuzzy Sets and Systems*, 88(1), 91–97, 1997.

80. Peña-Reyes, C.A. and Sipper, M., Fuzzy CoCo: a cooperative coevolutionary approach to fuzzy modeling. *IEEE Trans. Fuzzy Systems*, 9(5), 727–737, 2001.

81. Peña-Reyes, C.A. and Sipper, M., Applying fuzzy CoCo to breast cancer diagnosis. In *Proc. 2000 Congr. Evolutionary Computation*, 1168–1175, Piscataway, NJ, IEEE Press, 2000.

82. Potter, M.A. and De Jong, K.A., Cooperative coevolution: an architecture for evolving coadapted subcomponents. *Evolutionary Computation*, 8(1), 1–29, 2000.

83. Roubos, H. and Setnes, M., Compact and transparent fuzzy models and classifiers through iterative complexity reduction. *IEEE Trans. Fuzzy Systems*, 9(4), 516–524, 2001.

84. Sánchez, L. et al., Some relationships between fuzzy and random set-based classifiers and models. *Int. J. Approximate Reasoning*, 29(2), 175–213, 2002.

85. Setnes, M. et al., Similarity measures in fuzzy rule base simplification. *IEEE Trans. Systems, Man, and Cybernetics—Part B: Cybernetics*, 28(3), 376–386, 1998.

86. Setnes, M. and Hellendoorn, H., Orthogonal transforms for ordering and reduction of fuzzy rules. In *Proc. 9th IEEE Int. Conf. Fuzzy Systems*, San Antonio, TX, 700–705, 2000.

87. Setnes, M. and Roubos, H., GA-fuzzy modeling and classification: complexity and performance. *IEEE Trans. Fuzzy Systems*, 8(5), 509–522, 2000.

88. Shi, Y., Eberhart, R., and Chen, Y., Implementation of evolutionary fuzzy systems. *IEEE Trans. Fuzzy Systems*, 7(2), 109–119, 1999.

89. Söderström, T. and Stoica, P., *System Identification*. Prentice Hall, Englewood Cliffs, NJ, 1989.

90. Sugeno, M. and Kang, G.T., Structure identification of fuzzy model. *Fuzzy Sets and Systems*, 28, 15–33, 1988.

91. Sugeno, M. and Yasukawa, T., A fuzzy-logic-based approach to qualitative modeling. *IEEE Trans. Fuzzy Systems*, 1(1), 7–31, 1993.

92. Surmann, H., Kanstein, A., and Goser, K., Self-organizing and genetic algorithms for an automatic design of fuzzy control and decision systems. In *Proc. 1st European Congresss on Fuzzy and Intelligent Technologies*, Aachen, Germany, 1097–1104, 1993.

93. Suzuki, T. et al., Fuzzy modeling using genetic algorithms with fuzzy entropy as conciseness measure. *Information Sciences*, 136(1–4), 53–67, 2001.

94. Takagi, T. and Sugeno, M., Fuzzy identification of systems and its application to modeling and control. *IEEE Trans. Systems, Man, and Cybernetics*, 15, 116–132, 1985.

95. Thrift, P., Fuzzy logic synthesis with genetic algorithms. In *Proc. 4th Int. Conf. Genetic Algorithms*, R.K. Belew, and L.B. Booker, Eds., 509–513, San Mateo, CA, 1991. Morgan Kaufmann Publishers.

96. Valente de Oliveira, J., Semantic constraints for membership function optimization. *IEEE Trans. Systems, Man, and Cybernetics—Part A: Systems and Humans*, 29(1), 128–138, 1999.

97. Valente de Oliveira. J., Towards neuro-linguistic modeling: constraints for optimization of membership functions. *Fuzzy Sets and Systems*, 106(3), 357–380, 1999.

98. Wang, C.-H., Hong, T.-P., and Tseng, S.-S., Integrating fuzzy knowledge by genetic algorithms. *IEEE Trans. Evolutionary Computation*, 2(4), 138–149, 1998.

99. Wang, L.-X. and Mendel, J.M., Generating fuzzy rules by learning from examples. *IEEE Trans. Systems, Man, and Cybernetics*, 22(6), 1414–1427, 1992.

100. Xiong, N. and Litz, L., Fuzzy modeling based on premise optimization. In *Proc. 9th IEEE International Conference on Fuzzy Systems*, 859–864, San Antonio, TX, 2000.

101. Yam, Y., Baranyi, P., and Yang, C.-T., Reduction of fuzzy rule base via singular value decomposition. *IEEE Trans. Fuzzy Systems*, 7(2), 120–132, 1999.

102. Yen, J. and Wang, L., Simplifying fuzzy rule-based models using orthogonal transformation methods. *IEEE Trans. Systems, Man, and Cybernetics—Part B: Cybernetics*, 29(1), 13–24, 1999.

103. Yen, J., Wang, L., and Gillespie, C.W., Improving the interpretability of TSK fuzzy models by combining global learning and local learning. *IEEE Trans. Fuzzy Systems*, 6(4), 530–537, 1998.

104. Yu, W. and Bien, Z., Design of fuzzy logic controller with inconsistent rule base. *J. Intelligent and Fuzzy Systems*, 2, 147–159, 1994.
105. Yuan, Y. and Zhuang, H., A genetic algorithm for generating fuzzy classification rules. *Fuzzy Sets and Systems*, 84, 1–19, 1996.
106. Zadeh, L.A., A fuzzy-set theoretic interpretation of linguistic hedges. *J. Cybernetics*, 2(2), 4–34, 1972.
107. Zadeh, L.A., Outline of a new approach to the analysis of complex systems and desision processes. *IEEE Trans. Systems, Man, and Cybernetics*, 3, 28–44, 1973.
108. Zadeh, L.A., The concept of a linguistic variable and its application to approximate reasoning. Parts I, II and III, *Information Science*, 8, 8, 9, 199–249, 301–357, 43–80, 1975.
109. Zadeh, L.A., Soft computing and fuzzy logic. *IEEE Software*, 11(6), 48–56, 1994.

8

GA-FRB: A Novel GA-Optimized Fuzzy Rule System

W.J. Chen
Nanyang Technological University

Chai Quek
Nanyang Technological University

Abstract. Fuzzy logic was introduced to describe systems that are too complex or ill defined to allow the use of precise mathematical analysis. Fuzzy sets and fuzzy rules are two basic concepts of fuzzy logic. A fuzzy system with poorly designed fuzzy sets and rules will seldom perform well. Manual-construction of membership functions of fuzzy sets and fuzzy rules are very difficult. Although several algorithms have been proposed to automatically tune the membership functions or define the fuzzy rules, most of them are application dependent and require the manual selection and tuning of a set of preset parameters.

Genetic algorithms, on the other hand, are global search and optimization techniques exploring search space by incorporating a set of candidate solutions in parallel. Genetic algorithms, however, are

0-8493-1121-7/03/$0.00+$1.50
© 2003 by CRC Press LLC

normally used as an ad hoc method and cannot extract the distribution information about the input data features. The results produced by genetic algorithms are normally irreproducible.

However, the integration of these two techniques into a hybrid fuzzy genetic system, where genetic algorithms are used for auto-construction of membership functions in a fuzzy logic system, enables the system to extract the information about the input data. In this chapter, an algorithm based on genetic algorithms is proposed to automatically construct the membership functions and fuzzy rules of a fuzzy system. The resultant hybrid fuzzy-GA system is able to auto-construct the relevant fuzzy sets from the input data. It is application independent and uses very few preset parameters. The constructed membership functions are stored and retrievable, which makes the results of the system reproducible. The system was implemented and examined on Iris data sets, phoneme data, and traffic flow data. The results showed that these two techniques worked well together and the performance results were satisfactory.

Key words: Genetic algorithm, optimized fuzzy rule system, encoding, fitness function, traffic analysis and prediction, phoneme data classification, patterns, fuzzy membership construction.

8.1   Motivation

In the fifties, all real-world problems were solved using analytical methods. In most practical cases, however, the *a priori* data and the performance judgment criteria of these systems are far from being precisely specified or having accurately known probability distributions. These systems are too complex or ill defined to be handled by the traditional precise analytical methods. This led to the birth of fuzzy logic. Fuzzy logic aims to alleviate difficulties in developing and analyzing complex systems encountered by conventional mathematical technologies. This motivation requires fuzzy logic to work in quantitative and numeric domains. Fuzzy logic is also motivated by observing that human reasoning can utilize concepts and knowledge that do not have well-defined boundaries. This motivation enables fuzzy logic to have a descriptive and qualitative form. Hence, fuzzy logic is unique and different from other tools that only focus on either quantitative or qualitative domain. Furthermore, it becomes a natural bridge between the two domains. This unique feature allows fuzzy logic to provide not only a cost-effective way to model complex systems but also a comprehensive way to describe the systems.

Fuzzy sets, which represent the clusters of the input data, and fuzzy rules, which define the dynamics of the system, are the two basic concepts of fuzzy logic. A fuzzy system with poorly designed fuzzy sets and rules will seldom perform well. Fuzzy sets are defined by membership functions. Auto-construction of membership functions and fuzzy rules are very difficult. Although several algorithms have been proposed to automatically tune the membership functions or define the fuzzy rules, most of them are application dependent and rely heavily on the manual selection and tuning of a set of preset parameters.

Developed to mimic some of the processes observed in natural evolution, Genetic algorithms (GAs) are global search and optimization techniques exploring search space by incorporating a set of candidate solutions in parallel. They perform on the coding of the data and not on the exact data, which allows them to be independent of the continuity of the data and the existence of derivatives of the functions as needed in some conventional optimization algorithms. The coding method of GAs allows them to easily handle optimization problems with multi-parameters or multi-models, which is rather difficult or impossible to be treated by classical optimization methods.

Genetic algorithms, however, are normally used as an ad-hoc method and can only be used to produce a crisp partitioning of the input data space. They are unable to extract the information on the distribution of the input data.

When designing a fuzzy system, the identification of the parameters is the most important item. This can be viewed as an optimization problem—finding parameter values that optimize the model based on given evaluation criteria. This motivated the integration of fuzzy systems and genetic algorithms into a hybrid fuzzy-GA system, where genetic algorithms can be used for auto-construction of membership functions and fuzzy rules, while fuzzy logic enables the system to extract the information about the input data.

This chapter is organized as follows. Section 8.2 briefly describes current approaches in genetic algorithms and their applications. Clustering of data is covered in Section 8.3. The GA-based fuzzy system is described in Section 8.4. Extensive experimental results covering traffic data modeling, Iris data, and phoneme classifications are also presented for analysis. Section 8.5 summarizes the research results and provides directions for future work.

8.2 What Are Genetic Algorithms (GAs)?

The evolution of living beings is a process that operates on chromosomes—organic devices for encoding the structure of living beings. Natural selection is the link between chromosomes and the performance of their decoded structures. Chromosomes that encode successful structures will reproduce more often than those that do not, under the processes of natural selection. In addition to reproduction, mutations may cause the chromosomes of children to be different from those of their biological parents, and recombination processes may create quite different chromosomes in children by combining material from the chromosomes of their two parents. These features of natural evolution inspired the development of genetic algorithms.

Genetic algorithms (GAs) are global search and optimization techniques modeled from natural genetics, exploring search space by incorporating a set of candidate solutions in parallel. A genetic algorithm maintains a population of candidate solutions. Each candidate solution is usually coded as a string (typically, a binary string) called a chromosome, which is also referred to as a genotype. A chromosome encodes a parameter set for a set of variables being optimized. Each encoded parameter in a chromosome is called a gene. A decoded parameter set is called a phenotype. A set of chromosomes forms a population, which is evaluated and ranked by a fitness evaluation function. The fitness evaluation function is similar to the natural selection in the real world. It plays a critical role in GA because it provides information on how good each candidate solution is. This information guides the search of a genetic algorithm. More accurately, the fitness evaluation results determine the likelihood that a candidate solution is selected to produce candidate solutions in the next generation. A candidate solution with a better evaluation result will tend to reproduce more often than those with a worse result. The initial population is usually generated randomly.

8.2.1 The History of Genetic Algorithms

In 1962, the underlying principles of genetic algorithms were first published by Holland.[1] He continued to develop the mathematical framework in the 1960s. Since the birth of genetic algorithms, they have been employed primarily in two major areas: optimization and machine learning. In optimization applications, they have been used in many diverse fields, i.e., function optimization, image processing, system identification, and control. In machine learning, they have been used to learn syntactically simple string IF-THEN rules in an arbitrary environment. A lot of work has been carried out on machine learning using genetic algorithms since the late 1980s.[2,3]

8.2.2 Basic Concepts of Genetic Algorithms

The algorithm of a GA can be described as follows:

Step 1: Initialize a population of chromosomes.
Step 2: Evaluate each chromosome in the population.

Step 3: Create new chromosomes using different types of GA operators.

Step 4: Delete members of the population to make room for new ones.

Step 5: Insert the new chromosomes into the population.

Step 6: If the stopping criterion is satisfied, then stop and return the best chromosome; otherwise, go to step 3.

The total number of the candidates in the population will be kept unchanged throughout generations.

8.2.2.1 Encoding

As mentioned earlier, genetic algorithms will work on chromosomes coding from the parameter set of the variables being optimized, not directly on the real data. This feature makes the encoding process very important.

There are two types of encoding:

- Binary-Coded GA: In this type of encoding, a chromosome is represented by a binary string. The real parameter value will be mapped into an integer in the interval $[0, 2^n - 1]$, where n is the length of the binary string. Currently, this is the popular coding scheme due to its simplicity and good performance.
- Real-Coded GA: In this type of encoding, a chromosome is represented by a vector of floating-point numbers. It has been proven that in certain problems,[4] real-coded GAs have resulted in better performance than binary-coded GAs with some modifications of genetic operators.

8.2.2.2 Fitness Evaluation

A good design of fitness evaluation function is probably the most important factor in a successful application of GA, since this function is the only thing used to guide the search direction.

There are two major types of fitness evaluation functions:

- Fitness-based function: A fitness-based function calculates how well a model using the parameter set in a chromosome fits a set of input-output training data. In other words, the function computes the error between the target's output and the model's output.
- Performance-based function: A performance-based function evaluates how well a system using the parameter set in a chromosome achieves a set of performance objectives.

These two types of GA fitness evaluation functions are not mutually exclusive. A combination of them can be found in many applications.

8.2.2.3 Genetic Operators

There are three genetic operators used to create new generations:

- Selection: This is the process in which individual strings are selected for reproduction according to their fitness value. This operator is an artificial version of natural selection. Candidates with higher fitness value will be selected more often for future evolution than those without. The selected candidates will be copied directly into the next generation or be mated for crossover.
- Crossover: Selection directs the search toward the best existing individuals but does not create any new individuals. In nature, an offspring has two parents and inherits genes from both. It is the same in GAs. The main operator working on the parents is crossed over, which happens for a selected pair with a crossover probability p_c. Two candidates from the selected pool will be mated as parents to create new candidates. A crossover site will be randomly selected. Then the two parents are crossed and separated at the site. This can be shown in Figure II.8.1. This process produces two new strings, each of which takes after both parents. Selection and crossover provide GAs with considerable power by directing the search toward better areas using existing knowledge.

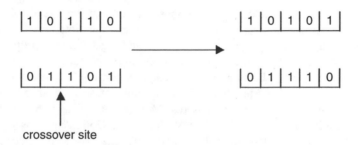

FIGURE II.8.1 An example of crossover in standard genetic algorithms. Two mating individuals (left) and their offspring (right).

- Mutation: Although selection and crossover produce many new strings, they do not introduce any new information into the population at the bit level. Mutation is introduced as the source of new bits. It inverts a randomly chosen bit on a string. Mutation should be used sparingly because it is a random search operator. Otherwise, with high mutation rates, the algorithm will become no more than a random search. That is the reason why mutation is often applied with a low probability p_m.

These three operators are applied repeatedly until the new candidates take over the entire population. The next generation is thus made up of candidates of three types:

- Mutated after crossover
- Crossover but not mutated
- Just selected, neither crossover nor mutated

8.2.3   Remarks on GAs

Although GAs are powerful tools for global search and optimization, if some aspects of GAs have not been taken good care of, they will easily become drawbacks of GAs:

1. Normally the initial population of a genetic algorithm is randomly generated. It is good that no manual predefinition is required, and it saves a lot of effort in the design of the system. However, the results that are achieved may not be reproducible, as normally different initialization will lead to different results. Therefore, GAs are more suitable for applications that are more interested in the statistical results than the results of every single execution.
2. GAs are tended to be used as an ad-hoc method and can only be used to produce a crisp partitioning of the input data space. They cannot extract the distribution information about the input data features. To resolve this problem, some other techniques, like fuzzy logic, have to be integrated into the same system.
3. Since the execution of a genetic algorithm is only guided by the fitness functions, if there are any preset parameters, the effect of these parameters will be unclear, and a lot of work has to be done to manually tune them. The solution to this problem is to use as few preset parameters as possible.

8.2.4   Current Research on Genetic Algorithms

Currently, researchers are applying genetic algorithms into many different areas, as well as introducing many algorithms or modifications on encoding, genetic operators, evaluation, etc.

Many new genetic algorithms or modifications have been developed. Krishna and Murty[5] proposed a novel hybrid genetic algorithm that finds a globally optimal partition of a given data into a specified number of clusters. Kemp, Porter, and Dawson[6] proposed a more efficient use of population members to allow a number of sub-populations to evolve separately and then interbreed. This is used when an optimization landscape is highly multi-modal. Roger and Prugel-Bennett[7] proposed a method for calculating genetic drift in terms of changing population fitness variance.

Control is one of the main application areas of genetic algorithms. Reformat et al.[8] proposed a new method of designing control system that relies on a combination of advanced system simulators and genetic computation. Shin and Park[9] proposed a GA-based optimization technique to handle the control problem of nonlinear processes. Vlachos, Williams, and Gomm[10] proposed a method for the tuning of decentralized PI controllers for multivariable processes, based on genetic algorithms.

Pattern recognition is another main application of genetic algorithms. Abutaleb and Kamel[11] proposed a genetic algorithm to find the ridges in fingerprints. Kabuka, Gudmundsson, and Essam[12] proposed an algorithm to detect well-localized, unfragmented, thin edges in medical images based on optimization of edge configuration using a genetic algorithm. Sankar, Bandyopadhyay, and Murthy[13] proposed a GA-based algorithm for pattern classification in R^n by fitting hyperplanes to model the decision boundaries in the feature space.

There are many other applications of genetic algorithms. Haydar, Dumirekler, and Yurtseven[14] proposed a feature selection using genetic algorithm, applying to speaker verification. Sun and Hu[15] proposed a new genetic algorithm for training hidden Markov models, which was applied to speech recognition. Haydar, Demirekler, and Yurtseven[16] further introduced the use of a genetic algorithm in the reduction of the parameter set for each speaker in text-independent speaker identification. Chang and Chen[17] presented novel solution algorithms and results based on a genetic algorithm to solve the hydrothermal generation scheduling problem. Gallego, Monticello, and Romero[18] presented and extended genetic algorithm for solving the optimal transmission network expansion planning problem. Shieh, Wang, and Tsai[19] presented a time-domain design methodology for optimal digital design of multivariable sampled-data parametric uncertain systems using genetic algorithms.

Although not very typical yet, there are also research efforts being done on integrating genetic algorithms with fuzzy systems. Some of them are on pattern recognition. Fung et al.[20] proposed a fuzzy genetic algorithm to solve feature selection problem. Rhee and Lee[21] presented an unsupervised feature selection method using a fuzzy-genetic approach. Kyung, Eung, and Sung[22] proposed an adaptive mesh grouping method based on a fuzzy-GA algorithm in Electrical Impedance Tomography technology for fast static image reconstruction.

Some of them are on control. White and Trebi-Ollennu[23] proposed multi-objective fuzzy GA optimization to provide an intuitive process for selecting free control parameters for nonlinear controllers with competing and fuzzy specifications. Kim, Kim, and Kim[24] used a genetic algorithm to determine membership functions under situations whereby fuzzy traffic control rules were provided by the human operators. Man[25] proposed a genetic algorithm to be applied for the simultaneous design of membership functions and rule sets for a fuzzy control method of water level in a steam generator. Qi and Chin[26] presented a fuzzy control algorithm for high-order processes. GAs are utilized for the fine tuning of the controller.

Some are on other issues. Streifel et al.[27] proposed an algorithm for adaptively controlling genetic algorithm parameter coding using fuzzy rules. Wang, Hong, and Tseng[28] proposed a GA-based fuzzy-knowledge integration framework that can simultaneously integrate multiple fuzzy rule sets and their membership function sets.

In the above applications, most of the algorithms have not taken care of some aspects of GAs, as mentioned in Section 8.2.2.4, especially (b) and (c). In these applications, very often GAs are still used as ad hoc methods and are unable to extract information of the input data. At the same time, these algorithms used quite a number of preset parameters, which required a lot of effort to manually tune them.

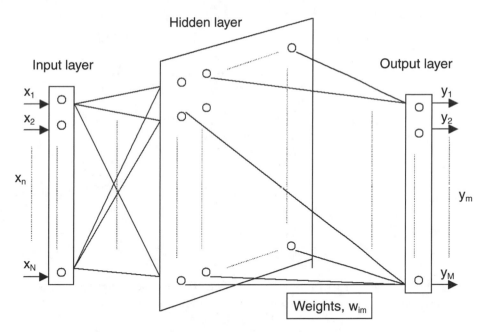

FIGURE II.8.2   The architecture of the NFLC based on the RBF neural network.

In Teo, Khalid, and Yusof,[29] the authors presented a neuro-fuzzy logic controller (NFLC), where all of its parameters can be tuned simultaneously by GA techniques. Since the algorithm they proposed is quite interesting and related closely to the research areas of this project, more details are given in the following.

The structure of the controller was based on the radial basis function neural network (RBF). This RBF neural network is usually used to approximate a continuous linear or nonlinear function mapping. The structure of the multi-input and multi-output NFLC can be shown as in Figure II.8.2. The input layer accepts the system state feedback $(x_1, x_2, \ldots, x_n, \ldots, x_N)$ (input vector) and the fuzzy inferencing is processed at the hidden layer. The strength of the control action for each of the fuzzy rules is given by the interconnected weights between the hidden and the output layers. The RBF structure can be used to implement the fuzzy control rules that are given as:

IF (X_1^i) and $\ldots (X_n^i) \ldots$ and (X_N^i) THEN (w_{i1}) and $\ldots (w_{im}) \ldots$ and (w_{iM})

where w_{im} is the singleton defined control action for the ith control rule of the mth output variable.

In their paper, Teo, Khalid, and Yusof[29] used this structure with two input variables, namely, the error (e) and the change of the error (Δe). Each of the variable takes five Gaussian type fuzzy membership functions that are labeled as positive big (PB), positive small (PS), zero (Z), negative small (NS), and negative big (NB). Each of the membership functions has two parameters, i.e., the center and width of the Gaussian functions. The multivariate Gaussian can also be viewed as the product of a single-variate Gaussian function. It performs a conjunctive operation on the "premise" part of the fuzzy rules in the hidden layer. The rule base matrix of the corresponding fuzzy basis units at the hidden layer of the controller is shown in Table II.8.1. Each of the kernel squares represents one control rule condition. Thus, the number of the hidden nodes for this network is exactly equal to the number of fuzzy control rules. The output from these units is the matching degree or inferred result (h_i) of the particular fuzzy control rules.

Rule 1 indicated in the rule map described by Table II.8.1 is "IF e is NB and Δe is PB" and the other rules are constructed in the same way. The input variable types rules are all Gaussian shape function.

TABLE II.8.1 The Fuzzy Basis Function at the Hidden Layer

		NB	NS	Z	PS	PB
				e		
	PB	Rule 1	Rule 2	Rule 3	Rule 4	Rule 5
	PS	Rule 6	Rule 7	Rule 8	Rule 9	Rule 10
Δe	Z	Rule 11	Rule 12	Rule 13	Rule 14	Rule 15
	NS	Rule 16	Rule 17	Rule 18	Rule 19	Rule 20
	NB	Rule 21	Rule 22	Rule 23	Rule 24	Rule 25

The matching degree of the input to the conditions of each of the fuzzy control rules were designed as:

$$h_i = \exp\left(-\left\|\frac{C_{x,n}^i - x_n}{D_{x,n}^i}\right\|^2\right) \quad \text{for} \quad i = 1 \text{ to } T \tag{II.8.1}$$

where T = the total number of fuzzy rules; $C_{x,n}^i$ and $D_{x,n}^i$ = the center and the width of nth input variable's membership assigned to the ith control rules, respectively; $\|.\|$ = the norm operator presented as either Euclidean, Hamming, Maximum, etc.

The matching degree process is simply an operation that returns the matching level $h_i \in [0, 1]$ between the inputs and the rule pattern for the ith rule. The higher the h_i, the better matching between the input pattern and the particular rule pattern.

The control output variables are then obtained by averaging the weights. Thus, the mth controller output (y_m) can be obtained by:

$$y_m = \sum_{i=1}^{p}(h_i^* w_{im}) \bigg/ \sum_{i=1}^{T}(h_i) \tag{II.8.2}$$

The controller output y_m is a crisp value that can be readily applied to the system.

A GA is then applied as an optimization algorithm to tune all the parameters of this NFLC.

Each Gaussian membership function has two variables, center and width. Therefore, there are in total 20 parameters for the 10 membership functions of the two input variables. Another 25 parameters are needed for the weights. Thus, a total of 45 parameters ($5_{\text{membership\_functions}} {}^*2_{\text{parameters}} {}^*2_{\text{variables}} + 25_{\text{weights}}$) are to be tuned by the GA.

The linear mapping method was used to encode, which can be expressed as:

$$g_p = G_{q\,\min} + (G_{q\,\max} - G_{q\,\min})^* A_q / (2^N - 1) \tag{II.8.3}$$

where g_p = the actual value of the qth parameter; A_q = the integer represented by a N-bit string gene; $G_{q\,\max}$ = user-defined upper limits of the gene; and $G_{q\,\min}$ = user-defined lower limits of the gene.

The encoded genes were concatenated to form a complete chromosome. Each of the parameters was encoded into 8-bit strings, resulting in a complete chromosome of 360 bits. Genes 1 to 10 were allocated to the sub-chromosome of the first controller input, where Genes 1, 3, 5, 7, 9 were for the membership functions centers and Genes 2, 4, 6, 8, 10 were for the width. Genes 11 to 20 were assigned in a similar way to the second controller input. The remaining 25 genes were allocated for the weights. This can be shown as:

Gene | 1 | 2 | ... | 9 | 10 | 11 | ... | 19 | 20 | 21 | ... | 45 |

Chromosome | sub-chr of X_1 | sub-chr of X_2 | sub-chr of weights |

Parameter | ... $(C_{x_1}^i, D_{x_1}^i$ ... | ... $(C_{x_2}^l, D_{x_2}^l)$ ... | w_{11} | w_{12} | ... | w_{il} | ... | w_{45} |

Where sub-chr stands for sub-chromosome, X_1 is the input vector 1, and X_2 is the input vector 2.

The initial population, consisting of 200 chromosomes, was randomly generated. This NFLC tuned by GA was tested on three different plants: a nonminimum and open loop plant, a nonlinear plant, and a car-parking mechanism. Each of the experiments used different GA objective function or performance indices. The performance index (F) is related to the fitness (f) using the following relationship:

$$f = A/(1 + F)^g \tag{II.8.4}$$

Where g = the constant that affects the performance curve, and A = a nonnegative constant and appropriately chosen so that f will not be too small.

According to the authors, the experiments' results are very encouraging. For more detail, please refer to Teo et al.[29]

This chapter proposes a novel way to automatically construct membership functions. It is applications independent since it used the fitness evaluation function to guide the evolution, not the actual objective function of the application. This makes it very easy for the system to be applied to different applications with only minor changes. The NFLC system, however, employs a lot of preset parameters. This not only requires a lot of effort on tuning the parameters, but also makes the system less application independent since different applications may require different values of the parameters.

8.2.5 Summary

This section summarizes the fundamentals of genetic algorithms. Genetic algorithms are global search and optimization techniques, exploring search space by incorporating a set of candidate solution in parallel. They start from an initial population, which is normally randomly initialized. The evolution process is guided by a fitness evaluation function with three genetic operators; namely: selection, crossover, and mutation.

Current research on genetic algorithm is studied and presented in this section. A growing interest in genetic algorithm is to integrate it with fuzzy logic, where genetic algorithms are used for auto-construction of membership functions and fuzzy rules in a fuzzy logic system. This enables the system to extract the information about the input data. Most existing algorithms, however, used too many preset parameters. This not only wastes a lot of effort to tune these parameters manually, but also makes the system less applications independent.

8.3 Data Clustering

Clustering essentially deals with the task of partitioning a data space or a data set into a number of more-or-less homogeneous classes (clusters) with respect to a suitable similarity measure, such that the patterns belonging to any one of the clusters are similar and the patterns of different clusters are as dissimilar as possible. The similarity measure used has an important effect on the clustering results, since it indicates which mathematical properties of the data set (e.g., distance, connectivity, and intensity) should be used and in what way they should be used to identify the clusters.

In crisp or hard clustering, one pattern is assigned to exactly one cluster. On the contrary, fuzzy clustering provides partitioning results with additional information supplied by the cluster membership values indicating different degrees of belongingness.

8.4 Fuzzy Clustering

Let a finite set of elements $X = \{x_1, x_2, \ldots, x_n\}$ be elements of the p-dimensional Euclidean space R^p, that is, $x_j \in R^p$, $j = 1, 2, \ldots, n$. The fuzzy clustering problem is to perform a partition of this collection of elements into c fuzzy sets with respect to a given criterion, where c is a given number of clusters. The criterion is usually to optimize an objective function that acts as a performance index of clustering. The end result of fuzzy clustering can be expressed by a partition matrix U such that:

$$U = [u_{ij}]_{i=1...c, j=1...n} \tag{II.8.5}$$

where u_{ij} is a numerical value in [0, 1] and expresses the degree to which the element x_j belongs to the ith cluster. However, there are two additional constraints on the value of u_{ij}. First, a total membership of the element $x_j \in X$ in all classes is equal to 1, i.e.:

$$\sum_{i=1}^{c} u_{ij} = 1 \quad \text{for all} \quad j = 1, 2, \ldots, n \tag{II.8.6}$$

Second, every constructed cluster is nonempty and different from the entire set, i.e.:

$$0 < \sum_{j=1}^{n} u_{ij} < n \quad \text{for all} \quad i = 1, 2, \ldots, c \tag{II.8.7}$$

A general form of the objective function is:

$$J(u_{ij}, v_k) = \sum_{i=1}^{c} \sum_{j=1}^{n} \sum_{k=1}^{c} g[w(x_i), u_{ij}] d(x_j, v_k) \tag{II.8.8}$$

where $w(x_i)$ is the *a priori* weight for each x_i and $d(x_j, v_k)$ is the degree of dissimilarity between the data x_j and the supplemental element v_k, which can be considered the central vector of the kth cluster. The degree of dissimilarity is defined as a measure that satisfies two axioms:

$$d(x_j, v_k) \geq 0,$$
$$d(x_j, v_k) = d(v_k, x_j), \tag{II.8.9}$$

and thus it is a concept weaker than distance measures.

With the above background, fuzzy clustering can be precisely formulated as an optimization problem:

$$\text{Minimize } J(u_{ij}, v_k), \quad i, k = 1, 2, \ldots, c; \ j = 1, 2, \ldots n \tag{II.8.10}$$

subject to Equations II.8.10 and II.8.15.

8.4.1   Current Research on Fuzzy Clustering

One of the widely used clustering methods based on Equation II.8.10 is the fuzzy c-means (FCM) algorithm developed by Bezdek.[30] The objective function of the FCM algorithm takes the form of

$$J(u_{ij}, v_k) = \sum_{i=1}^{c} \sum_{j=1}^{n} u_{ij}^{m} \|x_j - v_i\|^2, \quad m > 1 \tag{II.8.11}$$

where m is called the exponential weight which influences the degree of fuzziness of the membership matrix. The FCM algorithm provides an iterative approach to approximate the minimum of this objective function starting from a given position. It is summarized in Appendix A.

In this algorithm, there is no guideline to set the value of the parameter m, which is used to control the fuzziness of the membership functions. The degree of fuzziness increases as the value of m is increased. The iterative procedure may lead to any of the objective function's local minimal and does not guarantee global minimum.

Linear vector quantization (LVQ)[31] is another widely used clustering technique. It partitions a data set into c classes and represents each class using a prototype vector v_k, where c is a predetermined number. It is summarized in Appendix A.

The LVQ algorithm will converge if the sequence $\{\alpha_t\}$ satisfies the following two conditions:

$$\sum_{t=1}^{\infty} \partial_t = \infty \quad \text{and} \quad \sum_{t=1}^{\infty} \partial_t^2 < \infty \tag{II.8.12}$$

There are many preset parameters used in this algorithm, including the termination criterion ε, the number of clusters c, the initial V_0, and the choice of different strategies for α_t. The results of this algorithm vary a lot with changes of these parameters, which are normally manually set and require a lot of training.

Instead of creating visible clusters, the LVQ algorithm only produces c partitions of the input data space with a prototype vector representing each of the partitions. Therefore, strictly speaking, LVQ is not a clustering technique. Instead, it inspires the development of many clustering techniques, such as the modified linear vector quantization (MLVQ), fuzzy Kohonen partitioning (FKP), pseudo fuzzy Kohonen partitioning (PFKP),[32] and generalized linear vector quantization (GLVQ).[33]

8.5 Summary

This section summarizes the fundamentals of clustering and fuzzy clustering. Clustering essentially deals with the task of partitioning a data space or a data set into a number of classes with respect to a suitable similarity measure. In crisp clustering, one pattern is assigned to exactly one class, while fuzzy clustering provides partitioning results with additional information supplied by the cluster membership values indicating different degrees of belonging.

Current research on fuzzy clustering has been studied and two commonly used algorithms, namely, fuzzy c-means algorithm and linear vector quantization algorithm, are presented in this section. Most existing fuzzy clustering techniques employed quite a number of predefined parameters, which require a lot of manual training and make the system application dependent.

8.6 Auto MF Construction

Currently, the membership functions of most fuzzy systems are designed by interviewing those who are familiar with the underlying concept. They are subsequently tuned by trial-and-error. This is the major drawback of such fuzzy systems and makes them less cost-effective. Techniques to construct the membership functions automatically are in demand.

There are two main approaches to construct the membership functions automatically. One is to construct them automatically from the data, e.g., Holland,[1] and the other is to learn them based on the feedback from the system performance, e.g., Hong and Lee.[34]

Some of the existing membership function auto-construction techniques are summarized in Table II.8.2. In Table II.8.2, BT means Base Technique, AD means Application Dependent, UPP means Use of Preset Parameters, and No* means not fully application independent. From Table II.8.2, it is clear that most current membership function auto-construction techniques are application dependent and used preset parameters.

TABLE II.8.2   Some Existing Membership Function Auto-Construction Techniques

Paper	BT	Description	AD	UPP
Hong[34]		A general learning method as a framework to automatically derive membership functions and fuzzy rules.	Yes	Yes
Woodard[35]		A method to tune fuzzy controllers using numerical optimization.	Yes	Yes
Lee[36]	Neural Network	A feedforward, 5-layered fuzzy neural network model that can fine tune the parameters of a rough fuzzy model constructed using expert knowledge.	Yes	Yes
Furukawa[37]	Neural Network	Two design algorithm of membership functions for a fuzzy neuron using example-based learning with optimization of allocation of cross-detecting lines, namely, elimination method and selection method.	Yes	Yes
Sankar[38]	Neural Network	A way of formulating neuro-fuzzy approaches for both feature selection and extraction under unsupervised learning.	Yes	Yes
Teo[29]	Genetic Algorithm	A neuro-fuzzy logic controller where all of its parameters can be tuned simultaneously by GA.	No*	Yes
Kyung[22]	Genetic Algorithm	An adaptive mesh grouping method based on a fuzzy-GA algorithm for fast static image reconstruction.	Yes	Yes

In this section, an algorithm (GF) is proposed to automatically construct the membership functions based on the data input and system performance using genetic algorithms, which is application independent and uses no preset parameters.

8.7   Data

There are three data sets used to assess this technique. They are Fisher's iris data set,[39] phoneme data,[40] and traffic flow data.[41]

Fisher's Iris data set is used to demonstrate the structure of the system. This may be the most popular data set used in pattern recognition, machine learning, statistics, and data analysis. The data set contains 150 instances of iris flower belonging to three classes. The three classes are Setosa, Versicolor, and Virginica. Examples of them are shown in Figure II.8.3. Each instance of the iris flower has four physical attribute measurements. They are sepal length, sepal width, petal length, and petal width. These attributes of the Setosa class are linearly separable from the other two classes, while those of the other two classes (Versicolor and Virginica) are very similar to each other.

For easier reference, in Sections 8.8 to 8.11, the Setosa class is referred to as Class 1, the Versicolor class is referred to as Class 2, and the Virginica class is referred to as Class 3.

8.8   Theoretical Framework of the Fuzzy System

The strength of a good fuzzy rule-based system is built on the quality of the fuzzy sets and the fuzzy rules. Good fuzzy set definitions of a set of data should be consistent with the raw data that is associated with the information. This should complement a set of well-defined fuzzy rules. This section describes the set and rule definitions as well as a novel GA-based fuzzy membership construction technique.

FIGURE II.8.3   The three irises in Fisher's Iris data set.

8.8.1   Fuzzy Sets

Four linguistic variables are used in the systems, corresponding to the four attributes. They are sepal length (SL), sepal width (SW), petal length (PL), and petal width (PW). Each variable has three labels; therefore, there are in total 12 fuzzy sets used in the system ($3_{label}*4_{attribute}$). They are, namely,

- Setosa sepal length (SSL)
- Setosa sepal width (SSW)
- Setosa petal length (SPL)
- Setosa petal width (SPW)
- Versicolor sepal length (VSL)
- Versicolor sepal width (VSW)
- Versicolor petal length (VPL)
- Versicolor petal width (VPW)
- Virginica sepal length (ISL)
- Virginica sepal width (ISW)
- Virginical petal length (IPL)
- Virginical petal width (IPW)

The reason why the labels are defined to be "Setosa," "Versicolor," and "Virginica" instead of conventionally "Short," "Medium," and "Long" is to reduce the number of fuzzy rules used in the system (see Section 8.8.2).

The trapezoid membership function was chosen to be the membership function of these 12 fuzzy sets. They are constructed using the proposed genetic algorithms.

8.8.2   The Fuzzy Rules

Based on the fuzzy sets defined, there are three fuzzy rules used in the system:

- RULE 1: IF sepal length is SSL AND sepal width is SSW AND petal length is SPL AND petal width is SPL, THEN the flower belongs to Class 1
- RULE 2: IF sepal length is VSL AND sepal width is VSW AND petal length is VPL AND petal width is VPL, THEN the flower belongs to Class 2
- RULE 3: IF sepal length is ISL AND sepal width is ISW AND petal length is IPL AND petal width is IPL, THEN the flower belongs to Class 3

These rules are constructed *a priori*.

If the labels were instead defined as short, medium, and long, and with no prior knowledge of the input data as well as which class feature should be short, medium, or long, one would have to construct a rule space of $3^5 = 243$ rules. Using the proposed definition, the resultant membership functions can also be used to represent the distribution of the data. This can be considered as feature extraction. Some other clustering techniques, e.g., Sankar et al.[13] are able to perform clustering very well, but are unable to extract the features of the input data.

The multiplication operator is chosen as the conjunction operator. The matching degrees of the input data (the four attributes of a flower) to these three rules are defined as:

$$D_i = \prod_{j=1}^{4} D_{ij} \tag{II.8.13}$$

where $i = 1, 2, 3$, and D_{ij} is the matching degree of the jth condition in the ith rule. They will be compared, and if the matching degree of RULE i is the highest, the specimen is classified as belonging to Class i, where $i = 1, 2, 3$.

8.9 The Automatic Membership Functions Construction Technique

A proposed algorithm based on GA technique is used to construct the membership functions of the fuzzy sets. The details are described below.

8.9.1 Coding

Each trapezoid membership function has four parameters (l, lp, rp, and r). Therefore, there is a total of 48 parameters ($12_{\text{membership function}}*4_{\text{parameter}}$) needed to be tuned by the genetic algorithm.

The linear mapping method (Equation II.8.3) was used to encode the parameters. The parameters in Equation II.8.7 were chosen as:

- $N = 7$ (i.e., each membership parameter was encoded as a 7-bit string)
- $G_{q\,\text{min}} = 0$
- $G_{q\,\text{max}} = 12.7$ (the maximum value of the attributes is less than 10. Choosing 12.7 as the maximum value is simply because $12.7 = (2^7 - 1)/10$, which is easy for computation)

Therefore, Equation II.8.3 was simplified into:

$$g_q = 0 + (12.7 - 0)^* A_q/2^7 - 1 = A_q/10 \tag{II.8.14}$$

where g_q is the actual value of the qth parameter and A_q is the integer represented by the 7-bit string gene.

Since each membership parameter is encoded as a 7-bit string, the resultant complete chromosome is of 336 bits ($48_{\text{parameter}}*7_{\text{bit}}$). Gene i to Gene $i+3$ are allocated to the parameters of the membership function of the sepal length of Class i, Gene $i + 4$ to Gene $i + 7$ are allocated to the parameters of the membership function of the sepal width of Class i, Gene $i + 8$ to Gene $i + 11$ are allocated to the parameters of the membership function of the petal length of Class i, and Gene $i + 12$ to Gene $i + 15$ are allocated to the parameters of the membership function of the petal width of Class i, where $i = 1, 17, 33$. This arrangement of the coded parameters is shown as follows:

Gene | 1 | 2 | ... | 16 | 17 | 18 | ... | 32 | 33 | 34 | ... | 48 |

Chromosome | sub-chr of Class 1 | sub-chr of Class 2 | sub-chr of Class3 |

Parameter | l_1^i | lp_1^i | rp_1^i | r_1^i | l_2^i | lp_2^i | rp_2^i | r_2^i | l_3^i | lp_3^i | rp_3^i | r_3^i | l_4^i | lp_4^i | rp_4^i | r_4^i |

Where sub-chr stands for sub-chromosome and $i = 1, 2, 3$.

$$l \ \mathrel{<=} \ lp \ \mathrel{<=} \ rp \ \mathrel{<=} \ r \tag{II.8.15}$$

When $l = lp$, the left boundary of the membership function will become a sharp boundary. When $lp = rp$, the trapezoid membership function will become a triangular membership function. When $rp = r$, the right boundary of the membership function will become a sharp boundary.

In order to satisfy Equation II.8.15, after the birth of each generation of the population, the four genes for the same membership function in every chromosome will be sorted in ascending order.

8.9.2 Initialization

There are a total of 100 chromosomes used in this proposed algorithm to tune the parameters. They are all randomly initialized.

8.9.3 Fitness Function

The fitness function is probably the most important part in a genetic algorithm, because it is the only way to provide system information to guide the search.

In our experiments, a performance-based fitness function was used. The system is to classify the input data into different classes. Therefore, the fitness of the tuned parameter set is dependent on how well they can be used to classify the input data.

The fitness of a chromosome can be determined by the following steps:

Step 1: The chromosome is decoded into the actual parameters of the 12 membership functions.

Step 2: Each instance of the input data is classified by the systems using the 12 membership functions constructed from this chromosome. For every correct instant of classification, the fitness of the chromosome will increase by one, i.e.:

$$\text{Fitness} = \text{Number of successful classified instances.} \tag{II.8.16}$$

Since the system is to classify the input data, the more instances that are classified correctly, the better the membership functions. Equation II.8.16 directly relates the fitness of the chromosome to the number of successfully classified instances, which will provide simple and essential system information.

8.9.4 Selection

After the fitness evaluation, chromosomes will be selected for reproduction of the next generation. In the proposed algorithm, two types of selection were used. The details are shown in the following.

8.9.4.1 Selection Based on Fitness Value

A technique called *roulette-wheel parent selection*[42] is used to select chromosomes for reproduction. This is conducted by spinning a simulated biased roulette wheel whose slots have different sizes proportional to the fitness values of the individuals. Each spinning will yield a candidate for future reproduction. This technique can be described as:

- Sum the fitnesses of all the population members and call this result the total fitness
- Generate a random number, n, between 0 and the total fitness
- Return the first population member whose fitness, added to the fitness of the preceding population members (running total), is greater than or equal to n

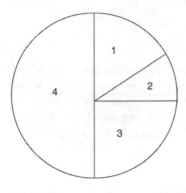

FIGURE II.8.4   An example of a weighted roulette wheel, where 1, 2, 3, and 4 refer to Chromosome 1, 2, 3, and 4, respectively.

TABLE II.8.3   The Selection Result Based on Figure II.8.4

Random Number	$0 <= n <= 10$	$10 < n <= 15$	$15 < n <= 30$	$30 < n <= 60$
Chromosome Chosen	Chromosome 1	Chromosome 2	Chromosome 3	Chromosome 4

This technique can be further illustrated using the following example: Consider a population of 4 chromosomes, whose fitness values are 10, 5, 15, and 30, respectively. The weighted roulette wheel is constructed as shown in Figure II.8.4. The selection result can be shown in Table II.8.3. It is clear that the higher the fitness value of a chromosome, the better chance this chromosome will be chosen.

8.9.4.2   Selection Based on Ranking

Again, the *roulette-wheel parent selection* technique is used for this type of chromosome selection. The only difference is that instead of using a fitness value, the Weighted Roulette-Wheel has been designed based on the ranking of the chromosome.

After the fitness evaluation, the chromosomes in the entire population will be ranked according to their fitness values. The chromosome with the highest fitness value will be ranked No.1, the one with the second highest fitness value will be ranked No. 2, etc.

In our algorithm, a value called ranking value was computed for each chromosome. It was based on the ranking number of the chromosome and can be expressed as:

$$\text{Ranking Value} = N - \text{Ranking Number} \qquad (\text{II}.8.17)$$

Where N is the total number of chromosomes in the population. Therefore, a chromosome with a high fitness value will have a small ranking number and result in a big ranking value.

The selection technique is expressed as:

- Sum the ranking values of all the population members and call this result the total ranking values
- Generate a random number, n, between 0 and the total fitness
- Return the first population member whose ranking value, added to the ranking value of the preceding population members (running total), is greater than or equal to n

Given the same example in Section 8.9.4.1 for the roulette-wheel parent selection based on ranking, the ranking values of the 4 chromosomes, whose fitness values are 10, 5, 15, and 30, will then become

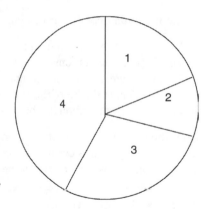

FIGURE II.8.5   An example of a weighted roulette wheel, where 1, 2, 3, and 4 refer to Chromosome 1, 2, 3, and 4, respectively.

TABLE II.8.4   The Selection Result Based on Figure II.8.5

Random Number	$0 <= n <= 2$	$2 < n <= 3$	$3 < n <= 6$	$6 < n <= 10$
Chromosome Chosen	Chromosome 1	Chromosome 2	Chromosome 3	Chromosome 4

2, 1, 3, and 4, respectively. The weighted roulette wheel will be constructed as shown in Figure II.8.5 and the selection result can be shown in Table II.8.4. It is clear that the higher the ranking value, the better the chance this chromosome will be chosen.

8.9.5   Genetic Operations

After the selection process, the selected chromosomes are used to create new chromosomes for the next generation using genetic operators.

8.9.5.1   Direct Copyover

In the proposed algorithm, the top 10% of the population on the basis of ranking is directly copied over to the next generation, i.e., $100 * 10\% = 10$ chromosomes are moved to the next generation. This is to ensure that the best solution in the current generation will be moved to the next generation and will not be lost. The rest, 90% of the population of the next generation, will be created by the crossover and mutation operations.

8.9.5.2   Crossover

Two chromosomes are selected using the roulette-wheel technique to crossover to create two new chromosomes. Then, a random number is created, from 2 to 336 (there is no point to crossover from the first bit, which will result in two same chromosomes), to be the crossover point. After that, all the bits to the right of the crossover point will exchange with each other. This process continues until $100 * 90\% = 90$ new chromosomes are created.

8.9.5.3   Mutation

The 90 newly created chromosomes will then proceed further to the mutation process. The probability of mutation is chosen to be 0.001, i.e., there are in total around $90_{\text{Chromosome}} * 336_{\text{bit}} * 0.001 = 30.24 \cong 30$ bits that will be changed.

The mutation operation is designed as:

Step 1: Create a random number, n, between 1 and 90.
Step 2: Select the chromosome for mutation based on the random number n created.
Step 3: Create another random number, m, between 1 and 336.
Step 4: Change the bit of the selected chromosome in the position m from 1 to 0 and 0 to 1.
Step 5: If this operation has been done 30 times, stop; otherwise, go to Step 1.

After the mutation, a new generation of the population has been created. They will go for the next round's sorting, fitness evaluation, and reproduction until the stop criteria is satisfied/matched. Then the best solution will be chosen to be the final solution.

8.9.6　Solution Storage

The final solution will then be output into a text file. This is to make sure that the results obtained from the current process will be reproducible. The system can simply retrieve the membership functions from the stored file and produce the same results at any time. This solves one of the GAs' drawbacks (see Section 8.2.3).

8.10　Implementation

To evaluate the performance of the proposed algorithm, the iris data set was broken up into two groups. One was used to train the system (training group), the other was used to test the performance of the system (test group). These two groups are disjointed, i.e., no data is repeated in both groups.
　In the experiments, two types of breakup of the iris data were used:

- 30–70% breakup, i.e., $150 * 30\% = 45$ instances of data were used for training and $150 * 70\% = 105$ instances of data were used for testing
- 50–50% breakup, i.e., $150 * 50\% = 75$ instances of data were used for training and 75 instances of data were used for testing

The other parameters used are:

- population size $= 100$
- binary coding
- chromosome length $= 336$ bit
- two types of selection implementation: fitness-based or ranking-value-based
- copyover rate $= 10\%$
- crossover probability, $p_c = 1$
- mutation probability, $p_m = 0.001$
- maximum number of training iterations $= 10000$
- stop criteria: either the best solution has reached 100% correct classification rate, or the maximum number of training iterations has been reached

The experiments were simulated on a 300-MHz Pentium-Pro IBM-compatible personal computer with 128 MB of RAM, under Microsoft Visual C++ environment.

8.11　Result Analysis

The two programs, with different selection implementations, were each run 10 times on the data. The results are shown in Tables II.8.5 to II.8.8, respectively. In these tables, C-C-R stands for correct classification rate and C-C-N stands for correct classification number. "Training Iterations"

TABLE II.8.5 Experiment Results on 30–70% Breakup Data with Fitness-Based Selection

Times (No.)	Training Iterations	Running Time (Sec)	Best Solution Fitness	C-C-R on Training Group (%)	C-C-N on Test Group	C-C-R on Test Group (%)
1	290	11	45	100	96	91.43
2	20	1	45	100	97	92.38
3	48	2	45	100	97	92.38
4	14	1	45	100	100	95.24
5	23	1	45	100	97	92.38
6	10,000	365	44	97.78	93	88.57
7	744	29	45	100	99	94.29
8	92	4	45	100	95	90.48
9	2554	87	45	100	99	94.29
10	5017	218	45	100	97	92.38
Average	1880.2	71.9	44.9	99.78	97	92.38

TABLE II.8.6 Experiment Results on 50–50% Breakup Data with Fitness-Based Selection

Times (No.)	Training Iterations	Running Time (Sec)	Best Solution Fitness	C-C-R on Training Group (%)	C-C-N on Test Group	C-C-R on Test Group (%)
1	10,000	468	74	98.67	72	96
2	1964	89	75	100	70	93.33
3	4953	232	75	100	69	92
4	267	13	75	100	70	93.33
5	10,000	479	74	98.67	69	92
6	1168	56	75	100	70	93.33
7	10,000	476	74	98.67	71	94.67
8	112	6	75	100	71	94.67
9	45	3	75	100	71	94.67
10	10,000	499	74	98.67	70	93.33
Average	4850.9	232.1	74.6	99.47	70.3	93.73

refers to the iterations needed for the program to stop training the input training data. "Running Time" refers to the total running time (sec) for the program to execute. "Best Solution Fitness" refers to the fitness of the best solution of the last generation, which is equal to the number of correct classification of the input training data. "C-C-R on Training Group" refers to the percentage of correct classification, i.e., the number of correct classification divided by the total number of training data (for 30–70% breakup case, divided by 45; for 50–50% breakup case, divided by 75). "C-C-N on Test Group" is the number of correct classification of the test data. "C-C-R on Test Group" is the percentage of correct classification rate, i.e., the number of correct classification divided by the total number of test data (for 30–70% breakup case, divided by 105; for 50–50% case, divided by 75).

Shown in Tables II.8.5 to II.8.8, on average, the program will take several thousand iterations and several min to converge. Compared to Teo, Khalid, and Yusof,[29] which takes roughly three hr for the GA process to converge (their experiments were simulated on a 133 MHz Pentium IBM-compatible

TABLE II.8.7   Experiment Results on 30–70% Breakup Data with Ranking-Value-Based Selection

Times (No.)	Training Iterations	Running Time (Sec)	Best Solution Fitness	C-C-R on Training Group (%)	C-C-N on Test Group	C-C-R on Test Group (%)
1	10,000	293	37	82.22	85	91.43
2	10,000	315	44	97.78	98	92.38
3	622	27	45	100	95	92.38
4	4228	118	45	100	96	95.24
5	42	2	45	100	100	92.38
6	25	1	45	100	97	88.57
7	185	6	45	100	99	94.29
8	10,000	299	30	66.67	65	90.48
9	17	1	45	100	99	94.29
10	815	26	45	100	100	92.38
Average	3593.4	108.8	42.6	94.67	93.4	88.95

TABLE II.8.8   Experiment Results on 50–50% Breakup Data with Ranking-Value-Based Selection

Times (No.)	Training Iterations	Running Time (Sec)	Best Solution Fitness	C-C-R on Training Group (%)	C-C-N on Test Group	C-C-R on Test Group (%)
1	10,000	386	74	98.67	71	94.67
2	3340	227	75	100	69	92
3	10,000	423	73	97.33	70	93.33
4	192	9	75	100	70	93.33
5	10,000	515	50	66.67	45	60
6	8	1	75	100	70	93.33
7	10,000	571	74	98.67	71	94.67
8	10,000	629	73	97.33	46	61.33
9	24	2	75	100	70	93.33
10	10,000	542	74	98.67	68	90.67
Average	6356.4	330.5	71.8	95.73	65	86.67

personal computer with 16 MB of RAM, running under Borland C+ + Version 3.1), the results are encouraging.

The correct classification rate can be up to 93.73% on the average for the 50–50% data breakup. Compared to Castro and Zurita[43] (the results are 90% of CART algorithm, 90% of C4 algorithm, and 93.33% of the proposed algorithm in their paper), which used up to 80% of the iris data for training, the results achieved here are better. It is not as good as the results of Sanker et al.,[13] which is 96%. However, the proposed algorithm is not only able to perform classification, it is also able to extract the data features to form the membership functions. This is an added advantage.

It has been shown that using fitness-based selection will converge earlier than using ranking-value-based selection. In fact, in some cases of ranking-value-based selection implementation, the program had not converged after 10,000 iterations. This is why the average correct classification rate of implementation using ranking-value-based selection is lower. However, excluding those implementations that did not converge, the average correct classification rate of implementation

TABLE II.8.9   The Detail Results of the Implementation of Row 1 in Table II.8.6

Class	True Acceptance (%)	False Acceptance (%)	False Rejection (%)	True Rejection (%)
1	100	0	0	100
2	91.42	10	8.58	90
3	80	4.29	20	95.71

using ranking-value-based selection is 93.33% for 30–70% data breakup case, which is even better than the fitness-based selection implementation, and 93.17% for 50–50% data breakup case, which is close to the fitness-based selection implementation.

Normally, fitness-based selection is not encouraged because it often results in premature convergence, especially when the difference of the fitness value is very large. In our case, since the difference of the fitness value is considered small (at most 45 for the 30–70% case and 75 for the 50–50% case), it is still working fine. However, if the fitness value can have a very big range, ranking-value-based selection should be a better choice.

Experiments had also been carried out using the same setting without the direct copyover operator. The results showed that although the final results are unrelated to whether or not this operator had been used, the experiments using it would converge faster.

Further analysis of the results also shows that they are affected by the input data. As mentioned before, Class 1 data are linearly separated from the other classes. The results also show that the misclassification rate of instances belonging to Class 1 is lower than those of the other classes. An example can be shown in Table II.8.9, where:

True Acceptance of Class i = 100% * (Number of Correct Classified Instances of Class i)/

(Total Number of Instances of Class i) $\qquad$ (II.8.18)

False Acceptance of Class i = 100% * (Number of Instances of Other Classes Classified

into Class i)/(Total Number of Instances of Other Classes) $\qquad$ (II.8.19)

False Rejection of Class i = 100% * (Number of Instances of Class i Classified into

Other Classes)/(Total Number of Instances of Class i) $\qquad$ (II.8.20)

True Rejection of Class i = 100% * (Number of Instances of Other Classes not Classified

into Class i)/(Total Number of Instances of Other Classes) $\qquad$ (II.8.21)

8.12   Phoneme Data

A second set of experiments based on phoneme data is used to evaluate the proposed GA-algorithm. This data set is available on-line at the ELENA website.[40] This database was in use in the European ESPRIT 5516 project: ROARS. There are two types of data in this Phoneme database, namely nasal and oral vowels. This database contains vowels coming from 1809 isolated syllables. Five different attributes were chosen to characterize each vowel. They are the amplitudes of the five first harmonics AHi, normalized by the total energy Ene (integrated on all the frequencies): AHi/Ene. Three observation moments have been kept for each vowel to obtain 5427 different instances. From these 5427 initial values, 23 instances for which the amplitude of the 5 first harmonics was zero were removed, leading to the 5404 instances of the present database. 3818 instances belong to the nasal class and 1586 instance belong to the oral class.

To benchmark the GA technique using the same structure, 20% of the data are used for training and 80% of the data are used for testing.

Since there are two classes and each class has five attributes, in total ten fuzzy sets are to be used in the system ($2_{label}*5_{attribute}$). They are:

- Nasal Attribute 1 (AN_1)
- Nasal Attribute 2 (AN_2)
- Nasal Attribute 3 (AN_3)
- Nasal Attribute 4 (AN_4)
- Nasal Attribute 5 (AN_5)
- Oral Attribute 1 (AO_1)
- Oral Attribute 2 (AO_2)
- Oral Attribute 3 (AO_3)
- Oral Attribute 4 (AO_4)
- Oral Attribute 5 (AO_5)

The reason why the labels are defined as "nasal" and "oral" instead of conventionally "short" and "long" is to reduce the number of fuzzy rules used in the system (see Section 8.2).

The trapezoid membership function is chosen to be the membership function of these ten fuzzy sets. They will be constructed by the proposed GA-based algorithm.

Based on the fuzzy sets defined, two rules are predefined in the system. They are:

- RULE 1: IF $Attr_1$ is AN_1 AND $Attr_2$ is AN_2 AND $Attr_3$ is AN_3 AND $Attr_4$ is AN_4 AND $Attr_5$ is AN_5, THEN the instance belongs to Class Nasal
- RULE 2: IF $Attr_1$ is AO_1 AND $Attr_2$ is AO_2 AND $Attr_3$ is AO_3 AND $Attr_4$ is AO_4 AND $Attr_5$ is AO_5, THEN the instance belongs to Class Oral

Where $Attr_i$ is the ith attribute of the input instance, AN_i and AO_i are the labels used for $Attr_i$, $i = 1, 2, 3, 4, 5$.

Other parameters used were:

- Population size $= 100$
- Binary coding
- Chromosome length $= 280$ bits ($2_{label}*5_{attribute}*4_{parameter}*7_{bit} = 280$)
- Ranking-value-based selection
- Copyover rate $= 10\%$
- Crossover probability, $p_c = 1$
- Mutation probability, $p_m = 0.001$
- Maximum number of training iterations $= 10,000$
- Stop criteria: either the best solution has reached 100% correct classification rate or the maximum number of training iterations has been reached

The results of the experiment are shown in Table II.8.10. In comparison, the results obtained using the K-NN classifier, which is obtained from ELENA[40] by using the leave-one-out method with k $= 20$, are shown in Table II.8.11; and the results of Quek and Tung[44] are shown in Table II.8.12.

It showed that the results of the proposed system were better than that of Quek and Tung.[44] Although the total error rate of the proposed system is higher than that of K-NN, the results were obtained using only 20% of the data for training in contrast to the K-NN, which used almost the

TABLE II.8.10 Phoneme Classification Results Using GF

	Confusion Matrix (avg)	
	Class Nasal (%)	Class Oral (%)
Class Nasal	90.6	9.4
Class Oral	53.4	46.6
	Error rate $= 22.3\%$	

TABLE II.8.11 Phoneme Classification Results Using K-NN ($k = 20$) and Leave-1-Out Method

	Confusion Matrix (avg)	
	Class Nasal (%)	Class Oral (%)
Class Nasal	91.4	8.6
Class Oral	27.8	72.2
	Error rate $= 14.2\%$	

TABLE II.8.12 Phoneme Classification Results of Quek and Tung[44]

	Falcon-FKP(CL)			Falcon-FKP (NCL)	
	Confusion Matrix			Confusion Matrix	
	Class Nasal (%)	Class Oral (%)		Class Nasal (%)	Class Oral (%)
Class Nasal	86.4	13.6	Class Nasal	86.8	13.2
Class Oral	49.1	50.9	Class Oral	50.1	49.9
	Error rate $= 24.12\%$			Error rate $= 24.11\%$	
	Falcon-PFKP (CL)			Falcon-PFKP (NCL)	
	Confusion Matrix			Confusion Matrix	
	Class Nasal (%)	Class Oral (%)		Class Nasal (%)	Class Oral (%)
Class Nasal	82.6	17.4	Class Nasal	92.1	7.9
Class Oral	42.1	57.9	Class Oral	71.5	28.5
	Error rate $= 24.76\%$			Error rate $= 26.73\%$	

whole set of data for training. In addition, using the proposed system, fuzzy membership functions can be achieved to represent the feature range.

8.13 Traffic Flow Data

This raw traffic flow data were obtained from Tan,[41] courtesy of School of Civil and Structural Engineering, NTU, Singapore. The data were collected at a site located at exit 15 along the eastbound Pan Island Expressway (PIE) in Singapore using loop detectors embedded beneath the road surface. There are a total of five lanes at the site; two exit lanes and three straight lanes for the main

FIGURE II.8.6   Location of Site 29 along PIE, Singapore.

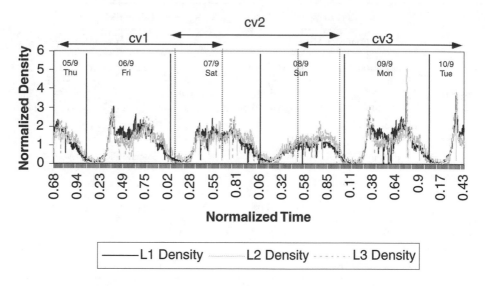

FIGURE II.8.7   Traffic density of three lanes along PIE (Site 29).

traffic. This is shown in Figure II.8.6. In our experiment, only the traffic flow data for the three straight lanes were used to assess the system. The density data of the three straight lanes were used to predict the density in lane 1, 2, or 3 in a short period of time. A plot of the traffic flow density data for the three straight lanes for a period of 6 days from 5th to 10th September 1996 is shown in Figure II.8.7.

TABLE II.8.13   The GF Prediction Results of the Lane 1 Density at $(t + 5)$ Min

	CV1_R2	CV2_R2	CV3_R2	Avg R2
Lane 1 Density in the Next 5 min	0.71325	0.70854	0.71139	0.71106

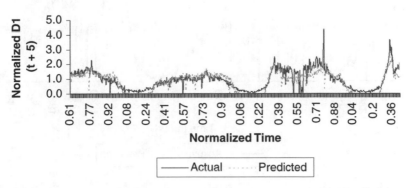

FIGURE II.8.8   A GF prediction result of the lane 1 density at $(t + 5)$ min. ($R^2 = 0.706622$).

In the experiment, three groups of training and test sets were used, namely, CV1, CV2, and CV3, as shown in Figure II.8.7. When CVI is used for training, the rest of the data will be used for testing, where $I = 1, 2, 3$. The square of the Pearson product-moment correlation value (R^2) was used to compute the accuracy of the result. Four inputs were used, namely, Time, Lane 1 Density, Lane 2 Density, and Lane 3 Density. Each of them is described by five labels. Five rules were predefined to predict the Lane 1 Density in the next 5 min. They are similar to those used in the experiments using iris data and phoneme data.

Other parameters used are:

- Population size $= 100$
- Binary coding
- Chromosome length $= 560$ bits ($5_{label} {}^* 4_{attribute} {}^* 4_{parameter} {}^* 7_{bit} = 560$)
- Ranking-value-based selection
- Copyover rate $= 10\%$
- crossover probability, $p_c = 1$
- Mutation probability, $p_m = 0.001$
- Maximum number of training iterations $= 10,000$
- Stop criteria: either the best solution has obtained 100% correct prediction or the maximum number of training iterations has been reached

The results are summarized in Table II.8.13 and one of the prediction results is plotted in Figure II.8.8. It shows a reasonable performance that allows the data to be predicted with a fairly high correlation. In comparison, the result using MLP, with 100 neurons in the hidden layer,[41] is plotted in Figure II.8.9. The proposed system produced a better R^2 value and the prediction results followed the actual data more closely.

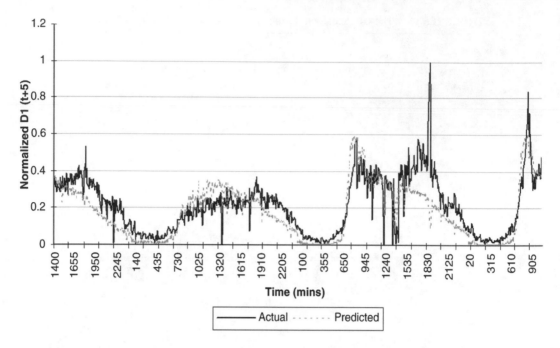

FIGURE II.8.9   The MLP prediction result of the lane 1 density at $(t + 5)$ min. ($R^2 = 0.615254$).

8.14   Conclusion

Fuzzy logic refers to all the theories and technologies that employ fuzzy sets, i.e., classes with unsharp boundaries, for reasoning under uncertainty. The performance of a fuzzy system depends on three factors: the fuzzy sets, the fuzzy rules, and the inference schemes. Fuzzy sets are defined by membership functions. Manual definition of membership functions and fuzzy rules is inefficient. Therefore, a good fuzzy system should be able to automatically construct membership functions and fuzzy rules. Currently there are many techniques on membership functions and fuzzy rules auto-construction. These techniques, however, normally are application dependent and use too many preset parameters. Fuzzy systems using such techniques are very difficult to use for different applications. Manually tuning those preset parameters also makes the design of such systems time-consuming and not cost-effective.

Genetic algorithms, on the other hand, are global search and optimization techniques exploring search space by incorporating a set of candidate solutions in parallel. The search process is only guided by a fitness function, which makes GAs application independent in nature. However, GAs are normally used as ad hoc methods and are unable to extract the distribution information about the input data features. In this chapter, the focus is on the integration of fuzzy logic and genetic algorithms into one system, where genetic algorithms can be used for the auto-construction of membership functions and fuzzy rules, while fuzzy logic enables the system to extract the information about the input data.

A GA-based algorithm (GF), to automatically construct membership functions has been proposed in this section. GF is capable of automatically constructing membership functions from the input training data. It utilizes the powerful search capability of genetic algorithm to pick up the best fuzzy sets for the input training data from a population pool. Guided by a fitness evaluation function, provided with sufficient running time, the system will converge to derive an optimal group of fuzzy sets.

The proposed system has the following features:

- It randomly generates the membership functions for the fuzzy rules, and subsequently optimizes them with the GA-based techniques.
- It operates on the fitness values instead of the real features, which makes it application independent and easily applicable to other applications.
- It requires very few complicated manually tuned preset parameters. In fact, the only preset parameter is the number of the fuzzy rules that should be used for the system. This makes the design of the system fast and cost-effective.
- The auto-generated membership functions and fuzzy rules are stored to a file. This makes the results reproducible. It also makes the migration of the system from working off-line to working one-line easier.

The proposed system has been implemented and assessed against the iris data set, phoneme data, and traffic flow data. The results are compared against other algorithms and the system is proven to be working well.

References

1. Holland, J.H., Outline for a logical theory of adaptive systems, *J. Assoc. Comput. Mach.* 3, 297–314, 1962.
2. Goldberg, D.E., *Genetic Algorithms in Search Optimization, and Machine Learning*, Reading, MA: Addison-Wesley, 1989.
3. Davis, L., *Handbook of Genetic Algorithms*, New York: Van Nostrand Reinhold, 1991.
4. Janikow, C.Z. and Michalewicz, Z., An experimental comparison of binary and floating point representation in genetic algorithms, *Proc. 4th Int. Conf. Genetic Algorithms*, 31–36, San Diego, CA, 1991.
5. Krishna, K. and Murty, M.N., Genetic K-Mean Algorithm, *IEEE Trans. on Systems and Man, and Cybernetics – Part B: Cybernetics*, 29(3), 433–439, 1999.
6. Kemp, B., Porter, S.J., and Dawson, J.F., Population partitioning in genetic algorithms, *Electronics Letters*, 34(20), 1928–1929, 1998.
7. Rogers, A. and Prugel-Bennett, A., Genetic drift in genetic algorithm selection scheme, *IEEE Trans. on Evolutionary Computation*, 3(4), 298–303, 1999.
8. Reformat, M., Kuffel, E., Woodford, D., and Pedrycz, W., Application of genetic algorithms for control design in power systems, *IEE Proc.-Gener. Transm. Distrib.*, 145(4), 345–354, 1998.
9. Shin, S.C. and Park S.B., GA-based predictive control for nonlinear processes, *Electronics Letters*, 34(20), 1980–1981, 1998.
10. Vlachos, C., Williams, D., and Gomm, J.B., Genetic approach to decentralized PI controller tuning for multivariable processes, *IEE Proc.—Control Theory Appl.*, 146(1), 58–64, 1999.
11. Abutaleb, A.S. and Kamel, M., A genetic algorithm for the estimation of ridges in fingerprints, *IEEE Tran. Image Processing*, 8(8), 1134–1139, 1999.
12. Kabuka, M.R., Gudmundsson, M., and Essam, A.E., Edge detection in medical images using a genetic algorithm, *IEEE Trans. Medical Imaging*, 17(3), 469–474, 1998.
13. Sankar, K.P., Bandyopadhyay, S., and Murthy, C.A., GA-based pattern classification: Theoretical and experimental studies, *Proc. 13th Int. Conf. Pattern Recognition*, vol. 3, 758–762, vol. 4, 1996.
14. Haydar, A., Demirekler, M., and Yurtseven, M.K., Feature selection using genetic algorithm and its application to speaker verification, *Electronics Letters*, 34(15), 1457–1459, 1998.
15. Sun, F. and Hu, G.R., Speech recognition based on genetic algorithm for training HMN, *Electronics Letters*, 34(16), 1563–1564, 1998.

16. Haydar, A., Dmirekler, M., and Yurtseven, M.K., Speaker identification through use of features selected using genetic algorithm, *Electronics Letters*, 34(1), 39–40, 1998.

17. Chang, H.C. and Chen, P.H., Hydrothermal generation scheduling package: a genetic based approach, *IEE Proc.-Gener. Transm. Distrib.* 145(4), 451–457, 1998.

18. Gallego, R.A., Monticelli A., and Romero, R., Transmission system expansion planning by an extended genetic algorithm, *IEE Proc.-Gener. Transm. Distrib.*, 145(3), 329–335, 1998.

19. Shieh, L.S., Wang, W., and Tsai, J.S.H., Optimal digital design of hybrid uncertain systems using genetic algorithms, *IEE Proc. – Control Theory Appl.*, 146(2), 119–130, 1999.

20. Fung, S.K., Liu, N.K., Chan, K.H., and Lau, W.H., Fuzzy genetic algorithm approach to feature selection problem, *Fuzz-IEEE' 97*, 441–446, 1997.

21. Rhee, C.H. and Lee, Y.J., Unsupervised feature selection using a fuzzy-genetic algorithm, *1999 IEEE Int. Fuzzy Systems Conf. Proc.*, III, 1266–1269, 1999.

22. Kyung, H.C., Eung, J.W., and Sung, T.K., Fast static image reconstruction using adaptive mesh grouping method in EIT, *Proc. 19th Int. conf.* IEEE/EMBS, 441–444, 1997.

23. White, B.A. and Trebi-Ollennu, A., Multiobjective fuzzy genetic algorithm optimization approach to nonlinear control system design, *IEE Proc. Control Theory Appl.*, 144(2), 137–142, 1997.

24. Kim, J.W., Kim, B.M., and Kim, J.Y., Genetic algorithm simulation approach to determine membership functions of fuzzy traffic controller, *Electronics Letters*, 34(20), 1982–1983, 1998.

25. Man, G.N., Design of a genetic fuzzy controller for the nuclear steam generator water level control, *IEEE Trans. Nuclear Science*, 45(4), 2261–2271, 1998.

26. Qi, X.M. and Chin, T.C., Genetic algorithms based fuzzy controller for high order systems, *Fuzzy Sets and Systems* 91: 279–284, 1997.

27. Streifel, R.J., Marks, R.J., Reed, R., Choi, J.J., and Healy, M., Dynamic fuzzy control of genetic algorithm parameter coding, *IEEE Trans. Systems, Man, and Cybernetics – Part B: Cybernetics*, 29(3), 426–433, 1999.

28. Wang, C.H., Hong, T.P., and Tseng, S.S., Integrating fuzzy knowledge by genetic algorithms, *IEEE Trans. Evolutionary Computation*, 2(4), 138–149, 1998.

29. Teo, L.S., Khalid, M.B., and Yusof, R., Tuning of a neuro-fuzzy controller by genetic algorithm, *IEEE Trans. Systems, Man, and Cybernetics – Part B: Cybernetics*, 29(2), 226–236, 1999.

30. Bezdek, J.C., *Pattern Recognition with Fuzzy Objective Function Algorithms*, Plenum Press, New York, 1981.

31. Kohonen, T.K., *Self-Organization and Associative Memory*, 3rd Ed. New York, Springer-Verlag, 1989.

32. Quek, H.C. and Ang, K.K., POPFNN-CRI(S): A fuzzy neural network based on the compositional rule of inference, Master Thesis, SCE, Nanyang Technological University, Singapore, 1998.

33. Pal, N.R., Bezdek, J.C., and Tsao, E.C.K., Generalized clustering networks and Kohonen's self-organizing scheme, *IEEE Trans. Neural Networks*, 4(4), 549–558, 1993.

34. Hong, T.P. and Lee, C.Y., Induction of fuzzy rules and membership functions from training examples, *Fuzzy Sets and Systems* 84: 33–47, 1996.

35. Woodard, S.E. and Gerg, D.P., A numerical optimization approach for tuning fuzzy logic controllers, *IEEE Trans. Systems, Man, and Cybernetics – Part B: Cybernetics*, 29(4), 565–569, 1999.

36. Lee, K.M., Kwak, D.H., and Hyung, L., Tuning of fuzzy models by fuzzy neural networks, *Fuzzy Sets and Systems* 76: 47–61, 1995.

37. Furukawa, M. and Yamakawa, T., The design algorithms of membership functions for a fuzzy neuron, *Fuzzy Sets and Systems* 71: 329–343, 1995.

38. Sankar, K.P., Rajat, K.D., and Jayanta, B., Unsupervised feature evaluation: A neuro-fuzzy approach, *IEEE Trans. Neural Networks*, 11(2), 366–376, 2000.

39. Fisher, R.A., The use of multiple measurement in taxonomic problems, *Ann. Eugenics*, 7, 179–188, 1936.

40. ELENA website: http://www.dice.ucl.ac.be/neural-nets/Research /Project /ELENA/elena.htm.

41. Tan, G.K., Feasibility of predicting congestion states with neural networks, Final Year Project, CSE, Nanyang Technological University, Singapore, 1997.

42. Lin, C.T. and Lee, C.S., *Nerual Fuzzy Systems*, Prentice-Hall, Inc, New York, 1996.

43. Castro, J.L. and Zurita, J.M., An inductive learning algorithm in fuzzy systems, *Fuzzy Sets and Systems* 89: 193–203, 1997.

44. Quek, H.C. and Tung, W.L., PACL-FNNS: A novel class of Falcon-like fuzzy neural networks based on positive and negative exemplars, in preparation, 2003.

Appendix A: Clustering Techniques

Algorithm FCM: Fuzzy c-Means Algorithm

- Step 1: Select a number of clusters $c(2 \leq c \leq n)$ and exponential weight $w(1 < m < \infty)$. Choose an initial membership (partition) matrix $U^{(0)}$ and a termination criterion ϵ. Set the iteration index l to 0.
- Step 2: Calculate the fuzzy cluster centers $\{v_i^{(l)} | i = 1, 2, \ldots, c\}$ by using $U^{(l)}$ and Equation II.A.1.

$$v_i = \frac{1}{\sum_{j=1}^{n}(u_{ij})^m} \sum_{j=1}^{n}(u_{ij})^m x_j \quad i = 1, 2, \ldots, c \qquad \text{(II.A.1)}$$

- Step 3: Calculate the new partition matrix $U^{(l+1)}$ by using $\{v_i^{(l)} | i = 1, 2, \ldots, c\}$ and Equation II.A.2.

$$u_{ij} = \frac{(1/\|x_j - v_i\|^2)^{1/m-1}}{\sum_{k=1}^{c}(1/\|x_j - v_k\|^2)^{1/m-1}} \quad i = 1, 2, \ldots, c; \ j = 1, 2, \ldots, n \qquad \text{(II.A.2)}$$

- Step 4: Calculate $\Delta = \|U^{(l+1)} - U^{(l)}\| = \max_{i,j} |u_{ij}^{(l+1)} - u_{ij}^{(l)}|$. If $\Delta > \epsilon$, then set $l = l + 1$ and go to step 2. Otherwise, stop.

End FCM

Algorithm LVQ: Linear Vector Quantization

- Step 1: Initialize the prototype vectors $v_k, k = 1, 2, \ldots, c$, such that $V_0 = \{v_1^0, \ldots, v_k^0, \ldots, v_c^0\} \in R^{cp}$ represents the prototype vectors for all the classes at time step $t = 0$. Set the initial learning rate $\alpha_1 \in (0, 1]$. Choose a termination criterion ϵ.
- Step 2: $t = t + 1$; For $i = [1, \ldots, n]$

  (i) Find the winning cluster the input element x_i is closest to by using Equation II.A.3:

$$Winner = \arg \min_{k=1}^{c} \|x_i - v_k^{t-1}\|, \ Winner = 1, 2, \ldots, c \qquad \text{(II.A.3)}$$

  where v_k^{t-1} is the prototype vector for class k at time step $t - 1$.

(ii) Update the prototype vector of the winning cluster by using Equation II.A.4:

$$v_{winner}^{t} = v_{winner}^{t-1} + \partial_t(x_i - v_{winner}^{t-1}) \qquad \text{(II.A.4)}$$

where v_{winner}^{t-1} is the prototype vector of the winning class at time step $t - 1$.

- Step 3: Calculate $\Delta = \|V_t - V_{t-1}\| = \sum_{k=1}^{c} \|v_k^t - v_k^{t-1}\|$. If $\Delta > \epsilon$, then set $\alpha_{t+1} = f(\alpha_t)$ and go to step 2, where $f()$ is a modifying function for the learning α. Otherwise, stop.

End LVQ

9

Fuzzy Cellular Automata and Fuzzy Sequential Circuits

Mraz Miha
University of Ljubljana

Zimic Nikolaj
University of Ljubljana

Virant Jernej
University of Ljubljana

Ficzko Jelena
University of Ljubljana

Preface. In this chapter we present the structure of fuzzy cellular automata. The structure can be derived with fuzzy generalization of ordinary (crisp) cellular automata. Fuzzy generalization is reached with the introduction of fuzzy logic into ordinary cellular automata. If the ordinary finite automaton can be used as an abstract model of a basic entity (a cell) of cellular automata structure, then the fuzzy finite automaton may be used as an abstract model for a fuzzy cellular automata structure. In our work we generalize the definition of fuzzy automaton for the needs of fuzzy cellular automata.

In the second part of the chapter we propose a new approach to the logic design of the FCA structure, based on fuzzy sequential circuits. Such systems are configured with fuzzy cellular fields of the dimensions 1×1, 1×4, 4×1, and 4×4, and each cell in such fields contains a fuzzy memory device of fuzzy delay and fuzzy trigger type. The problem to be solved is composed with fuzzy propositions. Such proposition(s) may be rewritten in the form of fuzzy switching functions, which are input functions of fuzzy memory devices. These switching functions could also be rewritten to the list of fuzzy rules, which are responsible for the life of a whole fuzzy cellular system.

All conclusions in the work are verified with MATLAB-based simulation. Using simulation results, we demonstrate attractive pattern formation and static and dynamic phenomena of the proposed fuzzy cellular systems. From the application point of view, we analyze results in the domain of a fuzzy inference engine. First of all, cellular systems with fuzzy trigger or fuzzy delay can take their place in a fuzzy inference engine, where we combine fuzzy memorizing and processing to achieve high-speed approximate reasoning.

Key words: Fuzzy automaton, fuzzy cell, fuzzy cellular automata, fuzzy cellular system, fuzzy memory device, fuzzy rule, fuzzy switching function, MATLAB simulation.

9.1 Introduction

The evolution of the literature in the field of cellular automata (CA) over the last 10 years reveals a decrease in useful applications. Cellular automata were one of the most promising interdisciplinary research areas in the 1980s.[1,2] From the computer science viewpoint they were very promising, due to the concepts of parallelism, interaction, locality and processing, homogeneity. All these characteristics gave hope of eliminating Von Neumann's narrow neck in the field of computer architecture. We presume that the main reason for decreased interest in cellular automata lies in the unavailability of exact or crisp knowledge of the behavior of real system dynamics, which we would like to model with the structure of cellular automata.

On the other hand, fuzzy logic theory has progressed greatly during the same period. With the generalized and continuous treatment of an element's membership in the observed set, it enables linguistic programming of systems on the basis of uncertain knowledge about real systems and also the use of uncertain, vague, and possible data, which enter the inference process. In the field of applications, the use of fuzzy logic lies mainly in simple controllers and much less in other applications, such as recognition or modeling applications. Fuzzy controllers can be found in very simple consumer products, as well as in very complex control applications, i.e., space ships, modern weapons.

In the present chapter the possibilities of involving fuzzy logic in the field of cellular automata are explored. Fuzzy logic is applied to obtain the generalization of cellular automata structure, which we will call *Fuzzy Cellular Automata* (FCA). Such a structure can be used for modeling in cases when the knowledge and data about the observed system are uncertain.

First, we explore the required characteristics of a cell, which represents the main entity of a cellular automata structure. We also analyzed most of the fuzzy automaton definitions and discovered that most of them do not suit the requirements of the desired cell's characteristics. Because of this, we have defined a new abstract model of *Fuzzy Finite Automaton*. The new definition enables fuzzy or uncertain states of the cell, fuzzy or uncertain input variables, and the fuzzy inference process, which leads to the formation of a new fuzzy state. Founded on the fuzzy automaton's definition, we have formed a definition of fuzzy cellular automata. In this form, the possibility of fuzzy neighboring or the possibility of different treatments of neighboring influential cells has been assumed. The fuzzy cellular automata definition was confirmed with application examples of fuzzy right shift, a fuzzy variant of the Conway game of life,[3] and finally with modeling the shape of fire spreading in a homogeneous natural environment.[4]

It was evident from these examples that involving fuzziness has brought some advantages. These are the human-friendly linguistic style of programming and the possibility of involving uncertain information, so that finally, richer values of behavior can be obtained by the time sequence of patterns from the fuzzy cellular automata. The example of the fire spread shape modeling demonstrates another advantage, which is the decreased time for model construction using fuzzy cellular automata, compared with traditional approaches, where a lot of measurements have to be made.

In the second part of the chapter a new approach to logic design and organization of homogeneous fuzzy cellular systems is proposed. The latter are configured with fuzzy cellular fields 1×1, 1×4, 4×1, and 4×4, and each cell in such fields contains a fuzzy memory device of delay (FD) and trigger type (FT, FTDC).[21,22] Some simulation results using MATLAB and Simulink tools are presented.

The rules of notation are as follows. The symbol c denotes an individual cell in the space of cells and the symbol c_c denotes the observed cell in the set of neighboring cells c_i, $i = 1, \ldots, k$; $k \in N$. The well-known fuzzy terms fuzzy set and fuzzy number are denoted with a symbol "~" (e.g., $\widetilde{A}, \widetilde{a}$).

New complex terms will be denoted with a symbol "∧". Fuzzified versions of well-known fuzzy memory cells (D, JK, RS, etc.) will acquire the prefix sign F (FD, FJK, etc). Other abbreviations are CA for cellular automata, FCA for fuzzy cellular automata, and FA for fuzzy automata. In all cases where fuzzy rules appear, the left side of the rule (conditional side) is observed in time t and the right side or a conclusion is formed for time $t + 1$. Some of the new introduced complex fuzzy terms are presented as vectors. For instance, a fuzzy state of a fuzzy automaton is presented as $\hat{q} = [\widetilde{Q}_1, \widetilde{Q}_2, \widetilde{Q}_3]$.

9.2   Basics of Cellular Automata

The fundamental part of a cellular automata is an individual cell c, which can also be treated as a basic modeling entity. Its main feature is a state which can be changed with time. The transition from one state to another is stimulated by the state of the observed cell and by the states of its neighboring cells in the past. The basic model for an entity cell is a finite automaton. Moore's automaton, which is a suitable realization of a finite automaton,[5] has the following definition:

Definition 1: Moore's automaton is defined as a quintuplet $A = \langle X, Q, Y, \delta, \lambda \rangle$, where X, Q, and Y represents nonempty and finite sets of input symbols, internal states, and output symbols, respectively. δ is a state translation function as given in Equation II.9.1 and λ is a function which translates into the output symbol Equation II.9.2.

$$\delta : Q \times X \to Q, \tag{II.9.1}$$

$$\lambda : Q \to Y. \tag{II.9.2}$$

Ordinary (or crisp) cellular automata can be defined as summarized from References 6 and 7.

Definition 2: The quadruplet $CA = \langle P, Q, N, F \rangle$ is a cellular automata (CA). Let the space P be equal to Z^m, where Z is a set of integer numbers and m the dimension of space. Q is a finite nonempty set of all possible states of a cell, F is a state translation function as given in Equation II.9.6, N is a function which returns the neighborhood of the observed cell c_c (Equation II.9.5) and T is a set of discrete time points.

$$E : P \times T \to Q, \tag{II.9.3}$$

$$\forall t \in T, \forall c_c \in P, \exists! q \in Q : q = E(c_c, t), \tag{II.9.4}$$

$$N : P \to P(P), \tag{II.9.5}$$

$$F : Q^{k+1} \to Q. \tag{II.9.6}$$

Every m-tuple from P represents an address of a cell (automaton). In this way, the space of addressable cells is formed. The number of cells neighboring the observed cell c_c is stored in k. $E(c_c, t)$ is a function which assigns a state q in time t to every observed cell c_c from space P. The function F given in Equation II.9.6 translates to a new state of cell in time $t + 1$. The translation depends on the states of the k neighboring cells and the cell's own state in time t. The most widely used method for the determination of neighborhood is given in Equation II.9.7. The term $d(c_c, c_j)$ represents the distance between two cells in m-dimensional space. Term R represents the criterion of radius of neighborhood in space P; if a cell lies within the radius of the observed cell c_c, then it is classified into the set $N(c_c)$.

$$\forall c_c \in P : N(c_c) = \{c_1, \ldots, c_k\} \Rightarrow d(c_c, c_j) \leq R, j = 1, \ldots, k; R, k \in N, \qquad (\text{II.9.7})$$

$$k = (2^*R + 1)^m - 1, \qquad (\text{II.9.8})$$

$$\forall c_c \in P : E(c_c, t + 1) = F(E(c_c, t), E(N(c_c, t))). \qquad (\text{II.9.9})$$

The main expansion of research into cellular automata theory took place in the eighties of the last century. In the same period the first hardware applications were carried out (CAM-6), enabled by increased processing speed.[2] At the beginning of the nineties research stopped, as seen from the decreasing number of publications. The research done in the field of CA may be divided in two main parts:

- An *Analytical* or forward approach of research where the authors are analyzing the behavior of dynamics of a CA structure built in advance; the problems observed are (a)periodicity, (ir)reversibility, etc.
- A *Synthesis* or inverse approach of research where the authors' goal was to build a CA structure on the base of desired dynamic; a CA model built in this way and the simulation should give results as similar as possible to the real ones.

9.3 Fuzzy Automaton Definitions for the Needs of Fuzzy Cellular Automata

9.3.1 Introduction to the Abstract Model of a Cell

The basic model for an entity cell of CA could be an ordinary finite automaton as mentioned in the previous section. In the present section we will try to generalize an abstract structure of a finite automaton, with the goal that it will be able to process on the basis of fuzzy variables. The term "generalization" is used because we expect the same behavior as that of the ordinary automaton when the crisp input values and the exact rules of behavior are used as a description of a fuzzy automaton structure. In other words, we can say that the fuzzy automaton will have the same behavior as the ordinary one if all memberships will be $1 (\forall i : \mu_i = 1, 1 \leq i \leq n)$. Compared with the ordinary automaton, the fuzzy automaton will also operate in the cases when $\mu_i < 1$. The Moore's automaton defined in the previous section does not meet the requirements for the fuzzy automaton mentioned above. In the following subsections, we first review some existing fuzzy automaton definitions, together with their deficiencies, and then analyze individual parameters of Moore's automaton from the fuzzifying point of view.

9.3.2 Review of Fuzzy Automaton Definitions

A lot of fuzzy automaton definitions exist in literature, but none of them satisfy the requirements for a fuzzy entity cell as an FCA basic entity. The definitions could be grouped into three categories. The typical representatives of definitions are cited from Wee and Fu,[8] Mizumoto, Toyoda, and Tanaka,[9] and Virant and Zimic.[10] They are presented in the following subsections with their main deficiencies. The usage of special characters in cited definitions, such as "~", is the same used in other works by the authors.

9.3.2.1 Fuzzy Automaton by Wee and Fu[8]

Definition 3:[8] The finite fuzzy automaton is a quintuplet $\widetilde{A} = \langle X, Q, Y, \widetilde{\delta}, \widetilde{\lambda} \rangle$, where X represents a finite nonempty set of inputs, Q represents a finite nonempty set of internal states, and Y represents the finite nonempty set of outputs of a finite automaton. $\widetilde{\delta}$ and $\widetilde{\lambda}$ are static membership functions of fuzzy sets $\widetilde{B}$ and $\widetilde{C}$. The first translates between states and the second translates to output space of symbols:

$$\widetilde{\lambda} : Q \times X \times Y \to [0, 1],$$
$$q_l, q_m \in Q, x_j \in X, k \in T. \tag{II.9.10}$$

$$\widetilde{\delta}_{\widetilde{B}}(q_l, x_j, q_m) = \widetilde{\delta}\{q(k) = q_l, x(k) = x_j, q(k+1) = q_m\},$$
$$q_l \in Q, x_j \in X, y_m \in Y, k \in T. \tag{II.9.11}$$

$$\widetilde{\lambda}_{\widetilde{C}}(q_l, x_j, y_m) = \widetilde{\lambda}\{q(k) = q_l, x(k) = x_j, y(k) = y_m\}. \tag{II.9.12}$$

The transition caused by the string of input symbols is defined in Equation II.9.13, where j is the length of the input string and X^* a set of all possible strings on the basis of X.

$$q_l, q_m \in Q, X_j(k) \in X^*, k \in T : \widetilde{\delta}_{\widetilde{B}}(q_l, X_j(k), q_m)$$
$$= \max_{q_0, q_q, \ldots, q_s \in Q} \min[\widetilde{\delta}_{\widetilde{B}}(q_l, x_1, q_0), \widetilde{\delta}_{\widetilde{B}}(q_l, x_2, q_1), \ldots, \widetilde{\delta}_{\widetilde{B}}(q_{m-1}, x_j, q_m)]. \tag{II.9.13}$$

The expression $\widetilde{\delta}_{\widetilde{B}}(q_l, x_j, q_m)$ is called "the grade of transition" and belongs to the interval $[0, 1]$. The transition between two states is made if $\widetilde{\delta}_{\widetilde{B}}(q_l, x_j, q_m) \geq \alpha$ and is not made if $\widetilde{\delta}_{\widetilde{B}}(q_l, x_j, q_m) \leq \beta$. The parameters $\alpha, \beta (0 < \beta \leq \alpha < 1)$ are chosen subjectively. The transition is uncertain in the case where $\beta < \widetilde{\delta}_{\widetilde{B}}(q_l, x_j, q_m) < \alpha$.

The definition is based on the Mealy automaton, whose output symbol informs the outer world about the last transition of states. X, Q, and Y are ordinary crisp sets, so this automaton is not able to deal with fuzzy states, fuzzy inputs, etc. The crisp states are obtained through state transitions by rounding up on the basis of the subjectively chosen parameters. The state transition is from exactly one state in time t to exactly one state in time $t + 1$. It can be concluded that this automaton cannot be used for FCA purposes.

9.3.2.2   Fuzzy Automaton by Mizumoto, Toyoda, and Tanaka[9]

Definition 4:[9]   The finite fuzzy automaton is defined as quintuplet $\widetilde{A} = \langle X, \widetilde{Q}, \widetilde{\pi}, \{\widetilde{F}(x) : x \in X\}, \eta^G \rangle$, where X represents a finite nonempty set of inputs, $\widetilde{Q}$ represents a finite nonempty fuzzy set of internal states, and $\widetilde{\pi}$ is a vector which represents the starting distribution of states $(0 \leq \widetilde{\pi}_j \leq 1, j, \ldots, n; |\widetilde{Q}| = n)$. G represents the set of final states and η^G a 1-column matrix with values 1, where $q_i \in G$, and values 0, where $q_i \notin G$. For every $x \in X$, there exists a fuzzy translation matrix of nth order $\widetilde{F}(x) = \| f_{q_k, q_l}(x) \|, 1 \leq k, l \leq n, f_{q_k, q_l}(x) = f_{\widetilde{B}}(q_k, x, q_l)$. X^* represents the set of all finite strings over set X. An empty string e also belongs to X^*.

$$f_{\widetilde{B}}(q_l, e, q_m) = \begin{cases} 1, & \text{if } q_l = q_m \\ 0, & \text{if } q_l \neq q_m \end{cases}, \tag{II.9.14}$$

$$\forall (x, y) \in X^* : f_{\widetilde{B}}(q_l, xy, q_m) = \max_{q \in Q} \min[f_{\widetilde{B}}(q_l, x, q), f_{\widetilde{B}}(q, y, q_m)] \tag{II.9.15}$$

$$\forall x \in X^* : f_{\widetilde{B}}(x) = \widetilde{\pi} \circ \widetilde{F}(x) \circ \eta^G, length(x) = m,$$
$$\widetilde{F}(x) = \widetilde{F}(x_1) \circ \widetilde{F}(x_2) \circ \cdots \circ \widetilde{F}(x_m). \tag{II.9.16}$$

The term xy used in Equation II.9.15 represents the concatenation of two strings, and Equation II.9.16 represents the method of calculation of a state transition in fuzzy automaton $\widetilde{A}$. The fuzzy automaton $\widetilde{A}$ "accepts" the string x with a degree of $f_{\widetilde{B}}(x)$. A fuzzy distribution of initial states is introduced into the definition. The states of the fuzzy automaton become fuzzy, but the input and output sets remain crisp. In this way, processing on the basis of uncertain knowledge is enabled, but processing with uncertain data is not possible.

9.3.2.3 Fuzzy Automaton by Virant and Zimic

Definition 5:[10] Finite fuzzy automaton is a sextuplet $\widetilde{A} = \langle \widetilde{X}, \widetilde{Q}, \widetilde{Y}, \widetilde{\delta}, \widetilde{\lambda}, \widetilde{q_0} \rangle$, where $\widetilde{X}$ represents a finite unempty fuzzy set of inputs, $\widetilde{Q}$ represents a finite unempty fuzzy set of internal states, $\widetilde{Y}$ represents a finite unempty fuzzy set of output symbols, and $\widetilde{q_0}$ represents a fuzzy initial state of the fuzzy automaton. $\widetilde{\delta}$ and $\widetilde{\lambda}$ are the membership functions of fuzzy sets $\widetilde{B}$ and $\widetilde{C}$ where the first one translates between the fuzzy states and the second one translates to output space.

$$\widetilde{\delta} : \widetilde{Q} \times \widetilde{X} \times \widetilde{Q} \times \widetilde{T} \to [0, 1], \tag{II.9.17}$$

$$\widetilde{\lambda} : \widetilde{Q} \times \widetilde{X} \times \widetilde{Y} \times \widetilde{T} \to [0, 1]. \tag{II.9.18}$$

Compared to previous definitions, functions from Equations II.9.17 and II.9.18 translate on the basis of fuzzy operands $\widetilde{Q}, \widetilde{X}$, and $\widetilde{Y}$. All basic sets are fuzzified. The expected values for fuzzy sets $\widetilde{Q}, \widetilde{X}$ are continuous intervals. The authors did not explain the definition with a concrete example of processing. The definition is done because of the research of relief of fuzzy automaton. From the definition it cannot be seen what is the fuzzy state of the automaton in time t. The definition of term $\widetilde{\lambda}$ is composed in a similar way to that in the Mealy automaton, which is therefore not suitable for CA or FCA definition.

9.3.2.4 Consistency of the Notation Used

The above-mentioned definitions use different notations. For the remainder of this chapter we use the term FA for fuzzy automaton, instead of $\widetilde{A}$ that is ordinarily used as a symbol for fuzzy set representation.

9.3.3 The New Definition of Fuzzy Automaton

9.3.3.1 Different Possibilities of Input Variable Values Set to Fuzzy Automaton

In all types of automata we have to deal with input variable values (symbols) and initial values (initial states). From Definition 1 of the Moore automaton, both sets of variable values should be finite. For applications we will deal with the situations described in the follow-up.

9.3.3.2 Finite Set of Crisp Variable Values

In most cases we take from the environment a value of variable x, which originates from the finite set of crisp values X. The possible set of values could then be an interval of discrete values ([0, n]) or, for example, $n + 1$ nonnumerical values of variable x. If internal processing of the observed automaton is fuzzy based, then we have to provide in advance membership functions $\mu_i, 1 \le i \le n$, which are used for fuzzification of the crisp variable's values. The pairs $(x, \mu_{\widetilde{A_i}}(x))$ are obtained with fuzzification where $\widetilde{A_i}$ is a fuzzy set and $\mu_i(x)$ a membership of an element x to fuzzy set $\widetilde{A_i}$.

9.3.3.3 Infinite Set of Crisp Variable Values

There are not many applications where we take from the environment a value of variable x, which originates from the infinite set of crisp values X. In this case, we are confronted with the problem of

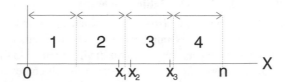

FIGURE II.9.1   Subinterval approach of finite set of entities forming ($m = 4$).

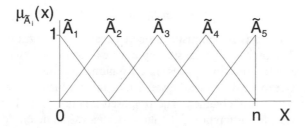

FIGURE II.9.2   An example of fuzzy sets distribution on the interval $[0, n]$, $m = 5$.

a translation from the infinite set of values (e.g., the continuous interval $[0, n]$), into the finite set of variable values, which is defined in the finite automaton. Two solutions are proposed:

- *Subinterval approach:* a continuous interval $[0, n]$ is divided into m subintervals; those represent the finite set of m entities (input symbols or states). Subintervals are of equal length or they can be formed logarithmically or even binary;[11] the value $i \in \{1, \ldots, m\}$ is chosen as an element of a finite set of values of input variable if the crisp value x captured from the environment falls into the ith subinterval (Equation II.9.19). Figure II.9.1 demonstrates the disadvantage of inadequate treatment of input values; the crisp pair of values (x_2, x_3) whose mutual distance is relatively big fall into the same subinterval, and the pair (x_1, x_2), whose mutual distance is smaller, fall into different intervals. This means that the pair of values (x_2, x_3) have the same influence on the processing of the automaton, and the pair of values (x_1, x_2) different influences on the processing. Another disadvantage, which can be seen from the Figure II.9.1, is in the decreased granulation of information representation. Both disadvantages are reduced with the increased number of subintervals, which, however, influences the complexity realization in the sense of processing time and place.

$$\forall x \in [0, n] : (x \in [0, j[\Rightarrow i = 1), (x \in [j, 2 \times j[\Rightarrow i = 2), \ldots,$$
$$(x \in [(m - 1) \times j, m \times j] \Rightarrow i = m). \tag{II.9.19}$$

- *Fuzzy approach:* m fuzzy sets (membership functions) are put into the continuous interval $[0, n]$ and they represent the finite set of m elements;[12] in this way, different crisp values from the continuous interval $[0, n]$ have different memberships to fuzzy sets. By distinguishing between different crisp values, we eliminate both problems from the interval approach. An example of m fuzzy sets distribution is presented in Figure II.9.2 and a formal expression of fuzzy sets in Equation II.9.20, where x represents a crisp value of input variable, $T(x)$ represents a linguistic variable, and $\tilde{A}_i$ a fuzzy set.

$$T(x) = \{\tilde{A}_1, \tilde{A}_2, \ldots, \tilde{A}_m\}, \forall x \in [0, n] : x \to (x, \mu_{\tilde{A}_i}(x)), 1 \leq i \leq m. \tag{II.9.20}$$

9.3.3.4 Infinite Set of Uncertain Crisp Variable Values

From the environment we take an uncertain crisp value a, which comes from the continuous interval $[0, n]$. Using a subjective approach a fuzzy number $\tilde{a}$ could be defined, where $a \in Supp(\tilde{a})$. The fuzzy number $\tilde{a}$ could then be said to be entering from the environment.[13] In the process of inference, the entering value is compared to some fuzzy criterion set $\tilde{A}$ which comes as a proposition part of a fuzzy rule. That means that the fuzzification process takes place over the fuzzy number. In the literature we could find only one approach of fuzzifying a fuzzy number.[14] It is presented in the expression below.

$$\mu_{\tilde{A}}(\tilde{a}) = \max[\min(\mu_{\tilde{A}}, \tilde{a})] \qquad (II.9.21)$$

Equation II.9.21 can be treated as an optimistic measure because of the max function used above the min operator. The result of processing with such optimistic measure is presented in Figures II.9.3 to II.9.5, where we can see a fuzzification process of crisp number c and fuzzy numbers $\tilde{a}$ and $\tilde{b}$ with respect to fuzzy set $\tilde{A}$. Due to the difference between $\tilde{a}$, $\tilde{b}$, and c, which can be seen from the figures, the same numerical result r was obtained. The proposed function from Equation II.9.21, which calculates the same results for different input values, decreases the quality of the fuzzy inference process. Instead of Equation II.9.21, therefore, we propose a new fuzzification approach of fuzzy number, formally written in Equation II.9.22. This distinguishes between different fuzzy values as presented in Figures II.9.4 and II.9.5.

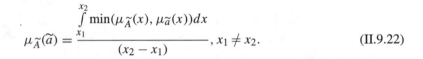

$$\mu_{\tilde{A}}(\tilde{a}) = \frac{\int_{x_1}^{x_2} \min(\mu_{\tilde{A}}(x), \mu_{\tilde{a}}(x))dx}{(x_2 - x_1)}, \, x_1 \neq x_2. \qquad (II.9.22)$$

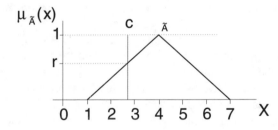

FIGURE II.9.3   Fuzzification of a crisp number $c(c = \frac{19}{7})$, with respect to $\tilde{A}$.

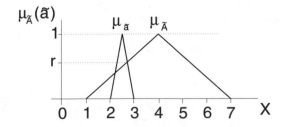

FIGURE II.9.4   Fuzzification of a fuzzy number $\tilde{a} = (2, 2.5, 2.5, 3)$, with respect to $\tilde{A}$.

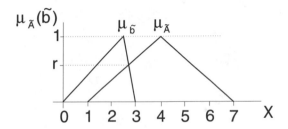

FIGURE II.9.5   Fuzzification of a fuzzy number $\tilde{b} = (0, 2.5, 2.5, 3)$, with respect to $\tilde{A}$.

The integration process takes place in the interval $[x_1, x_2]$, which is determined by the area where we achieved nonempty intersection between the supports of fuzzy entities $\tilde{a}$ and $\tilde{A}$.

Example 1:   Suppose parametrically (trapezoidal) determined fuzzy numbers $\tilde{a} = (2, 2.5, 2.5, 3)$ and $\tilde{b} = (0, 2.5, 2.5, 3)$ and a fuzzy set $\tilde{A} = (1, 4, 4, 7)$ (Figures II.9.4 and II.9.5).

Using Equation II.9.21 both memberships can be calculated as in Equation II.9.23. Using the calculation with Equation II.9.22, the results in Equation II.9.24 are obtained.

$$\mu_{\tilde{A}}(\tilde{a}) = \mu_{\tilde{A}}(\tilde{b}) = \frac{12}{21} \cong 0.57, \tag{II.9.23}$$

$$\mu_{\tilde{A}}(\tilde{a}) \cong 0.37, \mu_{\tilde{A}}(\tilde{b}) \cong 0.29. \tag{II.9.24}$$

Analyzing the numerical results from Equation II.9.24 and Figures II.9.4 and II.9.5 again, the membership of "narrow" fuzzy number $\tilde{a}$ to the fuzzy set $\tilde{A}$ can be seen to be higher than the membership of fuzzy number $\tilde{b}$, which is more dispersed to the left.

9.3.3.5   Fuzzy States of the Automaton

According to Definition 1, the set of states is finite and nonempty, and its elements are crisp. The last property is not suitable for integrating uncertain knowledge and data of a system's dynamics into the main entity of a cell that is modeled with an abstract term of fuzzy automaton. The following example is a real case study of collected knowledge from the observation process of the dynamics of a real system.

Example 2:   From observation of a real system we have determined three fuzzy states and the crisp values set of the infinite input variable which originates from interval $[0, n]$. Due to this we have defined three fuzzy input symbols (three fuzzy sets for input variable). Possible transitions from the state at time t to a new state at time $t + 1$ are shown in Figure II.9.6.

The membership functions used for the fuzzification process of input variables are first defined and presented in Figure II.9.7. Basic fuzzy automaton definitions are given in Equations II.9.25 to II.9.27.

$$\hat{\mathbf{q}} = [\tilde{Q}_1, \tilde{Q}_2, \tilde{Q}_3], \hat{\mathbf{x}} = [\tilde{X}_1, \tilde{X}_2, \tilde{X}_3], \tag{II.9.25}$$

$$\hat{q}(t) = [(\mu_{\tilde{Q}_i}(t), \tilde{Q}_i)], 1 \le i \le k, \tag{II.9.26}$$

$$\mu_{\hat{q}}(t) = [(\mu_{\tilde{Q}_i}(t))], 1 \le i \le k. \tag{II.9.27}$$

Two new terms are introduced in Equation II.9.25: $\hat{\mathbf{q}}$, which represents the vector of fuzzy sets of internal states of automaton and $\hat{\mathbf{x}}$, which represents the vector of fuzzy sets of input symbols. The

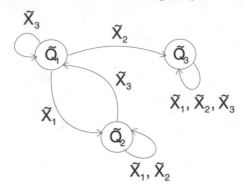

FIGURE II.9.6   An example of state transitions in fuzzy automaton.

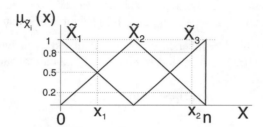

FIGURE II.9.7   The definition of input terms.

individual fuzzy set in a vector represents a single fuzzy state or an input symbol. Another new term is $\hat{q}(t)$ which represents the *global state* of the fuzzy automaton at time t. With respect to Equation II.9.26, it is a vector which consists of k pairs of type (membership/fuzzy set), where k is the number of fuzzy states defined in advance. Due to its vectorial nature, it allows an automaton to be in more than one state at time t. Every state in which the automaton is at time t is described with a membership to the corresponding fuzzy set. In Equation II.9.27 a vector of memberships $\mu_{\hat{q}}(t)$ is introduced representing the global state of the automaton at time t.

Let us also suppose for our example that the system starts in state $\hat{q}(t_0) = \{1/\widetilde{Q}_1, 0/\widetilde{Q}_2, 0/\widetilde{Q}_3]$. That means that it is certain that the state of the automaton belongs "fully" to the fuzzy set $\widetilde{Q}_1$ and to any other fuzzy set. What happens if we start to process such an automaton? Suppose that two input crisp values, x_1, x_2, come consecutively after time t_0. Both values are marked in Figure II.9.7. The starting conditions for processing are determined by Equation II.9.28. On the basis of these starting conditions, the initialized set of rules can be extracted from Figure II.9.6. They are presented in Equation II.9.29.

$$\hat{q}(t_0) = [1/\widetilde{Q}_1, 0/\widetilde{Q}_2, 0/\widetilde{Q}_3], \mu_{\hat{q}}(t_0) = [1, 0, 0],$$
$$x_1 \to (\mu_{\widetilde{X}_1}(x_1) = 0.5, \mu_{\widetilde{X}_2}(x_1) = 0.5, \mu_{\widetilde{X}_2}(x_1) = 0). \tag{II.9.28}$$

IF ($\hat{q}[1]$ is $\widetilde{Q}_1$) AND (X is $\widetilde{X}_1$), THEN ($\hat{q}[2]$ is $\widetilde{Q}_2$), $(\mu_{\widetilde{Q}_1}(t_0) = 1, \mu_{\widetilde{X}_1}(x_1) = 0.5)$,

IF ($\hat{q}[1]$ is $\widetilde{Q}_1$) AND (X is $\widetilde{X}_2$), THEN ($\hat{q}[3]$ is $\widetilde{Q}_3$), $(\mu_{\widetilde{Q}_1}(t_0) = 1, \mu_{\widetilde{X}_2}(x_1) = 0.5)$,

IF ($\hat{q}[1]$ *is* $\widetilde{Q}_1$) AND (X is $\widetilde{X}_3$), THEN ($\hat{q}[1]$ is $\widetilde{Q}_1$), $(\mu_{\widetilde{Q}_1}(t_0) = 1, \mu_{\widetilde{X}_3}(x_1) = 0.5)$.   (II.9.29)

In the present work all conditional terms (left side of the rule) are treated at time t and all conclusion terms (right side of the rule) at time $t + 1$. Using Mamdani's type of fuzzy inference,[15] which uses the transmission of minimal influence factor (membership) or relevancy from the conditional to the conclusion side of the rule, we get a new global state of the fuzzy automaton at time $t_0 + 1$, as presented in Equation II.9.30.

$$\hat{q}(t_0 + 1) = [0/\widetilde{Q}_1, 0.5/\widetilde{Q}_2, 0.5/\widetilde{Q}_3], \ \mu_{\hat{q}}(t_0 + 1) = [0, 0.5, 0.5]. \tag{II.9.30}$$

The result of a new global state at time $t_0 + 1$ can be interpreted as a state of the automaton where it has some membership to state $\widetilde{Q}_2$ and some other membership to state $\widetilde{Q}_3$. If we then process the second entering value x_2, we get the result in Equation II.9.31.

$$\hat{q}(t_0 + 2) = [0.5/\widetilde{Q}_1, 0.2/\widetilde{Q}_2, 0.5/\widetilde{Q}_3], \ \mu_{\hat{q}}(t_0 + 3) = [0.5, 0.2, 0.5]. \tag{II.9.31}$$

Due to the enabling of fuzzy input data, uncertainty about the input data is transformed into uncertainty about the state of the automaton during processing.

9.3.3.6 Definition of the New Fuzzy Automaton

We can now present a new extended definition of the fuzzy automaton, based partially on Virant and Zimic[10] and Adamtzky.[16]

Definition 6: A fuzzy automaton is defined as a sextuplet $FA = \langle \hat{\mathbf{x}}, \hat{\mathbf{q}}, Y, \widetilde{\delta}, \widetilde{\lambda}, \mu_{\hat{q}}(t_0) \rangle$, where $\hat{\mathbf{x}}$ and $\hat{\mathbf{q}}$ represent finite vectors of unempty fuzzy sets of inputs and internal states and Y represents a finite and crisp set of output symbols. $\widetilde{\delta}$ is a stationary function which translates to a new internal state and $\widetilde{\lambda}$ is a stationary function which translates to a new output symbol on the basis of the global state $\hat{q}(t)$. $\mu_{\hat{q}}(t_0)$ represents the vector of memberships to individual fuzzy states of the automaton.

Definition 7: For the needs of fuzzy cellular automata the fuzzy automaton is defined as above, with some additions. $\hat{\mathbf{x}}$ is defined as a concatenation of vectors of states of n neighboring cells in cellular automata $\hat{\mathbf{q}}^n$ and of vectors of nonempty fuzzy sets $\hat{\mathbf{x}}_c$, thus providing a method for introducing some global variable values into the decision process ($\hat{\mathbf{x}} = \hat{\mathbf{q}}^n \cdot \hat{\mathbf{x}}_c$). $\widetilde{\delta}^*$ is an extension of $\widetilde{\delta}$ and translates on the basis of finite strings over input fuzzy symbols.

Functions $\widetilde{\delta}, \widetilde{\lambda}$ are realized mainly through the sets of fuzzy rules. The above definitions lead to the valid relations in Equation II.9.32.

$$\widetilde{\delta} : \hat{\mathbf{q}} \times \hat{\mathbf{x}} \times \hat{\mathbf{q}} \to [0, 1] = \hat{\mathbf{q}} \times \hat{\mathbf{q}}^n \times \hat{\mathbf{x}}_c \times \hat{\mathbf{q}} \to [0, 1],$$

$$\widetilde{\lambda} : \hat{\mathbf{q}} \times Y \to [0, 1],$$

$$\hat{q}(t + 1) = \widetilde{\delta}(\hat{x}(t), \hat{q}(t)), \ \hat{x} \in \hat{\mathbf{x}}, \forall \hat{x}_1 \in \hat{\mathbf{x}}, \forall \hat{x}_2 \in \hat{\mathbf{x}}^*:$$

$$\widetilde{\delta}^*(\hat{q}_0, \hat{x}_1 \hat{x}_2, \hat{q}_n) = \max_{\forall \hat{q}_1 \in \hat{\mathbf{q}}} \min[\widetilde{\delta}(\hat{q}_0, \hat{x}_1, \hat{q}_l), \widetilde{\delta}^*(\hat{q}_l, \hat{x}_2, \hat{q}_n)]. \tag{II.9.32}$$

If the fuzzy automaton is introduced into the space of two-dimensional cellular automata space, then it is marked as $FA_{i,j} = \langle \hat{\mathbf{x}}, \hat{\mathbf{q}}, Y, \widetilde{\delta}, \widetilde{\lambda}, \mu_{\hat{q}}(i, j, t_0) \rangle$. The above definitions lead to a basic entity which changes its vague (fuzzy) state in time. The transition is influenced by the fuzzy input data and the fuzzy state of the entity in the previous step.

9.4   Fuzzy Cellular Automata

9.4.1   Introduction

The recent literature on the cellular automata and/or fuzzy logic contains little research in the field of fuzzy cellular automata. The first reference to the term "fuzzy cellular automata" is in Adamatzky.[16,17] The author describes approaches of classifying structures into different fuzzy cellular automata classes. Ordinary crisp cellular automata have been expanded to a structure with multi-value internal states which is not yet linguistically programmable.[18,19] They use such structures in applications of two-dimensional cellular automata to filtering and heat diffusion. In Cattaneo et al.,[20] a one-dimensional crisp cellular automata structure with a defective subpart (a fuzzy part of an array of cells) is presented, capable of processing under fuzziness. The most interesting part of this work is a study of an intersection of fuzzy and crisp areas. The structure is still not programmable and does not belong to the field of the synthesis approach. In the following subsections we will attempt a definition of a completely linguistically programmable structure of fuzzy cellular automata.

9.4.2   Definition of Fuzzy Cellular Automata

Definition 8: Fuzzy cellular automata are defined as a septuplet $FA = \langle P, \hat{\mathbf{q}}, \hat{N}, \hat{F}, \hat{T}, \widetilde{\lambda}, \hat{\mathbf{x}} \rangle$, where P is an m-dimensional space of cells (automata), $\hat{\mathbf{q}}$ is a vector of unempty and finite fuzzy sets of all possible cell states, $\hat{N}$ is a fuzzy definition of set of neighbors and $\hat{F}$ a finite set of fuzzy rules describing transitions between states of cells. $\hat{T}$ denotes the possibility of fuzziness of time and $\widetilde{\lambda}$ denotes a translation function which forms a new output crisp symbol on the basis of a global fuzzy state $\hat{q}$ for each cell. $\hat{\mathbf{x}}$ denotes a set of fuzzy variables, which have the significance of global variables. Space P is discrete.

Compared to the ordinary CA presented in Definition 2, the space P remains unaltered. The characteristics of vector $\hat{\mathbf{q}}$ and the set of fuzzy rules $\hat{F}$ coincide with the fuzzy state and the translation function from the fuzzy automaton definition. Determination of $\hat{N}$ depends mainly on the type application. The fuzziness of time is applicable in the area of referencing uncertain time points in fuzzy rules and also in the area of possible asynchronicity of processing.

Example 3: Suppose a fuzzy cellular automata structure which simulates the right-shift function in one-dimensional space. The shift is stimulated by input symbol x_0. From observation of the simulated system it is evident that the first left neighboring cell has a very large influence on the shifted summary and the next left neighbor half of that influence. The formal description of such an automaton is then given by Equation II.9.33.

$$m = 1, \hat{\mathbf{q}} = [\widetilde{Q}_1, \widetilde{Q}_2], \hat{\mathbf{x}} = [\widetilde{X}_0],$$

$$\hat{N}(c_i) = \{(c_{i-1}, \mu_{\hat{N}_{(c_i)}}(c_{i-1}) = 1), \ (c_{i-2}, \mu_{\hat{N}_{(c_i)}}(c_{i-2}) = 0.5)\},$$

$$\hat{F} = \{(\text{if } (X(t) \text{ is } \widetilde{X}_0) \text{ and } (c_{i-2}(t) \text{ is } \widetilde{Q}_j) \text{ and } (c_{i-1}(t) \text{ is } \widetilde{Q}_j) \text{ then } (c_i(t+1) \text{ is } \widetilde{Q}_j))\},$$

$$j \in \{0, 1\}. \tag{II.9.33}$$

9.4.3   Fuzzy Neighborhood

In Figures II.9.8 and II.9.9 are two widely used neighborhoods in 1-D and 2-D CA applications. In all cases we have observed cell $c_c(t)$, which forms its next state in time $t+1$ according to Equation II.9.9 in functional dependence on the gray-shaded neighbors.

For cases b) and d), it is typical that the influence of neighboring cells on the observed cell $c_c(t)$ is artificially equaled because, in the sense of real-world application, the cell is a model of a small part,

FIGURE II.9.8   Two 1-D CA neighborhoods: a) $R = 1, k = 3$, b) $R = 2, k = 5$.

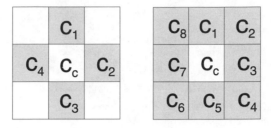

FIGURE II.9.9   Von Neumann and Moore neighborhoods: c) $k = 5$, d) $R = 1, k = 9$.

which has nearest neighbors, some neighbors with smaller influence, etc. For case c), the influence of diagonal neighbors is simply eliminated, which is not the case in real systems.

If we introduce the fuzzy approach of neighborhood, where the fuzzy set "Cell c_i is a neighbor of cell c_c" with one term is formed, then we can build more realistic models of neighbor influence. The main starting point for membership to a neighborhood set should be the distance between the observed central cell c_c and the candidate neighbor cell c_i. In this case we have, for 1-D CA, only one approach:

- *Euclidean* distance between cells in place

$$d(c_c, c_i) = |x_c - x_i|. \tag{II.9.34}$$

For 2-D CA, there are two basic approaches for distance evaluation between two points:

- *Euclidean* distance between cells in place

$$d(c_c, c_i) = \sqrt{(x_c - x_i)^2 + (y_c - y_i)^2}, \tag{II.9.35}$$

- *Manhattan* distance between two points

$$d(c_c, c_i) = |x_c - x_i| + |y_c - y_i|. \tag{II.9.36}$$

After obtaining these different distances between potential neighboring cells and the central cell, we also have to define the criteria for determining membership function. For this, two different simple approaches can be used:

- The *grading* method, where all distances are sorted in increasing order and marked with ascending numbers

$$f_{neighbor}(d(c_c, c_i)) = \frac{1}{position(d(c_c, c_i))}. \tag{II.9.37}$$

- The *inverse* approach, where we simply use the inverse function of distance

$$f_{neighbor}(d(N, C)) = \frac{1}{d(N, C)}. \tag{II.9.38}$$

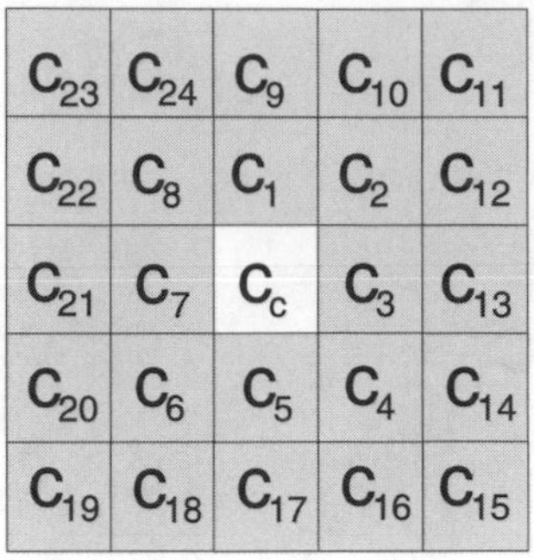

C_{23}	C_{24}	C_9	C_{10}	C_{11}
C_{22}	C_8	C_1	C_2	C_{12}
C_{21}	C_7	C_c	C_3	C_{13}
C_{20}	C_6	C_5	C_4	C_{14}
C_{19}	C_{18}	C_{17}	C_{16}	C_{15}

FIGURE II.9.10   Moore's neighborhood, $R = 2$, $k = 25$, $m = 2$.

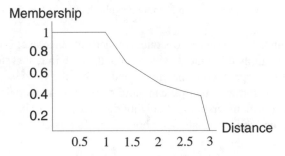

FIGURE II.9.11   Membership function, $m = 2$, $R = 2$, *Euclidean* distance, *inverse* approach.

By these two approaches, membership function shape can be obtained. In the next three figures we present an extended radius version ($R = 2$) of Figure II.9.9, case d). D-2 CA, with Moore definition of neighborhood, is presented first graphically (Figure.II.9.10), then with the *inverse* (Figure II.9.11), and third, with the *grading* approach (Figure II.9.12) to defining membership function of a fuzzy set with one term.

Selection of one of the methods shown above is assumed to depend mainly on the real system, which may be modeled with fuzzy cellular automata. More precisely, it depends on the difference between the influence of nearest and remote cells.

9.5   Logical Design of Fuzzy Cellular Automata Circuits

9.5.1   Introduction

This section deals with the fuzzy sequential (switching) circuit which represents the role of a fuzzy cell (FC) in fuzzy cellular automata structure (FCAS). In this case, FCAS is an array of $m \times n$ identical fuzzy cells which mutually interact. For the basic interactive cell we take a fuzzy trigger device, named FTDC.[22] The FTDC-device placed at location (i, j) in two-dimensional space has the state $\hat{q}_{i,j}(t)$, which is also a state of the fuzzy cell itself. In this way, the state of an FCAS at time t is represented by the set $FCAS = \{\hat{q}_{i,j}(t)\}, i = 1, \ldots, m, j = 1, \ldots, n$. FCAS is always observed through an individual cell $c_{i,j}$ with regard to its neighborhood. An example of cells neighboring to $c_{i,j}$ could be the set of cells $\{c_{i+1,j}, c_{i-1,j}, c_{i,j+1}, c_{i,j-1}\}$ (eastern, western, northern, and southern cells). This is called the Von Neumann's neighborhood. The treatment of just a four-cell neighborhood invokes no restrictions on the real design of FCAS. With the knowledge and skills attained in this section, the designer could also take, e.g., an eight-cell neighborhood.

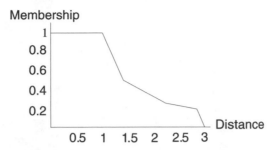

FIGURE II.9.12 Membership function, $m = 2$, $R = 2$, *Manhattan* distance, *grading* approach.

Processing a fuzzy rule is quite similar to the working of sequential circuits or automata. The general theory of sequential circuits or general theory of automata can therefore be used to represent the behavior of FCAS. In this section, we take sequential circuits as a basis for realizing the FCAS. A fuzzy memory device is used to interpret fuzzy rule. The first description of fuzzy memory device was by Hirota and Ozawa,[23] and with other later descriptions, takes into account only a JK memory device. Virant[22] presented in addition to the fuzzy JK memory device, FD (fuzzy delay), FT (fuzzy trigger), and FSR (fuzzy set-reset) fuzzy memory devices (cells) with correspondence to the binary memory devices D, T, and SR (flip-flops). Normally fuzzy devices such as FD, FT, or FSR are needed for fuzzy memory design, but it is shown in this section that they could be also used for designing cells in FCAS.

9.5.2 Logical Interpretation of the Fuzzy Cell

9.5.2.1 Basic Fuzzy Memory Devices

Consider the simple case of a fuzzy rule (Equation II.9.39) in the context of FCAS. Items $\hat{q}_{i,j}(t)$ and $\hat{N}_{i,j}(t)$ are not ordinary fuzzy sets or numbers, but represent, respectively, the fuzzy state of the observed cell and its neighborhood.

$$\text{IF } \hat{q}_{i,j}(t) \text{ is } \textbf{\textit{high}} \text{ AND } \hat{N}_{i,j}(t) \text{ is } \textbf{\textit{normal}} \text{ THEN } \hat{q}_{i,j}(t+1) \text{ is } \textbf{\textit{low}}. \tag{II.9.39}$$

Such a fuzzy rule is time sensitive because the variable t is a time step in the system behavior. This rule can be introduced with the fuzzy memory device FTDC,[22] where FT means fuzzy trigger cell (device), and characters D and C mean DOWN and CLEAR, respectively. Condition DOWN tells us that initial input signal T grows from 0–1 and condition CLEAR tells us that the output of FT can be forced to value 0. FT has the following output function:

$$\hat{q}(t+1) = \max(\min(\hat{q}(t), \overline{T}(t)), \min(\overline{\hat{q}}(t), T(t))),$$

$$\hat{q}(t+1) = \hat{q}(t)\overline{T}(t) + \overline{\hat{q}}(t)T(t),$$

$$\mathbf{D}^1 \hat{q} = \hat{q}\overline{T} + \overline{\hat{q}}T. \tag{II.9.40}$$

Here we have three different interpretations of the "next state" function. In the second case, the max( ) is represented with operator "+" and min( ) with "." (or nothing). From the theoretical point of view, the max( ) is an s-norm and the min( ) is a t-norm. Negation in fuzzy calculations is treated as $\overline{A} = 1 - A$. In the third case, a time operator "D" is introduced. It allows the simplest presentation of a time-dependent fuzzy switching function.[22] The variables T and $\hat{q}$ take values from fuzzy interval [0, 1], and can therefore be taken as membership functions of the corresponding

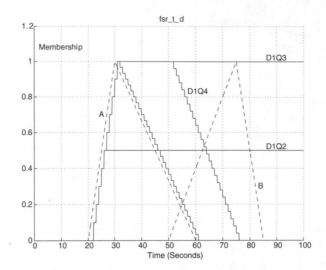

FIGURE II.9.13   Comparison of state behaviors of FD, FT, FTD, and FSR memory cells.

input and output fuzzy sets, respectively. In this way, output functions for FSR and FD may be presented as

$$\mathbf{FSR} : \mathbf{D}^1 \hat{q} = S + \overline{R}\hat{q},$$

$$\mathbf{FD} : \mathbf{D}^1 \hat{q} = D. \tag{II.9.41}$$

S, R, and D are input variables and $\hat{q}$ is an output variable. In the binary SR device (Set-Reset flip-flop) there exists the condition $SR = 0$. The possible value of this condition is extended to interval [0, 0.5] because in the FSR memory cell the relation $0 \leq SR \leq 0.5$ holds.[22] The first Equation II.9.41 holds only if $\max(\min(R, S, \hat{q}), \min(R, S, 1 - \hat{q})) = 0$. If this is not true, the output function in the FSR cell is

$$\mathbf{D}^1 \hat{q} = S + \overline{R}\hat{q} + SR\hat{q} + SR\overline{\hat{q}}. \tag{II.9.42}$$

The second Equation II.9.41 is valid for the memory device FD. Binary (delay) memory device D is equivalent to binary device SR if we take input equality $D = S = 1 - R$. Such equality is also useful in the case of FSR and FD.

Basic fuzzy cells FD, FSR, FT, FTD, and FTDC can be compared.[21,22] Figure II.9.13 shows the results obtained by execution of the small MATLAB simulation program FSR_T_D.MDL. Inputs A and B are two membership functions in the time domain of 100 sec. Annotations to the behavior in Figure II.9.13 are as follows:

- D1Q1 is the output function from an FD cell. Since $D = \widetilde{A}$, the next state is $\mathbf{D}^1 \hat{q} = \widetilde{A}$.
- Behavior D1Q2 is the result of the FT cell. In this case, the next state is given by the expression $\mathbf{D}^1 \hat{q} = \widetilde{A}\,\overline{\hat{q}} + \overline{\widetilde{A}}\hat{q}$. If $\hat{q} = 0.5$ then $\mathbf{D}^1 \hat{q} = 0.5$. The value $\hat{q} = 0.5$ is called the equilibrium (terminal) state from which it is impossible to come out.
- D1Q3 is output of the FTD cell. This fuzzy cell has an equilibrium point at maximum input $\widetilde{A}$ (core of fuzzy number). To move from this equilibrium point to $\hat{q} < 1$ a CLEAR possibility is needed in fuzzy cell FTD. The fuzzy memory cell with inputs T and C (Trigger and Clear) is named FTDC.[21,22] This cell can be very useful in the FIS (fuzzification, fuzzy inference, defuzzification) process. Output D1Q3 represents the membership function for fuzzy interval $[\mathbf{D}^1 \widetilde{A}, \infty]$.

- **D1Q4** is the behavior of an FSR cell if $S = \tilde{A}$ and $R = \tilde{B}$. In this case, membership function $\mathbf{D}^1 \hat{q}$ represents fuzzy interval $[\mathbf{D}^1 \tilde{A}, \tilde{B}$. If $\tilde{A}$ is the membership of fuzzy date $\tilde{a}$ and $\tilde{B}$ the membership of fuzzy date $\tilde{b}$, the corresponding time fuzzy interval is $(\tilde{a} + 1), \tilde{b}$. Boundary fuzzy set $(\tilde{a} + 1)$ is inside and boundary $\tilde{b}$ outside the time interval.

If the pair of variables $(T, \hat{q})$ is valid for memory device FT, the pair $(T^*, \hat{q}^*)$ is valid for device FTD with transformations

$$T = \frac{1}{G} \cdot T^*, \tag{II.9.43}$$

$$\hat{q}^* = G \cdot \hat{q}.$$

G is the gain with value $G > 2$ (from the point of view of approximation it is better to take a value of 100 or more). It is important that the memory device with pair $(T^*, \hat{q}^*)$ has a nonequilibrium (nonterminal) state $\hat{q} = 0.5$. The terminal state occurs at $\hat{q} = 1$. We can go out from this state or states $\hat{q} > T$ under constraint CLEAR. Thus, the memory device FTDC has two inputs, T and C.[21,22] Fuzzy trigger device as an automaton is shown in Virant, Zimic, and Mraz.[28]

9.5.2.2 Fuzzy Memory Cell and Its Rule

So far in this section we have discussed the concept of memorizing and we have also introduced devices which carry out memorizing functions. The basic memory device in any cell of FCAS has one or more inputs used in the inference process. The set of fuzzy rules is the same for all cells in FCAS. In previous discussions, we have introduced the two terms, *memory cell* and *memory device*. In some ways, they are synonymous, but to be more precise: *memory device* will be the basic item inside a cell. When input and output switching functions are added to the memory device, then it becomes a fuzzy cell with the capability of memorizing. Consequently, the *fuzzy memory device* together with its input and output functions is called *fuzzy memory cell*. In our case, the fuzzy rule set will be replaced with the fuzzy switching function(s) which uses input(s) of memory device in the observed cell. As mentioned at the beginning, there exists only one problem in FCAS design: how to find adequate fuzzy switching function(s) for given fuzzy proposition(s).

Let us take the single input T of a memory device. We define this input as a switching function with four independent variables, West (W), North (N), East (E), and South (S). The input function of the memory device is thus $T = f(W, N, E, S)$ in such a way that directions W, N, E, and S are input variables of the observed cell. The situation is inverted at the output of the memory device. This output is distributed, on the basis of certain criteria, into four outputs of the fuzzy cell. A cell $c_{i,j}$ in FCS therefore has four inputs and four outputs. In our MATLAB simulation programs, we take for those variables the following names InWij, InNij, InEij, InSij, and OuWij, OuNij, OuEij, OuSij. Examples of very simple input functions are shown in Equation II.9.44 and examples of very simple output functions in Equation II.9.45.

$$\text{Tij} = \max(\text{InWij}, \text{InNij}, \text{InEij}, \text{InSij}),$$

$$\text{Tij} = \min(\text{InWij}, \text{InNij}, \text{InEij}, \text{InSij}). \tag{II.9.44}$$

$$\text{OuWij} = \text{wW}*\hat{q}_{i,j}, \text{OuNij} = \text{wN}*\hat{q}_{i,j},$$

$$\text{OuEij} = \text{wE}*\hat{q}_{i,j}, \text{OuSij} = \text{wS}*\hat{q}_{i,j},$$

$$0 \leq \text{wX} \leq 1, 0 \leq \hat{q}_{i,j} \leq 1. \tag{II.9.45}$$

Here wX is some weight and $\hat{q}_{i,j}$ the output of the memory device. These descriptions allow us to draw, simulate, and design a cell which uses the model in Figure II.9.14.

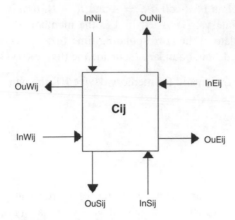

FIGURE II.9.14   The model of a fuzzy cell in FCS.

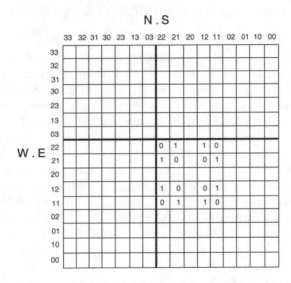

FIGURE II.9.15   Veitch diagram of fuzzy switching functions f_m (phrases labeled with 1) and f_e (phrases labeled with 0).

A fuzzy rule is placed between the fuzzy cell $c_{i,j}$ and the fuzzy memory device. To show the fuzzy rule setting, let us take an example to describe the working of an observed FCAS. The working is expressed with a fuzzy proposition as follows:

P_1: *'If odd number of input directions, then next state is equal odd number of directions.'*

The meaning of the phrase "odd number of directions" is given by the following fuzzy switching function.

$$f_m(S, N, W, E) = S\overline{N}\ \overline{W}\ \overline{E} + \overline{S}\ N\overline{W}\ \overline{E} + \overline{S}\ \overline{N}\ W\overline{E} + \overline{S}\ \overline{N}\ \overline{W}\ E$$
$$+ SNW\overline{E} + SN\overline{W}E + S\overline{N}WE + \overline{S}NWE. \tag{II.9.46}$$

The map of this function, shown in Figure II.9.15, is the fuzzy Veitch diagram (see Kandel's notation in Kandel and Lee).[24] Equation II.9.46 contains product-phrases where all of them are labeled with 1. All product-phrases labeled with 0 are in the Equation II.9.47.

$$f_e(S, N, W, E) = S\overline{N}\ \overline{W}\ E + \overline{S}\ N\overline{W}E + SN\overline{W}\ \overline{E} + \overline{S}NW\overline{E}$$
$$+ \overline{S}\ \overline{N}WE + S\overline{N}W\overline{E} + \overline{S}\ \overline{N}\ \overline{W}\ E + SNWE \qquad (\text{II}.9.47)$$

Functions f_m and f_e are in minimal product form, and therefore there is no possibility of minimizing their forms. All product-phrases in both functions are out of the rows/columns that are labeled with $3X$ or $X3$, $X = 1, 2, 3$. It means that the function does not include a switching variable of type $X\overline{X}$ that can lead to minimization with fuzzy consensus.[24] Still more, f_m is a fuzzy modulo-two function and f_e is fuzzy equivalence. Only these two fuzzy switching functions have the characteristic of direct complementary transformation, i.e., $f_m = 1 - f_e$.

Consider the Equation II.9.46 again. In Table II.9.1 we find function

$$f(S, N, W, E) = f_{num}(t, 0, 1, p, q). \qquad (\text{II}.9.48)$$

Here $S, N, W,$ and E are four independent directions, t an independent input variable, and $0, 1, p,$ and q four generic values. Suppose that each input direction has the possibility for value $t, 0, 1, p,$ or q. f_{num} means that the sum of a number of values $t, 0, 1, p,$ and q is 4. For example, if two directions have value t, only the remaining values 0 or 1, or p or q can be used later. Suppose that the next input direction has value 0. For the last input direction 1, p and q are available. If we take for this input direction the value p, then values 1 and q remain unused. Constants p and q are taken from fuzzy interval [0, 1].

Possible values of the function f_{num} are shown in the last column in Table II.9.1. The number 0 inside the table represents unused values $t, 0, 1, p,$ or q. The symbol $L_{\pm k}[\ ]$ represents the clipped rest of the function which depends on variable shown in [ ]. For example, the clipped function $0.5 + 0.5\sin(\ )$ is shown in Figure II.9.16. Clipping is performed with regard to value 0.5, and not perhaps to value 0, which is usually in use. The clipping factor k is p or $1 - q$ where $0 \leq p \leq 0.5$ and $0.5 \leq q \leq 1$. The symbol $\|$ represents the absolute value with regard to value 0.5 and not with regard to 0. Therefore, $\| = 0.5 + 0.5\ \text{abs}(\ )$.

Figure II.9.17 shows some generic fuzzy rule results in the cell $c_{i,j}$. Here $p = 0.4, q = 0.8,$ and $t = \sin(\ )$. The fuzzy rule result is not yet the state of the cell, while the fuzzy rule is just input to the fuzzy memory device in cell $c_{i,j}$. For the memory device in the cell, we can take FD, FT, FSR, FTD, FTDC, or FJK device. From the point of view of fuzzy hardware, the simplest way is FD. In this example we deal with the second expression of Equation II.9.41. That means, the next state of the memory device is the state of fuzzy cell $c_{i,j}$.

Going from proposition P_1 to the fuzzy rule domain, we define input and output variables of the cell. The name input is set for the input variable which has the value of $f(S, N, W, E)$ and output for the state of the fuzzy cell. In this way, for a given proposition P_1, the following rule is applied:

IF input is *odd number of directions*, THEN output is *odd number of directions*,

or in switching form

IF input is $f(S, N, W, E)$, THEN output is $f(S, N, W, E)$.

The output function of FD is the second Equation in II.9.41. For this device there is input $= D$ and output $= \mathbf{D}^1\hat{q} = D$. The last form of the rule can thus be written as

IF D is $f(S, N, W, E)$, THEN $\mathbf{D}^1\hat{q}$ is $f(S, N, W, E)$. $\qquad (\text{II}.9.49)$

TABLE II.9.1   Modulo-Two Fuzzy Switching Rule in a Fuzzy Cellular Cell with Memory Device FTDC

Number of t	Number of 0	Number of 1	Number of p	Number of q	$f(t, 0, 1, p, q)$
1	3	0	0	0	t
1	2	1	0	0	$-t$
1	1	2	0	0	t
1	0	3	0	0	$-t$
1	2	0	1	0	$L_{\pm p}[t]$
1	1	0	2	0	$L_{\pm p}[t]$
1	0	0	3	0	$L_{\pm p}[t]$
1	2	0	1	0	$L_{\pm p}[t]$
1	1	0	2	0	$L_{\pm p}[t]$
1	0	0	3	0	$L_{\pm p}[t]$
1	1	1	1	0	$-L_{\pm p}[t]$
1	0	2	1	0	$L_{\pm p}[t]$
1	2	0	0	1	$-L_{\pm(1-q)}[t]$
1	1	1	0	1	$L_{\pm(1-q)}[t]$
1	0	2	0	1	$-L_{\pm(1-q)}[t]$
1	1	0	1	1	$-L_{\pm(1-q)}[t]$
1	0	0	2	1	$-L_{\pm(1-q)}[t]$
2	0	0	2	0	$-L_{\pm p}[|t|]$
2	0	0	1	1	$L_{\pm p}[|t|]$
3	0	0	1	0	$L_{\pm p}[t]$
3	0	0	0	1	$-L_{\pm(1-q)}[t]$
4	0	0	0	0	$-|t|$
0	4	0	0	0	0
0	3	1	0	0	1
0	2	2	0	0	0
0	1	3	0	0	1
0	0	4	0	0	0
0	3	0	1 $p = 0.5$	0	0.5
0	3	0	0	1 $q = 0.5$	0.5
0	2	0	1 $p = 0.5$	1 $q = 0.5$	0.5
0	1	1	1 $p = 0.5$	1 $q = 0.5$	0.5
0	0	2	1 $p = 0.5$	1 $q = 0.5$	0.5

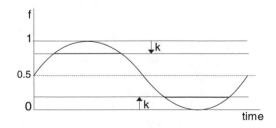

FIGURE II.9.16   Clipped function $t = 0.5 + 0.5 \sin(\ )$.

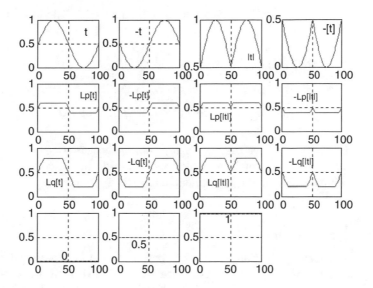

FIGURE II.9.17 Some generic behaviors for $t = 0.5 + 0.5 \cdot \sin(2\pi \cdot x/100)$ in some direction of $c_{i,j}$.

Here $f(S, N, W, E)$ is the fuzzy switching function (fuzzy algebraic/arithmetic expression), but proposition P_1, for which the $f(S, N, W, E)$ is written, linguistic expression. The difference between these two expressions is formal because each switching function can also be shown as a proposition. From this point of view, in the last rule 'is' is interchangable with '='.

Let us observe a more complicated fuzzy proposition. Instead of the proposition P_1 we now take

> P_2: *'If an odd number of directions are active with value a, and the state $\hat{q}$ is smaller than a, then the next state is equal to a, if an odd number of directions is active with value b and state $\hat{q}$ is greater than b, then the next state is equal to $\hat{q}$, or, if an odd number of directions are active with value c and state $\hat{q}$ equals c, then the next state is 0.5'.*

Proposition P_2 requires more than one fuzzy rule. So we have

> IF input is *odd number of directions* AND output is *smaller than odd number of directions*, THEN next state is equal to *odd number of directions*

> IF input is *odd number of directions* AND output is *greater than odd number of direction*, THEN next state is equal to *state*

> IF input is *odd number of directions* AND output is *equal to odd number of directions* THEN next state is 0.5.

Because proposition P_2 requires three (more than one as previously used) rules in fuzzy cell $c_{i,j}$, more than one memory device is needed. Besides this, it is also possible to activate some fuzzy logic characteristics of each memory device in the cell. To explain this, consider memory device FT. It is represented with output function from the second expression in Equation II.9.40. This expression in the domain of the above three rules is

$$
\begin{aligned}
D^1\hat{q} &= f_m(S, N, W, E)\overline{\hat{q}} + f_e(S, N, W, E)\hat{q} \\
&= f_m(S, N, W, E)\overline{\hat{q}} + \overline{f}_m(S, N, W, E)\hat{q} \\
&= f_m\overline{\hat{q}} + \overline{f}_m\hat{q}.
\end{aligned}
\tag{II.9.50}
$$

Consequently, for all three rules, we need only one memory device FT. The last form of the three fuzzy rules is as follows:

$$\text{IF } T \text{ is } f(S, N, W, E) \text{ AND } \hat{q} \text{ is } smaller\ than\ f(S, N, W, E), \text{ THEN } \mathbf{D}^1\hat{q} \text{ is } f(S, N, W, E),$$

$$\text{IF } T \text{ is } f(S, N, W, E) \text{ AND } \hat{q} \text{ is } greater\ than\ f(S, N, W, E), \text{ THEN } \mathbf{D}^1\hat{q} \text{ is } \hat{q},$$

$$\text{IF } T \text{ is } f(S, N, W, E) \text{ AND } \hat{q} \text{ is } equal\ to\ f(S, N, W, E), \text{ THEN } \mathbf{D}^1\hat{q} \text{ is } 0.5. \qquad (\text{II.9.51})$$

After terminal state $\mathbf{D}^1\hat{q} = 0.5$ we have no more changes of state because $\mathbf{D}^1\hat{q} = 0.5$ for all possible values of f_m. In this case, change of state is possible only by introducing extra input C with CLEAR characteristic. For such state clearing it is convenient to have the fuzzy memory device FTDC[21,22,28] in the fuzzy cell.

9.5.3 MATLAB Simulation

In this section we show some simulation results as a confirmation of FCS designed on the basis of fuzzy memory devices. To simulate FCS, we take the MATLAB environment that is very close to the design of fuzzy systems. As shown in Virant,[22] there are possible x.M MATLAB execution files, x.FIS MATLAB fuzzy inference systems, and x.MDL MATLAB modeling and simulating files. MATLAB also has Fuzzy Logic Toolbox that is very nice for the fuzzy system designer from an engineering point of view. In this section we use design skills from Virant.[22] So we have a base of fuzzy devices named TSNORM.MDL, from which we take fuzzy memory devices FDcell, FTcell, and FTDCcell, and nonmemory devices, e.g., max( ), min( ), and complements. SIMULINK library gives us many linear blocks, nonlinear blocks, sources, sinks, and binary switching blocks which enable us to create models which are used in this section.

In Figure II.9.18 are shown basic blocks for memory devices FD, FT, and FTDC and two kinds of fuzzy cells CA11 and CS11. CA11 is an asynchronous fuzzy cell in which there can be asynchronous memory devices FD or FT. CS11 is a synchronous cell with device FTDC. Input W of the FTDC device is needed to WRITE new value after CLEAR old output value. Simulation details of devices FD, FT, and FTDC and others can be found in Virant.[22] For FD we have output switching functions from the second expression of Equation II.9.41. The output switching function from Equation II.9.40 is valid for FT and that from Equation II.9.43 is valid for FTDC.

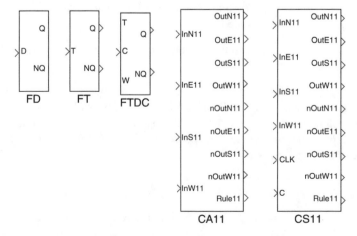

FIGURE II.9.18 MATLAB blocks for memory device: FD, FT, FTDC, and cells: CA11, CS11.

The main item in FCS design is the fuzzy rule represented in the fuzzy switching form. Function Equation II.9.46 is an example of a switching function with four input variables. From Figure II.9.15 we can conclude that with four independent variables there are approximately 1.158×10^{77} switching functions which are available in FCS design in our, or similar, designs.

Figure II.9.19 shows a simulation scheme for switching Equation II.9.50 in product-phrase form. Inputs InN11, InE11, InS11, and InW11 are inputs of the observed fuzzy cell, and output Rule11 is input to fuzzy memory device in this cell. Cell-state is the state of the memory device in the cell, or a part of this state. Properly speaking, the state of the memory device is split into directions, S, N, W, and E, depending on the case. For example, an application may dictate to take weighting factors $1.0 \times Q$ for S, $0.8 \times Q$ for N, $0.6 \times Q$ for W, and $0.3 \times Q$ for E. Negative outputs nOuS, nOuN, nOuW, and nOuE are also available. They can be used for some output direction. The numerical part of the cell labclcd 11 in Figure II.9.18 means that this cell is the first one in the cellular field. Actually in the cellular field there are labels containing numbers (i, j).

Now, let us consider the orientation of a fuzzy cellular cell in a cellular field. The left side of Figure II.9.20 shows that each direction of the cell has an input and an output. With these, the fuzzy cell $c_{i,j}$ is connected to the other equal cells in all four directions. The right side of Figure II.9.20 shows some auxiliary blocks needed in the simulation process. Viewer can plot four input/output signals. These signals are displaced by one y-unit in order to see better behavior of signals. XY scopes signal in plane $P_{xy} = [0, 1] \times [0, 1]$ in all simulation. Boundary signals are already performed in output-blocks SRL, BTT, and k. SRL includes sin( ), random, and linear membership functions, and BTT includes bell, trapezoidal, and triangle membership functions. k is any constant value from interval $[0, 1]$. All these functions can be changed with parameters, as is recommended within MATLAB. Clock Circuits provides signals needed within the FTDC device. Signals enable clearing and then writing input values into the memory device in the cell. Any data

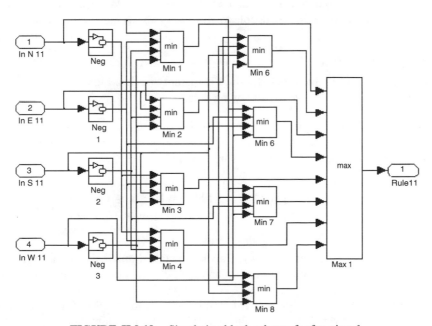

FIGURE II.9.19   Simulation block scheme for function f_m.

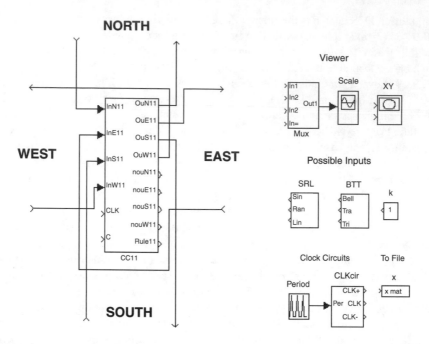

FIGURE II.9.20 Orientation of fuzzy cell (left) and auxiliary circuits needed in the simulation process (right).

can be stored to the file with block x. This enables the results of simulation to be analyzed (off-line analysis).

9.5.3.1 Experiments and Results

Using the proposed simulation items, we can construct cellular fields of dimension 1×1, 1×4, and 4×4. From these basic fields we are able to construct FCS with any dimension $m \times n$. Let us examine these three fields.

9.5.3.2 Fuzzy Cellular Field 1×1

First consider the fuzzy cell which includes the FD memory device. Figure II.9.21 shows the East output when four inputs of cell have the following data:

- InW11 represents the triangle membership function of fuzzy set $\widetilde{W} = (49,100,100,150)$
- InS11 represents the trapezoidal membership function of fuzzy set $\widetilde{S} = (40,60,80,100)$
- InN11 represents the half-sin( ) membership function

$$\widetilde{N} = \begin{cases} \sin(1.8^*t - 144) & \text{if } 80 \leq t \leq 180 \\ 0 & \text{otherwise} \end{cases}$$

- InE11 represents membership $\widetilde{E} = 0.1$. This means that through all time there exists a small membership $\mu > 0$ which reduces discrimination in the process of approximate reasoning.

The simulation process takes 200 time steps (1 step is 1 simulation sec). The input data is obtained with the help of Input Viewer shown in Figure II.9.21. Putting this input data (Figure II.9.22) into the fuzzy switching circuit in Figure II.9.18 we get D input of memory device FD in the observed cell. The plot of this function is performed using the input In2 in Output Viewer (Figure II.9.23).

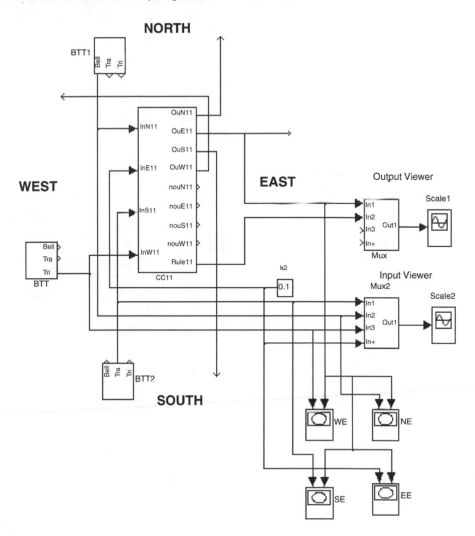

FIGURE II.9.21 Fuzzy cell with FD memory device with the North bell, West triangle, South trapezoidal, and East constant membership functions.

It is interesting to show the transition behaviors from all four inputs into one output. These behaviors are plotted for output OuE11 as follows:

WE direction from InW to OuE
NE direction from InN to OuE
SE direction from InS to OuE
EE direction from input InE to output OuE

In the interest of brevity only the second example, North-East is shown here. This relation is shown in Figure II.9.24. X-axis holds signal InN and Y-axis signal OuE. Here the simulation time is implicit. As the behavior N in Figure II.9.24 shows, the X-axis occurs twice, from 0–1 and then from 1–0.

Up to now we have observed the fuzzy cell with memory device FD which supports fuzzy rule of cellular field (Equation II.9.39). More complicated, there is a triplet of fuzzy rules (Equation II.9.51). To represent this triplet in the observed cell, we must exchange the memory device FD for FTDC.

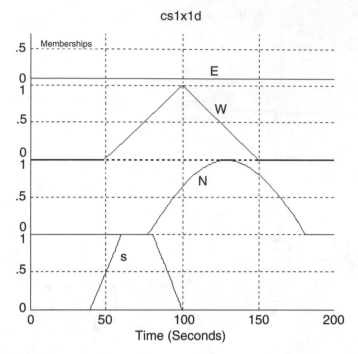

FIGURE II.9.22 *E*-, *W*-, *N*-, and *S*-inputs of fuzzy cell with fuzzy memory device FD.

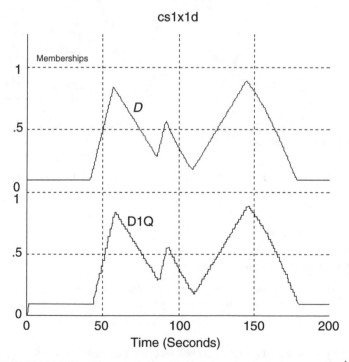

FIGURE II.9.23 Fuzzy rule as switching function D and next state $\mathbf{D}^1\hat{q}$.

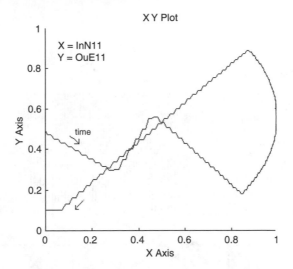

FIGURE II.9.24 Result of processing from the North to the East direction: X is N in Figure II.9.22 and Y is D1Q in Figure II.9.23.

This memory device is taken with $C = 1$ (see Figure II.9.20), i.e., no clearing situation. CLK input holds signal CLK- at maximal Period as shown in Figure II.9.25. CLK- is needed only to write membership value in the memory device. Figure II.9.25 also shows the unchanged behavior of input, now named T. The next state D1Q shows that at the end of simulation we have the maximum of all local maxima. Such replay to an input behavior is very important and useful in the FID (Fuzzification, Inference, Defuzzification) process.

The following procedure demonstrates how a fuzzy trigger memory device works from a logical point of view. From Equation II.9.40, Figure II.9.15, and Virant et al.[21] we can conclude that, in the case of output switching function $\mathbf{D}^1\hat{q}$, product-phrases exist which are denoted with the label 1 in Figure II.9.26. They are

$$\hat{q}\overline{\hat{q}}T\overline{T}, \hat{q}\overline{\hat{q}}T, \hat{q}\overline{\hat{q}}\overline{T}, \hat{q}T\overline{T}, \overline{\hat{q}}T\overline{T}, \hat{q}\overline{T}, \overline{\hat{q}}T. \tag{II.9.52}$$

Minimization of the fuzzy switching function enables us to write out the following simplifications.

$$\hat{q}\overline{\hat{q}}T\overline{T} + \hat{q}\overline{T} = \hat{q}\overline{T}(\overline{\hat{q}}T + 1) = \hat{q}\overline{T}$$

$$\hat{q}\overline{\hat{q}}T + \overline{\hat{q}}T = \overline{\hat{q}}T(\hat{q} + 1) = \overline{\hat{q}}T$$

$$\hat{q}\overline{\hat{q}}\overline{T} + \hat{q}\overline{T} = \hat{q}\overline{T}(\overline{\hat{q}} + 1) = \hat{q}\overline{T}$$

$$\hat{q}T\overline{T} + \hat{q}\overline{T} = \hat{q}\overline{T}(T + 1) = \hat{q}\overline{T}$$

$$\overline{\hat{q}}T\overline{T} + \overline{\hat{q}}T = \overline{\hat{q}}T(\overline{T} + 1) = \overline{\hat{q}}T \tag{II.9.53}$$

We see that all nonminimal product-phrases are covered with only two minimal, i.e., $\overline{\hat{q}}T$ and $\hat{q}\overline{T}$. Because $\max(x, 1) = x + 1 = 1$ and $\min(x, 1) = x1 = x$, we have no problems with fuzzy memory devices of FT-type. Negation of the fuzzy switching function $\mathbf{D}^1\hat{q} = \hat{q}\overline{T} + \overline{\hat{q}}T$ is simply $\mathbf{D}^1\overline{\hat{q}} = \hat{q}T + \overline{\hat{q}}\overline{T}$, consequently without any influence of $\overline{\hat{q}}\hat{q}$ and $T\overline{T}$. This in fact is very close to a binary memory trigger device where $\overline{\hat{q}}\hat{q} = 0$ and $T\overline{T} = 0$ are valid.

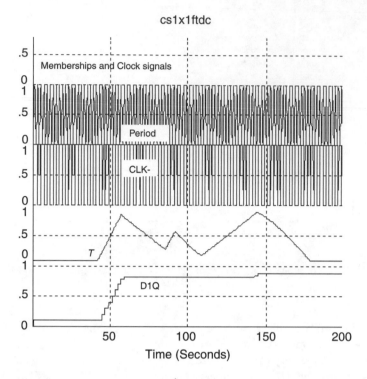

FIGURE II.9.25   Next state behavior $\mathbf{D}^1\hat{q}$ if memory device FD is changed with FTDC.

$$T$$

$$\mathbf{D}^1\mathbf{Q}:$$

	$T\overline{T}$	T	$\overline{T}$	1
$Q\overline{Q}$	1	1	1	
Q	1		1	
Q	1	1		
1				

FIGURE II.9.26   Fuzzy Veitch diagram for the next state $\mathbf{D}^1\hat{q}$.

Figure II.9.25 shows that for FT-type of cell the following expression holds.

$$\mathbf{D}^1\hat{q} = \begin{cases} T, \text{ if } \overline{\alpha} \geq T \\ \alpha, \text{ if } \alpha \leq \overline{T} \\ \alpha = C \cdot W, \\ \overline{\alpha}, \text{ if } \overline{\alpha} \leq T \\ \overline{T}, \text{ if } \alpha \geq \overline{T} \end{cases} \tag{II.9.54}$$

If $\hat{q} = \alpha$ then $\mathbf{D}^1\hat{q} = T, \overline{T}, \alpha, \overline{\alpha}$ respectively. Memory device TFDC has inputs C and CLK (see Figure II.9.20). Here, C is CLEAR C and CLK WRITE W. CLEAR takes place when $C = 0$. With inputs C and W we can obtain the dynamic behavior of output $\hat{q}$ or we can set the firing degree $0 \le \alpha \le 1$. This degree is written into the memory device with $C = 1$ and $W = \alpha$, or $C = W = \alpha$. If $\alpha = 0.5$, behavior of the memory device FTDC is equal to behavior of device FT.

In Figure II.9.27 the behaviors of FD, FT, FTDC (static), and FTDC (dynamic) are compared. The behaviors are as follows:

1. Input T:

$$x = 0.5 + 0.5 \sin(0.0235 \cdot t + 89), \, 0 \le t \le 200,$$

$$y = 0.5 + 0.5 \sin(0.0475 \cdot t + 29), \, 0 \le t \le 200,$$

$$T = \max(\min(\overline{x}, y), \min(x, \overline{y}))$$

2. $\mathbf{D}^1\hat{q}$ of memory device FD
3. $\mathbf{D}^1\hat{q}$ of memory device FT
4. $\mathbf{D}^1\hat{q}$ of memory device FTDC static ($\alpha = 0.8$ results from $C = W = 0.8$)
5. $\mathbf{D}^1\hat{q}$ of memory device FTDC dynamic ($C = W = CLK\text{-}$)

The difference between behaviors 2 and 1 is one time step delay. The difference between behaviors 3 and 2 is the memorizing of the last biggest maximum of 2 over past time and the firing $\alpha = 0.5$. The difference between 4 and 3 is the firing at $\alpha = 0.8$. We see that this function is under the control of inputs C and W. The difference between behaviors 5 and 2 is the dynamic behavior of 5 with regard to the static behavior of 2.

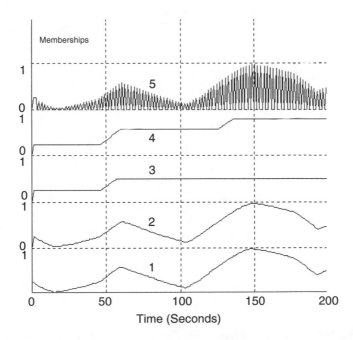

FIGURE II.9.27   Comparison between FD, FT, FTDC (static), and FTDC (dynamic).

9.5.3.3 Fuzzy Cellular Field 1×4

In contrast to 1×4 fields, cellular fields are a kind of shift register. To be brief, we will observe only the 1×4 field organized by the help of memory device FTDC. In Figure II.9.28 there is shown such a field in the direction West-East/East-West. Boundary values of North are (InN11, InN12, InN13, InN14) = (0, 0, 0, 0), and South boundaries are (0, 0.3, 0, 0). Each component can take any value from [0, 1]. West input InW11 is a bell membership function (positive half of $\sin(0.0235\ r)$) while the East input is a triangle membership function for fuzzy set $E = (49, 100, 100, 179)$. CLEAR and WRITE processes of the memory device require a time step vector $(s_0, s_1, s_2, s_3, s_4, s_5) = (0, 1, 0, 0, 0, 0)$ in block Period. The step s_1 enables CLEAR and the steps $s_2, \ldots, s_5$ enable asynchronous WRITE action.

The lower-half part of Figure II.9.28 shows Rule-Viewer, Dynamic Output-Viewer, Numeric Output-Viewer (dynamic), and Static Output-Viewer, which enable good data analysis to be made. MATLAB XY-Scope with label EW_i show the result for x-axis is input of the cell $CC1_i$ (it is equal to output of cell before), and for y-axis is the East output of the same cell. The result is a transfer function of cell in the direction East-West. Similarly, XY-Scope with label WE_i shows the transfer function of the memory device in the observed cell $CC1_i$. In the cells $CC1_i$ the fuzzy rule is shown in Figure II.9.19. Again, 200 time steps are taken for simulations time. The dynamic of a signal means that at each time period one CLEAR/WRITE action takes place. Therefore, a signal in the time step s_{2+i*5} takes value 0. Such a transition $a \rightarrow 0 \rightarrow b$ gives a very specific character, i.e., a feeling for looking at objects from one point (or a few points) in the coordinate system of an observed system.

To better understand the above-mentioned feeling, let us see the result of scope EW1 with the input pair (InE11, OuE11) = (OuW12, OuE11). Figure II.9.29 shows how the behavior of the function is

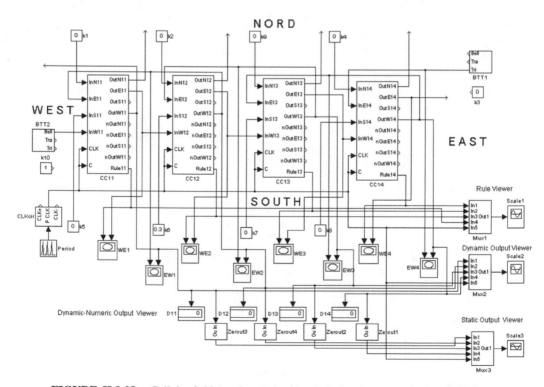

FIGURE II.9.28   Cellular field 1×4 organized by the help of memory devices FTDC.

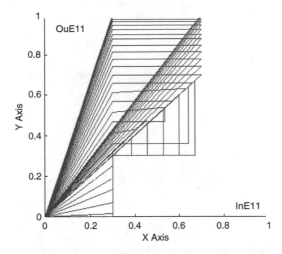

FIGURE II.9.29 (OuW12, OuE11)-view of the transfer function of cell CC11.

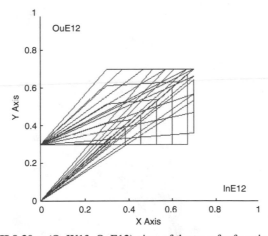

FIGURE II.9.30 (OuW13, OuE12) view of the transfer function of cell CC12.

closed into the space bounded with the various values of 0.3, $1 - 0.3 = 0.7$, as is required by the Modulo-Two rule of cell. If we go one step to the East we have the pair (InE12, OuE12) = (OuW13, OuE12). The result for this pair is shown in Figure II.9.30. The various values of 0.3, $1 - 0.3 = 0.7$ still have their impact but in another way.

Consider the results shown with scopes WE_i. In Figure II.9.28 it was seen that the WE1 x-axis is the Rule11 and y-axis the output OuE11. All cells have the same fuzzy switching rule, therefore Rule1i = Rule is the common point or common x-axis. Irrespective of this agreement, we use index i. The transfer function or, better, the FTDC output with regard to the given rule-value, is shown in Figure II.9.31. In this figure we see that various values of 0.3, $1 - 0.3 = 0.7$ still have their impact. Figure II.9.32 shows the impact of these values in the case of cell CC12.

XY-scopes show behaviors where time steps are unexpressed. We get the time-dependent behaviors with Rule Viewer, Dynamic Output Viewer, and Static Output Viewer. Consider first the dynamic behaviors. Through 200 time steps we have $200/5 = 40$ samples, but in the sequence of 5 time steps there is a possibility for four asynchronous WRITE actions. The Scale1 from Figure II.9.33 shows all

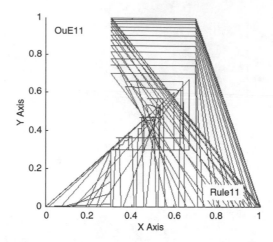

FIGURE II.9.31   (Rule11, OuE11) view of the cell CC11.

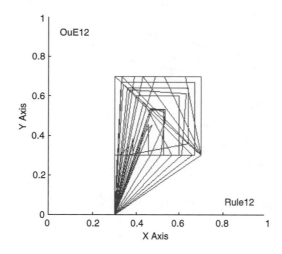

FIGURE II.9.32   (Rule12, OuE12) view of the cell CC12.

four rule behaviors and clock CLK. Two of them we have already seen, in XY-form, in Figures II.9.31 and II.9.32. Dynamic outputs as the consequence of input rule-values in Figure II.9.33 are shown in Figure II.9.34. In the last two figures can be seen synchronous and asynchronous processes of clearing and writing a value into the memory device.

   If there are inputs CLK and C constant values from interval [0, 1], FCS will be working, but after a certain time of simulation, all values of outputs become 0.5. In this case, results of FCS with FTDC are very simple. At the end of simulation, XY-scopes show only points (0.5, 0.5) or lines $y = 0.5$. Therefore, the attraction of results depends, first of all, on dynamic behaviors.

   There are many logical possibilities to reach different dynamics. For example, with the help of blocks Zerout we can perform only one WRITE-action into the memory device in such a way that

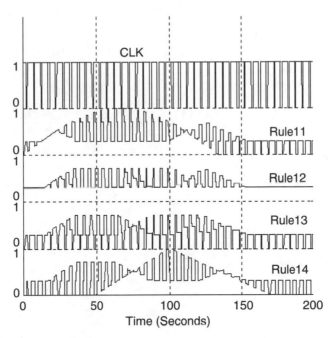

FIGURE II.9.33   Behavior of fuzzy switching rules of 1×4 FTDC FCS.

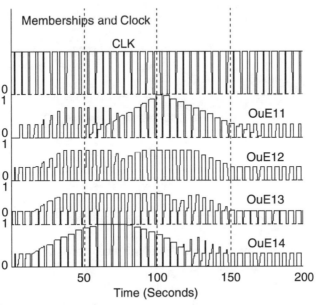

FIGURE II.9.34   Dynamic outputs in the cellular field 1×4 FTDC FCS.

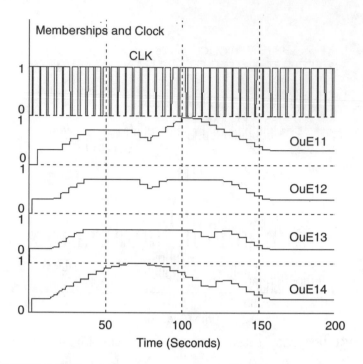

FIGURE II.9.35 Outputs as outcome for one WRITE per period of 6 Sec.

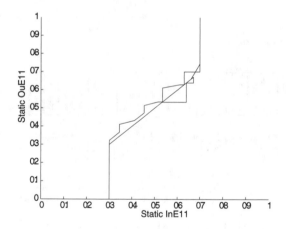

FIGURE II.9.36 Scope of static representation of Figure II.9.29 if InS12 = 0.3.

the signal represents only an envelope of a given dynamic signal. For example, in the block Zerout there is the following fuzzy switching function.

$$y = x + \mathbf{D}^{-1}x + \mathbf{D}^{-2}x + \mathbf{D}^{-3}x + \mathbf{D}^{-4}x \qquad (\text{II.9.55})$$

Here x is input and y output signal of the block. Figure II.9.28 shows *Zerout* implementation and adequate results obtained within *Static-Output Viewer*. These results are shown in Figure II.9.35.

The function (Equation II.9.55) can be implemented at any dynamic behavior. This implementation gives us static instead of dynamic behavior. The word "static" does not mean behavior without changes, but that the number of changes is considerably smaller than in the normal dynamic case with synchronous and asynchronous actions. For example, look at Figure II.9.29 in the domain of static signals. For transformations

$$\text{Static OuE11} = \text{Zerout(OuE11)}$$

$$\text{Static InE11} = \text{Zerout(InE11)}$$

we get Figure II.9.36. The dynamic in this figure is relatively poor in comparison with that shown in Figure II.9.29. It is also clear that static results show some horizontal and some vertical boundaries as a consequence of field-boundary constant values (e.g., in Figure II.9.28 we took $k_6 = 0.3$).

9.5.3.4 Fuzzy Cellular Field 4×4

We have collected enough information about behaviors in the 1×1 and 1×4 fuzzy cellular fields; therefore, we can now observe field with dimension 4×4. We shall observe the field with memory devices FTDC only. The simulation model is shown in Figure II.9.37. In this model only the following boundary inputs are different from zero:

InW41 = Negation of Bell = $1 - \sin(0.00235^*x) \geq 0$
InS43 = 0.6
InE24 = Bell = $\sin(0.00235^*x) \geq 0$
InN14 = 0.2

At the time of 200 steps we take 4.7 Radians. So one period requires $200^*6.28/4.7 = 267$ time steps and a half period of 133.5 steps. All inputs are plotted with the Input Viewer and they are shown in Figure II.9.38. This figure also shows clock CKL and Rule23. Simulation results are interpreted with the XY-scopes in two ways: (a) x-axis is the input of a selected direction; and y-axis is the output of this selected direction; and (b) x-axis is the rule of a selected direction and y-axis is the output of this selected direction. In Figure II.9.37 the second way is assumed.

There are some interesting results. For example, in Figure II.9.39 are OuS12(Rule12), OuS13 (Rule13), OuS22(Rule22), OuS21(Rule21), OuS23(Rule23), and OuS31(Rule31). The following conclusions can be drawn:

1. The output behaviors OuS12, OuS22, and OuS23 exhibit some similarities. All behaviors are formed with linear segments that break at some typical boundary value. For example, boundaries InS43 and InN14 given above have typical values: $0.6, 1 - 0.6; 0.2, 1 - 0.2$ (see Figure II.9.37).
2. A curve can be constructed from the linear segments.
3. There is no standard expression for any behavior (usual algebra or usual arithmetic), but the behaviors are strictly defined by fuzzy logic expressions, called fuzzy switching functions.
4. Input of fuzzy memory devices in all fuzzy cells (input T of FTDC device) can be used as a common point to draw attractive figures on XY-scope.
5. Dynamic behaviors of fuzzy fields with FTDC-conception are free of the equilibrium state 0.5.

It is interesting to compare the Rule23 with the left-down output behavior OuS23(Rule23) in Figure II.9.39. Going step by step through the simulation, we can see what form of the rule makes a curve with the linear segments, what part of the rule behavior makes changes between x-values 0.1 and 0.6, etc.

Homogeneous cellular arrays are used to perform image processing. For such processing, conclusion 5 is very important. If this conclusion is valid, we can control all points in the plane $X \times Y$ without terminal states. If conclusion 5 is not valid, the terminal state 0.5 exists. For example, conclusion 5 is not valid in the cases FD and FT.

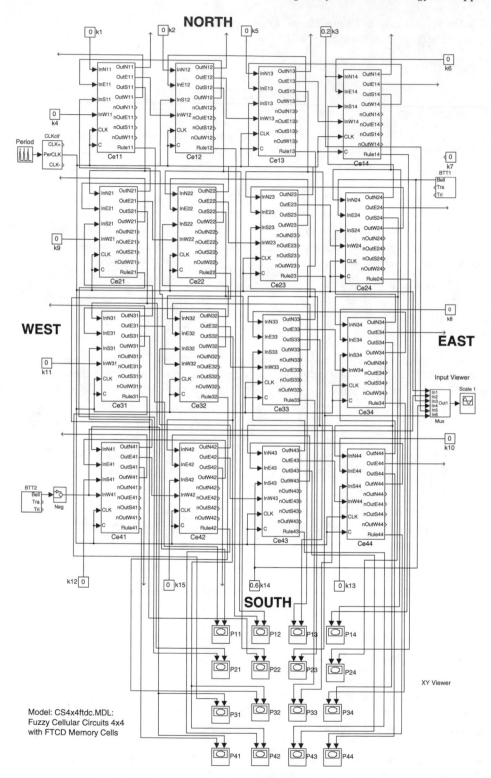

FIGURE II.9.37   Fuzzy cellular field 4 × 4 with FTDC memory devices.

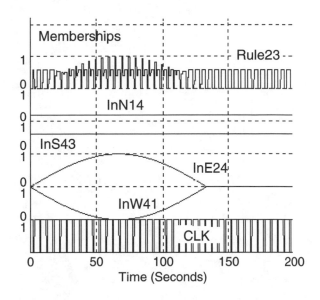

FIGURE II.9.38   Boundary inputs in simulation model shown in Figure II.9.37.

To show how a fuzzy cell goes into a terminal state, let us take in place of cs4 × 4ftdc the fuzzy cellular system cs4 × 4fd, but at the same boundaries shown in Figure II.9.38. The clock CLK in cs4 × 4fd does not exist, since the system is purely asynchronous. After simulation on cs4 × 4fd, we get the result shown in Figure II.9.40. OuS23(Rule23) relates the synchronous-asynchronous (mix) result in the case of the cs4 × 4ftdc in Figure II.9.39.

With respect to their asynchronous mode of memorizing, the clock CKL has a period that tolerates four memory devices in each of the directions W, N, E, and S. For example, the period $P = (0, 1, 0, 0, 0, 0)$ has, in step 2, a synchronous effect (CLEAR and WRITE), but in the steps 3, 4, 5, and 6 asynchronous memorizing can be achieved, if needed. In the ith period we have CLEAR and WRITE in step $2 \cdot i$ and asynchronous memorizing in steps $3 \cdot i, 4 \cdot i, 5 \cdot i$, and $6 \cdot i$. If only asynchronous memorizing is needed, CKL $= 1$ is taken.

9.5.4   Application of Fuzzy Cellular Fields

Let us see how the asynchronous mode of a 4×4 cellular field can be applied in fuzzy memory technique or in fuzzy inference processing. Fuzzy memory is an array of memory cells to store membership functions (fuzzy words[25]). Writing and reading from memory can be parallel or serial. There are two basic possibilities of realizing fuzzy memory: 1) The classical binary technique working on codes (classical computer words) for real numbers from range [0, 1];[26] and 2) a nonlinear analog technique where the memory cell can store any real number from [0, 1].[25,27]

In the first case, for one membership function with length n we need n computer words, but in the second case, only one computer word. Thus, the compression of memory for one membership function is $n : 1$. The processing time of one fuzzy word is also very high compared with the time processing of n binary words in the normal computer. On the other hand, a normal binary processor has high noise immunity and low sensitivity to the variance of device characteristics. In the case of a nonlinear analog technique this performance is not so good, but over the years the technique has proved quite useful.[27] Memory devices FD, FT, FTDC, FJK, and FRS, that are observed in this section, belong to the nonlinear analog techniques.

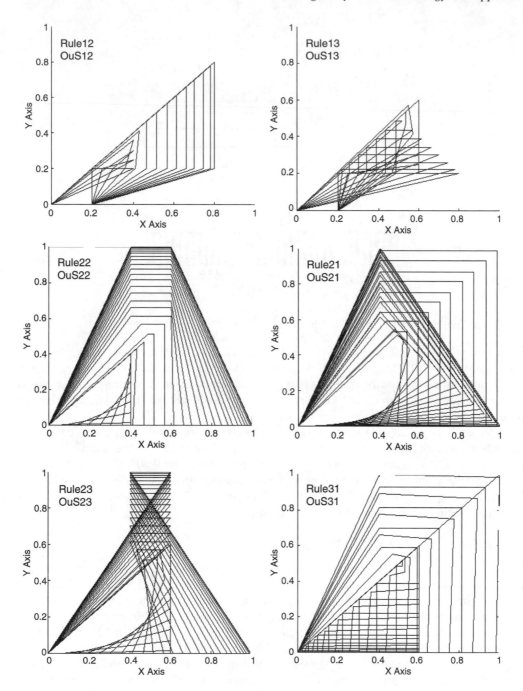

FIGURE II.9.39   Some attractive behaviors Rule(i, j)-Output(i, j).

Membership function is sampled with the highest density possible. This process transforms the continuous function to a vector with length n. If the vector has length $n = 16$, we need four fuzzy cellular fields 4×4, i.e., a 16×4 field to store four membership functions. Only 4 (serial

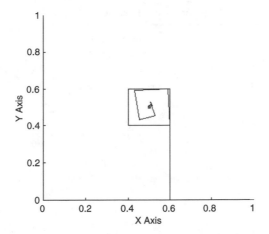

FIGURE II.9.40   Result OuS23(Rule23) in fuzzy field sc4 × 4fd.

type) or 16 (parallel type) boundary inputs are needed; others can be constants 0 or 1. Here, the problem is how to organize fuzzy cellular fields as a fuzzy memory or as a processor unit in a fuzzy inference engine.

Consider fragments of the mentioned organization in the frame of one 4 × 4 field only. If $n > 4$, we take an organization of more than one 4 × 4 fuzzy field. The space of input variables is divided into two parts: part I [0, 0.5] and part II [0.5, 1], because membership value $\mu = 0.5$ is the equilibrium value for the fuzzy trigger device. To represent a (segment of) membership function, we take values A, B, C, and D. In Part I, $0 \leq A \leq B \leq C \leq D \leq 0.5$ and in Part II, $0.5 \geq A \geq B \geq C \geq D \geq 1$, to be close to the monotonous membership behaviors. We take into account only asynchronous WRITE into 4 × 4 memory field (static working of fuzzy cells).

The state of the 4 × 4 memory field is represented by a matrix of 4 × 4 output values of fuzzy cells. These values are A, B, C, D, or a, b, c, d, where a, b, c, d are negated A, B, C, D, respectively. Table II.9.2 shows boundary input variables A, B, C, D which are taken only at the W-side of the field. Boundary inputs of other sides (N, E, S) are constants 0 or 1. We can exchange W-side with N- or E- or S-side, where the result will be the same, but in their respective directions. Because the cellular field is symmetrical, it is enough to observe one side only. Table II.9.2 shows only 16 states. This number is very small compared with the number of all states where A, B, C, A, a, b, c, d, 0, or 1 can be put on each boundary input. Number of all input combinations is 16^{10}. Some resulting states produced from input combinations are equal, however, the number of different states stays very high, but smaller than $16.^{10}$

The states in Table II.9.2 are results of simulation. Simulation is taken over one period only, $P = (1, 0, 1, 1, 1, 1)$. For fuzzy memory design, states numbered 1, 2, 9, and 11 are very attractive. For example, if the inputs $A = 0.51$, $B = 0.62$, $C = 0.73$, and $D = 0.84$ are used at the state 9, we have, after WRITE action into memory, the following memory state.

$$S = \begin{bmatrix} 0.51 & 0.51 & 0.51 & 0.51 \\ 0.62 & 0.62 & 0.62 & 0.62 \\ 0.73 & 0.73 & 0.73 & 0.73 \\ 0.84 & 0.84 & 0.84 & 0.84 \end{bmatrix}.$$

Here the vector $V = (A, B, C, D) = (0.51, 0.62, 0.73, 0.74)$ is in both the diagonals, and in all four columns. The first row of state S shows min(V) and the last row, max(V). With these conditions

TABLE II.9.2　Sixteen Generic States of a 4 × 4 Fuzzy Cellular Field

Boundary Inputs: I : $0 \leq A \leq B \leq C \leq D \leq 0.5$ II : $0.5 \geq A \geq B \geq C \geq D \geq 1$	State No. for I	State of cs4 × 4ftdc Value Label: X value x negated boundary	State No. for II	State of cs4 × 4ftdc Value Label: X value x negated boundary
0 0 0 0 A　0 B　0 C　0 D　0 0 0 0 0	1	D C B A D D C B D D D C D D D D	9	A A A A B B B B C C C C D D D D
0 0 0 0 A　0 B　0 C　0 D　0 1 1 1 1	2	d c b a d c b 1 d c 1 1 d 1 1 1	10	A A A a B B a 1 C b 1 1 c 1 1 1
0 0 0 0 A　1 B　1 C　1 D　1 0 0 0 0	3	d b 1 1 d c 1 1 d d 1 1 d d 1 1	11	A A 1 1 B B 1 1 C C 1 1 D D 1 1
1 1 1 1 A　0 B　0 C　0 D　0 0 0 0 0	4	a 1 1 1 c b 1 1 d d c 1 d d d d	12	a 1 1 1 B b 1 1 C C c 1 D D D d
0 0 0 0 A　1 B　1 C　1 D　1 1 1 1 1	5	D b 1 1 d D 1 1 d c C 1 d 1 1 0	13	A A 1 1 B B 1 1 C B C 1 c 1 1 0
1 1 1 1 A　0 B　0 C　0 D　0 1 1 1 1	6	a 1 1 1 c b 1 1 d d 1 1 d 1 1 1	14	a 1 1 1 B B 1 1 C B 1 1 c 1 1 1
1 1 1 1 A　1 B　1 C　1 D　1 0 0 0 0	7	d b 1 1 d c 1 1 d d 1 1 d d 1 1	15	a 1 1 0 B B B 1 B B 1 1 D D 1 1
1 1 1 1 A　1 B　1 C　1 D　1 1 1 1 1	8	a 1 1 1 c b b 1 d c c 1 d 1 1 0	16	a 1 1 0 B b b 1 C b b 1 c 1 1 0

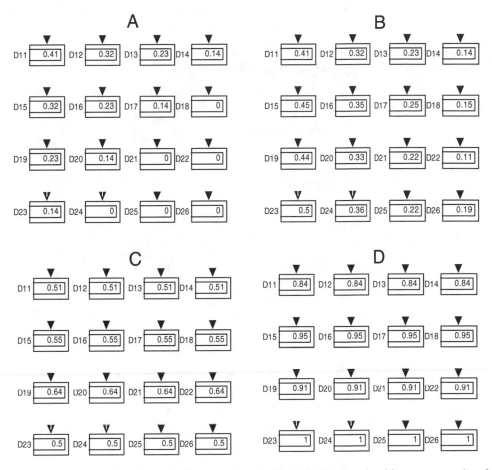

FIGURE II.9.41   Numerical view of end-states of 4 × 4 cellular field in domain of fuzzy memory in a fuzzy system.

and conditions of the 3 × 4 zeros in Table II.9.2 (see first row), the 4 × 4 fuzzy cellular field is placed into the fuzzy inference engine as a module of fuzzy memory or as a segment of the fuzzy processor.

State S was obtained using the constant values A, B, C, and D, as shown by the rows of the corresponding matrix. The opportunity exists to exchange these fixed values to the vectors $M_1 = (A_1, B_1, C_1, D_1), \ldots, M_4 = (A_4, B_4, C_4, D_4)$ over the clock period $P = (1, 0, 1, 1, 1, 1)$. At the value 0 CLEAR action of all fields takes place. The next four values "1" enable an asynchronous WRITE (SHIFT) action for inputs A, B, C, and D.

In Figure II.9.41 four field states are shown. A short analysis of them is as follows. The state labeled **A** corresponds to boundary situation

$$M_1 = (0.14, 0.23, 0.32, 0.41),$$

$$M_2 = (0, 0, 0, 0),$$

$$M_3 = (0, 0, 0, 0),$$

$$M_4 = (0, 0, 0, 0).$$

The whole memory function M_1 is written in the first row and in the first column. Other rows and columns are shortened in the process of shifting. Label **B** represents the following four different membership parts:

$$M_1 = (0.14, 0.23, 0.32, 0.41),$$

$$M_2 = (0.15, 0.25, 0.35, 0.45),$$

$$M_3 = (0.11, 0.22, 0.33, 0.44),$$

$$M_4 = (0.19, 0.22, 0.36, 0.5).$$

In the state with label **C** we have input data:

$$M_1 = (0.51, 0.51, 0.51, 0.51),$$

$$M_2 = (0.55, 0.55, 0.55, 0.55),$$

$$M_3 = (0.64, 0.64, 0.64, 0.64),$$

$$M_4 = (0.5, 0.5, 0.5, 0.5),$$

or constant vector

$$V = (0.51, 0.55, 0.64, 0.5)$$

or

$$M_1 = (0.51, 0.62, 0.73, 0.84),$$

$$M_2 = (0.55, 0.76, 0.85, 0.95),$$

$$M_3 = (0.64, 0.73, 0.82, 0.91),$$

$$M_4 = (0.5, 0.5, 0.5, 0.5).$$

In both diagonals and all four columns there is the vector

$$V = (\min(M_1), \min(M_2), \min(M_3), \min(M_4)) = (0.51, 0.55, 0.64, 0.5).$$

In the state with label **D** is the input data

$$M_1 = (0.84, 0.73, 0.62, 0.51),$$

$$M_2 = (0.95, 0.85, 0.76, 0.55),$$

$$M_3 = (0.91, 0.82, 0.73, 0.64),$$

$$M_4 = (1, 0.91, 0.77, 0.5).$$

In both diagonals and all four columns we have vector

$$V = (\max(M_1), \max(M_2), \max(M_3), \max(M_4)) = (0.84, 0.95, 0.91, 1).$$

So, we can conclude that FCS with FTDC conception can memorize and also perform operations that are necessary in fuzzy inference engine. Some readers may conclude that a major obstacle of the observed FCS is the state equilibrium 0.5, due to the use of fuzzy trigger conception. They may say that the equilibrium 0.5 requires that Table II.9.2 should be divided into two parts, I and II, resulting in difficulties with parts of the membership functions that go from $a \leq 0.5$ to $0.5 \leq b$. We have more possibilities to eliminate this problem. One of these possibilities is presented in the following example. Consider parts of four membership functions that can exceed the value 0.5

$$S = \begin{bmatrix} 0.82 & 0.64 & 0.46 & 0,28 \\ 0.9 & 0.7 & 0.5 & 0.3 \\ 0.88 & 0.66 & 0.44 & 0.22 \\ 1 & 0.72 & 0.44 & 0.38 \end{bmatrix}$$

For the first row (0.82, 0.64, 0.46, 0.28) we cannot select either part II or part I of Table II.9.2. If we divide this row by two, however, we get row (0.41, 0.32, 0.23, 0.14) for which part I is valid. The same is also valid for other rows of matrix S. In this way all state matrix S belongs to part I of Table II.9.2. In this part we carry out all the necessary operations. Clearly, we must multiply all the resulting membership values by two. Matrix S can be put into the form

$$M_1 = (0.82, 0.64, 0.46, 0.28),$$

$$M_2 = (0.9, 0.7, 0.5, 0.3),$$

$$M_3 = (0.88, 0.66, 0.44, 0.22),$$

$$M_4 = (1, 0.72, 0.44, 0.38).$$

It is then placed into FCS to memorize it. The result is shown as state, labeled **B**, in Figure II.9.41. During READ action, or just after, we multiply such contents from memory values by two. This may be done by $S = A + A$ or $S = 2 \cdot A$.

$$S = \begin{bmatrix} 0.41 & 0.32 & 0.23 & 0.14 \\ 0.45 & 0.35 & 0.25 & 0.15 \\ 0.44 & 0.33 & 0.22 & 0.11 \\ 0.5 & 0.36 & 0.22 & 0.19 \end{bmatrix} + \begin{bmatrix} 0.41 & 0.32 & 0.23 & 0.14 \\ 0.45 & 0.35 & 0.25 & 0.15 \\ 0.44 & 0.33 & 0.22 & 0.11 \\ 0.5 & 0.36 & 0.22 & 0.19 \end{bmatrix}$$

$$= \begin{bmatrix} 0.82 & 0.64 & 0.46 & 0.28 \\ 0.90 & 0.70 & 0.50 & 0.30 \\ 0.88 & 0.66 & 0.44 & 0.22 \\ 1 & 0.72 & 0.44 & 0.38 \end{bmatrix}$$

There is another fictional difficulty, if the observed membership function goes over its core. Consider a membership function with such a characteristic, $M = (0, 0.5, 0.6, 1, 0.8, 0.3, 0.1, 0)$. If we wish to memorize it in FCS, we need two 1×4 fields. In the first field we put segment $M_1 = (0, 0.5, 0.6, 1)$ and in the second field, segment $M_2 = (0.8, 0.3, 0.1, 0)$, $M = (M_1, M_2)$. This is presented in Figure II.9.42. The first fuzzy field uses the W-E and the second, the E-W shift (direction). Both M_1 and M_2 are put parallel, M_1 to input IW1 and M_2 to input IE4. Numerical Output Viewer shows contents in fuzzy cellular memory. Pro and Pro1 are multiply-by-two blocks. Input dividing-by-two is already taken into account in blocks M1 and M2. The length of the fuzzy word is 8 fuzzy bits in W-E and E-W directions. Both field blocks also get 8-bit words in N-S directions and 2×4 fuzzy rules.

The application described above shows that the $m \times n$ fuzzy fields can be very useful in the design and organization of fuzzy memories and processing units, both in the domain of a fuzzy inference engine. The many possibilities allow great freedom of ideas to designers.

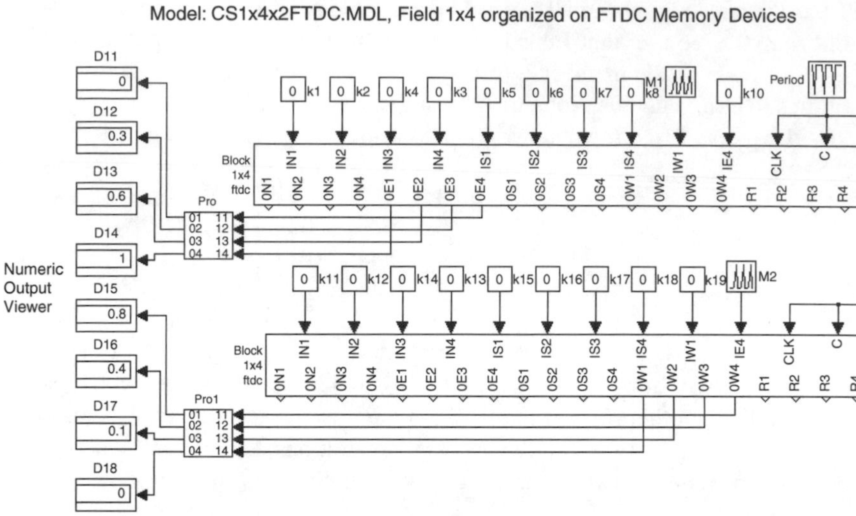

Model: CS1x4x2FTDC.MDL, Field 1x4 organized on FTDC Memory Devices

FIGURE II.9.42　Completing a membership vector over two fuzzy fields 1×4.

9.6　Conclusion

In the first four sections we have built basic definitions of fuzzy cellular automata. The structure is based on the entity of the finite fuzzy automaton. Basic approaches of designing and programming FCA structure are presented. The linguistic approach of programming makes the use of structure accessible to users who are not specialists in fuzzy logic or cellular automata. Some cases of FCA applications are presented in Mraz et al.[3] and Mraz, Zimic, and Virant.[4]

In the section of logical design of fuzzy cellular automata circuits, we have studied a new approach to the logical design of fuzzy cellular systems. The problem is stated in the form of fuzzy proposition(s) and the task of the designer is to go from this proposition(s) to a corresponding list of fuzzy rules that are responsible for the life of the fuzzy cellular system. The medium to use to go from the given propositions(s) to the behavior of a fuzzy cellular system is the fuzzy switching function which is, at the same time, the input function for the memory device in any cell of the cellular system.

In each cell there is a memory device which can memorize any value from fuzzy interval [0, 1]. The same domain has an input and output switching function before and after the memory device, which enables us to speak in the language of membership functions. Experiments take into account convex and nonconvex fuzzy sets.

The memory device used in a fuzzy cell is FTDC (fuzzy trigger with down-to-up signal and possibility of clearing). This device was designed and simulated in the laboratory with the intention of using it in sequential circuits in fuzzy inference systems and in designing fuzzy memories. The aim of this section is not exactly the theory and application of fuzzy cellular systems, but the consideration of the capability and power that FTDC has in the domain of such systems. Besides the FTDC, we have considered memory devices FT (common Fuzzy Trigger) and FD (Fuzzy Delay), first of all from a comparative point of view. FTDC, FT, and FD are explainable like the well-known fuzzy flip-flop FJK designed by Hirota and Ozawa.[22,23,25]

It has been pointed out that the implementation of FTDC devices in fuzzy cellular systems can be very attractive, and useful, if we take the same combination of synchronous and asynchronous system behaviors. Above all, FTDC cellular systems can take their place in a fuzzy inference engine where fuzzy memorizing and processing are combined to reach high-speed approximate reasoning.

Acknowledgment

The authors gratefully acknowledge the financial support of research project supported by the Ministry of Education, Science and Sport of Slovenia through grant Z2-3047.

References

1. Wolfram, S., *Theory and Applications of Cellular Automata*, World Scientific Publishing, Singapore, 1986.
2. Toffoli, T. and Margolus, N., *Cellular Automata Machines – A New Environment for Modelling*, MIT Press, Cambridge, MA, 1987.
3. Mraz, M. et.al., Notes on fuzzy cellular automata, *J. Chinese Inst. Industrial Engineering*, 17, 469, 2000.
4. Mraz, M., Zimic, N., and Virant, J., Intelligent bush fire spread prediction using fuzzy cellular automata, *J. Intelligent and Fuzzy Systems*, 7, 203, 1999.
5. Kohavi, Z., *Switching and Finite Automata Theory*, McGraw-Hill, New York, 1978.
6. Weimar, J., *Simulation with Cellular Automata*, Logos Verlag, Berlin, Germany, 1997.
7. Gutowitz, H.A., Introduction, in *Cellular Automata, Theory and Experiment*, Gutowitz, H.A., Ed., MIT Press, Cambridge, MA, 1991.
8. Wee, W.G. and Fu, K.S., A formulation of fuzzy automata and its application as a model of learning systems, *IEEE Trans. on Systems Science and Cybernetics*, 5, 215, 1969.
9. Mizumoto, M., Toyoda, J., and Tanaka, K., Some considerations on fuzzy automata, *J. Computer and System Sciences*, 3, 409, 1969.
10. Virant, J. and Zimic, N., Fuzzy automata with fuzzy relief, *IEEE Trans. on Fuzzy Systems*, 3, 69, 1995.
11. Virant, J., *The Use of Fuzzy Logic in Modern Systems* (in Slovene), Didakta, Radovljica, Slovenia, 1992.
12. Zimmerman, H.J., *Fuzzy Set Theory and Its Applications*, Kluwer Academic Publishers, Boston, 1991.
13. Dubois, D. and Prade, H., *Fuzzy Numbers: An Overview*, Prentice Hall, New York, 1994.
14. Yager, R.R. and Filev, D.P., *Essentials of Fuzzy Modelling and Control*, John Wiley and Sons Inc., New York, 1994.
15. Mamdani, E.H., *Fuzzy Reasoning and Its Applications*, Academic Press, London, 1987.
16. Adamatzky, A.I., Hierarchy of fuzzy cellular automata, *Fuzzy Sets and Systems*, 62, 167, 1994.
17. Adamatzky, A.I., Identification of fuzzy cellular automata, *Automatic Control and Computer Sciences*, 25, 75, 1991.
18. Caponetto, R. et al., Fuzzy Cellular System: Characteristics and Architecture, in *Fuzzy Hardware, Architectures and Applications*, Kandel, A. and Langholz, G., Eds., Kluwer Academic Press, Boston, 1998, Chapter 14.
19. Baglio, S. et al., Collective behaviour of cellular fuzzy systems for a new paradigm of computation, in *Proc. Int. Workshop on Artificial Intelligence in Real-Time Control*, Kocijan, J. and Karba, R., Eds., IFACS/IMACS, 1995.
20. Cattaneo, G. et al., Cellular automata in fuzzy backgrounds, *Physica D*, 105, 105, 1997.
21. Virant, J., Zimic, N., and Mraz, M., T-type fuzzy memory cells, *Fuzzy Sets and Systems*, 102, 175, 1999.
22. Virant, J., *Design Considerations of Time in Fuzzy Systems, Applied Optimization*, vol. 35, Pardalos, P.M. and Hearn, D. ser. Eds., Kluwer Academic Publishers, Dordrescht, 2000.

23. Hirota, K. and Ozawa, K., Fuzzy flip-flop as basis of fuzzy memory modules, in *Fuzzy Computing, Theory, Hardware, and Applications*, Gupta, M.M. and Yamakawa, T., Eds., North-Holland, Amsterdam, 1988.

24. Kandel, A. and Lee, S.C., *Fuzzy Switching and Automata: Theory and Applications*, Crane Russak, New York, 1979.

25. Yamakawa, T., Intrinsic fuzzy electronics circuits for sixth generation computers, in *Fuzzy Computing, Theory, Hardware, and Applications*, Gupta, M.M. and Yamakawa, T., Eds., North-Holland, Amsterdam, 1988.

26. Watanabe, H., Dettloff, W.D., and Yount, K.E., A VLSI fuzzy logic controller with reconfigurable, cascadable architecture, *IEEE J. Solid-state Circuits*, SC-25, 376, 1990.

27. Yamakawa, T., A fuzzy inference engine in nonlinear analog mode and its application to a fuzzy logic controller, *IEEE Trans. on Neural Networks*, 4, 496, 1993.

28. Virant, J., Zimic, N., and Mraz, M., Fuzzy sequential circuits and automata, in *Fuzzy Theory Systems: Techniques and Applications*, vol. 4, Leondes, C.T., Ed., Academic Press, 1999.

10

PACL-FNNS: A Novel Class of Falcon-Like Fuzzy Neural Networks Based on Positive and Negative Exemplars

W.L. Tung
Nanyang Technological University

Chai Quek
Nanyang Technological University

Abstract. Fuzzy neural systems are the integration of fuzzy systems and neural networks. The knowledge base of fuzzy neural systems consists of IF-THEN fuzzy rules. The fuzzy rules are intuitive to the human users and fuzzy neural systems have been widely implemented in real-life applications. However, most of the fuzzy neural networks proposed in the literature concentrate on the formulation of positive rules. That is, if x is A (the input matches and condition fulfills), then y is B (the consequence follows). As such, network outputs are computed using inference processes that place emphasis on the similarity or matching of the inputs and the input fuzzy labels of the rules. In this chapter, the notion of positive and negative exemplars in the inference process is proposed. The use of positive and negative exemplars (degree of matching as well as mis-matching) with respect to the fuzzy rules served to enhance the differentiating attributes between closely similar input classes and is thus able to produce a more stable network performance. Four novel *Falcon*-like networks are created to implement such a notion. They are named *Falcon-FCM, Falcon-MLVQ, Falcon-FKP*, and *Falcon-PFKP*. Their basic *Falcon*-like architecture is derived from the *Falcon-ART* network.

Falcon-ART uses fuzzy *adaptive resonance theory* (ART) as a clustering technique to partition the input-output space. The fuzzy rules of the system are subsequently derived through a mapping process. The evaluation of *Falcon-ART* network revealed several shortcomings:

1. Poor network performances when the classes of input data are closely similar to each other.
2. Weak resistance to noisy/spurious training data.
3. Termination of network training process depends heavily on a preset error parameter.
4. Learning efficiency may deteriorate as a result of using complementarily coded training data.

Falcon-MART (Modified ART) is proposed to handle the above deficiencies and is evaluated using three experiments. They are iris classification, traffic prediction, and phoneme classification. *Falcon-MART* is benchmarked against the *multi-layered perceptron* (MLP) network and the traditional *K-nearest neighbor* (K-NN) classifier. The simulation results are encouraging.

However, *Falcon-ART* and *Falcon-MART* both suffer from an inconsistent rule-base as new clusters are created when the resonance test fails, regardless of any duplication of fuzzy labels. Therefore, each of the four novel networks proposed uses a different clustering technique to derive the fuzzy sets of the problem domain. The performances of the proposed networks are evaluated using the same experiments. The simulation results showed that the contrast of positive and negative exemplars does bring about a more stable network performance and a more efficient recognition rate.

10.1   Introduction

"Soft computing" is the keyword coined by Lotfi A. Zadeh,[1] the founder of fuzzy logic, to describe approaches used to emulate the human style of reasoning in solving complex problems and the modeling of the human ability to tolerate uncertain, imprecise, and incomplete information in the decision-making process. These attributes are important in the physical sciences and engineering fields, as the measurements obtained are often not exact or precise due to the sensors or measuring tools used. Fuzzy logic was introduced as a mathematical framework to handle imprecise or ambiguous data encountered in daily life.[2] There are four major components constituting soft computing. They are *fuzzy system, neural network, evolutionary computing*, and *possibility reasoning*. Soft computing is concerned with the integration of these components to model the human intelligence in problem solving. A popular approach is the integration of neural network and fuzzy system to create a hybrid structure called *fuzzy neural network*. A fuzzy neural network is the realization of the operations of a traditional fuzzy system using neural network technology.

Fuzzy neural networks resolved some of the shortcomings that plagued the neural network and fuzzy system communities. Unlike the opaqueness of neural networks, the connectionist structure of such hybrid systems becomes transparent, as it is now possible to interpret the weights and the connections of the network using high-level IF-THEN fuzzy rules. The pitfalls of traditional fuzzy systems are also avoided, as the learning capabilities of a fuzzy neural network enable the network to automatically construct IF-THEN fuzzy rules from training data and tune the parameters of the embedded fuzzy system. In contrast, the help of experts has to be enlisted to formulate the expert knowledge base of a traditional fuzzy system in the form of IF-THEN fuzzy rules. Even then, the rules may be biased and inaccurate as they can vary with different experts. In addition, the parameters of a traditional fuzzy system are manually tuned to obtain the desired results. This may not be possible or is too difficult if the fuzzy system is large and has many variables. The performance of a fuzzy neural architecture depends on three factors:

1. The accuracy of the fuzzy sets in capturing the clustering nature of the training data
2. The accuracy of the fuzzy rules in modeling the dynamics of the problem domain
3. The type of reasoning/inference process used to compute the outputs of the fuzzy neural system in response to its inputs

In a fuzzy neural architecture, the accuracy of the fuzzy sets is directly related to the type of clustering technique used. Clustering is a process of assigning the elements in a data set into several clusters/classes such that elements within the same cluster/class are more similar to each other when compared against elements from another cluster/class. Clustering techniques can be grouped into two main categories: *Hierarchical-* or *partition-based* clustering. Under the branch of partition-based clustering techniques, there is *hard* and *fuzzy* clustering. Examples of hard clustering techniques are Kohonen's *Linear Vector Quantization* (LVQ)[3] and Ang's *Modified Linear Vector Quantization* (MLVQ).[4] On the other hand, fuzzy clustering techniques included *Fuzzy-C Means* (FCM)[5] and *Fuzzy Linear Vector Quantization* (FLVQ).[6] In hard clustering techniques, a data element is assigned to only one cluster using a metric measure such as the Euclidean distance. In fuzzy clustering, a data element can belong to more than one cluster with varying degrees or memberships. Pseudo partitioning is the constraint placed on fuzzy clustering techniques in which the membership values of a data element belonging to one or more clusters must have a sum of unity. Fuzzy sets are a by-product of clustering as fuzzy membership functions are used to represent the degree to which an element belongs to a cluster or class.

Once the fuzzy sets have been formed, IF-THEN fuzzy rules are formulated by determining the antecedents and consequent of the rules through a rule-learning process. Since fuzzy labels such as "tall," "short," and "heavy" in the antecedent and consequent sections of the rules represent fuzzy sets, any inaccuracy of the fuzzy sets in capturing the clustering nature of the training data would propagate to the fuzzy rules. This may lead to the formulation of incorrect rules and the dynamics of the problem domain not properly modeled. In addition, the learning algorithm employed to train the connectionist structure of a fuzzy neural network is important, as it directly affects the quality of the formulated rules. Such a learning algorithm may change the structure of the fuzzy neural network to formulate new fuzzy rules or delete redundant rules. This is known as *structural learning*. It may also adjust the weights of the links and hence update the fuzzy sets of the system. This is known as *parameter learning*. The aim of a learning algorithm is to optimally train the fuzzy neural network so that it gives the best performance. Such a learning algorithm can be classified as *supervised* or *unsupervised* learning.

Lastly, the performance of a fuzzy neural network is determined by its ability to draw valid or reasonable conclusions using its fuzzy rule base in response to its inputs. Hence, the inference process chosen to compute the outputs of a fuzzy neural network is an important consideration. Different inference processes may lead to different conclusions. The inference process used to compute the outputs must be structured and intuitive to the human reasoning process. Besides giving the fuzzy neural network a strong fuzzy logical foundation, it also ensures that the performance of the fuzzy neural network is comparable to that of a human expert solving the same problem under the same conditions. Examples of fuzzy inference techniques are *Compositional Rule of Inference* (CRI),[2] *Analogous Approximate Reasoning Schema* (AARS),[7] and *Truth-Value Restriction* (TVR).[8]

In this chapter, the contrast of positive and negative exemplars in the inference process is the emphasis. Most of the fuzzy neural networks proposed in the literature concentrate on the formulation of positive fuzzy rules. That is, if x is A then y is B. Hence, the inference process used inevitably computes the outputs based on the degree of matching or similarity of the inputs against the antecedents of the fuzzy rules. This may lead to undesirable or poor results if there are closely similar classes. In addition, the effects of attributes that give similar firings of the fuzzy rules in the rule base may obscure or weaken the effects of attributes that can differentiate closely similar inputs.

On the other hand, the contrast of positive and negative exemplars during the inference process may produce a more stable network performance and a better recognition rate in the presence of closely similar classes. Hence, four novel *Falcon-like* networks are proposed to examine the above hypothesis. Their *Falcon*-like architecture is derived from the *Falcon-ART* network, a fuzzy neural architecture developed by Lin and Lin.[9] *Falcon-ART* makes use of a variant of *Adaptive Resonance*

Theory (ART),[10] known as fuzzy ART, to derive the fuzzy sets in its fuzzy rule base. ART is a powerful model of the way human beings learn and classify concepts.

According to Grossberg, human beings learn through *match-based* and *error-based* (mismatch) learning. Existing concepts or knowledge is reinforced or resonated by repeated encounters (match-based learning), and new concepts are learned and absorbed into memories when a search or recall of existing memories produce no close match (error-based learning). As a clustering technique, ART works by creating a new concept when a new (or negative) exemplar to the existing knowledge is encountered. If an exemplar is not entirely new or is expected (positive exemplar), then it will be incorporated into the closest matched concept. However, *Falcon-ART* suffers from an inconsistent rule-base because new clusters are formed whenever resonance fails, regardless of any duplication of fuzzy labels. The four novel fuzzy neural network architectures proposed used four other clustering techniques to overcome this deficiency. They are FCM, MLVQ, *Fuzzy Kohonen Partitioning* (FKP), and *Pseudo Fuzzy Kohonen Partitioning* (PFKP),[4] hence, the four proposed networks were intuitively named *Falcon-FCM, Falcon-MLVQ, Falcon-FKP*, and *Falcon-PFKP*, respectively. Since the four proposed networks implement the contrast of positive and negative exemplars in their inference process, they are collectively termed the *Pseudo Adaptive Complementary Learning* (PACL) fuzzy neural networks.

This chapter is organized as follows. Section 10.2 of this chapter covers the literature review of neural networks, fuzzy systems, and clustering techniques. The different types of neural network learning algorithms available are presented. The concept of fuzzy logic and a brief introduction of the CRI and AARS inference schemes are also covered. The concept of clustering and the various clustering techniques available are introduced, and some of the problems encountered by the various clustering techniques are highlighted.

Section 10.3 introduces the concept of *Adaptive Resonance Theory* (ART)[10] and the *Falcon-ART*[9] fuzzy neural network that makes use of its clustering properties in identifying the fuzzy rules of a given problem domain. However, several shortcomings are found in the performance evaluation of the *Falcon-ART* network. Therefore, a modified version of the *Falcon-ART* network is proposed; it is termed the *Falcon-MART* (Modified ART). Simulation results of *Falcon-MART* are presented to justify the modifications. However, both *Falcon-ART* and *Falcon-MART* suffer from an inconsistent fuzzy rule base. This is a result of using ART (and fuzzy ART) as a clustering technique.

In Section 10.4, four novel *Falcon*-like fuzzy neural networks utilizing the concept of complementary learning (the contrast of positive and negative exemplars in the inference process) are presented. The *Fuzzy C-Means* (FCM), *Modified Linear Vector Quantization* (MLVQ), *Fuzzy Kohonen Partitioning* (FKP), and *Pseudo Fuzzy Kohonen Partitioning* (PFKP) clustering techniques are used to compute the fuzzy sets of the networks. Hence, the four proposed networks were appropriately named *Falcon-FCM, Falcon-MLVQ, Falcon-FKP*, and *Falcon-PFKP*.

Section 10.5 benchmarks the performances of these networks based on computer simulations, and Section 10.6 concludes the findings.

10.2 Background

10.2.1 Neural Networks

The research of neural networks (NN) begins in the 1940s, driven by the motivation to emulate human intelligence and implement such synthetic intelligence in a new generation of "thinking" machines and intelligent systems. Neural Networks are synthetic systems modeled on the brain architecture to capture the human learning capabilities. Artificial neural networks are good at tasks such as pattern matching and classification, function approximation, optimization, vector quantization, and data clustering. Over the years, many forms of learning algorithms and network architectures have been developed due to the intense interest in this area.

10.2.1.1 What Is a Neural Networks?

An artificial neural network is a parallel information processing structure with the following characteristics:

- It is a neural inspired mathematical model.
- It consists of a large number of highly interconnected processing elements (neurons).
- Its connections (weights) hold the knowledge. It emulates the operation of a neuron using aggregation and activation functions.
- A processing element can dynamically respond to its input stimulus, and the response completely depends on its local information; that is, the state of the neuron.
- It has the ability to learn, recall, and generalize from training data by assigning or adjusting the connection weights. Neural network can sensibly interpolate input patterns that are new to the network and automatically adjust its connection weights and even the network structure to optimize its performance.
- Its collective behavior demonstrates the computational power, and no single neuron carries specific information (distributed representation property). Hence, the performance of neural network degrades gracefully under faulty conditions such as damaged neurons or broken connections, as the neuron acts independently of all the others and relies only on local information.

Other commonly used names for artificial neural networks are *parallel distributed processing models, connectionist models, self-organizing systems, neuro-computing systems, and neuromorphic systems* or simply *neural networks*. For simplicity, the term neural network (NN) will henceforth be used.

Initial research on neural networks was inspired by interest in the neuro-physiological fundamentals of the human brain. The fundamental building block of a biological neural network is the elementary nerve cell called a *neuron*. The human brain consists of about 10^{11} neurons. Figure II.10.1 depicts three typical neurons.

Warren S. McCulloch and Walter Pitts in 1943[11] introduced a model for the functionality of such a neuron and had also shown in their paper how their proposed model can be used to implement logical connectives like AND and OR operations.

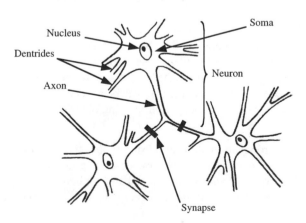

FIGURE II.10.1   A sketch of three connected neurons.

10.2.1.2 Development of Neural Networks

The major developments in the field of neural network are outlined in a chronological order as given below. The list is not exhaustive and only highlights the important milestones in the study of neural network.

1. **Hebbian rule:**[12] Donald O. Hebb developed the first learning rule used to update the weights of a neural network. The synaptic connection between two neurons is intensified or strengthened if it repeatedly fires the receptive neuron.

2. **Rosenblatt's perceptron:**[13] In 1958, the *perceptron* was developed by Frank Rosenblatt to make use of the Hebbian principle to learn its weights. It is an artificial neural network consisting of only a single neuron with one or more inputs and a single output. The perceptron is able to solve *linearly separable* classifying problems.

3. **ADALINE:**[14] Bernhard Widrow and Marcian E. Hoff developed the ADALINE (**Ada**ptive **Li**near **Ne**uron) model in 1960. The learning rule used in the training of the ADALINE is the *delta-learning rule* or the *least mean square* (LMS) rule. This is an improved learning process over that of the perceptron, as it uses the concept of *negative gradient descent*.

4. **Linear Associative Memory:**[15,16] During the 1970s, Teuvo Kohonen and James A. Anderson independently published their works in the implementation of *linear associated memory* using neural networks.

5. **Kohonen's network:**[17] In 1982, Kohonen published his work in self-organizing networks that make use of the unsupervised learning algorithm known as *competitive learning* or the *winner-takes-all* algorithm. Later, Kohonen also introduced the concept of *self-organizing feature maps* (SOFMs). These neural networks create a feature space in which other than the actual winning node, neighboring output nodes are also activated in response to the input stimuli.

6. **Hopfield network:**[18] At about the same time, John J. Hopfield published a paper on a model that is known as *Hopfield network*. Hopfield has proven in his paper that the stable states of the network have minimum energy levels and learning always converges. The Hopfield network has been used as memory associator and to optimize NP (Non-Polynomial) hard problems such as the *traveling salesman problem* (TSP).

7. **Boltzman machine:**[19] The *Boltzman machine* developed by David H. Ackley et al. uses the Hopfield network as a backbone structure. The problem of false wells is overcome by changing the activation function of the neurons to be of stochastic nature. Training of the Boltzman machine uses a scheme known as *simulated annealing*[20] to gradually reduce the randomness of the outputs of the neurons through the adjustment of a parameter called *temperature*, which is often denoted as T.

8. **Back-propagation:**[21] The discovery of the *back-propagation* learning algorithm leads to the prosperous age of neural network. The back-propagation learning algorithm extends the negative gradient-descent concept to update the weights of multi-layered networks. The back-propagation learning algorithm removes the limitation of the perceptron and ADALINE models so that more complex, multi-layered networks can be developed to solve real-world problems. Networks trained with the back-propagation learning algorithm are called *back-propagation networks*.

The discovery of the back-propagation algorithm brings about the developments of modern neural network study. More complex networks such as time-delayed neural network (TDNN)[22] and radial basis functions (RBF)[23] networks have been developed and applied to information-processing tasks such as pattern recognition, speech recognition, forgery detection, and control purposes.

10.2.1.3 Neural Network Fundamentals

Neural network consists of a large number of simple *processing elements* (PE for short) called *neurons, units, cells*, or *nodes*. It is characterized by the aggregation and activation functions of the neurons, its network architecture, and the learning (training) algorithm. Each neuron is connected to other neurons by means of directed communication links (each with an associated weight), and has an internal state called its activation or activity level that is a function of the inputs it has received.

- Aggregation and Activation Functions

  The basic operation of an artificial neuron involves combining its weighted input signals using an aggregation function, usually denoted as f, and to produce an output using its activation function a based on the aggregated inputs. Depending on the type of neural networks, the same activation function may be used for all neurons in the network (such as SOFMs), or each layer of a neural network has its own activating function (such as *Falcon-ART*[9]). Usually, the activation function a is nonlinear and differentiable so that the back-propagation algorithm can be applied to train the network.

- Network Architecture

  A neural network consists of a set of highly interconnected neurons such that each neuron's output is connected through weights to other neurons or back to itself. The operation of a neural network depends on the structure or topology that organizes the connections of the neurons in the network. Besides the simplest neural network that consists of only a single neuron, more complex neural networks consist of two or more neurons. These neurons are arranged in layers, and neurons within the same layer may behave in the same manner. Within the layer, the neurons may be fully interconnected or not interconnected at all. The arrangement of neurons into layers and the connection patterns within and between layers is called the *network architecture*. Neural networks are often classified as *single layer* or *multi-layer*. These layered networks can be *feedforward* networks in which signal flows from the input to the output or *recurrent* network in which there are closed-loop signal paths from a neuron back to itself or to its neighboring neurons in the same layer or the preceding layers. As the term suggests, single-layered network consists of only one layer of output nodes. There may be one or more nodes in the output layer. Here, the inputs are not considered as a layer of nodes. Figure II.10.2 depicts the structure of a single-layered neural network.

  Neural networks such as Kohonen's Self-Organizing Feature Maps (SOFM),[17] the Perceptron, and ADALINE models can be considered as single-layered networks.

  On the other hand, multi-layered networks consist of two or more layers of neurons. The first layer is known as the input layer that acts as a buffer between the network and the external environment, and usually performs no operation on the inputs. The last layer of neurons is the output layer and computes the network outputs to the external environment. Depending on the network architecture, there may be any number of hidden layers of neurons between the input and output layers. For example, if there are only two layers in a network, then there will be no hidden layers. Figure II.10.3 shows a three-layered network with one input, one hidden, and one output layer.

  Examples of multi-layered neural networks include RBF networks[23] and the counter-propagation network.[24–26]

  The neural network structures shown in Figure II.10.2 and II.10.3 are feedforward networks, since no neuron's output is an input to a neuron in the same or preceding layers. When outputs can be directed back as inputs to the same or preceding layer nodes, the network is called a recurrent or feedback network. Similar to feedforward networks, recurrent or feedback networks can be single-layered or multi-layered networks. The simplest recurrent network

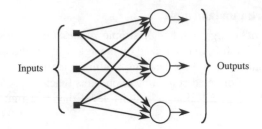

FIGURE II.10.2   A single-layered neural network.

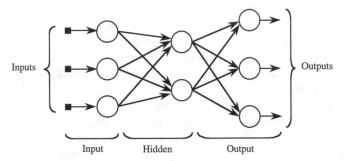

FIGURE II.10.3   A multi-layered network.

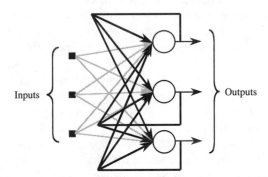

FIGURE II.10.4   A single-layered recurrent network.

consists of a single neuron with its output redirect as input to itself. Figure II.10.4 shows a single-layered recurrent network. The feedback paths are shown in bold.

Examples of recurrent networks are the Hopfield network and *Bi-directional Associative Memory* (BAM). However, in the Hopfield network, there is no feedback from the output of a neuron back to itself.

- Learning/Training Algorithms

The third important element of neural networks is the learning algorithms. There are two kinds of learning in neural networks: *parameter* learning and *structural* learning. Parameter learning involves the update of the weights connecting the neurons while structural learning is concerned with the modification of the connections between the neurons. Parameter and structural learning may be performed separately or simultaneously. An example in which both parameter and structural learning are performed simultaneously is the *Falcon-ART* network.[9] For a given input, structural learning is performed followed by the training of the weights using

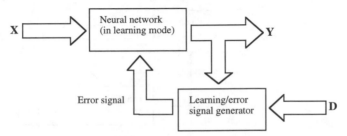

FIGURE II.10.5   Block diagram for supervised learning.

back-propagation before the next input is presented. In parameter learning, there are two types of training methods available—*supervised* and *unsupervised* training.

1. Supervised Training

A set of input and desired output patterns called a *training set* is required for this type of learning algorithm. Training is accompanied by presenting a sequence of training vectors or patterns, each with an associated target output vector. The weights are then updated and adjusted according to a learning algorithm (same as learning with a teacher). Examples of supervised learning are the back-propagation learning algorithm or the generalized delta-learning rule, Hebbian rule, and the delta-learning rule used to train the ADALINE model. Figure II.10.5 depicts the block diagram for supervised learning. The input vector is denoted as X, the output vector is denoted as Y, and the desired output vector is denoted as D.

During the training cycle, at each instant when the input is applied, the desired response of the neural network system is also provided. The distance between the actual and desired response serves as an *error measure* and is used to correct the network parameters. The learning algorithm tries to adapt the network's weights so that the error decreases. Sometimes, it may be impossible or too expensive to obtain the desired output vectors to train the neural network. However, some kind of training feedback is required to allow the network to be efficiently trained. This training feedback usually takes the form of a binary value, that is, 0 or 1 or it may be a real number in the range [0, 1]. Such feedback is known as a *critic signal*. 0 may be taken to represent that the network output is totally undesirable and 1 highly desirable. Hence, with this critic signal, the network can appropriately adjust its weights. This form of training is known as *reinforcement learning*. Reinforcement learning is a special type of supervised learning in which there is only a single training feedback and its value is constrained to a certain range. The critic signal provides an indication whether the network output is desirable or not, but gives no indication as to how to adjust the weights of the network.

2. Unsupervised Training

Unsupervised learning is the grouping of similar vectors together without the use of training data to specify what a typical member of each group looks like. A sequence of input vectors is provided, but no target output vectors are specified. The weights are modified such that the most similar input vectors are assigned to the same output. Examples of unsupervised learning are linear vector quantization (LVQ), Kohonen's competitive learning or winner-take-all algorithm, and SOFMs. Figure II.10.6 shows the block diagram for unsupervised learning. X and Y represent the input and output vectors, respectively.

In learning without supervision, the desired response is unknown. No information is available as to the correctness or incorrectness of responses. Learning must be accomplished based on observations of responses to the inputs. The network must discover for itself any possible existence of clusters, patterns, regularities, separating properties, etc. While discovering these properties, the network

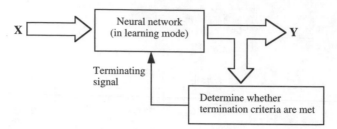

FIGURE II.10.6   Block diagram for unsupervised learning.

undergoes changes to its parameters. The unsupervised learning algorithm terminates when one or more of the preset termination criteria are met. Such criteria may include the accumulated error being smaller than a preset value or the learning parameter converging to 0, or that the number of training cycles exceed a preset limit. The terminating signal in Figure II.10.6 is used to terminate the neural network training cycle when one or more of the termination criteria are met.

10.2.1.4   Strengths and Weaknesses of a Neural Network

The strengths of a neural network lies in its ability to learn and adapt its weights, its fault-tolerance characteristic due to its distributed knowledge representational attribute, and ability to model a given problem domain and derive reasonable outputs in response to given inputs. However, there are certain weaknesses that restrict the usage of neural networks. They are listed as follows.

1. Black box
   It is common knowledge that it is difficult or almost impossible to interpret the weight matrix of a trained neural network in relation to the dynamics of the problem domain that it has modeled. Hence, a neural network cannot be verified that it performs as expected. To the user, a trained neural network acts as a black box as he/she cannot learn and assess the knowledge stored in the network.
2. Cannot make use of any *prior* knowledge
   A neural network always learns from scratch. Although *prior* knowledge on the problem domain such as the clustering of the data may be available, it cannot make use of them to shorten the learning cycle.
3. Stability and plasticity dilemma
   Once trained, a neural network has difficulty incorporating new data or information and in the extreme case, may even lead to the collapse of the network. Although stability of the network is important in computing consistent outputs, plasticity of the network is equally important, as it has to adapt to changes in the problem domain and may need to incorporate new information. This is especially so since the problem domain is often a physical process such as a control problem. Pure neural networks cannot incorporate new data once they are trained and a re-learning is inevitable if new data needs to be included.
4. Convergence of learning
   Although there are numerous learning algorithms available to train the network, there is no guarantee that the best results can be achieved. Supervised learning algorithms such as the back-propagation learning algorithm make use of a cost function to train the network. But such cost or error functions are often convex. Hence, there is a possibility that learning may terminate when the network is at a local minima rather than the global minima. Also, due to the complex surfaces of the cost functions, the training of a neural network may take a long time with no guaranteed results.

5. Optimization of network structure

Very often, the structure of a neural network is determined through past experience or in heuristic manner. There are no guidelines in determining the number of neurons in a network, how many layers there should be, or how the neurons should be connected. Hence, it is impossible to determine an optimized structure for a network for a given problem. Optimization is important for network performance, as it reduces the complexity of the network and the number of computations to be performed.

10.2.2 Fuzzy Systems

10.2.2.1 Fundamentals of Fuzzy Systems

Fuzzy system is a control paradigm that uses a linguistic approach to model a process or a problem that would otherwise be too complex to describe using mathematical models. The basis of fuzzy systems is fuzzy logic introduced by Zadeh A. Lotfi.[27] The theory of fuzzy logic provides a mathematical framework to capture the uncertainties associated with the human cognitive processes, such as thinking and reasoning. The concept of fuzzy sets is used to represent these uncertainties, and fuzzy systems are control or expert systems that make use of the fuzzy sets and the associated fuzzy rules to imitate the control actions or reasoning of an expert. Figure II.10.7 shows how a fuzzy system may be implemented to provide control to an arbitrary plant.

The main components of a fuzzy system consist of the following:

1. *Input fuzzifier module*

   Input status from a plant is captured, but because these readings are absolute values, they have to be converted to fuzzy values before the inference engine can make use of them.
2. *Inference engine*

   Derive appropriate outputs given the fuzzified inputs and the established facts in the fuzzy rule base using an inference/reasoning scheme.
3. *Fuzzy rule base*

   Store all the fuzzy rules associated with the system. Contains the knowledge base of the control actions in response to the fuzzified inputs.
4. *Output defuzzifier module*

   Convert the fuzzified outputs derived by the inference engine to absolute values so that they can be applied to the external plant.

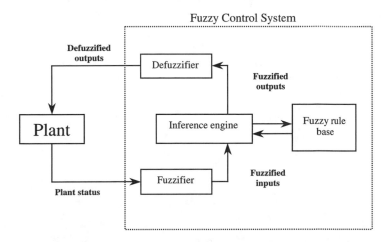

FIGURE II.10.7   Block diagram of a fuzzy control system.

Fuzzy systems model the dynamics of a problem domain using high-level IF-THEN fuzzy rules that are intuitive to the understanding of the human user. This enables the user to easily understand and learn from a fuzzy system and also ensures that the derived outputs of the system approximate the control actions of an expert. Any problem-specific *prior* knowledge can also be easily incorporated into a fuzzy system. The performance of a fuzzy system is based on its ability to draw or derive valid conclusions given the inputs and the established facts stored in its fuzzy rule base. The quality of the derived outputs depends on three factors:

- The fuzzy sets representing the underlying clustering structure of the problem domain
- The IF-THEN fuzzy rules reflecting the dynamics of the environment
- The fuzzy inference/reasoning scheme used to derive the outputs

However, traditional fuzzy systems such as ANFIS[28–30] have their limitations. An Achilles' heel of traditional fuzzy systems is the formulation of the fuzzy rule base. The fuzzy rules are manually specified and they may be inaccurate and biased, as opinions may differ with different experts and it may not be easy to verbally define the desired actions for a complex system. In addition, the parameters of a fuzzy system are manually tuned to produce the desired outputs. This may be too difficult or impossible for a complex system with many variables. Even if manual tuning is possible, the approach is heuristic, as there are no formal methods for tuning.

10.2.2.2 Fuzzy Logic and Fuzzy Sets

Fuzzy logic provides a mathematical framework to capture imprecise or fuzzy concepts. It uses fuzzy sets to represent these concepts by means of graded membership values. Fuzzy sets provide a mechanism for representing imprecise concepts such as "fat," "heavy," and "tall" that exist in human cognitive processes such as learning and reasoning. Fuzzy logic is an extension of classical set-based binary logic and serves as the basis for *approximate reasoning*. Approximate reasoning is the process of inferring or the deduction of a valid conclusion, given a set of premises that take the forms of IF-THEN fuzzy rules. Fuzzy sets are the extensions to classical crisp sets.

A fuzzy set is characterized by its membership function. The most commonly used fuzzy membership functions are Gaussian and trapezoidal functions. Gaussian membership function provides a smooth transition. Triangular-shaped membership functions are a subset of the trapezoidal-shaped membership functions. The trapezoidal-shaped and triangular-shaped membership functions are most commonly used in current fuzzy applications due to their simplicity in storage and computation.[31] The trapezoidal membership function (μ_A) can be described by a fuzzy interval formed by four parameters ($\alpha, \beta, \sigma, \tau$) and a centroid φ as shown in Figure II.10.8. This fuzzy interval is also known as a trapezoidal fuzzy number.

The interval $[\beta, \sigma]$ where $\mu_A(x) = 1$ is called the kernel of the fuzzy interval, and the interval $[\alpha, \tau]$ is called the support. If the kernel reduces to one single point, the fuzzy interval becomes a triangular fuzzy number.[32]

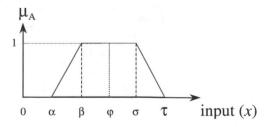

FIGURE II.10.8   Trapezoidal fuzzy membership function.

Depending on the form of the membership function chosen to represent a fuzzy set, one of the tasks in designing a fuzzy system is to determine the parameters of the membership function. In a traditional fuzzy system, the experts may determine the form and parameters of the membership function. In a fuzzy neural architecture, the learning ability of the neural network can be harnessed to learn the parameters of the membership function whose form is determined by the designer of the system. This automated learning of the parameters includes *clustering* of the numerical data and the self-adaptation of the parameters to achieve the desired results. More details on clustering will be covered later.

10.2.2.3 Fuzzy Rules

A fuzzy rule base is a linguistic model of a problem domain. It consists of numerous high-level IF-THEN fuzzy rules. It captures the dynamics of the problem domain and the associated response action/behavior of a human expert in handling the problem. A fuzzy rule base tries to model the problem domain from a human perspective (linguistic model) so as to imitate the human response rather than the physical perspective (mathematical models). For example, if the problem is to bring a cruising car to a stop in front of a traffic junction, the driver is concerned with mainly two things: speed of the car as it approaches the junction, and distance of the car to the junction. He is not concerned about the torque of the car engine, the drag on the car, or the friction of the road on the car, which would require complex mathematical models (to capture such physical characteristics). In his mind, a linguistic model of the problem would be something like:

- If *speed* is *fast* and *distance* is *short*, then *apply heavy brakes.*
- If *speed* is *fast* and *distance* is long, then *apply moderate brakes.*
- If *speed* is *slow* and *distance* is short, then *apply moderate brakes.*
- If *speed* is *slow* and *distance* is long, then *apply light brakes.*

Through visual feedback and the speedometer of the car, the driver receives information on the speed of the car and its distance to the junction and takes the necessary action (applying the brakes). The above four rules are examples of IF-THEN fuzzy rules and together they constitute a fuzzy rule base. In general, a fuzzy rule has the form shown in Equation II.10.1:

$$\text{If } x_1 \text{ is } \mathbf{A}_{x_i} \text{ and } \cdots \text{ and } x_i \text{ is } \mathbf{A}_{x_i} \cdots \text{ and } x_n \text{ is } \mathbf{A}_{x_n} \text{ then}$$
$$y_1 \text{ is } \mathbf{B}_{y_1} \text{ and } \cdots \text{ and } y_j \mathbf{B}_{y_j} \cdots \text{ and } y_m \text{ is } \mathbf{B}_{y_m} \tag{II.10.1}$$

where

n = number of input linguistic variable;
m = number of output linguistic variable;
x_i = the ith input linguistic variable;
y_j = the jth output linguistic variable;
$\mathbf{A}_{x_i}$ = the linguistic term (input term) of the ith input linguistic variable; and
$\mathbf{B}_{y_i}$ = the linguistic term (output term) of the jth output linguistic variable.

An IF-THEN fuzzy rule consists of five parts. The first two parts are the input linguistic variables or simply the inputs and their associated input terms. They form the antecedent section of the fuzzy rule. The last two parts are the output terms and the output linguistic variables. They form the consequent of the fuzzy rule. The third part of a fuzzy rule is the IF-THEN construct that links the antecedent and the consequent of the rule. In a Mamdani's fuzzy model, the input and output terms are represented as fuzzy sets. The definition of a linguistic variable is given in Lin and Lee.[33] The domain of possible values a linguistic variable can take is sometimes called the *universe of discourse*. A general fuzzy rule base of L rules may take the form:

$$\text{Rule 1 : If } x_1 \text{ is } \mathbf{A}^1_{x_1} \text{ and } \cdots \text{ and } x_i \text{ is } \mathbf{A}^1_{x_i} \cdots \text{ and } x_n \text{ is } \mathbf{A}^1_{x_n} \text{ then}$$

$$y_1 \text{ is } \mathbf{B}^1_{y_1} \text{ and } \cdots \text{ and } y_j \text{ is } \mathbf{B}^1_{y_j} \cdots \text{ and } y_m \text{ is } \mathbf{B}^1_{y_m} \tag{II.10.2}$$

$$\vdots$$

$$\text{Rule } l : \text{If } x_1 \text{ is } \mathbf{A}^l_{x_1} \text{ and } \cdots \text{ and } x_i \text{ is } \mathbf{A}^l_{x_i} \cdots \text{ and } x_n \text{ is } \mathbf{A}^l_{x_n} \text{ then}$$

$$y_1 \text{ is } \mathbf{B}^l_{y_1} \text{ and } \cdots \text{ and } y_j \text{ is } \mathbf{B}^l_{y_j} \cdots \text{ and } y_m \text{ is } \mathbf{B}^l_{y_m}$$

$$\vdots$$

$$\text{Rule } L : \text{If } x_1 \text{ is } \mathbf{A}^L_{x_1} \text{ and } \cdots \text{ and } x_i \text{ is } \mathbf{A}^L_{x_i} \cdots \text{ and } x_n \text{ is } \mathbf{A}^L_{x_n} \text{ then}$$

$$y_1 \text{ is } \mathbf{B}^L_{y_1} \text{ and } \cdots \text{ and } y_j \text{ is } \mathbf{B}^L_{y_j} \cdots \text{ and } y_m \text{ is } \mathbf{B}^L_{y_m}$$

The fuzzy rule given in Equation II.10.2 has multiple input and multiple output (MIMO). It is a composite conjunctive fuzzy rule with multiple propositions in the antecedent section connected by "AND." Fuzzy rules with multiple propositions in the antecedent section connected by "OR" will not be considered here, as they can be split into many simple rules.[7,34] In addition, a multiple output fuzzy rule can also be decomposed into simpler single output fuzzy rules by creating a rule for each of the connected output in the consequent section. Hence, the simplest type of fuzzy rule is a single input and a single output (SISO) fuzzy rule. In the literature, variations of fuzzy rule representation include specifying a certainty factor (CF) for a fuzzy rule, and a threshold, and a weight parameter for each of the propositions in the antecedent of the fuzzy rule. A certainty factor of [0,1] specifies the strength of the fuzzy rule given the supporting numerical data. The threshold for each proposition determines whether there is enough evidence for the proposition to be active, and contributes to the firing of the rule while the weight parameter determines the degree of influence that a proposition has in firing the fuzzy rule.

10.2.2.4 Fuzzy Inference/Reasoning Schemes

Reasoning capability plays a major role in human intelligence because in many situations, a human being has to make decisions based on incomplete, vague, and fuzzy information. Fuzzy reasoning of fuzzy rules is a mechanism to draw a conclusion from the given input that does not exactly match with the antecedents of the fuzzy rules. By modeling the human reasoning process using fuzzy inference/reasoning schemes and incorporating this capability into a fuzzy system, the intelligence of the system can be substantially enhanced.

There are two main branches of fuzzy inference schemes available: CRI (Compositional **R**ule of Inference)[2] based and *similarity-based* inference schemes. Inference schemes following the *generalized modus ponens* (GMP) inference rule are the most commonly used. The generalized modus ponens inference rule is defined as follows.

Definition 1: GMP is defined as the inference procedure shown in Equation II.10.3.

$$
\begin{array}{l}
\text{Premise 1: If } x \text{ is } \mathbf{A}, \text{ then } y \text{ is } \mathbf{B} \\
\underline{\text{Premise 2: } x \text{ is } \tilde{\mathbf{A}}} \\
\text{Conclusion: } y \text{ is } \tilde{\mathbf{B}}
\end{array} \tag{II.10.3}
$$

where

x = input linguistic variable taking values in universe of discourse $\mathbf{U}_1$;
y = output linguistic variable taking values in universe of discourse $\mathbf{U}_2$;
$\mathbf{A}, \tilde{\mathbf{A}}$ = fuzzy sets defined on $\mathbf{U}_1$; and
$\mathbf{B}, \tilde{\mathbf{B}}$ = fuzzy sets defined on $\mathbf{U}_2$.

Premise 1 can be considered as a typical fuzzy rule in the fuzzy rule base of a fuzzy system and premise 2 can be viewed as the input to the fuzzy system. If a fuzzy rule is interpreted as some forms of relation associating the antecedent (input data space) to the consequent (output data space), then CRI-based inference schemes result. For similarity-based inference schemes, a fuzzy rule is interpreted as a statement of truth, where the similarity between the conclusion to the consequent of the fuzzy rule correlates to the similarity between the input to the antecedent of the fuzzy rule.

10.2.2.4.1 *CRI-Based Inferences*

Consider the fuzzy rule shown as premise 1 in Equation II.10.3. In the framework of Zadeh's *composition rule of inference* (CRI),[2] such a fuzzy rule is interpreted as a fuzzy relation R[33] restricting the possible values of the ordered pair (x, y). This fuzzy relation R has a membership function μ_R (it is a fuzzy set) that can be resolved using a generalized implication operator I as defined in Equation II.10.4:

$$\mu_R(x, y) : \mathbf{U}_1 \times \mathbf{U}_2 \rightarrow [0, 1]; \text{ and} \qquad (\text{II}.10.4)$$

$$\mu_R(x, y) = I(\mu_{\mathbf{A}}(x), \mu_{\mathbf{B}}(y)) \quad \forall (x, y) \in \mathbf{U}_1 \times \mathbf{U}_2; \text{ and}$$

$$I : [0, 1]^2 \rightarrow [0, 1]$$

where

R = fuzzy relation used to interpret the fuzzy rule;
$\mathbf{U}_1, \mathbf{U}_2$ = universes of discourse;
μ_R = membership function of fuzzy relation R;
$\mu_{\mathbf{A}}$ = membership function of fuzzy set $\mathbf{A}$ defined on $\mathbf{U}_1$;
$\mu_{\mathbf{B}}$ = membership function of fuzzy set $\mathbf{B}$ defined on $\mathbf{U}_2$; and
I = implication operator used to resolve μ_R.

Since the introduction of the CRI inference scheme, many fuzzy implication relations have been proposed. The following lists six of them:

1. [2] : $I(\mu_{\mathbf{A}}(x), \mu_{\mathbf{B}}(y)) = \max[\min(\mu_{\mathbf{A}}(x), \mu_{\mathbf{B}}(y)), (1 - \mu_{\mathbf{A}}(x))]$
2. [35] : $I(\mu_{\mathbf{A}}(x), \mu_{\mathbf{B}}(y)) = [\mu_{\mathbf{B}}(y)]^{\mu_{\mathbf{A}(x)}}$
3. [36] : $I(\mu_{\mathbf{A}}(x), \mu_{\mathbf{B}}(y)) = \min(1, \frac{\mu_{\mathbf{B}}(y)}{\mu_{\mathbf{A}}(x)})$
4. [2] : $I(\mu_{\mathbf{A}}(x), \mu_{\mathbf{B}}(y)) = \min(1, 1 - \mu_{\mathbf{A}}(x) + \mu_{\mathbf{B}}(y))$
5. [37] : $I(\mu_{\mathbf{A}}(x), \mu_{\mathbf{B}}(y)) = \min(\mu_{\mathbf{A}}(x), \mu_{\mathbf{B}}(y))$
6. [38] : $I(\mu_{\mathbf{A}}(x), \mu_{\mathbf{B}}(y)) = \max[(1 - \mu_{\mathbf{A}}(x)), \mu_{\mathbf{B}}(y)]$

The most commonly used implication operator is the one proposed by Mamdani due to its simplicity in computation. Once a fuzzy rule is interpreted as a fuzzy relation R and its membership μ_R is resolved using an implication operator, it is then possible to derive a valid conclusion from the input to a fuzzy system and its fuzzy rule base with the GMP inference rule.

10.2.2.4.2   *Similarity-Based Inferences*

Similarity-based fuzzy reasoning methods do not require the construction of a fuzzy relation R between the antecedent and consequent of a fuzzy rule. Instead, they are based on the degree of similarity between the inputs to the fuzzy system and the antecedents of the fuzzy rules. Because the fuzzy rules are interpreted as statements of truth, similarity-based inference methods use a metric to measure the degree of similarity of the inputs to the antecedents of the fuzzy rules. The conclusion is subsequently derived by suitably modifying the consequent of the fuzzy rules based on the computed degree of similarity. Several similarity-based inference schemes are listed below.

1. Approximate Analogical Reasoning Schema (AARS)[7]
2. Chen's Matching Function (MF) Method[39]
3. Chen's Function T (FT) Method[40]
4. Yeung et al.'s Degree of Subsethood (DS) Method[34]
5. Yeung et al.'s Inclusion and Cardinality (IC) Method[41]
6. Yeung et al.'s Equality and Cardinality (EC) Method[41]

The inference schemes listed above are the same in principle. They differ only in their computation of the degree of similarity between the inputs to the fuzzy system and the antecedents of the fuzzy rules and the type of modifying functions used to derive the conclusion. The motivation for developing the various methods is to provide more flexibility when applying similarity-based reasoning methods in applications. The AARS reasoning scheme is as follows.

10.2.2.4.2.1   *Approximate Analogical Reasoning Schema (AARS)*   A fuzzy system with a simple rule and input is given below. A threshold τ is attached to the fuzzy rule so that the rule is fired only when the computed degree of similarity is greater than τ.

$$\frac{\text{Rule: If } x \text{ is } \mathbf{A} \text{ then } y \text{ is } \mathbf{B} \text{ (threshold } \tau)}{\text{Input: } x \text{ is } \tilde{\mathbf{A}} \text{ (such that similarity of } \tilde{\mathbf{A}} \text{ and } \mathbf{A} \text{ exceeds } \tau)}$$
$$\text{Conclusion: } y \text{ is } \tilde{\mathbf{B}} \tag{II.10.5}$$

where

$x =$ an element in universe of discourse $\mathbf{U}_1$;

$y =$ an element in universe of discourse $\mathbf{U}_2$;

$\mathbf{A}, \tilde{\mathbf{A}} =$ fuzzy sets defined on $\mathbf{U}_1$; and

$\mathbf{B}, \tilde{\mathbf{B}} =$ fuzzy sets defined on $\mathbf{U}_2$.

The AARS inference method modifies the consequent "y is **B**" based on the similarity (closeness) between the input "x is $\tilde{\mathbf{A}}$" and the antecedent of the fuzzy rule, "x is **A**." If the degree of similarity measure is greater than the predefined threshold τ of the fuzzy rule, the rule will be fired and the consequent is deduced by some modification functions. Since the input and the antecedent of the fuzzy rule are both fuzzy sets, a *distance measure* (DM) is required to compute the similarity between the two fuzzy sets. In Turksen and Zhong,[7,42] five good distance measures are introduced. The distance measures are Disconsistency Measure, Hausdorff measure (∞), Hausdorff measure (*), Kaufman and Gupta measure (∞), and Gupta measure (*). The similarity measure between the fuzzy sets **A** and $\tilde{\mathbf{A}}$, S_{AARS}, is defined as in Equation II.10.6:

$$S_{AARS} = \frac{1}{(1 + DM)}; S_{AARS} \in [0, 1] \tag{II.10.6}$$

where

DM = any of the distance measure listed above.

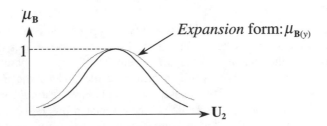

FIGURE II.10.9 μ_B and the *expanded* $\mu_{\tilde{B}}$ membership functions.

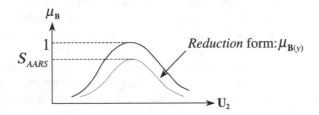

FIGURE II.10.10 μ_B and the *reduced* $\mu_{\tilde{B}}$ membership functions.

If S_{AARS} exceeds threshold τ, the fuzzy rule will be fired and the consequent of the fuzzy rule is modified by one of the modification functions below. The label $\mu_{\tilde{B}}$ is the membership function of the derived output of a fuzzy system employing the AARS inference method.

1. *Expansion* function:

$$\mu_{\tilde{B}}(y) = \min\left[1, \frac{\mu_B(y)}{S_{AARS}}\right] \qquad (\text{II.10.7})$$

where

$y =$ an element in universe of discourse U_2;
$B, \tilde{B} =$ fuzzy sets defined on U_2; and
$S_{AARS} =$ similarity measure of fuzzy sets B and $\tilde{B}$.

The effect of the *expansion* modifying function on μ_B is depicted in Figure II.10.9. It has the same modifying effect on μ_B as the *dilation* operation.[33]

2. *Reduction* function:

$$\mu_{\tilde{B}}(y) = S_{AARS} \times \mu_B(y) \qquad (\text{II.10.8})$$

where

$y =$ element in universe of discourse U_2;
$B, \tilde{B} =$ fuzzy sets defined on U_2; and
$S_{AARS} =$ similarity measure of fuzzy sets B and $\tilde{B}$.

The effect of the *reduction* modifying function on μ_B is depicted in Figure II.10.10. It has the same modifying effect on μ_B as the *concentration* operation.[33]

10.2.3 Clustering

The usefulness of a fuzzy system depends on its ability to derive valid conclusions in response to its inputs using a fuzzy inference scheme. The fuzzy inference scheme uses the fuzzy rule base and the inputs to compute the outputs of the fuzzy system. The fuzzy rule base consists of a set of high-level IF-THEN fuzzy rules. The fuzzy rules form a linguistic model of the problem domain. The derivation of the fuzzy rules consists of two phases: input and output feature/data space partitioning into fuzzy subspaces followed by the construction of the associated IF-THEN fuzzy rules by linking the appropriate input fuzzy subspaces to the appropriate output fuzzy subspaces. In a traditional fuzzy system, these two phases are performed manually. In fuzzy neural systems, the popular approach is to partition the input and output data spaces using clustering techniques and subsequently formulate the fuzzy rules by the rule-learning phase of the network training cycle.

Clustering is a process to partition a data space or a given data set into different classes/groups so that the data points within the same class are more similar to one another than to data points in other classes. This similarity is measured using a metric of which the Euclidean distance is commonly used. Clustering is an exploratory approach to analyze a given numerical data set by creating a structural knowledge representation of the data set. Cluster analysis attempts to discover the inherent organization structure in a set of data points. This structural knowledge representation is the grouping of the data points into classes.

Clustering techniques can be categorized into two main groups: *partition-based* and *hierarchical-based* clustering. In partition-based clustering techniques, the data space is partitioned or grouped into k classes where the value of k is known *a priori*:. The *k-means*[43,44] and *linear vector quantization* (LVQ)[3] are classical examples of partition-based clustering techniques. There are also partition-based clustering techniques[45,46] that try to predict or estimate the value of k before the actual clustering to overcome the limitation of having to know the value of k beforehand.

In hierarchical-based clustering, a data set of N data points is repeatedly partitioned into finer regions or clusters based on some objective functions and/or criteria. This results in a hierarchy (or inverted tree) of clusters being formed. At each level, if a cluster is not the leaf node of the partitioning tree, then the cluster is broken up into smaller clusters. *Single link*,[47] *complete link*,[48] and the algorithms proposed by Benkirane et al.[49] and Lin et al.[50] are examples of hierarchical-based clustering. In hierarchical-based clustering, the fuzzy rules modeling the problem domain are implicitly derived, since the clusters obtained are regions (may or may not be fuzzy regions) in the feature space such that data points in the same cluster (region) will be mapped to the same output.

In the design of fuzzy neural systems, partition-based clustering techniques are more suitable for implementation than hierarchical-based clustering techniques. This is because the former can be easily incorporated as an autonomous phase in the training cycle of the fuzzy neural networks and the parameters of the clusters can be fine-tuned using neural network learning algorithms. On the other hand, it can be very difficult to tune the parameters of the clusters (regions) of hierarchical clustering techniques, since any modifications may affect the integrity of the partitioning tree.

10.2.3.1 Different Types of Clustering Algorithms

In hierarchical-based clustering, a data point either belongs to a region or it is outside the region. If crisp sets are used to represent the regions, then the transition boundaries between regions will be crisp. On the other hand, if Gaussian- or trapezoidal-shaped fuzzy sets are used to represent the membership values of the data points in a region with respect to some known or reference point (i.e., centriode), then the transition boundaries between regions will be gradual. As mentioned earlier, clustering techniques can be classified as hard clustering or fuzzy clustering. Since fuzzy clustering is interpreted as having a data point belonging to one or more classes with different degrees of membership values, hierarchical-based clustering techniques are not considered as fuzzy clustering

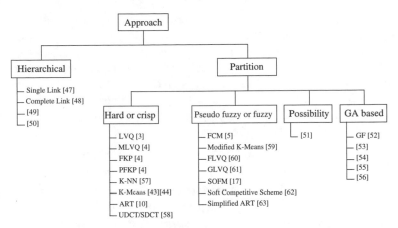

FIGURE II.10.11   Classification of various clustering techniques.

techniques. On the contrary, they are considered as hard clustering techniques since in most of the hierarchical-based clustering techniques, there is a clear partition or demarcation of regions.

In partition-based clustering techniques, the relationship between a data point and the class it belongs to is characterized by its membership value to the class. A C-class clustering problem for a data set $\mathbf{X}$ consisting of N data points, $\mathbf{X} = \{x_1, \ldots, x_p, \ldots, x_N\}$ where $p \in [1 \cdots N]$, is basically an assignment of the membership values $\mu_c(x_p)$ for each data point $x_p \in \mathbf{X}, \forall c \in [1 \cdots C]$ and $\forall p \in [1 \cdots N]$. The result of clustering on $\mathbf{X}$ (i.e., the membership values of all the data points with respect to C classes) can be presented as a $C \times N$ matrix, $\mathbf{S}$, as shown in Equation II.10.9.

$$\mathbf{S} = \begin{bmatrix} \mu_1(x_1) & \mu_1(x_2) & \cdots & \mu_1(x_N) \\ \mu_2(x_1) & \mu_2(x_2) & \cdots & \mu_2(x_N) \\ \cdots & \cdots & \ddots & \cdots \\ \mu_c(x_1) & \mu_c(x_2) & \cdots & \mu_c(x_N) \end{bmatrix} \tag{II.10.9}$$

where $\mu_c(x_N) = $ the membership value of the Nth data point to class C.

The value of matrix $\mathbf{S}$ varies with the type of clustering performed. There are three types of clustering, namely: *hard clustering, pseudo fuzzy clustering*, and *possibility clustering*.[51]

Besides the three types of clustering techniques mentioned, there is another branch of clustering algorithms based on or drawing inference from *Genetic Algorithms* (GAs). Examples are the GF classifier developed by Quek and Chen,[52] the work by Lin and Yang,[53] and Sankar.[54–56] Figure II.10.11 classifies some clustering techniques based on their approach (partition or hierarchical) and type of clustering adopted (hard, pseudo fuzzy or fuzzy, possibility, and GA-based). The list is not exhaustive and covers only the more popular and commonly used clustering techniques.

Many surveys of clustering techniques done by researchers have been published in the literature. They include Baraldi and Blonda.[64,65] In these two papers, the authors describe a detailed framework that they used to compare different clustering techniques and provide a comprehensive discussion of each of the clustering techniques reviewed. The clustering techniques reviewed include *self-organizing map* (SOM), *fuzzy linear vector quantization* (FLVQ), and fuzzy ART (F-ART).

10.2.3.2   LVQ-Inspired Clustering Techniques

LVQ[3] provides inspiration to the development of many clustering techniques such as the *modified linear vector quantization* (MLVQ),[4] *fuzzy Kohonen partitioning* (FKP), and *pseudo fuzzy Kohonen partitioning* (PFKP).[4] This is because the prototype vectors derived using LVQ can be seen as the centriodes of the clusters to be formed. Since these prototype vectors are representatives of the classes (clusters) they belong to, they can be assumed to have a membership value of unity to the class they represent. As a data point moves away from the prototype vector of a class, intuitively, its membership value to that class will decrease with distance. The nature of this membership value can be captured using either a Gaussian membership function with the prototype vector as its center or a trapezoidal membership function with the center of its kernel being the prototype vector.

Other variants of LVQ are *fuzzy linear vector quantization* (FLVQ)[60] and *generalized linear vector quantization* (GLVQ).[61] Both of these techniques incorporate the strengths of LVQ and SOFM to address issues of initialization in LVQ, sensitivity of LVQ in rugged decision surface, and underutilization of the LVQ network resources.[60]

10.2.3.3 FCM

In classical clustering analysis, cluster centers that characterize relevant classes of a finite set of data are computed. These classes form a crisp partition of the data set. When the crisp partition is replaced with a *fuzzy partition* or *fuzzy pseudo-partition*, it is referred to as *fuzzy clustering*.[31] Fuzzy pseudo-partitions are also called *fuzzy C-partitions* or *fuzzy C-means partitions*, where C designates the number of fuzzy classes in the data set. The clustering technique of obtaining the fuzzy pseudo-partition of a data set is hence known as *Fuzzy C-Means* (FCM). Consider a finite data set $\mathbf{X}$ consisting of N elements, $\mathbf{X} = \{x_1, \cdots, x_k, \cdots, x_N\}$ where $k \in [1 \cdots N]$ that is partitioned using FCM. The resultant fuzzy pseudo-partition P is defined by Equation II.10.10:

$$P = \{A_1, 1 \cdots NA_c, 1 \cdots NA_C\} \tag{II.10.10}$$

where

A_c = the fuzzy set for the cth cluster in the data set $\mathbf{X}$; and
C = total number of clusters for $\mathbf{X}$.

Each fuzzy set A_c is characterized by a membership function μ_c, where $c \in [1KC]$ and the pseudo-partition P has to satisfy the two conditions stated in Equations II.10.11 and II.10.12:

$$\sum_{c=1}^{C} \mu_c(x_k) = 1 \qquad \forall k \in [1 \cdots N] \tag{II.10.11}$$

$$0 < \sum_{k=1}^{N} \mu_c(x_k) < N \qquad \forall c \in [1 \cdots C] \tag{II.10.12}$$

where

C = number of clusters (linguistic labels) in the data set $\mathbf{X}$;
N = total number of elements in the data set $\mathbf{X}$; and
$\mu_c(x_k)$ = membership value of the kth element in the data set (denoted
          as x_k) with respect to the cth cluster.

An objective function for fuzzy clustering is formally expressed in Equation II.10.13 to partition the collection of data elements $\mathbf{X}$ into C fuzzy sets:[66]

$$J_M(P) = \sum_{k=1}^{N} \sum_{c=1}^{C} (\mu_c(x_k))^M |x_k - v_c|^2 \tag{II.10.13}$$

where

M = exponential value that influences the degree of fuzziness of the partition;

x_k = kth data element;

v_c = centriod for the cth cluster; and

$|x_k - v_c|$ = Euclidean distance between x_k and v_c.

The *fuzzy C-means* (FCM) algorithm was developed to derive fuzzy pseudo-partitions that minimize the above objective function. FCM requires that the number of cluster C is specified prior to clustering. In addition, all the data points in a training data set and their associated membership values are required to explicitly represent a cluster. This causes FCM to be memory intensive. The membership functions derived using FCM are not convex and there are no guidelines to set the value of the fuzzy parameter M. The fuzzy parameter M is used to control the fuzziness of the membership functions. The degree of fuzziness increases as the value of M is increased. The selection of the value of M remains open to research.

FCM belongs to a family of algorithms that are based on an iterative optimization of a fuzzy objective function. Due to their efficacy and computational efficiency, these are very popular clustering techniques. However, the users of these algorithms are usually required to specify the number of clusters C and some other parameters. These *a priori* assumptions are necessary, but do not guarantee the global minimum.[67] Different initial conditions may lead to different fuzzy C-partitions and consequently different clustering results. The heuristic algorithm proposed by Zahid et al.[45] performs fuzzy clustering without assumptions on initial guesses. In principle, it combines the favorable features of FCM and the K-Nearest Neighbors (K-NN) decision rule.[57] To estimate the optimal number of clusters C, the proposed KNN-one pass FCM algorithm incorporates a cluster-validity criterion. This cluster-validity criterion is based on two functions that measure the global compactness of a fuzzy C-partition[68] and the fuzzy separation of the fuzzy clusters.[51] These two functions assign a real number to the proposed cluster-validity criterion that determines the quality of the fuzzy C-partitions provided by the proposed KNN-one pass FCM algorithm.

10.2.3.4 GA-Based Clustering Techniques

Recently, there has been a surge in interest in GA-based clustering techniques. This is because the clustering process can be seen as a procedure to determine the optimal classes to represent a given data set and each of these classes is characterized by a membership function whose parameters need to be determined. Although a lot of algorithms have been proposed in the literature to automatically obtain the clustering of a given data set, most of them are application dependent and required the manual tuning of a set of pre-determined parameters to achieve optimal results. GA-based clustering techniques, such as the ones proposed by Quek and Chen[52] and Lin and Yang,[53] are attractive as they require very few manually tuned parameters and are application independent.

Genetic Algorithms (GAs) are global search and optimization techniques modeled from natural genetics, and explore the search space for an optimal solution by computing a set of candidate solutions in parallel. A GA maintains a population of candidate solutions. Each candidate solution is usually coded as a string (typically a binary string) called a *chromosome* (with reference to the similarity of GA and natural genetics), which is also referred to as a genotype. A chromosome encodes a set of parameters to be optimized. Each encoded parameter (usually a few bits long) is called a *gene*. A decoded parameter set is called a *phenotype*. A set of chromosomes forms a population that is evaluated and the chromosomes are ranked in order of merits by a fitness evaluation function. This fitness evaluation function encompasses a set of criteria specifying the characteristics of the optimal solution. Candidate solutions in the population are evaluated using these criteria and the fitness function will assign a real number to each chromosome to reflect its closeness to the desired solution. The fitness evaluation function is modeled after the natural selection process in the real world, where species with favorable genes live on while others become extinct. It plays an important role in GA because it provides information about how good each candidate solution is. This information guides the search for an optimal solution by determining the likelihood that a candidate solution is selected to produce candidate solutions in the next generation of possible solutions. A candidate solution with a better evaluation result will tend to reproduce more often than candidate solutions with poorer results.

An initial population of candidate solutions is usually randomly generated. Currently, GAs have been widely used in Shin and Park,[69] Kamel and Abutaleb,[70] Haydar, Demirekler, and Yurtseven,[71] and Gallego, Monticelli, and Romero.[72] With its powerful optimization and searching capability, GA can be used for clustering applications. Under GA, clustering has been transformed to the search for the optimized parameters of the membership functions characterizing each cluster. Once the parameters are found, the clusters are determined accordingly.

In Quek and Chen,[52] GA is proposed to automatically construct the trapezoidal membership functions of a fuzzy classifier system from a given data set and the performance responses of the system. Efforts have also been extended to the formulation of IF-THEN fuzzy rules that accurately reflect the dynamics of the problem domain using GA. In Lin and Shiueng,[53] GA is used to predict the optimal number of clusters C for clustering problems with no *prior* information on the value of C. There are also works on using GA for image processing by Sankar et al.[54] In addition, Sankar has extended GA for pattern classification.[55,56] However, the pattern classification implementation proposed has been limited to the search for hyper-planes that partition the data set. The results do not show the clustering nature of the data points in terms of fuzzy sets.

Although GA has a lot of potential in clustering applications, it also has its drawbacks. First, the initial population of candidate solutions in GA is randomly generated. Hence, the results that are achieved using GA may not be reproducible, as different initializations may lead to different results. Second, the results derived using GA tend to vary a lot and it may not be possible to identify the correlation between the results and the few preset parameters that the user has to specify.

10.2.3.5 Comparing FCM, MLVQ, FKP, and PFKP Clustering Techniques

As mentioned earlier, the FCM, MLVQ, FKP, and PFKP clustering techniques will be used to derive the fuzzy sets of four novel Falcon-like fuzzy neural networks. These four networks are intuitively named *Falcon-FCM, Falcon-MLVQ, Falcon-FKP*, and *Falcon-PFKP*. Hence, it is important to understand their characteristics and how they compare against other clustering techniques. The four clustering techniques are compared against the likes of LVQ, FLVQ, GLVQ, and fuzzy ART. Table II.10.1 shows the comparisons of the various techiniques.

From Table II.10.1, FCM, FKP, and PFKP perform clustering in the batch mode. FCM requires all the training data points to be present in order to generate a pseudo-partition of the training

TABLE II.10.1 Comparison of Various Clustering Techniques

Functional Features	Clustering Techniques							
	FCM	MLVQ	FKP	PFKP	LVQ	FLVQ	GLVQ	F-ART
Type of learning	Batch	On-line	Batch	Batch	On-line	On-line	On-line	On-line
Type of clustering	Fuzzy	Hard	Hard	Hard	Hard	Fuzzy	Fuzzy	Hard
Membership functions	Irregular	G	T	T	N.A.	N.A.	N.A.	T
A prior knowledge of C	Yes	Yes	Yes	Yes	Yes	Yes	Yes	No
Obj function minimization	Yes	No	No	No	No	Yes	Yes	No

Note: F-ART = fuzzy ART; G = Gaussian; T = Trapezoidal; C = number of classes; N.A. = Not Applicable.

data space. On the other hand, FKP and PFKP need to recycle the training data points in order to compute the trapezoidal membership functions. Fuzzy ART, in conforming to the concept of ART, is an on-line clustering algorithm that is proposed to handle the plasticity-stability dilemma. LVQ and its variants FLVQ, GLVQ, and MLVQ are all on-line clustering techniques because they update the prototype vectors in real time and do not need to recycle the training data points in order to compute membership functions.

According to the definition of hard and fuzzy clustering used in this manuscript, only FCM, FLVQ, and GLVQ are considered as fuzzy clustering techniques, because more than one cluster is being updated for each training data point. The rest of the clustering techniques in Table II.10.1 perform hard/crisp clustering as only one cluster is updated for each training data point. The membership functions of the clustering techniques are also examined. LVQ, FLVQ, and GLVQ represent the clusters using prototype vectors; hence, there are no membership functions. MLVQ, FKP, and PFKP, on the other hand, compute the membership functions in addition to performing clustering of the training data points. MLVQ derives Gaussian-shaped membership functions while FKP and PFKP derive trapezoidal-shaped membership functions. All the clustering techniques under review, with the exception of fuzzy ART, require that the number of classes be specified prior to clustering. This is one of the main limitations encountered by clustering techniques today, because not all data set used, especially those collected from real-time processes, have the number of classes C defined or it may be difficult to estimate the appropriate number of classes. Finally, only FCM, FLVQ, and GLVQ are known to optimize a well-defined objective function. The rest of the clustering techniques either do not optimize any objective functions or there is no clear definition of such functions.

FCM, MLVQ, FKP, and PFKP are chosen to be the clustering techniques of the four novel Falcon-like networks because of the following reasons. MLVQ, FKP, and PFKP are chosen for their ease of computation and the simplicity of their algorithms. Also, the Gaussian- and trapezoidal-shaped membership functions derived are commonly used in other fuzzy neural applications. Hence, it is easy to verify the accuracy of the fuzzy sets against other clustering techniques. FCM is chosen because the fuzzy sets it derived are of irregular shapes. This ensures that the training data set is not "fitted" with membership functions of pre-determined shapes.

10.3 Falcon-ART and Adaptive Resonance Theory (ART)

Stephen Grossberg introduced *adaptive resonance*[10] as a theory of human cognitive information processing. He concluded that the human learning process consists of *match-based* and *error-based* (mismatch) learning processes. Figure II.10.12 shows a simplified ART model. The memories are updated only when inputs from the external world match internal expectations or stored memories (match-based learning). For example, when we see a dog and hear someone saying that it is a dog, we recall our interpretation of a dog from memories. This reinforces and strengthens our understanding of a dog and the general attributes associated with it. Now, assume that there is a person who has never seen a Dalmatian before. When he encounters a Dalmatian, he does not know what breed of dog it is because there is no image of a Dalmatian in his memory. But he is able to tell that it is a dog from his past experiences. When he hears someone calling it a Dalmatian, he immediately learns what a Dalmatian looks like. At the same time, he also classifies Dalmatian under the category dog that he knows and forms relationships between the old memories and the new memory that he has just acquired. This is known as error-based learning. That is, new memories are acquired only if it cannot be found in existing memories.

The *adaptive resonance theory* (ART) clearly explains how human beings learn and classify things. As the aim of neural network is to emulate human reasoning and learning capabilities, ART becomes a natural tool to improve the performance of neural networks. The motivation behind the development of ART in neural networks is that in learning techniques such as Kohonen's competitive learning

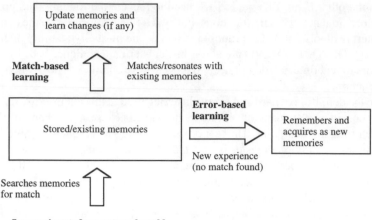

FIGURE II.10.12 Simplified ART model.

rule,[3] old memories tend to be washed away or eroded by new information. ART, on the other hand, offers a way of acquiring new information but at the same time preserving the old memories.

Cognitive theory has led to an evolving series of real-time ART neural network models for unsupervised and supervised category learning and pattern recognition. These models, including ART1, ART2, and ART3,[73–76] are capable of learning stable recognition categories in response to arbitrary input sequences.

The pure ART model is a set of design concepts. These concepts give rise to many ART models when we translate the concepts into mathematics. The models range from pattern classifiers and real-brain models to bi-directional associative memories and models of eye and robotic function. The fuzzy ART model is one such translation.

10.3.1 Fuzzy ART

Fuzzy ART is a model that incorporates fuzzy techniques into the basic ART model. It maps closely to the basic model shown in Figure II.10.12. Fuzzy sets are used to represent the stored/existing memories (stored categories). A memory entity consists of one or more dimensions or attributes. For example, a dog is an animal with four legs, wags its tail, has a sharp sense of smell, etc. If each dimension is represented by a fuzzy set, then each memory entity can be represented by a fuzzy hyper-cube of multiple dimensions. In the fuzzy ART model adopted by Lin in his *Falcon-ART* network,[9] trapezoidal fuzzy sets are used.

In Fuzzy ART, in order to emulate the memory-matching criterion of human cognitive learning (so as to determine when to create a new category to contain the new experience or to update an existing category for a recurring experience), a parameter called the vigilance parameter (ρ) is used. The vigilance parameter is used to determine the degree of matching before resonance occurs and is preset by the user. A high ρ means the input must be very compatible to a stored category before it can be admitted as an exemplar of that category. If two or more stored categories meet the matching criterion, then the one with the closest match will be updated to include the input. If no stored categories meet the matching criterion, then a new category is created to hold the input.

The fuzzy ART algorithm is implemented in the *Falcon-ART* architecture[9] to cluster the input and output training data and build its memory of patterns. This memory is subsequently used to classify new input patterns.

10.3.2 The Falcon-ART Architecture

The *Falcon-ART* network developed by Lin[9] is a highly autonomous system. It has five layers, as shown in Figure II.10.13, and generates fuzzy rules of the form shown by Equation II.10.14:

> **If** *input* 1 *is* $L_{1\phi}$ ... *and input* p *is* $L_{p\phi}$ ... *and input* n *is* $L_{n\phi}$ **Then**
>
> *output* 1 *is* $L'_{1\theta}$ ... *and output* t *is* $L'_{t\theta}$ ... *and output* m *is* $L'_{m\theta}$ (II.10.14)

The fuzzy rule in Equation II.10.14 has five elements, namely the input linguistic variables and terms, the output linguistic variables and terms, and the IF-THEN rule construct. The labels *input p*, where $p \in [1 \cdots n]$, denote the input linguistic variables, and the labels *output t*, where $t \in [1 \cdots m]$, denote the output linguistic variables. The variables, n and m denote the number of inputs and outputs, respectively. They represent entities such as height, speed, and weight. The labels $L_{1\phi}, \ldots, L_{n\phi}$ denote the input linguistic terms and the labels $L'_{1\theta}, \ldots, L'_{m\theta}$ denote the output linguistic terms.

In *Falcon-ART*, the input and output linguistic terms are represented as trapezoidal fuzzy sets. The linguistic terms represent fuzzy concepts such as tall and short. The input linguistic variables and terms constitute the antecedent (condition) section of a fuzzy rule while the output linguistic variables and terms make up the consequent section of the rule. The IF-THEN construct is used to join the condition section to the consequent section. Each of the five layers in *Falcon-ART* is mapped to the respective elements of the fuzzy rule as shown in Figure II.10.13.

Prior to training, *Falcon-ART* has only the input and output layers to represent the input and output linguistic variables, respectively. There are n inputs and m outputs. The hidden layers for the input

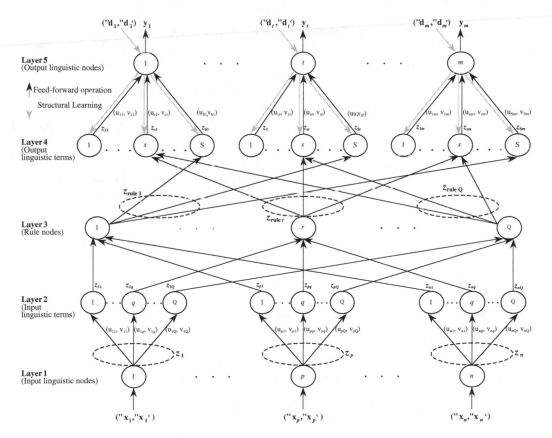

FIGURE II.10.13   Structure of *Falcon-ART*.

and output term nodes (layers 2 and 4), and the fuzzy rules (layer 3) are created and begin to grow as the learning cycle progresses. *Falcon-ART* dynamically partitions the input-output data spaces into trapezoidal fuzzy sets, tunes the trapezoidal membership functions representing the linguistic terms, and determines the proper network connections (fuzzy rules) in a single pass of the training data set. The fuzzy ART algorithm is employed to perform the fuzzy clustering of the input-output spaces into fuzzy hyper-boxes (hyper-cubes). *Falcon-ART* then dynamically determines the proper fuzzy rules by connecting the appropriate input and output clusters (input and output hyper-boxes) through a mapping process. The back-propagation learning scheme is subsequently used to tune the input-output membership functions. Thus, *Falcon-ART* effectively combines the fuzzy ART algorithm for structural learning (formulation of the fuzzy rules) and the back-propagation algorithm for parameter learning (tuning of the membership functions).

With reference to Figure II.10.13, layer 1 nodes represent the input linguistic variables and each node in layer 2 acts as a one-dimensional trapezoidal fuzzy set for an input node. The label Q denotes the number of term nodes for each input variable. In *Falcon-ART*, each input node has the same number of term nodes. This is the inherent characteristic of fuzzy ART. The nodes in layer 3 are the rule nodes and they form the fuzzy rule base of the *Falcon-ART* network. Layer 5 represents the output linguistic variables, and layer 4 the respective output term nodes. The label S denotes the number of term nodes for each output node. After training, the number of rule nodes in layer 3 is determined by the number of term nodes for each input node. That is, there will be Q fuzzy rules. This is because in *Falcon-ART*, each input term node is connected to only one rule node and only one term from each input node contributes to the antecedent of a fuzzy rule.

The output of a node is denoted by z. The subscript denotes the origin of the output. That is, the label z_p denotes the output from the pth input node in layer 1 and z_{pq} denotes the output of the qth term of the pth input node. The weights connecting each layer are unity unless otherwise shown. The trapezoidal membership functions of the term nodes in layer 2 and 4 are represented by a tuple consisting of the left (u) and right (v) flat points of the kernel. The tuple is implemented as the weights of the links in layers 2 and 4. The membership function of the qth term node of the pth input node is denoted by (u_{pq}, v_{pq}). Similarly, for the sth term node of the tth output node, its membership function is denoted by (u_{st}, v_{st}). During training, the inputs to the *Falcon-ART* network are the *complementarily coded*[33] input vector $\overline{\mathbf{X}} = [(\text{"}x_1\text{,"}x_1^c), \cdots, (\text{"}x_p\text{,"}x_p^c), \cdots, (\text{"}x_n\text{,"}x_n^c)]^T$, where "$x_p^c = 1 - \text{"}x_p$, and the complementarily coded desired output vector $\overline{\mathbf{D}} = [(\text{"}d_1\text{,"}d_1^c), \cdots, (\text{"}d_t\text{,"}d_t^c), \cdots, (\text{"}d_m\text{,"}d_m^c)]^T$, where "$d_t^c = 1 - \text{"}d_t$. The output of *Falcon-ART* is denoted as $\mathbf{Y} = [y_1 \cdots, y_t \cdots, y_m]^T$.

Based on the five layers of the *Falcon-ART* network shown in Figure II.10.13, Lin has developed an on-line learning algorithm **FALCON_ART**[9] to train the network. Refer to Lin and Lee[33] for the detailed functionality of the various layers in the Falcon-ART network.

10.3.3 Performance of Falcon-ART

To evaluate the performance of the *Falcon-ART* network, a simple classification experiment using the *Fisher's Iris Data* set[77] is conducted. The data set is partitioned into a training set and a test set. The training set consists of one third of the data points, (approx. 35%, i.e., 17 data points for each class) while the test set contains the remaining 65% (33 data points for each class). The results are cross-validated by using three different groups of training and test sets. They are denoted as CV1, CV2, and CV3. For the experiment, the following *Falcon-ART* parameters are used.

- Learning constant in the *back-propagation* algorithm (η) = 0.005
- In-vigilance parameter in the fuzzy ART algorithm (ρ_{in}) = 0.80
- Out-vigilance parameter in the fuzzy ART algorithm (ρ_{out}) = 0.80
- Targeted cost error E_{max} = 0.00005
- Sensitivity parameter of the trapezoidal membership function (γ) = 30.00

TABLE II.10.2 Hardware Configuration for the Simulations

System Configuration	
CPU: Intel Pentium III 450Mhz	File system: Fat 32
OS: Win95 (4.00.950 B)	Virtual memory: 32 bit
Memory: 128 Mbytes	Disk compression: Not installed
Hard disk: Seagate 4.0 GBytes	

- Train set $= 35\%$
- Test set $= 65\%$
- Number of input (input linguistic variables) $= 4$
- Number of output (output linguistic variables) $= 3$
- Maximum number of training iterations $= 1000$

The parameter 'Maximum number of training iterations' is used to ensure that the training cycle stops if the targeted cost error is unrealized. When it is realizable, then the training may terminate before the specified 1000 iterations. The hardware configuration on which the simulations are performed is listed in Table II.10.2.

Three outputs are used to represent the three classes of irises for the experiment. They are: Class 1-Setosa, Class 2-Virginica, and Class 3-Versicolor. A cost function E defined in Equation II.10.15 is used to measure the convergence of the back-propagation learning algorithm of *Falcon-ART* during training:

$$E^{(b)} = 1/2 \sum_{t=1}^{m} (d_t^{(b)} - y_t^{(b)})^2, \quad \text{for } b \in [1 \cdots \Omega] \tag{II.10.15}$$

where

m = number of outputs from the *Falcon-ART* network;
$d_t^{(b)}$ = desired output for the t th output node for the bth training pair;
$y_t^{(b)}$ = actual output for the t th output node for the bth training pair; and
Ω = number of training pairs in the training set.

Cost function E is computed for each training pair in the training set and the total error (denoted as TE) for one training epoch is obtained using Equation II.10.16:

$$\text{TE} = \sum_{b=1}^{\Omega} E^{(b)} \tag{II.10.16}$$

where

TE = total error for one training epoch;
Ω = number of training pairs in the training set; and
$E^{(b)}$ = cost function for the bth training pair in the training set.

Training stops when TE is less than the user preset $E_{\max}$ or the number of training cycle exceeds the maximum number of iterations specified. Figure II.10.14 shows the convergence of the back-propagation algorithm during training (CV1).

TE remains unchanged throughout the 1000 training iterations except for the initial rise. This is far from the preset targeted cost error of 0.00005. This problem can be explained when the fuzzy sets

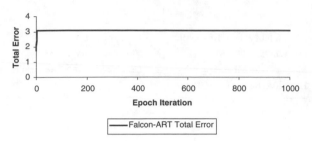

FIGURE II.10.14   A plot of total error (TE) against the number of training iterations.

that are derived using *Falcon-ART* are examined (Figure II.10.15). The result shows that Virginica and Versicolor are very much alike from the large regions of overlap in their respective fuzzy sets. In each of the numeric attributes, there is only a slight difference between the two classes. The total error (TE) does not converge (Figure II.10.14) because the fuzzy ART algorithm is unable to distinguish minute differences between fuzzy sets of highly similar classes as it examines the membership values of the fuzzy sets together as a class. For example, a class 2 training data has membership values of $\{0.9, 0.8, 0.9, 0.9\}$ for class 2 fuzzy sets. Hence, it gives a resemblance of 0.875 (i.e., $0.9/4 + 0.8/4 + 0.9/4 + 0.9/4 = 0.875$) to class 2. However, the same training data has membership values of $\{1.0, 0.6, 1.0, 0.6\}$ for class 3 fuzzy sets. This gives a resemblance of 0.8 to class 3 and thus gives rise to a large error at the outputs.

Learning by back-propagation causes the corners of the membership functions to oscillate. This occurs when class 2 and 3 training data are fed into the network. The network initially tunes itself toward class 2 outputs. When a class 3 training data is presented, the correction is reversed. Therefore, the total error TE remains constant. The initial rise of TE occurs between the first and the second epoch when the fuzzy sets have not been completely formed. Hence, the generated error was much lesser than when all three classes of fuzzy sets had been formed.

This assessment reveals significant drawbacks of the *Falcon-ART* architecture. First, the network is able to provide a satisfactory classification performance only when the classes of the input data are very different from each other (e.g., between Setosa and Virginica). If they only differ by a slight difference (like in the case of Virginica and Versicolor), then this difference is likely to be missed by the network, because the input iris data on the whole closely resembles both classes and hence, gives a similar firing strength for both classes. Because of this drawback, the classification results using the test sets are poor. Each test set consists of 65% of the original iris data set. There are 99 data points in a test set. The classification results for CV1, CV2, and CV3 are summarized in Table II.10.3.

The learning and classification performance of the *Falcon-ART* is poor because class 2-Virginica and class 3-Versicolor are still inseparable even with the complementary coding of the input data. Figure II.10.16 shows the plot for all 150 instances of irises after complementary coding.

It is clear that class 1-Setosa is linearly separable from the other two classes, and classes 2 and 3 are inseparable. When one examines the sepal length attribute, both classes 2 and 3 have large overlaps. This also occurs in the sepal width, petal width, and petal length attributes. Figure II.10.15 shows why back-propagation in *Falcon-ART* is unable to converge for the iris data set used and the poor classification performances of the network for the test groups. This is not limited only to *Fisher's Iris Data* set but also for any data set with clusters (classes) that are inseparable.

Another problem of *Falcon-ART* is its weak resistance to noisy/spurious training data points in the train sets. The fuzzy ART algorithm incorporates a new training data point into an existing category as long as that category is the closest to the training data point and passes the resonance test. However, this training data may be far from most of the data in the category and is incorporated

TABLE II.10.3 Iris Classification Using
Falcon-ART

	Network : Falcon-ART		
	CV1	CV2	CV3
C	69	83	73
M	30	16	26
U	0	0	0
C rate	69.70%	83.84%	73.74%
Mean C rate = 75.76%		Std Dev = 7.54%	

Note: (C = Classified; M = Misclassified; U = Unclassified; C rate = Classification rate).

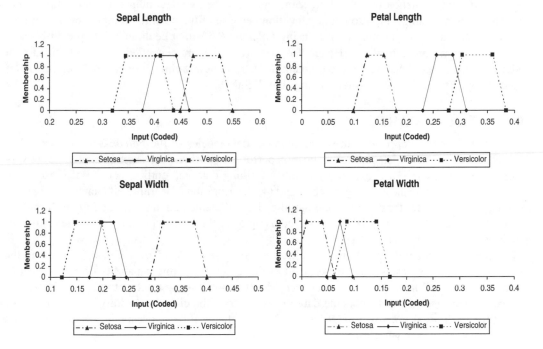

FIGURE II.10.15 Fuzzy sets derived using the *Falcon-ART* network.

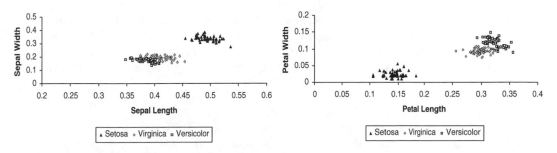

FIGURE II.10.16 Complementary coded iris data set.

because the resonance parameter is set small enough to accommodate such an inclusion. Inclusion is done through the *fast learning* scheme in the structural learning step of the *Falcon-ART* architecture.

The learning cycle in *Falcon-ART* very much depends on the setting of the targeted cost error. If this is set to a large value, then learning will terminate quickly. On the other hand, if it is set to too small a value, then learning may not cease. This is likely to occur for data sets like *Fisher's Iris Data* since there are classes that are inseparable from one another. This makes comparison of results difficult since the optimal value of E_{max} is often obtained through trial-and-error and is different for different data sets. This makes *Falcon-ART* application dependent. An alternative is to terminate the learning process when the difference in the total error (TE) for consecutive training epochs is sufficiently small.

In addition, the use of complementary coding in *Falcon-ART* can cause the learning efficiency of the network to deteriorate. Consider when two different training data points with magnitudes that are proportional to each other, but belonging to different classes, enter the network. Complementary coding under this situation will yield the same input for the two data points. In addition, *Falcon-ART* cannot handle single input/output application because of the need for complementary coding. Moreover, the fuzzy sets that are derived using *Falcon-ART* cannot be directly interpreted because all the values used are normalized. Hence, the fuzzy rules that are generated are nonintuitive to our understanding of the dynamics of the problem domain. A new architecture called *Falcon-MART*[78] was developed to overcome the shortcomings highlighted.

10.3.4   Falcon-MART

Falcon-MART is developed to address the inherent deficiencies of the *Falcon-ART* network. The deficiencies are: (1) unsatisfactory classification performance when the classes of input data are very similar to each other; (2) susceptibility to noisy training data and outlier; (3) use of preset error parameter to terminate the learning process; and (4) deterioration of learning efficiency. Please refer to Quek and Tung[78] for the details of the modifications implemented in the *Falcon-MART* network. The performance of *Falcon-MART* is evaluated using three different experiments and the results are benchmarked against *Falcon-ART*.

The first experiment demonstrates the efficiency of *Falcon-MART* over *Falcon-ART* using the *Fisher's iris data* set. The second experiment evaluates the modeling capability of *Falcon-MART* against the classical *multi-layered perceptron* (MLP) network using a set of traffic flow data. The last experiment uses a set of phoneme data to demonstrate the clustering ability of *Falcon-MART* against the traditional *K-nearest-neighbor* (KNN) classifier. All the experiments are carried out on the same hardware configuration as listed in Table II.10.2.

10.3.4.1   Experiment 1: Classification of Iris Data

To benchmark against the performance of *Falcon-ART*, the *Fisher's iris data* set with the same training and test sets are used. The network parameters common to both architectures are kept constant with the exception of the sensitivity parameter, γ. The parameter γ is the gradient of the slope of the trapezoidal membership function. This is set to 2.0 in *Falcon-MART* to provide a buffer of 0.5 on either side of the kernel of the trapezoidal function. The following *Falcon-MART* parameters are used in the assessment.

- Learning constant in the back-propagation algorithm $(\eta) = 0.005$.
- In-vigilance parameter in the fuzzy ART algorithm $(\rho_{in}) = 0.80$.
- Out-vigilance parameter in the fuzzy ART algorithm (ρ_{out}) is 0.80.
- Termination criterion $\varepsilon = 0.00005$.
- Sensitivity parameter of the trapezoidal membership function $(\gamma) = 2.00$.
- Training set $= 35\%$.

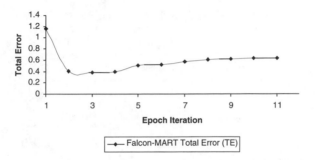

FIGURE II.10.17 Total error (TE) against the number of training iterations.

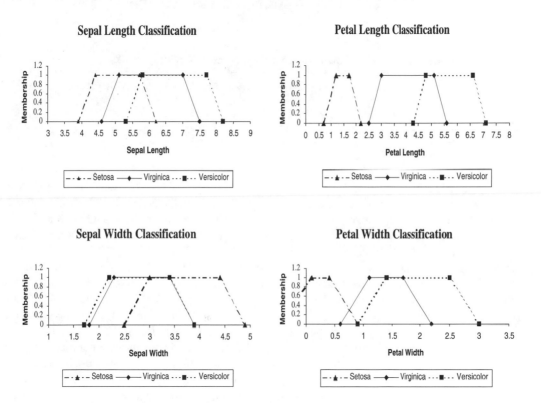

FIGURE II.10.18 Fuzzy sets of the three classes of irises derived using *Falcon-MART*.

- Test set = 65%.
- Number of input (input linguistic variables) = 4.
- Number of output (output linguistic variables) = 3.
- Maximum number of training iterations = 1000.

Figure II.10.17 shows the convergence of the back-propagation algorithm in *Falcon-MART* for the training set of CV1.

Falcon-MART completes the learning cycle in 11 iterations. Moreover, comparing against the *Falcon-ART* results in Figure II.10.14, the total error (TE) has converged and is much lower than that of the original *Falcon-ART* architecture. The fuzzy sets derived using the *Falcon-MART* network are shown in Figure II.10.18.

TABLE II.10.4 Iris Classification Using *Falcon-MART*

| | Network : Falcon-MART | | |
	CV1	CV2	CV3
C	98	95	89
M	1	4	10
U	0	0	0
C rate	98.99%	95.96%	89.90%
Mean C rate = 94.95%		Std Dev = 4.64%	

Note: C = Classified; M = Misclassified; U = Unclassified; C rate = Classification rate.

TABLE II.10.5 CPU Timings and Memory Usage of Various Networks

Network (parameter)	CPU Time (sec)	Recall Time (sec)	Memory Usage (h)
Falcon-ART ($\eta = 0.005$)	98.75 ± 8.57	0.07 ± 0.051	24
Falcon-MART ($\eta = 0.005$)	1.46 ± 0.39	0.07 ± 0.045	24

From Figure II.10.18, it can be seen that Class 2-Virginica and Class 3-Versicolor are still very much alike. In the fuzzy sets for the sepal width and petal width attributes, the two classes are almost identical. Hence, the distinguishing properties of these two classes lie in the sepal length and petal length attributes. The classification results using the training and test sets of CV1, CV2, and CV3 are summarized in Table II.10.4.

Comparing against the classification results of *Falcon-ART* in Table II.10.3, *Falcon-MART* offers a much-improved performance for the classification of the iris data in the test sets. In addition to the smaller number of training epochs required, it also has a higher mean classification rate of 94.95% for correct classification as compared to the *Falcon-ART* (75.76%). It also has a higher noise tolerance than *Falcon-ART*, as shown by the smaller value in the standard deviation of the classification rate. In addition to the classification rate, three other benchmark criteria are used to compare the two architectures. They are the training time of the network, recall time, and memory usage. The CPU timing for training and recall, and the memory usage for the two network architectures are shown in Table II.10.5. The variable h denotes the size of a *long double* variable type. Here, only the memory space used by the networks to store the fuzzy sets of the input space partition is used in the comparison.

The timings shown in Table II.10.5 are obtained by simulating the experiments ten times each for CV1, CV2, and CV3. The mean and standard deviation of the timings are shown in the table. *Falcon-ART* requires a longer CPU training time because of its failure to converge to the preset termination criterion ($E_{max} = 0.00005$); and hence, training continued to the preset limit of 1000 cycles before terminating. On the other hand, *Falcon-MART* requires a mean CPU training time of just 1.46 sec. This is less than 2% of the average training time for *Falcon-ART*. The memory usage of both networks is $24h$. They are used essentially for the storage of the two corners of the kernel of the membership functions in the input term nodes of hidden layer 2. The iris data set has four numeric inputs, and both networks derived three fuzzy rules (therefore, three terms per input node). Hence, the total memory requirement is four inputs × three input terms × two corners × size of long double variable type = $24h$. The recall time for both architectures are similar.

10.3.4.2 Experiment 2: Modeling of Traffic Flow Data

This experiment is conducted to evaluate the effectiveness of the *Falcon-MART* network in universal approximation and data modeling using a set of traffic flow data. The raw traffic flow data for the experiment was obtained from Tan,[79] courtesy of the School of Civil and Structural Engineering (CSE), NTU, Singapore. The data were collected at a site (Site 29) located at exit 15 along the eastbound Pan Island Expressway (PIE) in Singapore, using loop detectors embedded beneath the road surface. These inductive loop detectors were pre-installed by the Land Transport Authority (LTA) of Singapore in 1996 along major roads to facilitate traffic flow data collection.

There are a total of five lanes at the site; two exit lanes and three straight lanes for the main traffic. For this experiment, only the traffic flow data for the three straight lanes were considered. The purpose of this experiment is to model the traffic flow trend at the site using four inputs, i.e., time and traffic density of the three lanes using *Falcon-MART*. The trained network is subsequently used to obtain prediction for the traffic density of a particular lane at a time $t + \tau$, where $\tau = 5, 15, 30, 45$, and 60 min. Figure II.10.19 shows a plot of the traffic flow density data for the three straight lanes spanning a period of six days from September 5–10, 1996.

The following *Falcon-MART* parameters are used for the simulation.

- Learning constant in the back-propagation algorithm (η) = 0.005.
- In-vigilance parameter in the fuzzy ART algorithm (ρ_{in}) = 0.70.
- Out-vigilance parameter in the fuzzy ART algorithm (ρ_{out}) = 0.95.
- Termination criterion ε = 0.005.
- Sensitivity parameter of the trapezoidal membership function (γ) = 5.00.
- Training set = 40%.
- Test set = 60%.
- Number of input (input linguistic variables) = 4.
- Number of output (output linguistic variables) = 1.
- Maximum number of training iterations = 1000.

For the experiment, three groups of training and test sets are used. They are CV1, CV2, and CV3. The training windows are labeled as such in Figure II.10.19. To compute the accuracy of the predictions by *Falcon-MART*, the square of the Pearson product-moment correlation value (R^2) is used. The prediction made by *Falcon-MART* for $\tau = 60$ min for lane 1 traffic density (based on CV3) is shown in Figure II.10.20.

The simulation is repeated using the *Falcon-ART* network. The benchmarking criterion for both the *Falcon-ART* and *Falcon-MART* networks is the "Avg R^2" index. "Avg R^2" is the mean R^2 value

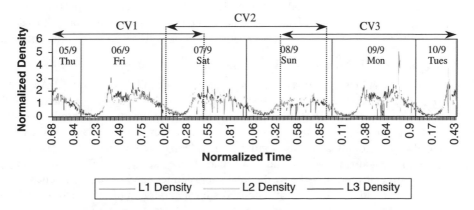

FIGURE II.10.19   Traffic density of three straight lanes along PIE.

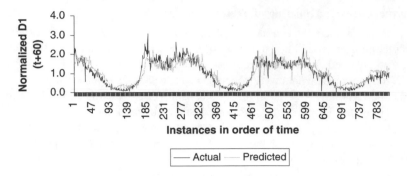

FIGURE II.10.20   Prediction of lane 1 density for $\tau = 60$ mins using *Falcon-MART*.

TABLE II.10.6   Prediction of Traffic Density for Lane 1 at $\tau = 5$ min by MLP

Lane 1	τ	$cv1\_R^2$	$cv2\_R^2$	$cv3\_R^2$	Avg R^2
	5	0.834817	0.847881	0.881109	0.854602

across the three cross-validation sets for each value of τ. The variation of "Avg R^2" against τ for the three straight lanes using *Falcon-ART* and *Falcon-MART* are shown in Figure II.10.21.

Figure II.10.21 shows that *Falcon-MART* consistently has better prediction accuracy than *Falcon-ART*. This applies across all three different lanes for different values of τ. The same set of experiments is repeated using the *Multi-Layered Perceptron* (MLP) network with ten input nodes, five hidden nodes, and a single output node. The bipolar sigmoid function with an output range of $[-1, 1]$ is used as the activation function for the hidden and output nodes. The input nodes simply relay the input signals to the hidden nodes. The MLP network is trained with the back-propagation algorithm and the training data has been normalized to unity to fit into the range of the bipolar sigmoid activation function. During the prediction phase, the MLP network functions in a feedforward mode. The results of the MLP network for the prediction of lane 1 traffic density for $\tau = 5$ min across the three cross-validation groups are given in Table II.10.6.

Comparing against the results of *Falcon-MART* in Figure II.10.21, the MLP can achieve better prediction accuracy. However, the trained MLP is a black box and the linguistic rules defining the traffic flow pattern cannot be extracted from the MLP network. Hence, *Falcon-MART* may not be as efficient an architecture as MLP in data modeling and universal approximation, but its advantage lies in its ability to formulate a set of understandable fuzzy rules to describe the problem domain.

10.3.4.3   Experiment 3: Clustering of Phoneme Data

This experiment is conducted to determine the clustering capability of *Falcon-MART* against a traditional classifier such as the K-NN classifier. The data set used is a set of phoneme data that is available on-line at the ELENA[80] Web site. The aim of the phoneme database is to distinguish between nasal and oral vowels. There are thus two different classes, namely Class 0-Nasals and Class 1-Orals. This database contains vowels coming from 1809 isolated syllables (for example: pa, ta, pan, etc). Five different attributes were chosen to characterize each vowel. They are the amplitudes of the five first harmonics, normalized by the total energy (integrated on all the frequencies). Each harmonic is signed: positive when it corresponds to a local maximum of the spectrum and negative otherwise.

Traffic Prediction for Lane 1 against Time Interval

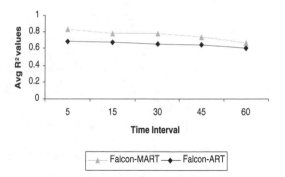

Traffic Prediction for Lane 2 against Time Interval

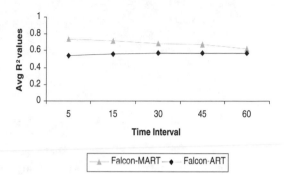

Traffic Prediction for Lane 3 against Time Interval

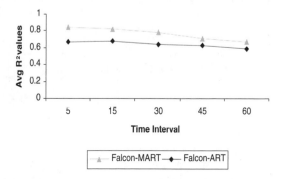

FIGURE II.10.21 Avg R^2 variations against τ for the three lanes.

Three observation moments have been kept for each vowel to obtain 5427 different instances:

Observation corresponding to the maximum total energy.
Observations taken 8 millisec before and after the observation corresponding to this maximum total energy.

From these 5427 initial values, 23 instances for which the amplitude of the first five harmonics was zero were removed, leading to the 5404 instances of the present database. The patterns are presented in a random order. The entire phoneme data set consists of 70.65% Class 0 data points and 29.35% Class 1 data points. Due to the size of the data set, it is partitioned into two sets: 20% for the training

TABLE II.10.7   Phoneme Classification Results Using *Falcon-MART* and *Falcon-ART*

	Falcon-MART Confusion Matrix			Falcon-ART Confusion Matrix	
	Class 0 (%)	Class 1 (%)		Class 0 (%)	Class 1 (%)
Class 0	93.17	6.83	Class 0	100.00	0.00
Class 1	56.21	43.79	Class 1	100.00	0.00
	Mean error rate = 21.32%			Mean error rate = 29.35%	

TABLE II.10.8   Phoneme Classification Result Using K-NN ($k = 20$) and Leave-One-Out Method

K-NN Classifier ($k = 20$ with Leave-One-Out) Confusion Matrix		
	Class 0 (%)	Class 1 (%)
Class 0	91.40	8.60
Class 1	27.80	72.20
	Error rate = 14.20%	

set and the remaining 80% for the test set. A total of five groups of training and test sets are used for the experiment. They are labeled as CV1, CV2, CV3, CV4, and CV5. The data for the training sets do not repeat among the five groups. For this experiment, the *Falcon-MART* network has five inputs and one output. The following *Falcon-MART* parameters are used.

- Learning constant in the back-propagation algorithm (η) = 0.005.
- In-vigilance parameter in the fuzzy ART algorithm (ρ_{in}) = 0.70.
- Out-vigilance parameter in the fuzzy ART algorithm (ρ_{out}) is 0.70.
- Termination criterion ε = 0.0005.
- Sensitivity parameter of the trapezoidal membership function (γ) = 8.00.
- Training set = 20%.
- Test set = 80%.
- Number of input (input linguistic variables) = 5.
- Number of output (output linguistic variables) = 1.
- Maximum number of training iterations = 1000.

The classification results using *Falcon-MART* is benchmarked against that of *Falcon-ART* in Table II.10.7. The results shown are the average classification results across the five cross-validation groups.

In comparison, the confusion matrix obtained with the K-NN classifier (result obtained from ELENA and tested with the Leave-One-Out method with $k = 20$) is shown in Table II.10.8.

From Table II.10.7, it is obvious that *Falcon-MART* offers better classification performance than *Falcon-ART*. The classification rate for Class 1 data points significantly improved from 0.00–43.97% when *Falcon-MART* is used instead of *Falcon-ART*. In addition, there is also a significant drop in the mean error rate from 29.35–21.32%. Comparing the results in Table II.10.7 and II.10.8, K-NN seems to have the better clustering capability. The K-NN classifier has a classification rate of 91.40% for

Class 0 data points and a classification rate of 72.20% for Class 1 data points, which is significantly better than the result obtained with *Falcon-MART*. However, this is achieved with nearly the whole data set as the training set except for one data point that is used to perform the classification test. The classification test for K-NN is performed a total of 5404 times to derive the results shown in Table II.10.8. On the other hand, *Falcon-MART* is able to achieve a reasonable result by using just 20% of the data set for training. Moreover, IF-THEN fuzzy rules can be extracted from the *Falcon-MART* network to describe the clustering behavior of the phoneme data set. This cannot be achieved using the K-NN classifier.

10.4   PACL-FNNS (Pseudo Adaptive Complementary Learning-Based FNNs)

10.4.1   Positive and Negative Exemplars (Complementary Learning)

As described in the last section, the ART model (including fuzzy ART) uses positive and negative exemplars to create a clustering of the input and output data space. However, besides the creation of a new cluster, the negative exemplars can also be used in the inference process. Most of the fuzzy neural network architectures described in the literature, e.g., Jang's ANFIS (Adaptive Network-based Fuzzy Inference System),[28–30] and Quek's POPFNN (Pseudo Outer-Product-based FNN)[81,82] uses positive fuzzy rules to infer the network outputs. An example of a positive fuzzy rule is shown in Equation II.10.17:

$$\text{If } x \text{ is } \boldsymbol{A} \text{ Then } y \text{ is } \boldsymbol{B} \tag{II.10.17}$$

where

x, y are linguistic variables; and

$\boldsymbol{A}$ and $\boldsymbol{B}$ are fuzzy sets.

This fuzzy rule adopts only positive exemplars for its inference. That is, if input matches and condition fulfills, then the consequent follows. This is equivalent to teaching a child about the concept of "car." The best way to teach the child what is a car is to show him/her photos of different makes of cars. It is then hoped that the child will be able to recognize a car whenever he/she sees one. However, very often, the learning process and the ability to recognize a vehicle as a car or not a car are often enhanced if we also show the child photos of trucks, buses, etc. This is the use of negative exemplars. By teaching the child both positive and negative exemplars, the results of a recognition test will be better because the negative exemplars serve as a contrast to the positive exemplars. Hence, the same notion applies to fuzzy neural networks. If negative exemplars are used in addition to positive exemplars during the inference process, then more accurate outputs can be derived and this leads to a more stable network performance. The concept of positive and negative exemplars (complementary learning) is formalized as follows.

Definition 2:   Given a universe of discourse $\mathbf{U}$ consisting of N elements, $\mathbf{U} = \{x_1, \ldots, x_p, \ldots, x_N\}$, a crisp set $\mathbf{C}$ representing a concept C is defined. Hence, elements in $\mathbf{U}$ belonging to the concept C will have a membership of unity in the set $\mathbf{C}$. That is:

$$\mu_c(x_p) = \begin{cases} 1, \text{if } x_p \in \mathbf{C} \\ 0, \text{otherwise} \end{cases} \tag{II.10.18}$$

where μ_c = membership function of set $\mathbf{C}$.

Definition 3: Following Definition 2, a crisp set ¬**C** representing elements not belonging to the concept C is defined. Hence, elements in ¬**C** will have the memberships as defined in Equation II.10.19.

$$\mu_{\neg\mathbf{C}}(x_p) = \begin{cases} 1, \text{if } x_p \notin \mathbf{C} \\ 0, \text{otherwise} \end{cases} \tag{II.10.19}$$

where $\mu_{\neg\mathbf{C}}$ = membership function of set ¬**C**.

The above two definitions are for crisp concepts (crisp sets). If they are extended to fuzzy concepts (fuzzy sets) such as "high," "tall," etc., then each element x_p will have a membership value of [0, 1] for **C** and ¬**C**. Hence, for a given element x_p, the following cases may be expected:

Case 1: $\mu_{\mathbf{C}}(x_p) \to 0$ and $\mu_{\neg\mathbf{C}}(x_p) \to 0$ i.e., x_p is not similar to **C** and x_p is not similar to ¬**C**.
Case 2: $\mu_{\mathbf{C}}(x_p) \to 0$ and $\mu_{\neg\mathbf{C}}(x_p) \to 1$ i.e., x_p is not similar to **C** and x_p is similar to ¬**C**.
Case 3: $\mu_{\mathbf{C}}(x_p) \to 1$ and $\mu_{\neg\mathbf{C}}(x_p) \to 0$ i.e., x_p is similar to **C** and x_p is not similar to ¬**C**.
Case 4: $\mu_{\mathbf{C}}(x_p) \to 1$ and $\mu_{\neg\mathbf{C}}(x_p) \to 1$ i.e., x_p is similar to **C** and x_p is similar to ¬**C**.

In Case 1 since x_p belongs to neither **C** nor ¬**C**, we can only conclude that x_p does not belong to the concept "C." The element x_p may belong to another concept defined on the same universe of discourse **U**. In Case 2, the possibility of x_p not belonging to concept "C" is high, since there is strong evident that x_p belongs to the concept "not C." For Case 3, the element x_p is expected to belong to "C" and in Case 4, the result is inconclusive, as x_p has a high possibility of belonging to "C" and "not C" at the same time. Hence, the element x_p should not be a deciding factor if it is used in a classification system.

10.4.2 CL-Based Fuzzy Neural Networks

To investigate the effect of using complementary learning (i.e., the contrast between negative and positive exemplars) in fuzzy neural systems, four novel Falcon-like fuzzy neural architectures are proposed. They are named *Falcon-FCM, Falcon-MLVQ, Falcon-FKP*, and *Falcon-PFKP*. The naming of the networks is based on the clustering techniques used to derive the fuzzy sets in the rule base. The new networks use the basic *Falcon* structure as the backbone architecture.

10.4.2.1 Falcon Network

The basic *Falcon* network has a 5-layer structure as shown in Figure II.10.22.

The *Falcon* is a feedforward multi-layer network that integrates the basic elements and functions of a traditional fuzzy logic controller into a connectionist structure that has distributed learning abilities.[33] In the *Falcon* structure, the input and output nodes represent the input states and output control signals or decisions, respectively. In the hidden layers 2 and 4, there are nodes functioning as input and output membership functions respectively, and the nodes in hidden layer 3 act as fuzzy logic rules. In the *Falcon* architecture, each input term node is connected to one rule node only. In addition, only one term node from each input node contributes to the antecedent of a fuzzy rule that conforms to the ART notion of positive and negative exemplars. The *Falcon* can be contrasted with a traditional fuzzy logic control and decision system in terms of its network structure and learning abilities. The *Falcon* network can be constructed from training exemplars using neural learning techniques and the connectionist structure can be trained to determine proper input-output membership functions. The *Falcon* model provides human-understandable IF-THEN fuzzy reasoning to the normal feedforward multi-layer network in which the internal structure is often opaque to the users. The connectionist structure also avoids the rule-matching time of the inference engine in the traditional fuzzy control systems.

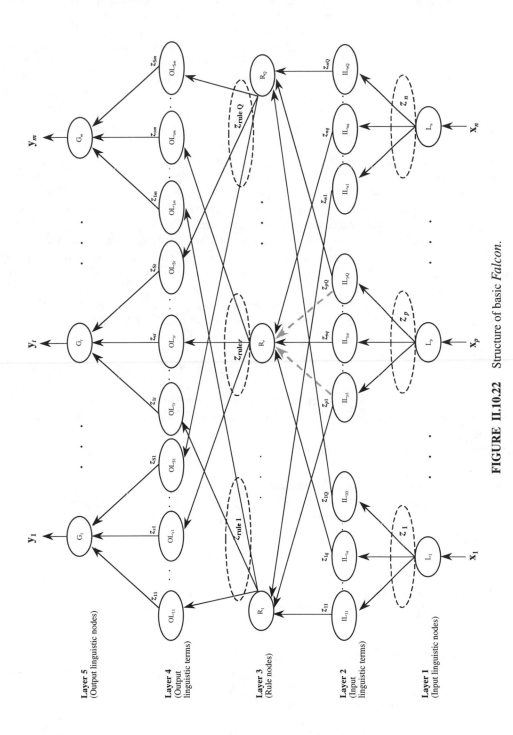

FIGURE II.10.22 Structure of basic *Falcon*.

Figure II.10.22 shows the structure of the *Falcon* network. The system has a total of five layers. The network has n input and m outputs. The input to the network is represented by the vector $\mathbf{X} = [x_1, \ldots, x_p, \ldots, x_n]^T$ and the output is represented by the vector $\mathbf{Y} = [y_1, \ldots, y_t, \ldots, y_m]^T$. The nodes in layer 1 are input nodes (linguistic nodes) that represent the input linguistic variables, and layer 5 is the output layer representing the output linguistic nodes. The label L_p denotes the pth input node and the label G_t denotes the tth output node. Nodes in layers 2 and 4 are *term nodes* that act as membership functions representing the fuzzy terms of the input and output linguistic variables respectively. The qth term node of the pth input node is denoted by $\mathbb{L}_{pq}$ and the label OL_{st} denotes the sth term of the tth output node. Each node in layer 3 is a rule node that represents one fuzzy rule. The label R_r denotes the rth rule node. Thus, all the layer 3 nodes form a fuzzy rule base. Links in layers 3 and 4 function as a *connectionist inference engine*, which avoids the rule-matching process. Layer 3 links define the preconditions of the rule nodes, and layer 4 links define the consequent of the rule nodes. The label z defines the output of an arbitrary node in Figure II.10.22. The subscripts specified the origin of that output. That is, z_p denotes the output of the pth input node while z_{pq} denotes the output of the qth term of the pth input node, etc.

The membership functions of the term nodes are derived using clustering techniques such as fuzzy ART that make use of the input and desired output training pair to locate the membership function and formulate the fuzzy rules. This serves to allocate network resources efficiently by placing domains of the membership function covering only those regions of the input-output space where data are present. Linking the input terms to the output terms then forms the fuzzy rule nodes. One main characteristic of *Falcon* is that a fuzzy rule is formed only when there is a training exemplar from the external environment that supports the existence of the rule. The term nodes and rule nodes are formed simultaneously, and this is a great contrast to some other fuzzy neural systems in which all the memberships are first determined before the training exemplars are recycled to derive the fuzzy rules. The *Falcon* network is not a rule selection network, as no redundant rules without supporting exemplars are created. To implement the contrast between the positive and negative exemplars, layer 2 nodes are fully connected to layer 3 nodes. That is, each of the input term nodes is fully connected to the rule nodes. The gray dashed arrows in Figure II.10.22 show these connections. For simplicity, only two such arrows are drawn.

10.4.2.1.1 *Positive and Negative Exemplars*

To implement the notion of positive and negative exemplars in the inference process, the basic functionality of the rule nodes (layer 3 nodes) has to be defined. The aim of using positive and negative exemplars is to contrast the latter with the former so those attributes that separate one class (category) from the other classes (categories) are enhanced. This can be achieved by computing the output of the rule nodes as follows.

10.4.2.1.1.1 *Computing the Output of a Rule Node*   With respect to rule node R_r in Figure II.10.22, the net synaptic input Net_{R_r} and the net synaptic output z_{R_r} are defined as:

$$\text{Net synaptic input for node } R_r, \ Net_{R_r} = f^{(3)}(z_{1q}, \ldots, z_{pq}, \ldots, z_{nq})$$

$$= \{z_{1q}, \ldots, z_{pq}, \ldots, z_{nq}\} \tag{II.10.20}$$

$$\text{Net output for node } R_r, \ z_{R_r} = a^{(3)}(Net_{R_r})$$

$$= \frac{\sum_{p=1}^{n} w_p \times [z_{pq} - \max_{\theta \in \{1 \cdots Q\}; \theta \neq q}\{z_{p\theta}\}]}{\sum_{p=1}^{n} w_p}$$

where

z_{pq} = net output of the input term $\mathbb{L}_{pq}$ which is connected to rule R_r;

w_p = weighting parameter for $[z_{pq} - \max_{\theta \in \{1 \cdots Q\}; \theta \neq q} \{z_{p\theta}\}]$;

$f^{(3)}$ = aggregation function for nodes of layer 3; and

$a^{(3)}$ = activation function for nodes of layer 3.

The weighting parameter w_p is used to compute the weighted averaging of the n terms of $[z_{pq} - \max_{\theta \in \{1 \cdots Q\}; \theta \neq q} \{z_{p\theta}\}], \forall p \in \{1 \cdots n\}$, where n is the number of input nodes. All the n terms are sorted in ascending order and the largest weight is assigned to the largest element using Equation II.10.21:

$$w_{pos} = base^{pos} \tag{II.10.21}$$

where

pos = the position of an element in the sorted list; and

base = a real number greater than 0 (In the proposed networks, base is 1.5).

For the pth input node, and with respect to the rule node R_r, the term $[z_{pq} - \max_{0 \in \{1 \cdots Q\}; \theta \neq q} \{z_{p\theta}\}]$ computes the contrast between the positive and negative memberships. Positive membership is the membership induced by the input at the pth node with respect to the rule node R_r, and negative memberships refer to the memberships induced at rule nodes other than R_r. The term $\max_{0 \in \{1 \cdots Q\}; \theta \neq q} \{z_{p\theta}\}$ returns the largest of the negative membership values. Hence, for rule node R_r, the output will reflect the extent of the contrast. A large output (as z_{R_r} tends to one) implies there are distinguishing attributes in the input vector, and the possibility of correct classification is enhanced. The inverse is true when z_{R_r} tends to zero.

10.4.2.2 Augmented Training Vectors

Augmented training vectors are used to train the four proposed networks. The inputs and outputs of the four Falcon-like networks are represented as vectors $\mathbf{X} = [x_1, \ldots, x_p, \ldots, x_n]^T$ and $\mathbf{Y} = [y_1, \ldots, y_t, \ldots, y_m]^T$, respectively, where n denotes the number of input and m denotes the number of output linguistic variables. Due to the clustering techniques used, the number of fuzzy rules in the proposed networks is determined by a user-specified parameter C, where C is the number of clusters. That is, the number of rules is equal to C. Prior to training of the proposed networks, C rule nodes are created and connected to their respective input and output labels that are created at the same time. These input and output label nodes are subsequently connected to the respective input and output nodes.

In the four proposed networks, the input vector $\mathbf{X}$ and its corresponding output vector $\mathbf{Y}$ are grouped and treated as augmented data vector for the clustering phase. The rationale behind this approach is that if there exists a rule that maps an input space $\mathbf{U}_1(\mathbf{X} \in U_1)$ to an output space $U_2(\mathbf{Y} \in U_2)$, then there will exist a corresponding representative data sample in the joint input-output space. As a result, the input and output spaces are frequently regarded as joint space, and the data pairs in this joint space are treated as augmented data. Hence, given a group of p data samples:

$$(\mathbf{X}^{(1)}, \mathbf{Y}^{(1)}), (\mathbf{X}^{(2)}, \mathbf{Y}^{(2)}), \ldots, (\mathbf{X}^{(r)}, \mathbf{Y}^{(r)}), \ldots, (\mathbf{X}^{(p)}, \mathbf{Y}^{(p)}) \quad \text{for } \mathbf{X} \in U_1, \mathbf{Y} \in U_2,$$

that are associated with the rule "*If x is A Then y is B*," one can rewrite these ordered pairs into an augmented data set described as follows:

$$\mathbf{U}^{(1)}, \mathbf{U}^{(2)}, \ldots, \mathbf{U}^{(r)}, \ldots, \mathbf{U}^{(p)}; \quad \mathbf{U}^{(r)} = (\mathbf{X}^{(r)}, \mathbf{Y}^{(r)}) \in U_1 \times U_2$$

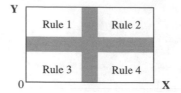

FIGURE II.10.23   Clusters in input-output space as rules.

This augmented data set is subsequently used to discern clusters in the joint space $U_1 \times U_2$. Each cluster in the joint space $U_1 \times U_2$ corresponds to a rule that maps the input space U_1 to the output space U_2 as shown in Figure II.10.23. The checkered area in Figure II.10.23 represents the fuzzy boundaries between clusters.

After the clustering phase, the input and output labels of the rule nodes will contain the respective fuzzy sets of the clusters. The detailed functionality of each of the proposed networks is as follows.

10.4.2.3   Falcon-FCM

The FCM clustering technique[5] presented in Section 10.2.3 is incorporated into the basic *Falcon* network. The resultant hybrid architecture is called *Falcon-FCM*. Like *Falcon-ART*, *Falcon-FCM* has five layers to implement each of the five elements of a fuzzy rule. The difference is that *Falcon-ART* uses the fuzzy ART algorithm to obtain its partition of the input-output space whereas *Falcon-FCM* uses the FCM algorithm. The membership functions of *Falcon-FCM* are no longer trapezoidal as in *Falcon-ART*, and the shapes of the membership functions depend on the exponential parameter M. The *Falcon-FCM* network architecture is as shown in Figure II.10.24.

Similar to the *Falcon-ART* and *Falcon-MART* networks, the input and output vectors to the *Falcon-FCM* network are nonfuzzy vectors. Fuzzification of the inputs and defuzzification of the outputs are automatically accomplished in the input and output layers, respectively. The label L_p denotes the pth input node in layer 1 that represents the pth input linguistic variable. Output of node L_p is denoted by z_p. The qth input term of the pth input linguistic variable is denoted as node $\mathbb{L}_{pq}$ in layer 2. The output of node $\mathbb{L}_{pq}$ is z_{pq}. In *Falcon-FCM*, each input variable in layer 1 has Q input-labels in layer 2. This number is determined by the user-defined parameter C that partitions the input space into C clusters. That is, $Q = C$. All the input nodes have the same number of input-label nodes; therefore, the number of input term nodes in layer 2 is $n \times Q$ where n denotes the number of input nodes.

To implement the notion of positive and negative exemplars, each input-label node in layer 2 is fully connected to the layer 3 nodes. When the contrast of positive and negative exemplars is not implemented, each of the input-label nodes in layer 2 is connected to only one rule node in the rule base layer (layer 3). There is Q rule node in layer 3. The label R_r denotes the rth rule node. The output z_{R_r} of rule node R_r is fed into a set of m output-label nodes in layer 4. Each output variable (output node) in layer 5 has exactly one output label connecting to a particular rule in layer 3. The output-label nodes in layer 4 represent the consequent parts of the fuzzy rules in *Falcon-FCM*. The label OL_{st} denotes the sth output-label of the tth output node and its output is z_{st}. Each output node in layer 5 has S output-label nodes and acts as a defuzzifier that produces an actual network output based on the inputs to the network in layer 1. In *Falcon-FCM*, each output linguistic variable has the same number of output-label as the number of input-label that each input linguistic variable has. That is, $S = Q$. The label G_t denotes the tth output node in layer 5. The detailed functionality of the *Falcon-FCM* network is as follows.

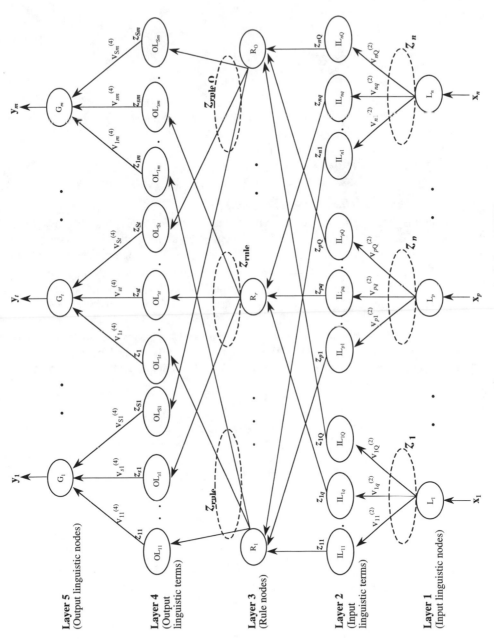

FIGURE II.10.24 Structure of *Falcon-FCM*.

10.4.2.3.1 *The Input Layer*

This layer represents the input variables in a fuzzy rule. The net synaptic input and the net synaptic output of node L_p are given in Equation II.10.22:

$$\text{Net synaptic input for node } L_p, \, Net_p = f^{(1)}(x_p)$$
$$= x_p \qquad \text{(II.10.22)}$$
$$\text{Net synaptic output for node } L_p, \, z_p = a^{(1)}(Net_p)$$
$$= x_p$$

where

$x_p = $ the pth element of the input vector $\mathbf{X}$;
$f^{(1)} = $ aggregation function for nodes of layer 1; and
$a^{(1)} = $ activation function for nodes of layer 1.

10.4.2.3.2 *The Condition (Input Linguistic Term) Layer*

This layer contains the input-label nodes. Input-label nodes in the condition layer constitute the condition section of the fuzzy rules in *Falcon-FCM*. The membership function of each input-label node is derived using the FCM clustering technique.[5] The centriod of the membership function for node $\mathbb{L}_{pq}$ is denoted as $v_{pq}^{(2)}$ in Figure II.10.24. Since the membership functions form a pseudo-partition of the data space and depend on the training data, they cannot be represented using mathematical expressions. Hence, each input-label node has to locally store the membership values of the entire set of training data with respect to the cluster it is representing. The set of membership values is sorted with respect to the magnitude of the training data. This makes the implementation of the *Falcon-FCM* network memory intensive. The net output z_{pq} of the input-label node $\mathbb{L}_{pq}$ is also not directly computed. The net synaptic input Net_{pq} and the net synaptic output z_{pq} of the input-label node $\mathbb{L}_{pq}$ are given in Equation II.10.23:

$$\text{Net synaptic input for node } \mathbb{L}_{pq}, \, Net_{pq} = f^{(2)}(z_p)$$
$$= z_p \qquad \text{(II.10.23)}$$
$$\text{Net synaptic output for node } \mathbb{L}_{pq}, \, z_{pq} = a^{(2)}(Net_{pq})$$

where

$z_p = $ output of node L_p in layer 1;
$f^{(2)} = $ aggregation function for nodes of layer 2; and
$a^{(2)} = $ activation function for nodes of layer 2.

Function $a^{(2)}$ is defined by Equation II.10.24:

$$a^{(2)}(z_p) = \begin{cases} 0, & \text{for } z_p < x_1 \\ \mu_{pq}(x_i) + \frac{z_p - x_i}{x_{i+1} - x_i} \times [\mu_{pq}(x_{i+1}) - \mu_{pq}(x_i)], & \text{for } x_i \le z_p \le x_{i+1} \text{ and } \mu_{pq}(x_i) \\ & \le \mu_{pq}(x_{i+1}) \\ \mu_{pq}(x_{i+1}) + \frac{x_{i+1} - z_p}{x_{i+1} - x_i} \times [\mu_{pq}(x_i) - \mu_{pq}(x_{i+1})], & \text{for } x_i \le z \le x_{i+1} \text{ and } \\ & \mu_{pq}(x_i) > \mu_{pq}(x_{i+1}) \\ 0, & \text{for } z_p > x_N \end{cases}$$

$$\text{(II.10.24)}$$

where

x_1 = first element after sorting the input training data of N elements;
x_i = ith element in the sorted list;
x_{i+1} = $(i + 1)$th element in the sorted list;
x_N = last element in the sorted list; and
μ_{pq} = fuzzy membership function as derived by FCM.

Interpolating the sorted membership values derives the membership value for an input that falls within the minimum and maximum training inputs. For any input that falls outside this range, a membership value of zero is returned by the function $a^{(2)}$.

10.4.2.3.3 *The Rule-Based Layer*

Each node in this layer represents a fuzzy rule in the *Falcon-FCM* network. The qth input-label of each input node is connected to the rule node R_r. The net synaptic input Net_{R_r} and the net synaptic output z_{R_r} of the rule node R_r are given in Equation II.10.20.

10.4.2.3.4 *The Consequence Layer*

The neurons in this layer are known as output-label nodes. The net synaptic input Net_{st} and the net synaptic output z_{st} of the output-label node OL_{st} are given in Equation II.10.25:

$$\text{Net synaptic input for node } OL_{st}, Net_{st} = f^{(4)}(z_{R_r})$$
$$= z_{R_r} \qquad (\text{II.10.25})$$
$$\text{Net synaptic output for node } OL_{st}, z_{st} = a^{(4)}(Net_{st})$$
$$= Net_{st}$$

where

z_{R_r} = net output of the rule node R_r;
$f^{(4)}$ = aggregation function for nodes of layer 4; and
$a^{(4)}$ = activation function for nodes of layer 4.

10.4.2.3.5 *The Output Layer*

The net synaptic input Net_t and the net synaptic output y_t of output node G_t are given in Equation II.10.26:

$$\text{Net synaptic input for node } G_t, Net_t = f^{(5)}(Z_1, z_{2t}, \dots, z_{st} \dots, Z_{st})$$
$$= \sum_{s=1}^{S} v_{st}^{(4)} \times z_{st} \qquad (\text{II.10.26})$$

$$\text{Net synaptic output for node } G_t, y_t = a^{(5)}(Net_t)$$
$$= \frac{Net_t}{\sum_{s=1}^{S} Z_{st}}$$

where

z_{st} = net synaptic output of the node OL_{st};

$v_{st}^{(4)}$ = centriod of the node OL_{st} (sth label of the tth output node);

$f^{(5)}$ = aggregation function for nodes of layer 5; and

$a^{(5)}$ = activation function for nodes of layer 5.

10.4.2.4 Falcon-MLVQ

The MLVQ clustering technique[4] is incorporated into the *Falcon* network and the proposed *Falcon-MLVQ* architecture has a 5-layered structure as shown in Figure II.10.25. The layers are characterized by the fuzzy operations they perform. They are: layer 1—the input layer (input linguistic variable); layer 2—the condition layer consisting of all the input linguistic terms; layer 3—the rule base layer; layer 4—the consequence layer consisting of all the output linguistic terms; and layer 5—the output layer (output linguistic variables).

The inputs and outputs of *Falcon-MLVQ* are represented as vectors $\mathbf{X} = [x_1, \cdots, x_p, \cdots, x_n]^T$ and $\mathbf{Y} = [y_1, \cdots, y_t, \cdots, y_m]^T$, respectively, where n denotes the number of input and m denotes the number of output linguistic variables. It must be noted that the input and output vectors are nonfuzzy vectors. That is, each element in $\mathbf{X}$ and $\mathbf{Y}$ has a nonfuzzy value. The fuzzification of the input data and defuzzification of the output data are automatically accomplished in the *Falcon-MLVQ* network.

In layer 1, L_p denotes the pth input linguistic node. Its input and output are denoted by x_p and z_p, respectively. Q denotes the number of input term nodes (linguistic labels) for each input node. The label $\mathbb{L}_{pq}$ denotes the qth term node of the pth input node. In *Falcon-MLVQ*, all input nodes have the same number of input term nodes Q. There are a total number of $n \times Q$ term nodes in layer 2. The membership value of the input data is computed in layer 2. The output of the node $\mathbb{L}_{pq}$ is denoted by z_{pq}. Similar to *Falcon-FCM*, each layer 2 node in *Falcon-MLVQ* is fully connected to all the rule nodes in layer 3, if the contrast between positive and negative exemplars is implemented. Otherwise, each layer 2 node will be connected to only one rule node.

Layer 3 has Q rule nodes. That is, the number of term nodes for each input variable determines the number of fuzzy rules in *Falcon-MLVQ*. On the other hand, the variable Q is determined by the number of clusters C that is specified prior to training. That is, $Q = c$. In layer 3, R_r denotes the rth rule node. The output of rule R_r is denoted by z_{R_r} and can be interpreted as a degree of matching between the input data and the stored exemplars of the rth fuzzy rule.

Every output term node (output linguistic label) in layer 4 receives input from one rule node only. This input is subsequently transmitted to the respective output node. Again, the number of output term nodes for each output variable is Q, since the output space is also partitioned like the input space. That is, $S = Q$. The sth term node (label) of the tth output variable is denoted by OL_{st} and its output is z_{st}.

Each of the output nodes (output variable) in layer 5 receives inputs from all its S term nodes and performs defuzzification of the data. The output of the node G_t is denoted by y_t, and is the response by the *Falcon-MLVQ* network with respect to the input stimulus $\mathbf{X}$. A detailed description of the functionality of the nodes in each of the five layers is given as follows.

10.4.2.4.1 *The Input Layer*

The input linguistic nodes directly transmit the nonfuzzy input values to the second layer. The net synaptic input Net_p and the net synaptic output z_p of node L_p are given in Equation II.10.27:

$$\text{Net synaptic input for node } L_p, Net_p = f^{(1)}(x_p)$$

$$= x_p \qquad (\text{II}.10.27)$$

$$\text{Net output for node } L_p, z_p = a^{(1)}(Net_p)$$

$$= x_p$$

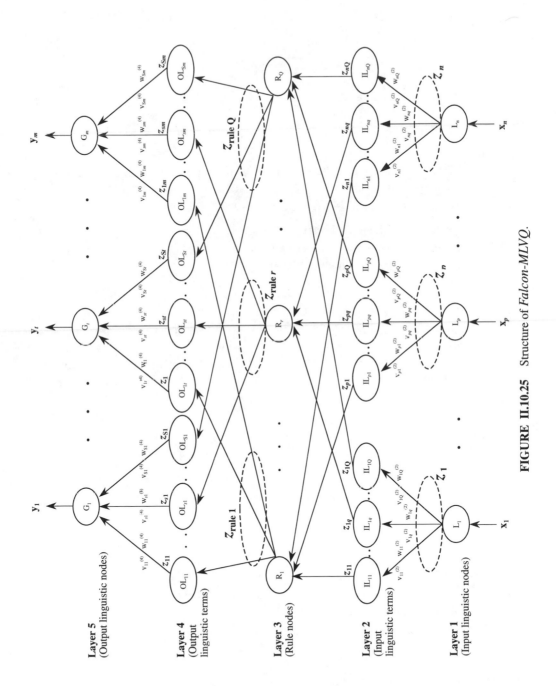

FIGURE II.10.25   Structure of *Falcon-MLVQ*.

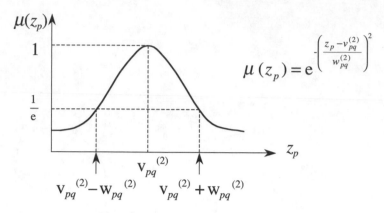

FIGURE II.10.26   Bell-shaped membership function.

where

$\quad x_p \quad = pth$ element of the input vector **X**;
$\quad f^{(1)} \quad$ = aggregation function for nodes of layer 1; and
$\quad a^{(1)} \quad$ = activation function for nodes of layer 1.

10.4.2.4.2 *The Condition (Input Linguistic Term) Layer*

Input-label nodes in the condition layer constitute the antecedents of the fuzzy rules in *Falcon-MLVQ*. Each input-label node $\mathbb{L}_{pq}$ is represented by a bell-shaped membership function as shown in Figure II.10.26. The net synaptic input Net_{pq} and the net synaptic output z_{pq} of the node $\mathbb{L}_{pq}$ are given in Equation II.10.28:

$$\text{Net synaptic input for node } \mathbb{L}_{pq}, Net_{pq} = f^{(2)}(z_p)$$

$$= -\left(\frac{z_p - v_{pq}^{(2)}}{w_{pq}^{(2)}}\right)^2$$

$$\text{Net synaptic output for node } \mathbb{L}_{pq}, z_{pq} = a^{(2)}(Net_{pq})$$

$$= e^{Net_{pq}} \tag{II.10.28}$$

where

$\quad v_{pq}^{(2)} \quad$ = centriod of the qth cluster of the pth input node;
$\quad w_{pq}^{(2)} \quad$ = width of the Gaussian membership function of qth cluster of the
$\qquad\qquad$ pth input node;
$\quad f^{(2)} \quad$ = aggregation function for nodes of layer 2; and
$\quad a^{(2)} \quad$ = activation function for nodes of layer 2.

10.4.2.4.3 *The Rule-Based Layer*

The net synaptic input and the net synaptic output of the rule node R_r are given in Equation II.10.20.

10.4.2.4.4   *The Consequence Layer*

Output-label node OL_{st} represents the sth label of the defuzzification node G_t in the output layer. The net synaptic input Net_{st} and the net synaptic output z_{st} are as given in Equation II.10.29.

$$\text{Net synaptic input for node } OL_{st}, Net_{st} = f^{(4)}(z_{R_r})$$

$$= z_{R_r} \tag{II.10.29}$$

$$\text{Net synaptic output for node } OL_{st}, z_{st} = a^{(4)}(Net_{st})$$

$$= Net_{st}$$

where

z_{R_r} = net synaptic output of the rule node R_r;
$f^{(4)}$ = aggregation function for nodes of layer 4; and
$a^{(4)}$ = activation function for nodes of layer 4.

10.4.2.4.5   *The Output Layer*

The net synaptic input Net_t and the net synaptic output y_t of the output node G_t are as given in Equation II.10.30:

$$\text{Net synaptic input for node } G_t, Net_t = f^{(5)}(z_{1t}, \ldots, z_{st} \ldots, Z_{St})$$

$$= \sum_{s=1}^{S} v_{st}^{(4)} \times z_{st} \tag{II.10.30}$$

$$\text{Net synaptic output for node } G_t, y_t = a^{(5)}(Net_t)$$

$$= \frac{Net_t}{\sum_{s=1}^{S} z_{st}}$$

where

z_{st} = net synaptic output of the node OL_{st};
$v_{st}^{(4)}$ = centriod of the sth label of the tth output node;
$f^{(5)}$ = aggregation function for nodes of layer 5; and
$a^{(5)}$ = activation function for nodes of layer 5.

The training cycle of *Falcon-MLVQ* consists of two phases: namely, self-organization and membership formation. Self-organization uses the MLVQ algorithm[4] for the clustering of all the training data to form the centriods of the respective clusters. Membership formation creates the bell-shaped membership function of each cluster using the pre-defined width constant ζ and the centriods that are computed by the MLVQ algorithm. A small ζ causes the Gaussian membership to be more spread out and vice versa.

10.4.2.5   Falcon-FKP

The *fuzzy Kohonen partitioning* (FKP) and pseudo FKP (PFKP) clustering algorithms developed by Ang[4] are inspired by the original LVQ algorithm. The main difference between the two algorithms is that FKP uses supervised learning to identify the winning cluster (prototype vector) to update while PFKP uses competitive (unsupervised) learning. In addition, FKP is not restricted by the need to generate a pseudo-partition of the data space that is imposed on PFKP. When the FKP algorithm is incorporated into the basic *Falcon* network, a new hybrid network is created. This network is henceforth referred to as *Falcon-FKP* and has the structure as shown in Figure II.10.27.

The *Falcon-FKP* network has the same structure as the *Falcon-MLVQ* network. There are five layers that represent the input linguistic variables, input linguistic labels, fuzzy rules, output linguistic labels, and the output linguistic variables, respectively. The connections in the network are the same for both *Falcon-MLVQ* and *Falcon-FKP*. Only the training phase and the functionality of the nodes for the two networks differ. The input to the *Falcon-FKP* network is represented by the vector $\mathbf{X} = [x_1, \cdots, x_p, \cdots, x_n]^T$ and the output by vector $\mathbf{Y} = [y_1, \cdots, y_t, \cdots, y_m]^T$, where n and m denotes the number of input and output variables, respectively. The detailed functionality of the *Falcon-FKP* network is as follows.

10.4.2.5.1 *The Input Layer*

The neurons of this first layer are the input nodes, and each node represents a linguistic variable of the fuzzy rules. Referring to Figure II.10.27, L_p denotes the pth input node of the layer. The net synaptic input Net_p and net synaptic output z_p of node L_p are given in Equation II.10.31:

$$\text{Net synaptic input for node } L_p, Net_p = f^{(1)}(x_p)$$

$$= x_p \tag{II.10.31}$$

$$\text{Net synaptic output for node } L_p, z_p = a^{(1)}(Net_p)$$

$$= Net_p$$

where

$x_p = P$th element of the input vector $\mathbf{X}$;

$f^{(1)} = $ aggregation function for nodes of layer 1; and

$a^{(1)} = $ activation function for nodes of layer 1.

10.4.2.5.2 *The Condition Layer*

This layer contains all the input-label nodes associated with the input nodes of the first layer. Similar to the *Falcon-MLVQ* network, every input node in *Falcon-FKP* has the same number of input-label nodes in the condition layer. Hence, the total number of input-label nodes in layer 2 is $Q \times n$ where Q is the number of input-label nodes per input variable. Similar to *Falcon-MLVQ*, for the *Falcon-FKP* network, $Q = C$, where C is the number of clusters defined by the user. The label $\mathbb{L}_{pq}$ denotes the qth input-label of the pth input node. The net synaptic input Net_{pq} and net synaptic output z_{pq} of node $\mathbb{L}_{pq}$ are given in Equation II.10.32:

$$\text{Net synaptic input for node } \mathbb{L}_{pq}, Net_{pq} = f^{(2)}(z_p)$$

$$= z_p \tag{II.10.32}$$

$$\text{Net synaptic output for node } \mathbb{L}_{pq}, z_{pq} = a^{(2)}(Net_{pq})$$

$$= membership(z_p)$$

where

$z_p = $ synaptic output of the node L_p;

$f^{(2)} = $ aggregation function for nodes of layer 2; and

$a^2 = $ activation function for nodes of layer 2.

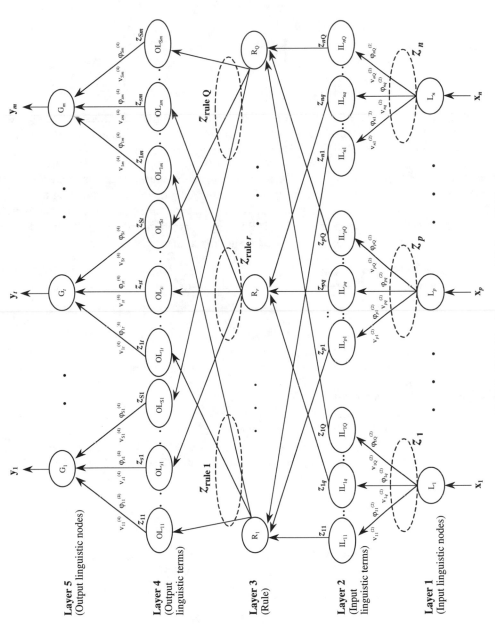

FIGURE II.10.27 Structure of *Falcon-FKP*.

The function *membership* is as defined in Equation II.10.33:

$$membership(z_p) = \begin{cases} 0 & \text{for } z_p < \alpha_{pq} \\ \frac{z_p - \alpha_{pq}}{\beta_{pq} - \alpha_{pq}} & \text{for } \alpha_{pq} \leq z_p \leq \beta_{pq} \\ 1 & \text{for } \beta_{pq} < z_p < \sigma_{pq} \\ \frac{\tau_{pq} - z_p}{\tau_{pq} - \sigma_{pq}} & \text{for } \sigma_{pq} \leq z_p \leq \tau_{pq} \\ 0 & \text{for } z_p > \tau_{pq} \end{cases} \qquad (II.10.33)$$

where

$\alpha_{pq}, \beta_{pq}, \sigma_{pq}$, and τ_{pq} are the four parameters of the trapezoidal membership function of the input-label node $\mathbb{L}_{pq}$. The variables α_{pq} and β_{pq} are the left support and left kernel points, while σ_{pq} and τ_{pq} are the right kernel and right support points, respectively.

The outputs of the nodes in layer 2 are the fuzzy membership values of the input data with respect to the different clusters. The four parameters of the trapezoidal membership function of node $\mathbb{L}_{pq}$ are derived using ν_{pq} and φ_{pq}. The variable ν_{pq} is the centriod of the cluster represented by $\mathbb{L}_{pq}$ after the supervised clustering phase of the FKP training algorithm and φ_{pq} is the pseudo weight used to form the kernel of the trapezoidal membership function of cluster $\mathbb{L}_{pq}$.

10.4.2.5.3 *The Rule-Based Layer*
Neurons in this layer are the rule nodes. They are responsible for mapping the input space to the output space. Each rule node in layer 3 represents a fuzzy rule in *Falcon-FKP*. In Figure II.10.27, the label R_r denotes the rth fuzzy rule. The net synaptic input Net_{R_r} and net synaptic output z_{R_r} of the rule node rth are as given in Equation II.10.20. The output of the rule node R_r can be considered as a result of matching the input vector to the trained exemplars of the conditional part of the fuzzy rule R_r. Hence, the function of the rule nodes in layer 3 is to perform precondition matching.

10.4.2.5.4 *The Consequent Layer*
Nodes in this layer are the output-labels to the output nodes in layer 5. They represent the consequent section of the fuzzy rules. They partition the output space into clusters. In *Falcon-FKP*, the input and output nodes have the same number of linguistic labels. This is because the input and output training vectors are combined as an augmented training vector. Therefore, $S = Q$ and the number of output-labels per output node is Q and hence, the total number of nodes in layer 4 is $Q \times m$ where m is the number of output variables. The label OL_{st} denotes the sth fuzzy term of the tth output variable. The net synaptic input Net_{st} and net synaptic output z_{st} of the node OL_{st} are given in Equation II.10.34:

$$\text{Net synaptic input for node } OL_{st}, Net_{st} = f^{(4)}(z_{R_r})$$
$$= z_{R_r} \qquad (II.10.34)$$
$$\text{Net synaptic output for node } OL_{st}, z_{st} = a^{(4)}(Net_{st})$$
$$= Net_{st}$$

where

z_{R_r} = net synaptic output of the rule R_r that is feeding into OL_{st};

$f^{(4)}$ = aggregation function for nodes of layer 4; and

$a^{(4)}$ = activation function for nodes of layer 4.

10.4.2.5.5   *The Output Layer*

The function of an output node in layer 5 is to perform defuzzification of the inputs it receives from all its term nodes in layer 4, and present the computed result as the network output. The label G_t denotes the tth output node in layer 5. The net synaptic input Net_t and net synaptic output y_t of node G_t are given in Equation II.10.35:

$$\text{Net synaptic input for output node } G_t, \; Net_t = f^{(5)}(Z_{1t}, \ldots, z_{st}, \ldots, z_{St})$$

$$= \sum_{s=1}^{S} v_{st}^{(4)} \times z_{st} \tag{II.10.35}$$

$$\text{Net synaptic output for output node } G_t, \; y_t = a^{(5)}(Net_t)$$

$$= \frac{Net_t}{\sum_{s=1}^{S} z_{st}}$$

where

z_{st} = net synaptic output of the node OL_{st};

$v_{st}^{(4)}$ = centriod of the sth label of the tth output node;

$f^{(5)}$ = aggregation function for nodes of layer 5; and

$a^{(5)}$ = activation function for nodes of layer 5.

Equation II.10.35 above approximates the *center-of-area* defuzzification method[33] that is popularly used in fuzzy neural networks.

Similar to the *Falcon-MLVQ* architecture, the *Falcon-FKP* creates the rule nodes and their respective condition and consequent nodes in layer 2 and layer 4 prior to the start of training. The number of rules that is created depends on the user-specified parameter C that states the number of clusters in the training data set. That is, the number of rules in *Falcon-FKP* is equal to C (and $Q = c$). Each of the input and output label nodes represents a cluster of the linguistic variable it belongs to. The input and output training data pair are grouped into an augmented vector for training. After the supervised clustering phase of the FKP algorithm[4] is performed, the training data are recycled to train the pseudo weight to form the kernel and support of the respective trapezoidal membership functions. Due to the nature of the training algorithm, *Falcon-FKP* is unsuitable for on-line learning application because the training data needs to be recycled to compute the trapezoidal membership functions.

10.4.2.6   Falcon-PFKP

The PFKP algorithm[4] is synthesized into the basic *Falcon* network to create a new hybrid network named *Falcon-PFKP*. It has the same structure as the *Falcon-FKP* network (Figure II.10.27). The functionality of the *Falcon-PFKP* network is exactly the same as *Falcon-FKP*. The main differences between the two networks are that Falcon-PFKP uses unsupervised competitive learning and the membership functions of the data clusters form a pseudo-partition for the data space. The membership values of an input data with respect to all the clusters give a sum of unity. This is more intuitive to the human cognition than the MLVQ and FKP algorithms, as a data will either belong entirely to one cluster (i.e., has a membership of unity for that cluster and zeros for the rest of the clusters), or belong partially to two or more clusters. In other words, an element in a given data space belongs to either only one cluster, or two or more clusters that partition the data space. However, its membership values with respect to all the clusters must have a total sum of unity to show that it lies within the boundaries of the data space.

The training cycle of the *Falcon-PFKP* network consists of two phases: the self-organization phase using unsupervised learning, and the membership phase in which the trapezoidal membership functions of the clusters are being formed. The resultant trapezoidal membership functions of the clusters form a pseudo-partition of the training space. The input and desired output training data pair is grouped together to form an augmented training vector. Prior to training, the *Falcon-PFKP* network creates the required number of rule nodes and their associated input and output terms. The number of rules in *Falcon-PFKP* is again determined by a user-preset parameter C, which denotes the number of clusters in the training data set. That is, the number of rules is equal to C. The input and output terms are then seeded with evenly spaced values to reduce unnecessary movement of the centriods.[4] Upon training, the winning cluster that has the smallest Euclidean distance from the input training vector is identified and its weight updated accordingly. This is repeated for all the augmented training vectors. After the self-organization clustering phase, the *Falcon-PFKP* network subsequently recycles the training data and uses a pseudo weight φ and its corresponding learning constant ψ to determine the trapezoidal membership functions. The kernels and supports of the trapezoidal membership functions are then adjusted to form a pseudo-partition of the input and output data space.

This concludes the discussion on the four proposed fuzzy neural architectures. The performance of the four new networks are evaluated and benchmarked against *Falcon-ART* and *Falcon-MART* in the next section.

10.5   Results and Analysis

This section presents the simulation results for the four proposed networks: *Falcon-FCM, Falcon-MLVQ, Falcon-FKP*, and *Falcon-PFKP*. To evaluate the performances of the proposed networks, three different experiments are conducted. They are 1) iris classification, 2) traffic prediction, and 3) phoneme classification. The simulations are performed using *complementary learning* (CL)-based networks as well as non-CL-based networks. To differentiate between the two sets of networks, CL-based networks will be denoted as *Falcon-FCM*(CL), *Falcon-MLVQ*(CL), *Falcon-FKP*(CL), and *Falcon-PFKP*(CL), respectively. Non-CL-based networks are denoted as *Falcon-FCM, Falcon-MLVQ, Falcon-FKP*, and *Falcon-PFKP*. The background on the data sets and the objectives of the experiments are given in Section 10.3.

10.5.1   Iris Classification

The same training and test sets are used as in Section 10.3. There are three groups of training and test sets and they are labeled as CV1, CV2, and CV3, respectively. This is to cross-validate the experimental results to avoid any inconsistencies and to reduce any experimental errors. The two indicators used to benchmark the efficiency of a network in the iris classification experiment are the *mean classification rate* (denoted as Mean c rate) and the *standard deviation* of the classification rate across the three cross-validation sets (Std Dev). The higher the "mean c rate," the better the classification result in terms of the percentage of correct classification across the three cross-validation sets. The indicator "Std Dev" reflects the performance stability of a network across different training and test sets. Hence, the smaller the value of "Std Dev," the stronger the network tolerance to variations across the different cross-validation sets. The simulation results of the four proposed networks using complementary learning for the iris classification experiment are summarized in Table II.10.9.

From Table II.10.9, several observations are made. First, the standard deviations of the classification rates for all four networks across the three cross-validation sets (Std Dev) are less than 5%, with the maximum Std Dev at 4.06% reported by *Falcon-MLVQ*(CL). Second, all the networks have 0% for unclassification rate, indicating that the general clustering nature of the iris data set

TABLE II.10.9 Simulation Results for Iris Classification Experiment (CL-Based)

	Network : Falcon-FCM(CL)				Network : Falcon-MLVQ(CL)		
	CV1	CV2	CV3		CV1	CV2	CV3
c rate (%)	89.90	91.18	89.90	c rate (%)	91.92	87.25	83.84
m rate (%)	10.10	8.82	10.10	m rate (%)	8.08	12.75	16.16
u rate (%)	0.00	0.00	0.00	u rate (%)	0.00	0.00	0.00
Mean c rate = 90.33%		Std Dev = 0.74%		Mean c rate = 87.67%		Std Dev = 4.06%	

	Network : Falcon-FKP(CL)				Network : Falcon-PFKP(CL)		
	CV1	CV2	CV3		CV1	CV2	CV3
c rate (%)	93.94	90.20	90.91	c rate (%)	91.92	85.29	86.87
m rate (%)	6.06	9.80	9.09	m rate (%)	8.08	14.71	13.13
u rate (%)	0.00	0.00	0.00	u rate (%)	0.00	0.00	0.00
Mean c rate = 91.68%		Std Dev = 1.99%		Mean c rate = 88.03%		Std Dev = 3.46%	

Note: (c rate = classification rate; m rate = misclassification rate; u rate = unclassification rate. All are expressed in terms of percentage).

TABLE II.10.10 Simulation Results for Iris Classification Experiment (non-CL-Based)

	Network : Falcon-FCM				Network : Falcon-MLVQ		
	CV1	CV2	CV3		CV1	CV2	CV3
c rate (%)	89.90	91.18	88.89	c rate (%)	91.92	87.25	81.82
m rate (%)	10.10	8.82	11.11	m rate (%)	8.08	12.75	18.18
u rate (%)	0.00	0.00	0.00	u rate (%)	0.00	0.00	0.00
Mean c rate = 89.99%		Std Dev = 1.15%		Mean c rate = 87.00%		Std Dev = 5.06%	

	Network : Falcon-FKP				Network : Falcon-PFKP		
	CV1	CV2	CV3		CV1	CV2	CV3
c rate (%)	92.93	90.20	87.88	c rate (%)	90.91	84.31	84.85
m rate (%)	7.07	9.80	12.12	m rate (%)	9.09	15.69	15.15
u rate (%)	0.00	0.00	0.00	u rate (%)	0.00	0.00	0.00
Mean c rate = 90.33%		Std Dev = 2.53%		Mean c rate = 86.69%		Std Dev = 3.66%	

Note: c rate = classification rate; m rate = misclassification rate; u rate = unclassification rate. All are expressed in terms of percentage.

has been effectively learned by all four networks. The same set of experiments is repeated for the non-CL-based networks, and the results are summarized in Table II.10.10.

The results from Table II.10.10 shows that the CL-based networks have better performance than the non-CL-based networks, both in terms of higher mean classification rates (mean c rate) and smaller standard deviation (Std Dev) values. Comparing the results in Table II.10.9 and II.10.10 with that of *Falcon-ART* (Table II.10.3), one can conclude that all four proposed architectures performed better than *Falcon-ART*, regardless of whether the networks are CL- or non-CL-based.

All four proposed networks have a higher "mean c rate" than *Falcon-ART*, and better stability in network performances as indicated by the low "Std Dev" readings. The difference in performances is due to the inherent shortcomings of the *Falcon-ART* architecture as discussed in Section 10.3. When the results of the four proposed networks (CL-based) are compared against that of *Falcon-MART* (Table II.10.4), it is apparent that *Falcon-MART* has a better classification rate. However, the benefits of the contrast of positive and negative exemplars can be seen on a closer look at the two sets of results. Using the notion of positive and negative exemplars, all four proposed architectures (Table II.10.9) have a smaller deviation in performance than *Falcon-MART* (at 4.64%). Hence, the results of the iris classification experiment clearly indicate that the hypothesis of improved performance in terms of network stability using the notion of positive and negative exemplars is correct.

10.5.2   Traffic Prediction

The same setup as in Section 10.3 is used to perform the traffic prediction experiment. There are three groups of training and test sets, labeled as CV1, CV2, and CV3, respectively. They are used to cross-validate the results obtained. Again, the square of the Pearson product-moment correlation value (R^2) is used to measure the accuracy of the predicted traffic trend. Since for all four proposed networks, the number of classes has to be specified prior to training (a constrain faced by the FCM, MLVQ, FKP, and PFKP clustering algorithms), it is therefore decided that C is preset as 5. Hence, the joint input-output space is partitioned into five clusters. The simulation results for the four CL-based fuzzy neural networks are presented as the average R^2 value (across the three cross-validation sets) of each lane (lanes 1, 2, and 3) with respect to time interval τ. The results are shown as Figure II.10.28.

Comparing the results of the traffic experiment (Figure II.10.28) for the four proposed CL-based networks, it is observed that the value of R^2 decreases as τ increases. This is expected, as it becomes less accurate to predict the traffic conditions at longer intervals (say $\tau = 60$ min) than shorter intervals (e.g., $\tau = 15$ min). In addition, *Falcon-PFKP*(CL) consistently provides more accurate predictions than the other three networks. On the other hand, *Falcon-FCM*(CL) has the worst performance, probably because of the operations of its layer 2 nodes. As discussed in Section 10.4, the fuzzy memberships of the inputs to *Falcon-FCM*(CL) are computed through the interpolation of the learned data points. As the membership value between two points is assumed to be linear, this could lead to unexpected errors. Hence, this may be the reason why the prediction results of *Falcon-FCM*(CL) are consistently the worst among the four proposed. The same simulations are repeated using the non-CL-based networks, that is, *Falcon-FCM, Falcon-MLVQ, Falcon-FKP*, and *Falcon-PFKP*. The results are as shown in Figure II.10.29.

The results for the non-CL-based networks in Figure II.10.29 reinforced what we have observed from the results of the CL-based networks in Figure II.10.28. That is, the accuracy of the predictions degrades as the time interval increases. The facts that *Falcon-PFKP* once again consistently provides better predictions than the other three networks, and that *Falcon-FCM* has the worst performance, demonstrate that the FCM clustering algorithm is unsuitable for the traffic experiment as compared to the PFKP algorithm.

The changes in the average R^2 values as τ increases from 5–60 min (denoted as Var and expressed as a percentage of the former) for each lane based on the different network architectures (both CL- and non-CL-based) are tabulated in Table II.10.11. The mean Var value across the three different lanes is denoted as Avg Var.

From Table II.10.11, it is obvious that the contrast of positive and negative exemplars brings about better network performances for all four proposed networks. This is demonstrated by the smaller degradation in accuracy of the predictions by the CL-based networks when the time interval increases from 5–60 min against that of the non-CL-based networks (as shown by the Avg Var values in the last column of Table II.10.11). Average variation can be taken to be a measure of consistency in the predictions across the various time intervals. Comparing the results of the four proposed networks

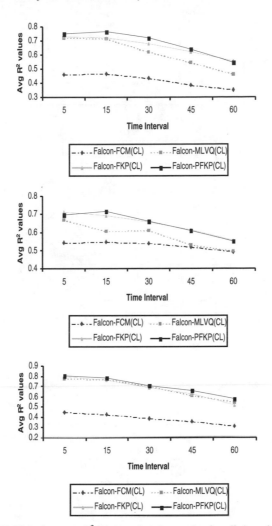

FIGURE II.10.28   Average R^2 against time interval τ for all three lanes (CL-based).

(both CL- and non-CL-based) against that of *Falcon-MART* (Figures II.10.20 and II.10.21), it is clear that *Falcon-MART* has a much superior performance. However, *Falcon-MART* uses 83 fuzzy rules to model the traffic conditions of the three lanes, while the four proposed networks used only five fuzzy rules (because $C = 5$). As more fuzzy rules can partition the data space into finer regions, it is not surprising that *Falcon-MART* should have a better performance.

10.5.3   Phoneme Classification

The phoneme data set is obtained from the ELENA[80] Web site. The same training and test sets as in Section 10.3 are used for the evaluations of the four proposed networks. Hence, there are five groups of training and test sets, and they are labeled as CV1, CV2, CV3, CV4, and CV5. This is to cross-validate the simulation results obtained. The simulation results are presented in order of the proposed networks and summarized in Table II.10.12. For each of the four proposed networks, two sets of results are presented. The first set (on the left) is labeled "CL" and the second set (on the right) is labeled "NCL." The set of results labeled "CL" is obtained when the notion of positive and negative exemplars in the inference process is implemented (complementary learning), while the set

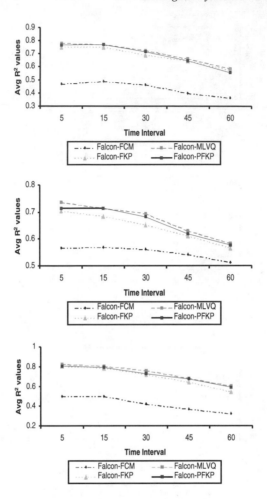

FIGURE II.10.29   Average R^2 against time interval τ for all three lanes (non-CL-based).

TABLE II.10.11   Variation in Average R^2 Values for $\tau = 5$–60 Min

Network	Lane 1—Var (Avg R^2 variation in %)	Lane—Var (Avg R^2 variation in %)	Lane 3—Var (Avg R^2 variation in %)	Avg Var (Average variation in %)
Falcon-FCM (CL)	30.47	9.32	24.17	21.32
Falcon-MLVQ (CL)	24.75	20.58	26.63	23.99
Falcon-FKP (CL)	22.73	19.29	32.09	24.70
Falcon-PFKP (CL)	27.34	19.14	26.93	24.47
Falcon-FCM	22.73	9.60	35.84	22.72
Falcon-MLVQ	30.21	25.94	36.41	30.85
Falcon-FKP	35.19	22.09	23.87	27.05
Falcon-PFKP	28.25	21.05	27.81	25.70

TABLE II.10.12 Phoneme Classification Results Using the Proposed Networks

	CL				NCL			
	Falcon-FCM	Falcon-MLVQ	Falcon-FKP	Falcon-PFKP	Falcon-FCM	Falcon-MLVQ	Falcon-FKP	Falcon-PFKP
MC-C0 (%)	66.25	80.56	86.42	82.56	66.28	96.58	86.86	92.13
MC-C1 (%)	87.93	70.79	50.89	57.87	87.93	11.92	49.86	28.52
Std C0 (%)	0.38	3.02	5.27	4.22	0.41	6.31	4.95	8.30
Std C1 (%)	0.91	4.81	10.65	9.54	0.91	23.42	10.94	25.19

of results labeled "NCL" is obtained when such a contrast is not implemented (noncomplementary learning).

The label "MC-C0" denotes the mean classification rate for Class 0 phoneme data. It refers to the average classification rate for Class 0 data points across the five cross-validation test sets. Similarly, the label "MC-C1" denotes the mean classification rate for Class 1 phoneme data. It refers to the average classification rate for Class 1 data points across the same five cross-validation test sets. Besides the mean classification rates for Class 0 and Class 1 data points, the standard deviations of the classification rates across the five cross-validation sets (for both Class 0 and Class 1 classifications) are also presented. They are denoted as "Std C0" and "Std C1," respectively. The standard deviation can be viewed as an indicator to the consistency in performance of the proposed networks across each of the five groups of training and test sets.

Comparing the phoneme results of the four proposed networks (Table II.10.12), several observations can be made. First, when the notion of positive and negative exemplars is implemented in the four networks (CL-based), it leads to a more stable network performance across different cross-validation sets in terms of smaller deviations in the classification rates of Class 0 and Class 1 data points. This can be seen from the results of the "CL" and "NCL" learning schemes. The change in standard deviation for *Falcon-MLVQ* and *Falcon-PFKP* using "CL" and "NCL" scheme is quite significant. However, this improvement in performance is not as well defined in *Falcon-FCM* and *Falcon-FKP* as compared to *Falcon-MLVQ* and *Falcon-PFKP*. This is because for the FCM clustering technique, it minimizes an objective function[5] and the pseudo-partition of the data space results when the algorithm terminates. Hence, it is not sensitive to the amount of data points each class has. As highlighted in Section 10.3, Class 0 data points are dominant in the phoneme data set, consisting of 70.35%, and the two classes are hardly separable.

Regardless of whether the two classes are separable and the number of data points each class has, the FCM algorithm creates two partitions of the data space that minimizes the objective function. On the other hand, FKP is a supervised learning algorithm. Hence, class information of the training data points is known beforehand and the two prototype vectors are updated accordingly. As a result, these two clustering techniques already produce a near optimal partition of the data space. Hence, the effect of the contrast between positive and negative exemplars is not so obvious.

On the other hand, MLVQ and PFKP are unsupervised clustering techniques. Because Class 0 data points are dominant in the phoneme data set, they also form the majority in the training sets. Hence, during training, Class 0 data points have an "overwhelming" effect on Class 1 data points, and because both sets of data points are very similar, the movement of the prototype vector representing Class 0 overshadows that of Class 1. Hence, even with the initial seeding of the prototype vectors (based on the MLVQ and PFKP algorithms), the two vectors will move toward one another, resulting in almost identical clusters. Therefore, when the notion of positive and negative exemplars is not implemented (NCL), both *Falcon-MLVQ* and *Falcon-PFKP* have very high classification rates for

Class 0, while the classification rates for Class 1 are very low. Referring to Table II.10.12, the mean classification rate for Class 1 using *Falcon-MLVQ* is 11.92%, and the mean classification rate for Class 1 using *Falcon-PFKP* is 28.52%. This is because the two networks "perceived" the majority of Class 1 data as Class 0 data.

Table II.10.12 shows that with the use of the contrast between positive and negative exemplars (CL), the classification rates for Class 0 drops as compared to the set of results obtained without such notion (NCL). However, the classification rates for Class 1 data points increases significantly, indicating a more balanced overall classification result with respect to Class 0 and Class 1 classification rates. In other words, the results are not lopsided, and the possibility of a Class 1 data point being classified as a Class 0 data point is greatly reduced. When the results of the proposed networks are compared against that of *Falcon-MART* (Table II.10.7), it can be seen that when the notion of positive and negative exemplars is implemented (CL), the four proposed networks give higher classification rates for Class 1 data points than *Falcon-MART*. In fact, *Falcon-FCM* and *Falcon-MLVQ* have classification rates (for Class 1) of 87.93 and 70.79%, respectively. The classification rate for Class 1 obtained using *Falcon-FCM* is even better than that of the K-NN classifier, which is 72.20% (see Table II.10.8).

Therefore, the effect of using the contrast between positive and negative exemplars is evident in the results of the phoneme classification. Not only does the contrast between positive and negative exemplars increase the stability of the network performance, it also prevents the "overwhelming" effect of Class 0 (with a majority of data points) over Class 1 in this case.

10.6   Conclusions

In this chapter, the concept of positive and negative learning is examined. Most of the fuzzy neural networks proposed in the literature use fuzzy rules derived from positive exemplars. They compute their outputs based on the degree of matching or similarity of the inputs to the antecedents of their fuzzy rules. Hence, the result of classifying/recognizing closely similar classes may not be desirable. As such, four Falcon-like fuzzy neural networks are proposed to examine the hypothesis that the contrast of positive and negative exemplars in the inference process will lead to a more stable network performance as well as better recognition rate. The four new fuzzy neural networks are to make use of positive and negative exemplars in their inference process. They are based on the basic *Falcon* architecture, and used different clustering techniques to compute the fuzzy sets of the problem domain. The clustering techniques used are FCM,[5] MLVQ, FKP, and PFKP.[4] Hence, the four proposed networks were intuitively named as *Falcon-FCM, Falcon-MLVQ, Falcon-FKP*, and *Falcon-PFKP*, respectively. Together, they are classified as *Pseudo Adaptive Complementary Learning* (PACL) fuzzy neural networks.

In addition, a fuzzy neural architecture, *Falcon-ART*[9] that is developed by Lin is studied in detail. The *Falcon-ART* network uses the fuzzified version of *Adaptive Resonance Theory* (ART)[10] to obtain a partition of the input-output space, and subsequently derives the fuzzy rules using a mapping process. In the evaluation of the *Falcon-ART* network using the iris data set,[77] several shortcomings have been uncovered. They are:

1. Poor classification performance when the classes of input data are closely similar.
2. Weak resistance to noisy/spurious training data.
3. Termination of network training process depends heavily on a preset error parameter.
4. Learning efficiency may deteriorate as a result of complementary coding of the training data.

Hence, a variant of *Falcon-ART*, named *Falcon-MART* (Modified ART) is proposed to overcome the above deficiencies. The proposed *Falcon-MART* network is evaluated using three experiments: Iris classification, Traffic prediction, and Phoneme classification; the results are encouraging. Meanwhile, the performances of the four proposed networks were evaluated using the same experiments.

The results of the simulations supported the hypothesis that the contrast of positive and negative exemplars do improve network performance, and in some cases, the performances of the four proposed networks are better than that of *Falcon-MART*, even though the latter uses more fuzzy rules to partition the input-output space.

References

1. Zadeh, L.A., Fuzzy logic, neural networks and soft computing, *Communications of ACM*, 37(3), 77–84, 1994.
2. Zadeh, L.A., et al., *Calculus of Fuzzy Restrictions, Fuzzy Sets and Their Applications to Cognitive and Decision Processes*, Academic Press, New York, 1975, 1–39, 1975.
3. Kohonen, T.K., *Self-Organization and Associative Memory*, 3rd ed., Springer-Verlag, New York, 1989.
4. Ang, K.K., POPFNN-CRI(S): A Fuzzy Neural Network based on the Compositional Rule of Inference, M.Phil Project, NTU Singapore, 1998.
5. Bezdek, J.C., Hathaway, R.J., and Tucker, W.T., An improved convergence theory for the fuzzy C-means clustering algorithms. In *Analysis of Fuzzy Information* 3, Bezdek, J.C. (Ed.) Chap. 8, 123–129, CRC Press, Boca Raton, 1987.
6. Tsao, E.C.K., Bezdek, J.C., and Pal, N.R., Fuzzy Kohonen clustering networks, *Pattern Recognition*, 27(5), 757–764, 1994.
7. Turksen, I.B. and Zhong, Z., An approximate analogical reasoning scheme based on similarity measures and interval valued fuzzy sets, *Fuzzy Sets and Syst.*, 34, 323–346, 1990.
8. Mantaras, R.L., Approximate reasoning models, Ellis Horwood Limited, New York, 1990.
9. Lin, C.J. and Lin, C.T., An ART-based fuzzy adaptive learning control network, *IEEE Trans. Fuzzy Syst.*, 5(4), 477–496, 1997.
10. Grossberg, S., Adaptive pattern classification and universal recoding: II. Feedback, expectation, olfaction, illusions. *Bio. Cybern.* 23, 187–202, 1976.
11. McCulloch, W.S. and Pitts, W., A logical calculus of ideas immanent in nervous activity, *Bull. Math. Biophys.*, 5, 115–133, 1943.
12. Hebb, D.O., *The Organization of Behavior: A Neuro-Psychological Theory*, John Wiley and Sons, New York, 1949.
13. Rosenblatt, F., The perceptron, a probabilistic model for information storage and organization in the brain, *Psychological Review*, 62, 589, 1958.
14. Widrow, B. and Hoff, M.E., Adaptive switching circuits, *Proc. 1960 IRE West. Elect. Show Conv. Rec.*, Part 4, 96–104, New York, 1960.
15. Anderson, J.A., A simple neural network generating an interactive memory, *Mathematical Biosciences*, 14, 197–220, 1972.
16. Kohonen, T.K., Correlation matrix memories, *IEEE Trans. Computers*, 21, 353–359, 1972.
17. Kohonen, T.K., Self-organized formation of topologically correct feature maps, *Biological Cybernetics*, 43, 59–69, 1982.
18. Hopfield, J.J., Neural networks and physical systems with emergent collective computational abilities, *Proc. Nat. Acad. Sci.*, 79, 2554–2558, 1982.
19. Ackley, D.H., Hinton, G.E., and Sejnowski, T.J., A learning algorithm for Boltzmann machines, *Cognitive Sci.*, 9, 147–169, 1985.
20. Kirkpatrick, S., Gelatt, C.D., and Vecchi, M.P., Optimization by simulated annealing, *Science*, 220, 671–680, 1983.
21. Rumelhart, D.E., Hinton, G.E., and Williams, R.J., Learning internal representations by error propagation. In Rumelhart, D.E., McClelland, J.L. et al., Eds., *Parallel Distributed Processing*, vol. 1, Chap. 8. Cambridge, MA, MIT Press, 1986.

22. Waibel, A., Modular construction of time-delay neural networks for speech recognition, *Neural Comput.*, 1, 39–46, 1989.

23. Moody, J. and Darken, C., Fast learning in networks of locally-tuned processing units, *Neural Comput.*, 1, 281–294, 1989.

24. Hecht-Nielson, R., Counterpropagation networks, *Appl. Opt.*, 26, 4979–4984, 1987.

25. Hecht-Nielson, R., Nearest matched filter classification of spatiotemporal patterns, *Appl. Opt.*, 26(10), 1892–1899, 1987.

26. Hecht-Nielson, R., Applications of counter-propagation networks, *Neural Networks*, 1, 131–139, 1988.

27. Zadeh, L.A., Fuzzy sets, *Inf. Cont.*, 8, 338–353, 1965.

28. Jang, J.S.R., Fuzzy modeling using generalized neural networks and Kalman filter algorithm, In *Proc. 9th Nat. Conf. Artificial Intelligence* (AAAI-91), 762–767, 1991.

29. Jang, J.S.R., Self-learning fuzzy controller based on temporal back-propagation, *IEEE Trans. Neural Networks*, 3, 714–723, 1992.

30. Jang, J.S.R., ANFIS: Adaptive-network-based fuzzy inference systems, *IEEE Trans. Systs., Man, and Cyberns.*, 23, 665–685, 1993.

31. Klir, G.J. and Yuan, B., *Fuzzy Sets and Fuzzy Logic, Theory and Applications*, Prentice Hall, New York, 1995.

32. Kaufmann, A. and Gupta, M.M., *Introduction to Fuzzy Arithmetic: Theory and Applications*, Van Nostrand Reinhold Company Inc., New York, 1985.

33. Lin, C.T. and Lee, C.S.G., *Neural Fuzzy Systems—A Neuro-Fuzzy Synergism to Intelligent Systems*, Prentice Hall, New York, 1996.

34. Yeung, D.S. and Tsang, E.C.C., Improved fuzzy knowledge representation and rule evaluation using fuzzy Petri nets and degree of subsethood, *Intell. Syst.*, 9(12), 1083–1100, 1994.

35. Yager, R., An approach to inference in approximate reasoning, *Man-Machine Studies*, 13, 323–338, 1980.

36. Mizumoto, M. and Zimmermann, H.J., Comparison of fuzzy reasoning methods, *Fuzzy Sets and Syst.*, 8, 253–283, 1982.

37. Mamdani, E.H., Application of fuzzy logic to approximate reasoning using linguistic systems, *IEEE Trans. Comput.*, 26, 1182–1191, 1977.

38. Bandler, W. and Kohout, L.J., Fuzzy power sets and fuzzy implication operators, *Fuzzy Sets Syst.*, 4, 13–30, 1980.

39. Chen, S.M., A new approach to handling fuzzy decision-making problems, *IEEE Trans. Syst., Man, Cybern.*, 18, 1012–1016, Nov/Dec., 1988.

40. Chen, S.M., A weighted fuzzy reasoning algorithm for medical diagnosis, *Decision Support Syst.*, 11, 37–43, 1994.

41. Yeung, D.S. and Tsang, E.C.C., A comparative study on similarity-based fuzzy reasoning methods, *IEEE Trans. Syst., Man, and Cyberns, part B: Cyberns*, 27(2), 216–227, April 1997.

42. Turksen, I.B. and Zhong, Z., An approximate analogical reasoning approach based on similarity measures, *IEEE Trans. Systems, Man and Cybernetics*, 18, 1049–1056, Nov/Dec., 1988.

43. Forgy, E.W., Cluster analysis of multivariate data efficiency versus interpretability of classification, Biometric Society Meetings, Riverside, California. Abstract in *Biometrics*, 213, 768, 1965.

44. MacQueen, J.B., Some methods for classification and analysis of multivariate observations, *Proc. Symp. Math. Statist. and Probability*, 5th Berkley 1, AD 669871, University of California Press, Berkeley, 281–297, 1967.

45. Zahid, N. et al., Unsupervised fuzzy clustering, *Pattern Recognition Letters*, 20, 123–129, 1999.

46. Lee, H.S., On system identification via fuzzy clustering for fuzzy modeling, *IEEE Trans. Fuzzy Sets and Syst.*, 1956–1961, 1998.

47. Gower, J.C. and Ross, G.J.S., Minimum spanning trees and single linkage cluster analysis, *Appl. Stat.*, 18, 54–64, 1969.

48. Johnson, S.C., Hierarchical clustering schemes, *Psychometrika*, 323, 241–254, 1967.
49. Benkirane, H. et al., Hierarchical fuzzy partition for pattern classification with fuzzy if-then rules, *Pattern Recognition Letters*, 21, 503–509, 2000.
50. Lin, Y.H., Cunningham G.A., and Coggeshall, S.V., Using fuzzy partitions to create fuzzy systems from input-output data and set the initial weights in a fuzzy neural network, *IEEE Trans. Fuzzy Syst.*, 5(4), 614–621, Nov 1997.
51. Pal, N.R. and Bezdek, J.C., On cluster validity for the fuzzy C-means model, *IEEE Trans. Fuzzy Syst.*, 3(3), 370–379, 1995.
52. Quek, H.C. and Chen, W.J., *GF, A GA-Optimized Fuzzy Classifier System*, NTU Singapore, 2000.
53. Lin, Y.T. and Shiueng, B.Y., A genetic clustering algorithm for data with non-spherical-shape clusters, *Pattern Recognition*, 33, 1251–1259, 1999.
54. Sankar, K.P., Bhandari, D., and Kundu, M.K., Image enhancement incorporating fuzzy fitness function in genetic algorithms, 2nd *IEEE Int. Conf. Fuzzy Syst.*, 2, 1408–1413, 1993.
55. Sankar, K.P., Bandyopadhyay, S., and Murthy, C.A., GA-based pattern classification: Theoretical and experimental studies, *Proc. 13th Int. Conf. Pattern Recognition*, 3, 758–762, 1996.
56. Sankar, K.P., Bandyopadhyay, S., and Murthy, C.A., Genetic algorithms for generation of class boundaries, *IEEE Trans. Syst., Man, Cyberns-Part B: Cyberns*, 28(6), 816–828, 1998.
57. Keller, J.M., Gray, M.R., and Givens, J.A., A fuzzy K-nearest neighbors algorithm, *IEEE Trans. Syst., Man, Cyberns.*, 15, 580–585, 1985.
58. Singh, A., Yager Fuzzy Neural Networks, Honors Year Project, School of Applied Science, NTU Singapore, 2000.
59. Chaudhuri, B.B. and Bhowmik P.R., An approach of clustering data with noisy or imprecise feature measurement, *Pattern Recognition Letters*, 19, 1307–1317, 1998.
60. Chung, F.L. and Tong, L., Fuzzy learning vector quantization, *Proc. 1993 Int. Joint Conf. Neural Networks*, 2739–2742, 1993.
61. Pal, N.R., Bezdek, J.C., and Tsao, E.C.K., Generalized clustering networks and Kohonen's self-organizing scheme, *IEEE Trans. Neural Networks*, 4(4), 549–558, 1993.
62. Yair, E., Zeger, K., and Gersho, A., Competitive learning and soft competition for vector quantizier design, *IEEE Trans. Signal Processing*, 40, 294–309, 1992.
63. Baraldi, A. and Parmiggian, F., Fuzzy combination of Kohonen's and ART neural network models to detect statistical regularities in a random sequence of multi-valued input patterns, *Int. Conf. Neural Networks*, 1, 281–286, 1997.
64. Baraldi, A. and Blonda, P., A survey of fuzzy clustering algorithms for pattern recognition—Part I, *IEEE Trans. Syst., Man, Cybern.-Part B: Cybern*, 29(6), 718–785, 1999.
65. Baraldi A. and Blonda, P., A survey of fuzzy clustering algorithms for pattern recognition—Part II, *IEEE Trans. Syst., Man, Cybern.-Part B: Cybern*, 29(6), 786–801, 1999.
66. Bezdek, J.C., *Pattern Recognition with Fuzzy Objective Function Algorithms*. Plenum Press, New York, 1981.
67. Al Sultan, K.S. and Selim, S.Z., Global algorithm for fuzzy clustering problem, *Pattern Recognition*, 26, 1357–1361, 1993.
68. Bensaid, A.M. et al., Validity guided (re)clustering with applications to image segmentation, *IEEE Trans. Fuzzy Syst.*, 4, 112–123, 1996.
69. Shin, S.C. and Park, S.B., GA-based predictive control for nonlinear processes, *Electronics Letters*, 34(20), 1980–1981, 1998.
70. Kamel, M. and Abutaleb, A.S., A genetic algorithm for the estimation of ridges in fingerprints, *IEEE Trans. Image Processing*, 8(8), 1134–1139, 1999.
71. Haydar, A., Demirekler, M., and Yurtseven, M.K., Feature selection using genetic algorithm and its application to speaker verification, *Electronics Letters*, 34(15), 1457–1459, 1998.

72. Gallego, R.A., Monticelli, A., and Romero, R., Transmission system expansion planning by an extended genetic algorithm, *IEEE Proc. Gen. Transm. Distr.*, 145(3), 329–335, 1998.

73. Carpenter, G.A. and Grossberg, S., A massively parallel architecture for a self-organizing neural pattern recognition machine. *Comput. Vision Graphics Image Process*, 37, 54–115, 1987.

74. Carpenter, G.A. and Grossberg, S., ART2: Self-organization of stable category recognition codes for analog input patterns. *Appl. Opt.*, 26, 4919–4930, 1987.

75. Carpenter, G.A. and Grossberg, S., The ART of adaptive pattern recognition by a self-organizational neural network. *Computer*, 21(3), 77–88, 1988.

76. Carpenter, G.A. and Grossberg, S., ART3, Hierarchical search using chemical transmitters in self-organizing pattern recognition architectures. *Neural Networks*, 3(2), 129–152, 1990.

77. Fisher, R.A., The use of multiple measurements in taxonomic problems. *Annals of Eugenics* 7, Part II, 179–188, 1936.

78. Quek, H.C. and Tung, W.L., A novel approach to the derivation of fuzzy membership functions using the Falcon-MART architecture, *Pattern Recognition Letters*, 22(9), 941–958, July 2001.

79. Tan, G.K., Feasibility of predicting congestion states with neural networks, Final Year Project, Nanyang Technological University, CSE, 1997.

80. ELENA website: http://www.dice.ucl.ac.be/neural-nets/Research/Projects/ELENA/elena.htm.

81. Quek, H.C. and Zhou, R.W., POPFNN-TVR: A pseudo outer-product-based fuzzy neural network, *Neural Networks*, 9(9), 1569–1581, 1996.

82. Quek, H.C. and Zhou, R.W., POPFNN-AARS(S): A pseudo outer-product-based fuzzy neural network, *IEEE Trans on Syst., Man, Cybern.*, 29(6), 859–870, 1999.

11

Maximum Margin Fuzzy Classifiers with Ellipsoidal Regions

Shigeo Abe
Kobe University

Abstract. In a fuzzy classifier with ellipsoidal regions, a fuzzy rule is defined by calculating the center and the covariance matrix using the training data included in the associated cluster. In classification, the membership function for each fuzzy rule is calculated using the inverse of the covariance matrix. Thus, when the number of training data is small, the covariance matrix becomes singular and the generalization ability is degraded. In this chapter, we improve the generalization ability using the concept of support vector machines. Namely, when the covariance matrix is found to be singular by the symmetric Cholesky factorization, we replace the diagonal element of the associated lower triangular matrix with a small positive value. Further, we tune the slopes of the membership functions so that the margins are maximized. We demonstrate the validity of our method by computer simulations.

Key words: Fuzzy classifiers, maximum margins, pattern classification, support vector machines, symmetric Cholesky factorization.

11.1 Introduction

In training neural networks or fuzzy systems, the sum of square errors for the training data is usually minimized. Thus, the generalization ability depends on the training method and the initial conditions. Support vector machines, on the other hand, minimize the generalization error as well as the training data error. The difference in the training methods affects the generalization ability, especially when classes do not overlap. Namely, according to the simulations for some benchmark data sets, the support vector machines outperform other classifiers when the training data are scarce and classes do not overlap.[1]

The generalization ability of conventional classifiers can be improved by introducing the support vector machine concept. In this chapter we discuss one such approach: we improve generalization ability of a fuzzy classifier with ellipsoidal regions,[1-4] by maximizing margins between class regions.

In the fuzzy classifier with ellipsoidal regions, we define a fuzzy rule calculating the center and the covariance matrix using the training data; and in classification, we need to calculate the inverse of the covariance matrix. When the covariance matrix is singular, usually we calculate the pseudo-inverse[5,6] and control singular values.[1] But the improvement is still not sufficient.[1] Thus, to realize high generalization ability, we use the symmetric Cholesky factorization in calculating the inverse of the covariance matrix, and when the covariance matrix is singular we replace the diagonal element of the associated lower triangular matrix with a small positive value. Furthermore, at the initial stage of tuning slopes of the membership functions, we maximize the slope margins to improve the generalization ability.

In the following, in Sections 11.2 and 11.3 we briefly explain the support vector machine and fuzzy classifier with ellipsoidal regions, respectively, and in Section 11.4 we discuss the generalization enhancement by the symmetric Cholesky factorization. Then in Section 11.5 we discuss how to maximize margins by tuning membership functions. Finally, in Section 11.6, the validity of the proposed method is evaluated for four data sets.

11.2 Support Vector Machines

11.2.1 Optimal Hyperplane

In the following we discuss support vector machines for a two-class problem. Then we discuss extensions to n-class problems.

Let m-dimensional inputs x_i $(i = 1, \ldots, M)$ belong to Class 1 or 2 and the associated labels be $y_i = 1$ for Class 1 and -1 for Class 2. If these data are linearly separable, we can determine the decision function:

$$D(\mathbf{x}) = \mathbf{w}^t \mathbf{x} + b, \tag{II.11.1}$$

where $\mathbf{w}$ is an m-dimensional vector, b is a scalar, and for $i = 1, \ldots, M$:

$$\mathbf{w}^t \mathbf{x}_i + b \begin{cases} \geq +1 & \text{for} \quad y_i = +1, \\ \leq -1 & \text{for} \quad y_i = -1. \end{cases} \tag{II.11.2}$$

Equation II.11.2 is equivalent to:

$$y_i \, (\mathbf{w}^t \mathbf{x}_i + b) \geq 1 \quad \text{for} \quad i = 1, \ldots, M. \tag{II.11.3}$$

The hyperplane:

$$D(\mathbf{x}) = \mathbf{w}^t \mathbf{x} + b = c \quad \text{for} \quad -1 < c < 1 \tag{II.11.4}$$

forms a separating hyperplane that separates x_i $(i = 1, \ldots, M)$. When $c = 0$, the separating hyperplane is in the middle of the two separating hyperplanes with $c = 1$ and -1. The distance between the separating hyperplane and the training datum nearest to the hyperplane is called the margin. Assuming that the hyperplanes $D(\mathbf{x}) = 1$ and -1 includes at least one training datum, the hyperplane $D(\mathbf{x}) = 0$ has the maximum margin for $-1 < c < 1$. The region $\{\mathbf{x} \mid -1 \leq D(\mathbf{x}) \leq 1\}$ is

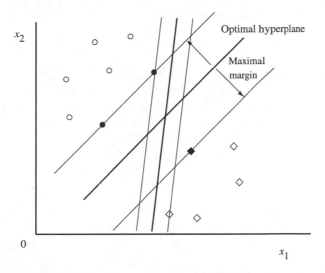

FIGURE II.11.1   Optimal separating hyperplane in a two-dimensional space.

the generalization region for the decision function. The hyperplane with the largest margin is called the optimal separating hyperplane (Figure II.11.1).

Figure II.11.1 shows two decision functions that satisfy Equation II.11.3. Thus, there are an infinite number of decision functions that satisfy Equation II.11.3. The generalization ability depends on the location of the separating hyperplane.

Here, we must bear in mind that the optimal separating hyperplane realizes the highest generalization ability from the standpoint of the VC dimension.[1] The only assumption made is that the training and test data are generated by a single unknown distribution. Thus, if the outliers are included in the training data or the training data are biased from the unknown distribution, the optimal separating hyperplane may not realize the highest generalization ability.

Now consider determining the optimal separating hyperplane. The Euclidean distance from a training datum $\mathbf{x}$ to the separating hyperplane is given by $|D(\mathbf{x})|/\|\mathbf{w}\|$. Thus, assuming the margin δ, all the training data must satisfy:

$$\frac{y_k D(\mathbf{x}_k)}{\|\mathbf{w}\|} \geq \delta \quad \text{for} \quad k = 1, \ldots, M. \tag{II.11.5}$$

Now if $\mathbf{w}$ is a solution, $a\mathbf{w}$ is also a solution where a is a scalar. Thus, we impose the following constraint:

$$\delta \, \|\mathbf{w}\| = 1. \tag{II.11.6}$$

From Equations II.11.5 and II.11.6, to find the optimal separating hyperplane, we need to find $\mathbf{w}$ with the minimum Euclidean norm that satisfies Equation II.11.3.

The data that satisfy the equality in Equation II.11.3 are called support vectors. In Figure II.11.1, the data corresponding to the filled circles and the filled rectangle are support vectors. These data are nearest to the separating hyperplanes and thus are difficult to be classified.

Now the optimal separating hyperplane can be obtained by minimizing:

$$\frac{1}{2} \, \|\mathbf{w}\|^2 \tag{II.11.7}$$

with respect to $\mathbf{w}$ and b subject to the constraints:

$$y_i \left(\mathbf{w}^t \, \mathbf{x}_i + b \right) \geq 1 \qquad \text{for} \qquad i = 1, \dots, M. \tag{II.11.8}$$

The number of variables for the convex optimization problem given by Equations II.11.7 and II.11.8 is the number of features plus 1: $m + 1$. When the number of features is small, we can solve Equations II.11.7 and II.11.8 by the quadratic programming technique. When the number of features is large, we can convert Equations II.11.7 and II.11.8 into the equivalent dual problem whose number of variables is the number of training data.

First we convert the constrained problem given by Equations II.11.7 and II.11.8 into the unconstrained problem:

$$Q(\mathbf{w}, b, \alpha) = \frac{1}{2} \mathbf{w}^t \, \mathbf{w} - \sum_{i=1}^{M} \alpha_i \left\{ y_i \left(\mathbf{w}^t \, \mathbf{x}_i + b \right) - 1 \right\}, \tag{II.11.9}$$

where $\alpha = (\alpha_1, \dots, \alpha_M)^t$ is the Lagrange multiplier. The optimal solution of Equation II.11.9 is given by the saddle point where Equation II.11.9 is minimized with respect to $\mathbf{w}$ and b and it is maximized with respect to α_i (≥ 0).

The optimal solution $\mathbf{w}^*$, b^*, and α^* of Equation II.11.9 must satisfy:

$$\frac{\partial Q(\mathbf{w}^*, b^*, \alpha^*)}{\partial b} = 0, \tag{II.11.10}$$

$$\frac{\partial Q(\mathbf{w}^*, b^*, \alpha^*)}{\partial \mathbf{w}} = 0. \tag{II.11.11}$$

Using Equations II.11.9, II.11.10, and II.11.11 reduce, respectively, to:

$$\sum_{i=1}^{M} \alpha_i^* \, y_i = 0, \qquad \alpha_i^* \geq 0 \quad \text{for} \quad i = 1, \dots, M, \tag{II.11.12}$$

$$\mathbf{w}^* = \sum_{i=1}^{M} \alpha_i^* \, y_i \, \mathbf{x}_i, \qquad \alpha_i^* \geq 0 \quad \text{for} \quad i = 1, \dots, M. \tag{II.11.13}$$

According to the Kuhn-Tucker theorem, in Equation II.11.3 the equality holds for the training input-output pair $(\mathbf{x}_i, y_i)$ only if the associated α_i^* is not 0. In this case, the training data $\mathbf{x}_i$ are the support vectors.

Substituting Equations II.11.12 and II.11.13 into Equation II.11.9, we obtain the following dual problem. Namely, maximize:

$$Q(\alpha) = \sum_{i=1}^{M} \alpha_i - \frac{1}{2} \sum_{i, j=1}^{M} \alpha_i \, \alpha_j \, y_i \, y_j \, \mathbf{x}_i^t \, \mathbf{x}_j \tag{II.11.14}$$

with respect to α_i subject to the constraints:

$$\sum_{i=1}^{M} y_i \, \alpha_i = 0, \quad \alpha_i \geq 0 \quad \text{for} \quad i = 1, \dots, M. \tag{II.11.15}$$

Solving Equations II.11.14 and II.11.15 for α_i ($i = 1, \ldots, M$), we can obtain the support vectors for Classes 1 and 2. Then the optimal hyperplane is placed at the equal distances from the support vectors for Classes 1 and 2, and b^* is given by:

$$b^* = -\frac{1}{2}(\mathbf{w}^t \, \mathbf{s}_1 + \mathbf{w}^t \, \mathbf{s}_2), \tag{II.11.16}$$

where $\mathbf{s}_1$ and $\mathbf{s}_2$ are, respectively, arbitrary support vectors for Classes 1 and 2. From Equation II.11.13, Equation II.11.16 is rewritten as follows:

$$b^* = -\frac{1}{2} \sum_{k=1}^{M} y_k \, \alpha_k^* \, (\mathbf{s}_1^t \, \mathbf{x}_k + \mathbf{s}_2^t \, \mathbf{x}_k). \tag{II.11.17}$$

In the above discussion, we assumed that the training data are linearly separable. In the following we consider determining optimal hyperplane when the training data are not linearly separable. In this case, we want to determine the optimal hyperplane with both the maximum margin and the minimum classification error.

To allow the data that do not have the maximum margin to exist, we introduce the nonnegative slack variables ξ_i (> 0) into Equation II.11.3:

$$y_i \, (\mathbf{w}^t \, \mathbf{x}_i + b) \geq 1 - \xi_i \qquad \text{for} \quad i = 1, \ldots, M. \tag{II.11.18}$$

For the training data $\mathbf{x}_i$, if $0 < \xi_i < 1$, the data do not have the maximum margin but are still correctly classified. But if $\xi_i \geq 1$, the data are misclassified by the optimal hyperplane. To obtain the optimal hyperplane in which the number of training data that do not have the maximum margin is minimum, we need to minimize:

$$Q(\mathbf{w}) = \sum_{i=1}^{n} \theta(\xi_i),$$

where

$$\theta(\xi_i) = \begin{cases} 1 & \text{for } \xi_i > 0, \\ 0 & \text{for } \xi_i = 0. \end{cases}$$

But this is a combinatorial optimization problem and is difficult to solve. Instead, we consider minimizing:

$$\frac{1}{2} \, \|\mathbf{w}\|^2 + C \sum_{i=1}^{M} \xi_i \tag{II.11.19}$$

subject to the constraints:

$$y_i \, (\mathbf{w}^t \, \mathbf{x}_i + b) \geq 1 - \xi_i \qquad \text{for} \qquad i = 1, \ldots, M, \tag{II.11.20}$$

where C is the upper bound that determines the trade-off between the maximization of margin and minimization of classification error, and is set to a large value. We call the obtained hyperplane soft margin hyperplane.

Similar to the linearly separable case, introducing the Lagrange multipliers α and β, we obtain:

$$Q(\mathbf{w}, b, \xi, \alpha, \beta) = \frac{1}{2} \, \|\mathbf{w}\|^2 + C \sum_{i=1}^{M} \xi_i$$

$$- \sum_{i=1}^{M} \alpha_i \, (y_i \, (\mathbf{w}^t \, \mathbf{x} + b) - 1 + \xi_i) - \sum_{i=1}^{M} \beta_i \, \xi_i. \tag{II.11.21}$$

The conditions of optimality are given by:

$$\frac{\partial Q(\mathbf{w}^*, b^*, \xi^*, \alpha^*, \beta^*)}{\partial b} = 0, \tag{II.11.22}$$

$$\frac{\partial Q(\mathbf{w}^*, b^*, \xi^*, \alpha^*, \beta^*)}{\partial \mathbf{w}} = 0, \tag{II.11.23}$$

$$\frac{\partial Q(\mathbf{w}^*, b^*, \xi^*, \alpha^*, \beta^*)}{\partial \xi} = 0. \tag{II.11.24}$$

Using Equations II.11.21 and II.11.22–II.11.24 reduce, respectively, to:

$$\sum_{i=1}^{M} \alpha_i^* y_i = 0, \quad \alpha_i^* \geq 0 \quad \text{for} \quad i = 1, \ldots, M, \tag{II.11.25}$$

$$\mathbf{w}^* = \sum_{i=1}^{M} \alpha_i^* y_i \mathbf{x}_i, \quad \alpha_i^* \geq 0 \quad \text{for} \quad i = 1, \ldots, M, \tag{II.11.26}$$

$$\alpha_i + \beta_i = C, \quad \alpha_i^*, \beta_i^* \geq 0 \quad \text{for} \quad i = 1, \ldots, M. \tag{II.11.27}$$

Thus we obtain the following dual problem. Namely, find α_i ($i = 1, \ldots, M$) that maximize:

$$Q(\alpha) = \sum_{i=1}^{M} \alpha_i - \frac{1}{2} \sum_{i,j=1}^{M} \alpha_i \alpha_j y_i y_j \mathbf{x}_i^t \mathbf{x}_j \tag{II.11.28}$$

subject to the constraints:

$$\sum_{i=1}^{M} y_i \alpha_i = 0, \qquad 0 \leq \alpha_i \leq C, \tag{II.11.29}$$

which is similar to the linearly separable case.

According to Kuhn-Tucker's condition, the optimal solution satisfies:

$$\alpha_i \left(y_i \left(\mathbf{w}^t \mathbf{x}_i + b \right) - 1 + \xi_i \right) = 0, \tag{II.11.30}$$

$$\beta_i \xi_i = (C - \alpha_i) \xi_i = 0. \tag{II.11.31}$$

Therefore, there are three cases for α_i:

1. $\alpha_i = 0$. Then $\xi_i = 0$. Thus $\mathbf{x}_i$ is correctly classified.
2. $0 < \alpha_i < C$. Then $y_i \left(\mathbf{w}^t \mathbf{x}_i + b \right) - 1 + \xi_i = 0$ and $\xi_i = 0$. Therefore, $y_i \left(\mathbf{w}^t \mathbf{x}_i + b \right) = 1$ and $\mathbf{x}_i$ is a support vector.
3. $\alpha_i = C$. Then $y_i \left(\mathbf{w}^t \mathbf{x}_i + b \right) - 1 + \xi_i = 0$ and $\xi_i > 0$. Thus $\mathbf{x}_i$ is a support vector and if $0 < \xi_i < 1$, $\mathbf{x}_i$ is correctly classified, and if $\xi_i \geq 1$, $\mathbf{x}_i$ is misclassified.

The decision function is the same for separable and nonseparable cases and is given by:

$$D(\mathbf{x}) = \sum_{i=1}^{M} \alpha_i^* y_i \mathbf{x}_i^t \mathbf{x} + b^*. \tag{II.11.32}$$

Since α_i are nonzero for the support vectors, the summation in Equation II.11.32 is added only for the support vectors.

Then unknown datum $\mathbf{x}$ is classified as follows:

$$\mathbf{x} \in \begin{cases} \text{Class 1} & \text{if } D(\mathbf{x}) > 0, \\ \text{Class 2} & \text{otherwise.} \end{cases} \tag{II.11.33}$$

Thus, when training data are separable, the region $\{\mathbf{x} \mid 1 > D(\mathbf{x}) > -1\}$ is a generalization region.

11.2.2   Mapping to a High-Dimensional Space

In support vector machines for a two-class problem, the optimal hyperplane is determined to maximize the generalization ability. Therefore, if the original input $\mathbf{x}$ are not sufficient to guarantee linear separability of the training data, the obtained classifier may not have high generalization ability, although the hyperplanes are determined optimally. Thus, to enhance linear separability, in the support vector machines, the original input space is mapped into a high-dimensional dot product space called feature space.

Now using the nonlinear vector function $\mathbf{g}(\mathbf{x}) = (g_1(\mathbf{x}), \ldots, g_l(\mathbf{x}))^t$ that map the m-dimensional input vector $\mathbf{x}$ into the l-dimensional feature space, the linear decision function in the feature space is given by:

$$D(\mathbf{x}) = \mathbf{w}^t \mathbf{g}(\mathbf{x}) + b, \tag{II.11.34}$$

where $\mathbf{w}$ is the l-dimensional vector and b is a constant.

According to the Hilbert-Schmidt theory, the dot product in the feature space can be expressed by a symmetric kernel function $H(\mathbf{x}, \mathbf{x}')$:

$$H(\mathbf{x}, \mathbf{x}') = \mathbf{g}(\mathbf{x})^t \mathbf{g}(\mathbf{x}'), \tag{II.11.35}$$

if:

$$\iint H(\mathbf{x}, \mathbf{x}')\, h(\mathbf{x})\, h(\mathbf{x}')\, d\mathbf{x}\, d\mathbf{x}' \geq 0 \tag{II.11.36}$$

is satisfied for all the square integrable functions $h(\mathbf{x})$ in the compact subset of the input space ($\int h^2(\mathbf{x})\, d\mathbf{x} < \infty$). This condition is called Mercer's condition.

Using the kernel function, the dual problem in the feature space is given as follows. Find α_i ($i = 1, \ldots, M$) that maximize:

$$Q(\alpha) = \sum_{i=1}^{M} \alpha_i - \frac{1}{2} \sum_{i,j=1}^{M} \alpha_i\, \alpha_j\, y_i\, y_j\, H(\mathbf{x}_i, \mathbf{x}_j) \tag{II.11.37}$$

subject to the constraints:

$$\sum_{i=1}^{M} y_i\, \alpha_i = 0, \qquad 0 \leq \alpha_i \leq C. \tag{II.11.38}$$

The decision function is given by:

$$D(\mathbf{x}) = \sum_{i=1}^{M} \alpha_i^*\, y_i\, H(\mathbf{x}_i, \mathbf{x}) + b^*. \tag{II.11.39}$$

Then unknown data are classified using the kernel function as follows.

$$\mathbf{x} \in \begin{cases} \text{Class 1} & \text{if } f(\mathbf{x}) = +1, \\ \text{Class 2} & \text{if } f(\mathbf{x}) = -1, \end{cases} \tag{II.11.40}$$

where:

$$f(\mathbf{x}) = \text{sign}\left(\sum_{\text{support vectors}} y_i \, \alpha_i^* \, H(\mathbf{x}, \mathbf{x}_i)\right). \tag{II.11.41}$$

In the following we discuss some of the kernel functions that are used in support vector machines.

11.2.2.1   Polynomial Kernels

The polynomial kernel with the degree of d is given by:

$$H(\mathbf{x}, \mathbf{x}') = (\mathbf{x}^t \, \mathbf{x}' + 1)^d. \tag{II.11.42}$$

As a special case, we call $H(\mathbf{x}, \mathbf{x}') = \mathbf{x}^t \, \mathbf{x}'$, the dot product kernel.

The class boundaries of the support vector machine with $d = 2$ are similar to those given by the fuzzy classifier with ellipsoidal regions.

11.2.2.2   Radial Basis Functions

The radial basis function (RBF) kernel is given by:

$$H(\mathbf{x}, \mathbf{x}') = \exp\left(-\gamma \, \|\mathbf{x} - \mathbf{x}'\|^2\right), \tag{II.11.43}$$

where γ is a positive parameter for slope control.

From Equation II.11.39, the resulting decision function is given by:

$$D(\mathbf{x}) = \sum_{i=1}^{M} \alpha_i^* \, y_i \, \exp\left(-\gamma \|\mathbf{x}_i - \mathbf{x}\|\right) + b^*. \tag{II.11.44}$$

Since α_i^* are nonzero only when $\mathbf{x}_i$ are support vectors, support vectors are the centers of the radial basis functions.

11.2.2.3   Three-Layer Neural Networks

The three-layer neural network kernel is given by:

$$H(\mathbf{x}, \mathbf{x}') = \frac{1}{1 + \exp(\nu \, \mathbf{x}^t \, \mathbf{x}' - a)}, \tag{II.11.45}$$

where ν and a are constants. The values of ν and a need to be determined so that Equation II.11.36 is satisfied.

From Equation II.11.39, the resulting decision function is given by:

$$D(\mathbf{x}) = \sum_{i=1}^{M} \frac{\alpha_i^* \, y_i}{1 + \exp(\nu \|\mathbf{x}_i - \mathbf{x}\| - a)} + b^*. \tag{II.11.46}$$

Since α_i^* are nonzero only when $\mathbf{x}_i$ are support vectors, the weights between the input and hidden neurons correspond to the support vectors. Since Mercer's condition is not always satisfied for three-layer neural networks, several approaches are made to overcome this problem.[7]

11.2.3 Support Vector Machines for Multi-Class Problems

In the conventional support vector machines, an n-class problem is converted into n two-class problems and for the ith two-class problem, class i is separated from the remaining classes. But by this formulation, unclassifiable regions exist. To solve this problem, Kreßel[8] converts the n-class problem into $n(n-1)/2$ two-class problems which covers all pairs of classes. Bennett[9] uses piecewise-linear separating hyperplane instead of the separating hyperplane. To resolve unclassifiable regions, Inoue and Abe[10] propose fuzzy support vector machines for one-to-$(n-1)$ formulation.

11.3 A Fuzzy Classifier with Ellipsoidal Regions

In this section, we overview the fuzzy classifier with ellipsoidal regions based on Abe,[1] Abe and Thawonmas,[3] and Abe.[4]

Consider classifying an m-dimensional input vector $\mathbf{x}$ into one of n classes using fuzzy rules with ellipsoidal regions. To simplify discussions, we assume that each class consists of one cluster and define a fuzzy rule for each class. We define the following fuzzy rule for class i:

$$R_i : \text{If } \mathbf{x} \text{ is } \mathbf{c}_i \text{ then } \mathbf{x} \text{ is class } i \tag{II.11.47}$$

where $\mathbf{c}_i$ is the center of class i and is calculated by the training data included in class i:

$$\mathbf{c}_i = \frac{1}{|X_i|} \sum_{\mathbf{x} \in X_i} \mathbf{x}, \tag{II.11.48}$$

where X_i is the set of training data included in class i, and $|X_i|$ is the number of data included in X_i.

For the center $\mathbf{c}_i$, we define the membership function $m_i(\mathbf{x})$ which defines the degree to which $\mathbf{x}$ belongs to $\mathbf{c}_i$:

$$m_i(\mathbf{x}) = \exp(-h_i^2(\mathbf{x})), \tag{II.11.49}$$

$$h_i^2(\mathbf{x}) = \frac{d_i^2(\mathbf{x})}{\alpha_i}, \tag{II.11.50}$$

$$d_i^2(\mathbf{x}) = (\mathbf{x} - \mathbf{c}_i)^t Q_i^{-1} (\mathbf{x} - \mathbf{c}_i), \tag{II.11.51}$$

where $h_i(\mathbf{x})$ is a tuned distance, $d_i(\mathbf{x})$ is a weighted distance between $\mathbf{x}$ and $\mathbf{c}_i$, α_i is a tuning parameter for class i, Q_i is an $m \times m$ covariance matrix for class i. And Q_i^{-1} denotes the inverse of the covariance matrix Q_i. Here we calculate the covariance matrix Q_i using the data belonging to class i as follows:

$$Q_i = \frac{1}{|X_i|} \sum_{\mathbf{x} \in X_i} (\mathbf{x} - \mathbf{c}_i)(\mathbf{x} - \mathbf{c}_i)^t. \tag{II.11.52}$$

For an input vector $\mathbf{x}$ we calculate the degrees of membership for all the classes. If $m_k(\mathbf{x})$ is the maximum, we classify the input vector into class k.

Recognition performance is improved by tuning fuzzy rules, i.e., by tuning α_i one at a time. When α_i is increased, the slope of $m_i(\mathbf{x})$ is decreased and the degree of membership is increased. Then, misclassified data may be correctly classified and correctly classified data may be misclassified. Based on this, we calculate the net increase of the correctly classified data. Likewise, by increasing the slope, we calculate the net increase of the correctly classified data. Then, allowing new misclassification, we tune the slope so that the recognition rate is maximized. In this way we tune fuzzy rules successively until the recognition rate of the training data is not improved.[1]

11.4   Generalization Improvement by the Symmetric Cholesky Factorization

The covariance matrix Q_i is guaranteed to be positive semi-definite, and to be positive definite, the number of the training data belonging to class i needs to be at least larger than the number of input variables.[12] Namely:

$$|X_i| \geq m + 1. \tag{II.11.53}$$

When the covariance matrix Q_i is singular, usually Q_i is decomposed into singular values.[5] But this slows down training. Therefore, to speed up training even when Q_i is singular, we use the symmetric Cholesky factorization.

When Q_i is positive definite, each diagonal element is positive and the value is the maximum among the column elements. Namely:

$$q_{ii} > |q_{ij}| \quad \text{for} \quad i,\, j = 1, 2, \ldots, m, \quad i \neq j. \tag{II.11.54}$$

Thus Q_i can be decomposed into the two triangular matrices by the symmetric Cholesky factorization without pivot exchanges as follows:[5]

$$Q_i = L_i L_i^t, \tag{II.11.55}$$

where L_i is the real-valued regular lower triangular matrix and each element of L_i is given by:

$$l_{op} = \frac{q_{op} - \sum_{n=1}^{p-1} l_{pn} l_{on}}{l_{pp}} \quad \text{for} \quad o = 1, \ldots, m, \quad p = 1, \ldots, o - 1, \tag{II.11.56}$$

$$l_{aa} = \sqrt{q_{aa} - \sum_{n=1}^{a-1} l_{an}^2} \quad \text{for} \quad a = 1, 2, \ldots, m. \tag{II.11.57}$$

Using L_i, Equation II.11.51 is written as:

$$d_i^2(\mathbf{x}) = (L_i^{-1}(\mathbf{x} - \mathbf{c}_i))^t L_i^{-1}(\mathbf{x} - \mathbf{c}_i). \tag{II.11.58}$$

Now we define the vector $\mathbf{y}_i (= (y_{i1}, \ldots, y_{im})^t)$ by:

$$\mathbf{y}_i = L_i^{-1}(\mathbf{x} - \mathbf{c}_i). \tag{II.11.59}$$

Then solving the following equation for $\mathbf{y}_i$:

$$L_i \mathbf{y}_i = \mathbf{x} - \mathbf{c}_i, \tag{II.11.60}$$

we calculate $d_i^2(\mathbf{x})$ without calculating the inverse of Q_i:

$$d_i^2(\mathbf{x}) = \mathbf{y}_i^t \mathbf{y}_i. \tag{II.11.61}$$

When the number of training data is small, the value in the square root of Equation II.11.57 is nonpositive. To avoid this, if:

$$q_{aa} - \sum_{n=1}^{a-1} l_{an}^2 \leq \eta, \tag{II.11.62}$$

where $\eta \, (> 0)$, we set:

$$l_{aa} = \sqrt{\eta}. \tag{II.11.63}$$

This increases the variances of the variables.

11.5   Maximizing Margins of α_i

11.5.1   Concept

In fuzzy classifier with ellipsoidal regions, if there are overlaps between classes, the overlaps are resolved by tuning the membership functions. But if there is no overlap, the membership functions are not tuned. When the number of training data is small, usually the overlaps are scarce. Thus, the generalization ability is degraded. To tune membership functions even when the overlaps are scarce, we use the idea used in training support vector machines. As discussed in Section 11.2, in training support vector machines for a two-class problem, the separating margin between the two classes is maximized to improve the generalization ability. In the fuzzy classifier with ellipsoidal regions, we tune the slopes of the membership functions so that the slope margins are maximized.

Initially, we set the values of α_i to be 1. Namely, the fuzzy classifier with ellipsoidal regions is equivalent to the classifier based on Mahalanobis distance if each class consists of one cluster. Then, we tune α_i so that the slope margins are maximized. Here, we maximize the margins without causing new misclassification. When the recognition rate of the training data is not 100% after tuning, we tune α_i so that the recognition rate is maximized as discussed in Abe.[1] Here we discuss how to maximize slope margins.

Unlike the support vector machines, tuning of α_i is not restricted to two classes, but for ease of illustration we explain the concept of tuning using two classes. In Figure II.11.2 the filled rectangle and circle show the training data, belonging to classes i and j, that are nearest to classes j and i, respectively. The class boundary of the two classes is somewhere between the two curves shown in the figure. We assume that the generalization ability is maximized when it is in the middle of the two.

In Figure II.11.3(a), if the datum belongs to class i, it is correctly classified since the degree of membership for class i is larger. This datum remains correctly classified until the degree of membership for class i is decreased as shown in the dotted curve. Similarly, in Figure II.11.3(b), if the datum belongs to class j, it is correctly classified since the degree of membership for class j is larger. This datum remains correctly classified until the degree of membership for class i is increased as shown in the dotted curve. Thus, for each α_i, there is an interval of α_i that makes

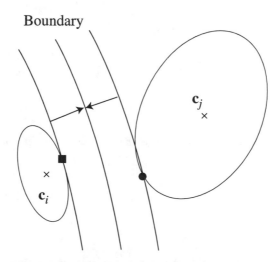

FIGURE II.11.2   Concept of maximizing margins.

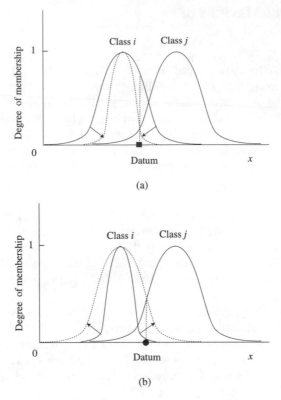

FIGURE II.11.3   Range of α_i that does not cause misclassification; (a) upper bound of α_i (b) lower bound of α_i.

correctly classified data remain correctly classified. Therefore, if we change the value of α_i so that it is in the middle of the interval, the slope margins are maximized.

In the following we discuss how to tune α_i.

11.5.2   Upper and Lower Bounds of α_i

Let X be the set of training data that are correctly classified for the initial α_i. Let $\mathbf{x}\ (\in X)$ belong to class i. If $m_i(\mathbf{x})$ is the largest there is a lower bound, $L_i(\mathbf{x})$, of α_i to keep $\mathbf{x}$ correctly classified:

$$L_i(\mathbf{x}) = \frac{d_i^2(\mathbf{x})}{\min_{j \neq i} h_j^2(\mathbf{x})}.$$ (II.11.64)

Then the lower bound $L_i(1)$ that does not cause new misclassification is given by:

$$L_i(1) = \max_{\mathbf{x} \in X} L_i(\mathbf{x}).$$ (II.11.65)

Similarly, for $\mathbf{x}\ (\in X)$ belonging to a class other than class i, we can calculate the upper bound of α_i. Let $m_j(\mathbf{x})\ (j \neq i)$ be the largest. Then the upper bound $U_i(\mathbf{x})$ of α_i that does not cause misclassification of $\mathbf{x}$ is given by:

$$U_i(\mathbf{x}) = \frac{d_i^2(\mathbf{x})}{\min_{j \neq i} h_j^2(\mathbf{x})}.$$ (II.11.66)

The upper bound $U_i(1)$ that does not make new misclassification is given by:

$$U_i(1) = \min_{\mathbf{x} \in X} U_i(\mathbf{x}). \tag{II.11.67}$$

In Abe,[1] $L_i(l)$ and $U_i(l)$ are defined as the lower and the upper bound in which $l - 1$ correctly classified data are misclassified, respectively. Thus, $L_i(1)$ and $U_i(1)$ are the special cases of $L_i(l)$ and $U_i(l)$.

11.5.3 Tuning Procedure

The correctly classified data remain correctly classified even when α_i is set to some value in the interval $(L_i(1), U_i(1))$. The tuning procedure of α_i becomes as follows. For α_i we calculate $L_i(1)$ and $U_i(1)$ and set the value of α_i with the middle point of $L_i(1)$ and $U_i(1)$:

$$\alpha_i = \frac{1}{2}(L_i(1) + U_i(1)). \tag{II.11.68}$$

We successively tune one α_i after another. Tuning results depend on the order of tuning α_i, but in the following simulation study, we tune from class 1 to class n.

11.6   Performance Evaluation

Using numeral data and hiragana data, we evaluate the recognition improvement for the test data by controlling the values of the diagonal elements in the symmetric Cholesky factorization and maximizing the slope margins. Numeral data and hiragana data were collected to classify numerals and hiragana characters on Japanese vehicle license plates.[1,13] Table II.11.1 shows the numbers of inputs, classes, training data, and test data.

Hiragana data gathered from the vehicle license plates were transformed into $5 \times 10 \,(= 50)$ and $7 \times 15 \,(= 105)$ gray-scale grid data. The hiragana-13 data were generated by calculating the 13 two-dimensional central moments of the hiragana-105 data.

For the fuzzy classifier with ellipsoidal regions, we set the parameter $\delta = 0.1,$[3] and the maximum number of allowable misclassifications l_M to be 10,[3] and we assumed that each class consisted of one cluster. We used a Sun UltraSPARC-IIi workstation (335MHz). To compare the results with other methods, we used the evaluation results using the singular value decomposition method and the support vector machine.[1] For evaluation by the support vector machine, we used the software developed by Royal Holloway, University of London.[11] We assumed that all the benchmark data sets were not linearly separable, i.e., we set the upper bound $C = 5000$ and we used the polynomial and RBF kernels.

Table II.11.2 shows the recognition rates of the test (training) data of the numeral data when the covariance matrices were diagonalized and when the η in Equation II.11.62 was changed. The column

TABLE II.11.1   Feature of Benchmark Data

Data	Inputs	Classes	Train.	Test
Numeral	12	10	810	820
Hiragana-50	50	39	4610	4610
Hiragana-105	105	38	8375	8356
Hiragana-13	13	38	8375	8356

"Initial" shows the recognition rates for $\alpha_i = 1$ and the column "Final" shows the recognition rates when α_i were tuned to resolve misclassification. The final recognition rate of the test data increased as η was increased and when $\eta = 10^{-2}$, the recognition rate was the highest.

Table II.11.3 shows the results when η was changed and α_i were tuned to maximize the margins. The column "Initial" shows the recognition rates when α_i were tuned to maximize the margins, but were not tuned to resolve misclassification. Comparing Tables II.11.2 and II.11.3, the recognition rates were almost the same for each value of η.

Table II.11.4 shows the best performance of the numeral data using the proposed methods and the support vector machine. In the table, CF means Cholesky factorization and CF + MM means that maximizing the slope margins was combined with the Cholesky factorization. There was not so much difference between the three. The best recognition rates by the singular value decomposition were the same with those listed in the first row of Table II.11.2.

Table II.11.5 shows the recognition rates of the hiragana-50 test (training) data when the covariance matrices were diagonalized and when the η in Equation II.11.62 was changed. The final recognition rate of the test data increased as η was increased and when $\eta = 10^{-2}$, the recognition rate was the highest.

Table II.11.6 shows the results when η was changed and α_i were tuned to maximize the margins. The column "Initial" shows the recognition rates when α_i were tuned to maximize the margins but were not tuned to resolve misclassification. Comparing Tables II.11.5 and II.11.6, the maximum

TABLE II.11.2   Recognition Rates for Numeral Data

η	Initial (%)	Final (%)	Time (s)
10^{-5}	99.63 (99.75)	99.39 (99.88)	0.5
10^{-4}	99.51 (99.75)	99.51 (99.88)	0.5
10^{-3}	99.51 (99.75)	99.51 (99.88)	0.5
10^{-2}	98.90 (99.63)	99.76 (100)	0.5
10^{-1}	98.29 (98.77)	99.02 (99.51)	0.6

TABLE II.11.3   Recognition Rates for Numeral Data by Maximizing Margins

η	Initial (%)	Final (%)	Time (s)
10^{-5}	99.15 (99.75)	99.27 (99.88)	1.2
10^{-4}	99.27 (99.75)	99.51 (99.88)	1.2
10^{-3}	99.76 (99.88)	99.76 (99.88)	1.2
10^{-2}	99.39 (99.63)	99.88 (100)	1.3
10^{-1}	98.29 (98.89)	98.66 (99.38)	1.3

TABLE II.11.4   Best Performance for Numeral Data

Method	Rates (%)	Time (s)
CF	99.76 (100)	0.5
CF + MM	99.88 (100)	1.2
SVM	99.76 (100)	27

TABLE II.11.5   Recognition Rates for Hiragana-50 Data

η	Initial (%)	Final (%)	Time (s)
10^{-5}	77.51 (100)	—	35
10^{-4}	81.39 (100)	—	35
10^{-3}	95.53 (99.98)	95.64 (100)	43
10^{-2}	98.79 (99.65)	98.85 (100)	101
10^{-1}	94.43 (96.14)	94.69 (97.57)	310

TABLE II.11.6   Recognition Rates for Hiragana-50 Data by Maximizing Margins

η	Initial (%)	Final (%)	Time (s)
10^{-5}	90.50 (100)	—	431
10^{-4}	92.17 (100)	—	458
10^{-3}	96.25 (100)	—	575
10^{-2}	98.83 (99.87)	98.85 (100)	488
10^{-1}	94.64 (96.59)	94.84 (97.51)	996

TABLE II.11.7   Best Performance for Hiragana-50 Data

Method	Rates (%)	Time (s)
CF	98.85 (100)	101
CF + MM	98.85 (100)	488
SVD	94.51 (99.89)	946
SVM	99.07 (100)	7144

TABLE II.11.8   Recognition Rates for Hiragana-105 Data

η	Initial (%)	Final (%)	Time (s)
10^{-5}	93.59 (100)	—	304
10^{-4}	97.93 (100)	—	305
10^{-3}	100 (100)	—	305
10^{-2}	99.99 (99.99)	100 (100)	380

recognition rates of the test data were the same, but for $\eta = 10^{-5}$ and 10^{-4}, the recognition rates by maximizing the slope margins were much better.

Table II.11.7 shows the best performance of the hiragana-50 data using the proposed methods, the singular value decomposition (SVD), and the support vector machine. Except for the SVD, there was not so much difference of recognition rates among the remaining three.

Table II.11.8 shows the recognition rates of the hiragana-105 test (training) data when the covariance matrices were diagonalized and when the η in Equation II.11.62 was changed. The final recognition rate of the test data increased as η was increased and when $\eta = 10^{-3}$. The recognition rates of the test and training data were 100%.

Table II.11.9 shows the results when η was changed and α_i were tuned to maximize the margins. Comparing Tables II.11.8 and II.11.9, the maximum recognition rates of the test data were the

TABLE II.11.9 Recognition Rates for Hiragana-105 Data by Maximizing Margins

η	Initial (%)	Final (%)	Time (s)
10^{-5}	99.23 (100)	—	7459
10^{-4}	99.90 (100)	—	4284
10^{-3}	100 (100)	—	3284
10^{-2}	99.99 (99.99)	100 (100)	3312

TABLE II.11.10 Best Performance for Hiragana-105 Data

Method	Rates (%)	Time (s)
CF	100 (100)	305
CF+MM	100 (100)	3284
SVD	99.99 (100)	2467
SVM	100 (100)	10443

TABLE II.11.11 Recognition Rates for Hiragana-13 Data

η	Initial (%)	Final (%)	Time (s)
10^{-6}	98.36 (99.84)	98.79 (99.99)	26
10^{-5}	98.59 (99.86)	99.02 (99.99)	26
10^{-4}	99.41 (99.86)	99.59 (99.96)	31
10^{-3}	99.49 (99.62)	99.44 (99.77)	42
10^{-2}	95.22 (95.52)	96.48 (97.00)	110

TABLE II.11.12 Recognition Rates for Hiragana-13 Data by Maximizing Margins

η	Initial (%)	Final (%)	Time (s)
10^{-6}	99.34 (99.94)	99.34 (99.99)	154
10^{-5}	99.46 (99.94)	99.45 (99.99)	155
10^{-4}	99.41 (99.90)	99.43 (99.96)	156
10^{-3}	99.56 (99.67)	99.49 (99.79)	185
10^{-2}	95.64 (96.08)	96.27 (96.93)	240

same, but for $\eta = 10^{-5}$ and 10^{-4}, the recognition rates by maximizing the slope margins were much better.

Table II.11.10 shows the best performance of the hiragana-105 data using the proposed methods, the singular value decomposition, and the support vector machine. The recognition performance was almost the same for four methods.

Table II.11.11 shows the recognition rates of the hiragana-13 test (training) data when the covariance matrices were diagonalized and when the η in Equation II.11.62 was changed. The final recognition rate of the test data increased as η was increased and when $\eta = 10^{-4}$, the recognition rate was the maximum.

Table II.11.12 shows the results when η was changed and α_i were tuned to maximize the margins. Comparing Tables II.11.11 and II.11.12, the maximum recognition rate of the test data, by only the

TABLE II.11.13   Best Performance for Hiragana-13 Data

Method	Rates (%)	Time (s)
CF	99.59 (99.96)	31
CF +MM	99.49 (99.79)	185
SVD	98.79 (99.99)	36
SVM	99.77 (100)	23,216

Cholesky factorization was better, but for $\eta = 10^{-6}$ and 10^{-5}, the recognition rates by maximizing the slope margins were better.

Table II.11.13 shows the best performance of the hiragana-13 data using the proposed methods, the singular value decomposition, and the support vector machine. Singular value decomposition showed the lowest recognition rate for the test data, but the recognition performances of the remaining three methods were comparable.

11.7   Conclusions

To improve the generalization ability of the fuzzy classifier with ellipsoidal regions when the number of training data is small, we proposed to replace the diagonal elements of the covariance matrix with a small positive value when they become nonpositive during the symmetric Cholesky factorization and to maximize the slope margins.

For the numeral data and hiragana data, we showed that by controlling the diagonal elements during the Cholesky factorization, the generalization ability of the classifier was improved, and that by combining this method with maximizing slope margins, the generalization ability became more robust.

Acknowledgment

We are grateful to Mr. Y. Kobayashi of Hitachi Research Laboratory, Hitachi. Ltd. for providing the hiragana data.

References

1. Abe, S., *Pattern Classification: Neuro-Fuzzy Methods and Their Comparison*, Springer-Verlag, 2001.
2. Abe, S., *Neural Networks and Fuzzy Systems: Theory and Applications*, Kluwer Academic Publishers, Boston, 1996.
3. Abe, S. and Thawonmas, R., A fuzzy classifier with ellipsoidal regions, *IEEE Trans. Fuzzy Systems*, vol. 5, no. 3, pp. 358–368, 1998.
4. Abe, S., Dynamic cluster generation for a fuzzy classifier with ellipsoidal regions, *IEEE Trans. Systems, Man, and Cybernetics–Part B*, vol. 28, no. 6, pp. 869–876, 1998.
5. Golub, G.H. and Van Loan, C.F., *Matrix Computations*, 3 ed., John Hopkins University Press, 1996.
6. Press, W.H. et al., *Numerical Recipes in C, The Art of Scientific Computing*, 2 ed., Cambridge University Press, 1996.
7. Suykens, J.A.K. and Vandewalle, J., Training multilayer perceptron classifiers based on a modified support vector method, *IEEE Trans. Neural Networks*, vol. 10, no. 4, pp. 907–911, 1999.

8. Kreßel, U.H.-G., Pairwise classification and support vector machines, in B. Schölkopf, C.J.C. Burges, and A.J. Smola, Ed., *Advances in Kernel Methods: Support Vector Learning*, pp. 255–268. The MIT Press, Cambridge, MA, 1999.

9. Bennett, K.P., Combining support vector and mathematical programming methods for classification, in B. Schölkopf, C.J.C. Burges, and A.J. Smola, Eds., *Advances in Kernel Methods: Support Vector Learning*, pp. 307–326. The MIT Press, Cambridge, MA, 1999.

10. Inoue, T. and Abe, S., Fuzzy support vector machines for pattern classification, in *Proc. Int. J. Conf. Neural Networks (IJCNN '01)*, vol. 2, pp. 1449–1454, July 2001.

11. Saunders, C. et al., Support Vector Machine Reference Manual, Technical Report CSD-TR-98-03, Royal Holloway, University of London, London, 1998.

12. Duda, R.O. and Hart, P.E., *Pattern Classification and Scene Analysis*, John Wiley & Sons, pp. 67–68, 1973.

13. Takenaga, H., Abe, S., Takatoo, M., Kayama, M., Kitamura, T., and Okuyama, Y., Input layer optimization of neural networks by sensitivity analysis and its application to recognition of numerals, *Electrical Engineering in Japan*, vol. 111, no. 4, pp. 130–138, 1991.

12

Neural Networks: Techniques and Applications in Telecommunications Systems

Si Wu
Sheffield University

K.Y. Michael Wong
Hong Kong University of Science and Technology

12.1 Introduction

As their name suggests, neural networks represent a class of intelligent techniques which derive their inspiration from neuroscience in biology.[1] Despite the slowness of signal transmission and processing in the human brain when compared with the digital computer, the brain remains superior in many everyday tasks such as retrieving memories, recognizing objects, and controlling body motion. This superiority is attributed to the fundamental difference in the way information is processed in the two systems. In conventional approaches, information is processed serially and with mathematical precision. Explicit algorithms for computation have to be prespecified. In contrast, the human brain consists of a network of interconnecting neurons which receive (often unprecise) information fed by other neurons or by external stimuli in a parallel fashion. The neurons then transmit output signals in an apparently stochastic manner, with probabilities determined by the received input. Amazingly, this endows them with the advantages of being robust, fault tolerant, flexible, and easy to adapt to environmental changes, when compared with conventional information processors. They are able to learn from experience even when no precise mathematical models are available. Artificial neural networks are introduced as systems which try to capture these features so that they can be utilized in intelligent applications.

With their well-known complexities, telecommunications systems become a natural niche for neural network applications. Studies in telecommunications systems are often classified into layers. The bottommost layers deal with physical connections and data links, addressing such issues as how to minimize the bit error rate, or to cancel interferences in the channel. Intermediate layers deal with the network as a whole, addressing such issues as how to allocate resources according to the demand of individuals or groups of customers, while generating maximum revenue for the network at the same time. Topmost layers deal with customer applications, addressing such issues as speech and text recognition, data mining, and image processing.

In terms of the temporal nature of the studies, these issues may arise in the design stage, in which one has to consider a static optimization problem. Other issues may be found in real-time control of the network, and hence the dynamical aspects of the problem have to be considered, and the computational output has to be fast.

In terms of the scope of the computation, network design problems and some real-time control problems may involve the entire network. Network management problems, such as fault diagnosis, may involve monitoring networkwide events which are highly correlated among themselves. Other control issues may involve a group of locally optimal controllers, which nevertheless are expected to collectively attain a networkwide optimal state.

Taking these factors into consideration, *complexity* is the best single word to summarize the situation of modern telecommunications systems. They are characterized by their openness to a growing population of customers with an expanding demand on bandwidths and new applications, a highly competitive market, and ever-evolving technology. Concerning the last point on technological advances, high-speed networks decimating the current ones will emerge, heightening demands for faster controls on more volatile situations.

Research literature, mainly (but by no means exclusively) found in several workshop proceedings and special issues, confirmed the rich variety of areas on which researchers have embarked.[2-6] At the bottommost layers of telecommunications, researchers have considered neural network techniques of channel equalization of digital signals,[7-17] switching,[18-22] and adaptive flow control[23] for data transmissions. On the network level, there is much interest in using neural networks in dynamic routing of teletraffic,[24] rapid configuration of the intercity backbone network (technically, the synchronous digital hierarchy, or SDH),[25,26] and overload control for distributed call processors.[27,28]

At the topmost layer of applications, neural network techniques are employed in speech processing,[29-32] smoothing of video delay variations,[33] data mining,[34,35] image coding,[36-38] clone detection,[39,40] growth prediction,[41] and marketing support.[42,43]

Network designers and planners also found support in neural network techniques, using them in network dimensioning,[44] network clustering and topology design,[45-47] optimal broadcast scheduling and channel assignment in wireless networks,[48-51] optimal routing,[52] optimal trunk reservation,[53] and expert system design tools.[54]

In network management, neural networks are used in monitoring,[55] path failure prediction,[56] fault detection,[57,58] error message classification and diagnosis,[59-66] alarm correlation,[67] and fraud detection.[68-71]

With the recent growth in mobile wireless networks and asynchronous transfer mode (ATM) networks carrying multiple classes of traffic, new and more complicated design, control, and management issues arise, stimulating an upsurge in the development of the neural network as a tool to deal with them. In mobile wireless networks, neural networks are considered in the handover decision process,[71] location detection,[72] dynamic channel allocation,[73-75] and admission control.[68]

In ATM networks, excitement about neural networks is raised in the areas of link allocation and call admission,[76–89] flow and congestion control,[90–92] dynamic routing,[93] capacity dimensioning,[94,95] traffic trends analysis,[96] generation and modeling and prediction of ATM traffic,[97–100] and multicast routing.[101]

As the neural network and other intelligent techniques became more mature, more recent applications in telecommunications began to appear in systems integrating neural networks with conventional approaches, fuzzy logic, and genetic algorithms. Neural networks may be combined with fuzzy logic to ensure media synchronization in a Web environment.[102] In another two-level attempt to combine call admission control and bandwidth allocation problems when accessing a multiservice network, neural networks are used to decide the balance of access of real-time and non-real-time traffic from each single user, whereas a dynamic programming algorithm is used to decide the share of each user in the common resource.[103]

For more sophisticated applications, neural and fuzzy systems may come in modules which cooperate to perform complex tasks. For example, a fuzzy/neural congestion controller used for multimedia cellular networks may consist of three modules: (1) a neural predictor of microwave interference, and (2) a fuzzy performance indicator, both feeding their outputs to (3) a neural controller assigning access probability to network users.[104]

Two features of neural networks will be apparent from the examples discussed in this chapter. First, when the function used to implement the task can be generated by conventional methods, it may be too complicated to be computed in real time. In these cases, neural networks are ideal substitutes for the conventional methods, since their outputs can be computed in a small number of passes through their structures. It has been proved that arbitrary functions can be approximated by neural networks with suitable structures.[105] Hence, neural networks can play the role of approximating, interpolating, or extrapolating complex functions. They undergo a preparatory learning stage, during which examples are generated by the conventional method and used to train the neural networks. After learning, the neural networks are expected to imitate the functions in general network situations. In this chapter we will consider the example of overload control of call processors in telecommunications networks. Here, the networkwide optimal function is known, but is too complex to be computed in real time. The neural network controllers are supposed to implement local control in a distributed manner, with only partial information of the network status available to them. In this way, they can be considered as extrapolators of complex functions.

Second, when the task function cannot be obtained by conventional methods, but learning examples of the target function are available, neural networks can play the role of a rule extractor, capitalizing on their capability to learn from experience without knowing explicitly the underlying rules. The learning examples may be generated off-line, so that the learning and operating stages of the neural network can be separated.

Alternatively, the learning examples may be generated directly from the network during its real-time operations. In many cases, such as those involving the prediction of temporal information, the current states of the neural networks affect the outcomes of the examples, which in turn affect the learning process of the neural networks themselves. The entire system can then be considered as a dynamical system. This approach works when the consequences of the control actions can be assessed shortly afterward, so that training examples can be collected for supervised learning. An example is the call admission control in ATM networks based on the steady-state behavior of the calls in the network.[76–79]

Even more complicated are the cases in which no error signals are available for the neural networks to formulate their learning processes. Instead, the environment only provides punishing or awarding feedbacks to the neural networks, which are less specific and less informative. In these cases, the neural networks may achieve their target tasks through a process of reinforcement learning, in much the same way that circus animals are trained through carrot-and-stick processes. In telecommunications applications, the complexity of the system often renders the long-term consequences of the control actions difficult to be assessed and collected into sets of training examples for supervised learning. For example, the acceptance of a call into the network may prevent the access of a more valuable call in the future. Thus, reinforcement learning approaches have been used in applications with this nature, such as dynamic packet routing in networks with many nodes,[106–108] dynamic channel allocation in mobile networks,[74,75] adaptive call admission control and routing in ATM networks,[81–85] and power management in wireless channels.[109] Reinforcement learning is often also considered as dynamic programming problems, which are outside the scope of this chapter.

In this chapter, we will use as an example a problem of fault diagnosis in telecommunications management, in which learning examples are available off-line, but no conventional mathematical models are available, and strong unknown correlations exist in the various symptoms used as the inputs.

The chapter is organized as follows. In Section 12.2 we will introduce the techniques of neural networks to be used in subsequent examples. In Sections 12.3 and 12.4, we will illustrate the applications of these techniques via two examples, respectively being overload control and fault diagnosis. A data model will be introduced in Section 12.5, in an attempt to understand the choice of appropriate network algorithms for general diagnostic tasks. Concluding remarks are given in Section 12.6.

12.2   Overview of Neural Network Methods (NNM)

There are many different ways to interpret the working principles of NNM, such as from the perspectives of statistical inference[110] or information geometry.[111] Here, we adopt the perspective of function approximation. Simply speaking, NNM can be considered as a way to construct a function by associating it with a network structure. The neural networks in Figures II.12.1 and II.12.2 are two common examples. The network weights are the adjustable parameters. Given examples of the function to be learned (often referred to as the teacher), a neural network adjusts the weights, so that its outputs can imitate those of the teacher. The power of neural networks comes from their ability of universal approximation and the ease of training them. For a two-layer neural network with sufficient number of units, any continuous function can be learned well.[105] It is remarkable that this learning process can be done without the knowledge of the real function, but instead, by using examples only. These properties make NNM extremely useful in applications when real functions are unknown.

12.2.1   Examples of Neural Networks

Below, we will first introduce two standard feedforward networks, namely, the single layer perceptron (SLP) and the multilayer perceptron (MLP), as shown in Figures II.12.1 and II.12.2, respectively. Various forms of neural networks exist in the literature, either for the purpose of simplifying the

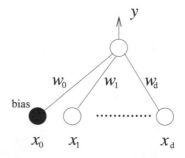

FIGURE II.12.1   The single layer perceptron.

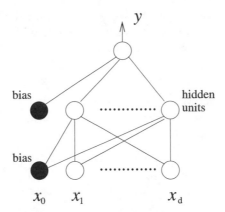

FIGURE II.12.2   The multiplayer perceptron.

training process, or as designs for specific applications. One such variation, called the radial basis function neural network (RBFNN), will be introduced next. Finally, a modern learning method, called the support vector machine (SVM), will be described and may be understood as an advanced version of RBFNN.

Apart from the feedforward network, another widely used neural network model is the Hopfield network, in which nodes of the same layer interact with each other (instead of acting only in the feedforward direction).[112] This kind of network is often used to solve difficult optimization problems, utilizing the property of its recurrent dynamics which evolves to a steady optimal state. They will not be discussed in this chapter.

12.2.1.1 Single Layer Perceptron (SLP)

Figure II.12.1 shows the structure of SLP, where $\mathbf{x} = \{x_i\}$, for $i = 1, \ldots, d$, represents the input; $\mathbf{w} = \{w_i\}$, for $i = 1, \ldots, d$, the network weights; and y the network output. w_0 is the bias, and x_0 is permanently set to $+1$. The function expressed by this network is:

$$y(\mathbf{x}) = f\left(\sum_{i=1}^{d} w_i x_i + w_0\right),$$
(II.12.1)

where f is a function chosen for the network. Given a set of examples of the teacher, $\{\mathbf{x}_l, O_l\}$ for $l = 1, \ldots, N$, where O_l is the output of the teacher for $\mathbf{x}_l$, SLP optimizes the weight by minimizing a cost function, which is often chosen to be the mean square error:

$$E = \frac{1}{N} \sum_{l=1}^{N} (O_l - y(\mathbf{x}_l))^2.$$
(II.12.2)

The simplest method to minimize Equation II.12.2 is the standard gradient descent method, which updates the weights as:

$$\Delta \mathbf{w} = -\gamma \frac{\partial E}{\partial \mathbf{w}},$$
(II.12.3)

where γ is a chosen learning rate.

The most serious limitation of SLP lies in the range of functions it can represent. For example, if SLP is used for a classification task, it outputs the class index of an input ($+1$ or -1), i.e., $y(\mathbf{x}) = sign(\mathbf{w} \cdot \mathbf{x} + w_0)$, where the function $sign(\cdot)$ is the sign of its argument. The separating boundary generated by SLP is determined by $\mathbf{w} \cdot \mathbf{x} + w_0 = 0$, which is a linear hyperplane in the d dimensional space. To allow for more general mappings (e.g., nonlinear boundaries), SLP should be extended to have more than one layer, which leads us to MLP.

12.2.1.2 Multilayer Perceptron (MLP)

Figure II.12.2 shows a two-layer MLP. The nodes in the second layer are called the hidden units, which receive inputs from the lower layer and feed signals to the upper one. The function expressed by this two-layer network has the form:

$$y(\mathbf{x}) = f\left(\sum_{j=0}^{M} w_j^{(2)} g\left(\sum_{i=0}^{d} w_{ji}^{(1)} x_i\right)\right), \tag{II.12.4}$$

where the function $g(\cdot)$ must be nonlinear, since the network reduces to SLP otherwise. The standard gradient descent method, when adapted to MLP, is called Back-Propagation (BP).[113] According to BP, MLP updates weights by:

$$\Delta w_j^{(2)} = -\gamma \frac{\partial E}{\partial w_j^{(2)}}, \tag{II.12.5}$$

$$\Delta w_{ji}^{(1)} = -\gamma \frac{\partial E}{\partial w_{ji}^{(1)}} = -\gamma \frac{\partial E}{\partial g} \frac{\partial g}{\partial w_{ji}^{(1)}}, \tag{II.12.6}$$

where E is a suitably chosen cost function. In Equation II.12.6, the chain rule is used to calculate the derivative of $w_{ji}^{(1)}$, which can be understood as back-propagating an error $(\partial E / \partial g)$ from the second layer to the first one.

12.2.1.3 Radial Basis Function Neural Network (RBFNN)

Though MLP with sufficient number of hidden units has guaranteed approximation ability, it is still not preferable in many cases. One complication in training MLP arises from the fact that the cost function is a nonlinear function of the network weights, and hence has many local minima. In this case, it is difficult to find the optimal solution by using any method based on gradient descent. To overcome this shortcoming, a new network model, RBFNN, is proposed.[114]

As shown in Figure II.12.3, RBFNN can be expressed as a two-layer network. In contrast to MLP, it has no adaptive weights in the first layer. The activations of the hidden units are radial basis

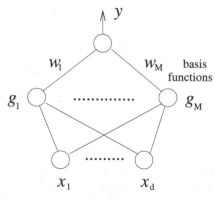

FIGURE II.12.3 The radial basis function neural network.

functions (RBF) of the inputs, centered at different locations. For example, when the RBF is chosen to be Gaussian:

$$g_j(\mathbf{x}) = \exp\left[-\frac{(\mathbf{x}-\mu_j)^2}{2\sigma^2}\right], \qquad (\text{II}.12.7)$$

where μ_j is the RBF center of the jth hidden unit, and σ the RBF width. The final output of RBFNN is a weighted summation of all RBFs:

$$y(\mathbf{x}) = \sum_{j=1}^{M} \mathbf{w}_j g_j(\mathbf{x}). \qquad (\text{II}.12.8)$$

The universal approximation ability of RBFNN with adequate RBFs is also guaranteed.[115] The adjustable parameters in the case of Gaussian RBFs include the centers and widths, and the network weights in the second layer. The efficiency of RBFNN comes from splitting the training process into two simple steps. The first step is an unsupervised procedure aiming to find out the centers and widths, which is often done through modeling the density distribution of input data by using methods such as K-means clustering[114] or EM algorithm.[116] (Here, supervised or unsupervised learning refers to whether the teacher outputs the examples used.) The second step is to optimize the network weights by using supervised learning analogous to that in SLP. Note that in the second step of training, the cost function is linear in weights, whose minimum can be easily obtained.

12.2.1.4 Support Vector Machine (SVM)

SVM is currently the most advanced technique for classification and regression.[117] It was originally developed by Vapnik and co-authors from the point of view of structure risk minimization,[118] and later connected with the regularization networks.[119] As far as we are aware, the application of SVM to telecommunications networks has not appeared in the literature yet. However, its popularity in the near future is anticipated. It is therefore worthwhile to give an introduction here. We will not go into the details of the SVM implementation, but will rather compare it with RBFNN to show its power. To get more knowledge of SVM, the reader can refer to Christianini and Shawe-Taylor.[120]

SVM looks for a solution of $y(\mathbf{x}) = \mathbf{w}\cdot\psi(\mathbf{x})$ by minimizing the cost function (the regression problem is used here as illustration):

$$E = C \sum_{l=1}^{N} |O_l - y(\mathbf{x}_l)|_{\varepsilon} + \frac{1}{2}\|\mathbf{w}\|^2, \qquad (\text{II}.12.9)$$

where $|f|_{\varepsilon}$ denotes the ε-*insensitive* function, which takes value of f when $|f| > \varepsilon$, and 0 otherwise. The solution of SVM turns out to have the form:

$$y(\mathbf{x}) = \sum_{l=1}^{M} h_l K(\mathbf{x}, \mathbf{x}_l), \quad \text{for } h_l \neq 0, \qquad (\text{II}.12.10)$$

where $K(\mathbf{x}, \mathbf{x}_l) = \psi(\mathbf{x}) \cdot \psi(\mathbf{x}_l)$ is called the kernel function, and $h_l = 0$ for $M + 1 \leq l \leq N$, after relabeling the examples. Due to the specific form of the cost function, the attractive property of SVM is that normally only a very few number of coefficients h_l are nonzero, i.e., $M \ll N$. These data points are called support vectors. Many efficient learning algorithms have been developed for SVM.

Let us compare SVM with RBFNN. The network structure is the same as in Figure II.12.3. The kernel function is now the corresponding RBF, the support vectors are the RBF centers, and the coefficients h_l are the network weights in the second layer. The advantage of SVM is that the RBF centers are automatically optimized during a one-step training, overcoming the weakness of RBFNN of choosing centers in an unsupervised manner. Therefore it is not surprising that SVM generally outperforms RBFNN.

12.2.2   Procedures for Applying NNM

While we have reviewed several neural network models, using them in practice is not as simple as randomly choosing one and then training it with examples. The real application involves many subtleties. Some issues affecting the final performance of the network may not be even related to the training step, but play very important roles. We summarize five essential steps of setting up NNM, irrespective of their orders of being considered.

12.2.2.1   Preprocessing the Data

This refers to the preprocessing of the representations of input and output variables before they are used in training. This step is often crucial in practice. It aims at simplifying the unknown function to be learned by using all available side information, such that the function can be easily learned at latter stages. In some cases, when the dimensionality of input is high, dimensionality reduction is essential to avoiding "the curse of dimensionality." Applications presented in this chapter give good examples of how to preprocess inputs.

12.2.2.2   Selecting a Model

This refers to the choice of a suitable neural network model, and is often problem-dependent. The most important issue is the so-called *bias-variance trade-off* in choosing a model.[121] On one hand, a high bias means that one does not have much confidence in the training data, and would thus resort to the predisposition of a simple model. This may not be adequate to approximate the desired function, and the training result becomes bad. On the other hand, a high variance means that one has much confidence in the training data, and is thus ready to fit the data by using a more flexible model. Though flexibility results in the good learning of the examples, it may generalize badly when new inputs are presented. In practice, prior knowledge of the task is often used to help us in model selection.

For example, if the teacher function is known to be linear, SLP is sufficient. Otherwise, more complex network structures, such as MLP or RBFNN, are needed. In many cases, the process of model selection can be interleaved with step 4, using the feedback obtained from intermediate training steps to assist the selection. Examples are the growing algorithms[122] and pruning algorithms for MLP.[123]

12.2.2.3   Choosing a Cost Function

This refers to the definition of a suitable criterion measuring how the teacher function is learned. The mean square error is often used and is adequate in many cases. From the statistical point of view, however, choosing a cost function corresponds to assuming a noise model for the process of generating data. Hence it should be problem dependent.

12.2.2.4   Optimizing the Free Parameters

This is the training process, during which the neural network adjusts the weights by minimizing the cost function. The standard learning algorithm is the gradient descent, which is simple, but is often slow and may stop at a plateau of the cost function in the parameter space. More advanced algorithms are subsequently developed, including natural gradient,[124] conjugate gradient,[125] and Newton's method.[110]

12.2.2.5   Implementing the Neural Network

This is the last step, which is the overall outcome of the previous steps and makes the neural network ready for application.

12.3   Neural Networks for Overload Control

In this section, we give an example of neural networks in their use for overload control in telecommunication networks. The motivation of using neural networks here is to overcome the shortcomings of traditional local and centralized control strategies, and combine their advantages. The idea is to use a group of student neural networks to learn the control actions of an optimal centralized teacher, yet operating in a distributed manner in their implementation. As a result, the control is simple, robust, and near-optimal. Below is a brief outline of the work. For more details, please refer to Wu and Wong.[28]

12.3.1 Overload Control in Telecommunications Networks

In telecommunications systems, overload control is critical to guarantee good system performances of the call setup and disconnection processes. Overload events occur in heavy traffic, when the number of call setup jobs exceeds the capacity of call processing computers. These events, if left uncontrolled, will cause the system to break down and bring disasters to the network performance. Some control actions are therefore required to protect the limited system resources from excessive load, based on a throttling mechanism for new arriving requests.

In general, there are two kinds of control strategies, i.e., local or centralized, according to the amount of information on which the control decisions are based. *Centralized control* consists of one main networkwide controller, which collects all the information through the signaling network, and hence can make the globally optimal decisions. The shortcoming of a centralized control is that it can be very complex and time-consuming when the network size is large. Also, the load in the signaling network is high, often rendering it impractical. Furthermore, centralized control is sensitive to network breakdown. On the other hand, *local control* makes decisions based on locally available information only. It has the advantages of easy implementation and robustness to partial network failure. However, local control has the shortcoming that the control decisions are generally not optimal, since they are based on local information.

In reality, centralized control is used in smaller networks, while localized control is preferred in larger networks. In the latter case, the challenge is to coordinate the control steps taken by each local controller to achieve performances approaching globally optimal ones.

For traditional hierarchical networks, centralized versions of overload control strategy have been well developed. There is a main controller located at the central call processor, which takes control actions in response to all call setup requests. An example is the STATOR method.[126]

For networks of distributed architecture, where the role of each processor is equivalent, the situation is much more complex and difficult. Some local control methods have been suggested for this situation,[127–129] in which each processor makes a decision depending only on its own status and there is no cooperation between them. In this case, the system is either over-controlled (wasting resources) or under-controlled (failing to curb overload events), and does not reach the optimal performance.

To improve the local control methods, some information on the traffic status among local processors needs to be exchanged. However, given that such information is available, it is still not easy to design a good local control method, the reason being that the teletraffic is stochastic and the mapping from traffic input to optimal decision is complex. To solve this problem, our attention turns to neural networks, bearing in mind its ability of learning unknown functions from a large number of examples and its implementation in real time once trained.

So, the first step of the work is to find a teacher of optimal performance, which generates examples for the training purpose. We find such a teacher by solving a sequence of linear programming problems. The second step is to train a group of decentralized neural controllers, each located on one processor node. After training, the neural controllers cooperate to infer the control decisions of the teacher based on locally available information.

12.3.1.1 Objectives of Overload Control

A processor is overloaded if its work load averaged over a period exceeds a predefined threshold. Overload control is implemented by gating new calls. The gate values, i.e., the fraction of admitted calls, are updated periodically. An effective control is to find out the optimal gate values for each period. To measure and compare the performances of control strategies, the objectives of the control need to be clarified first. An ideal control algorithm should satisfy the following requirements: 1) maximum throughput, therefore avoiding unnecessary throttling; 2) balance between stations; 3) fairness to every node; 4) robustness against changing traffic profiles and partial network breakdown; and 5) easy implementation.

12.3.1.2 A Simplified Call Processing Model

For the purposes of theoretical studies and simulations, we adopt a simplified call processing model, which captures the essential features of real processes. Figure II.12.4(a) shows a distributed telecom network which consists of N fully connected switch stations. Call requests between two stations are assumed to arrive as Poisson processes. Each call setup request initiates five jobs, referred to as jobs 1 to 5, respectively. They represent the jobs of sending dial tones, receiving digits, routing, connecting path, etc. Jobs 1 to 3 are processed on the original node, and jobs 4 and 5 on the terminating node. They generate different work load, measured in milliseconds of service times. Time delay between successive jobs are assumed to be stochastic and uniformly distributed within a certain range. The parameters used in this work are shown in Table II.12.1.

TABLE II.12.1 The Simplified Call Processing Model

Call Processing on the Originating Node	Call Processing on the Terminating Node
Job 1 (load = 50 ms)	Waiting
1–3 sec delay	1–3 sec delay
Job 2 (load = 150 ms)	Job 4 (load = 100 ms)
2–8 sec delay	2–8 sec delay
Job 3 (load = 50 ms)	Job 5 (load = 50 ms)

Source: Wu, S.I. and Wong, K.Y.M., *IEEE Trans. Neutral Networks*, 9, 6, 1998. With permission.

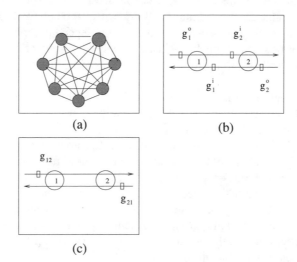

(a)   (b)

(c)

FIGURE II.12.4   (a) A seven-node fully connected network of switch stations. (b) The local control method. (c) The centralized control method. *Source*: Wu, S.I. and Wong, K.Y.M., *IEEE Trans. Neutral Networks*, 9, 6, 1998. With permission.

12.3.2 Conventional Control Strategies

12.3.2.1 The Local Control Method (LCM)

In the local control method, each processor node monitors its own load and makes decisions independent of all others. As shown in Figure II.12.4(b), there are two kinds of gate representing where throttling takes place. The gate values g_i^0 and g_i^i denote, respectively, the acceptance rates of calls outgoing from and incoming to node i. They are updated periodically. At the beginning of a control period, the control action at a node is computed from the gate values which satisfy the capacity constraint, which is obtained from the predefined capacity threshold, after subtracting the leftover load carried forward from the previous periods.

When a node is overloaded, priority should be given to the terminating calls to maximize the throughput, since they have already consumed processing resources in their originating nodes. Hence, the local controller should first reject the outgoing call requests. If this is still not effective, the controller should further adjust the incoming gate. Transcribing this into the control algorithm, the local controller first maximizes the incoming gate values, and next the outgoing gate values, while satisfying the capacity constraint.

LCM is not an optimal control, for there is no cooperation between different nodes. However, it has the advantages of being simple and robust.

12.3.2.2 The Optimal Centralized Control Method (CCM)

In the centralized control algorithm, networkwide information is available to the controller. Therefore, through cooperative control on each node, only outgoing calls need to be throttled, as shown in Figure II.12.4(c). CCM is able to take into account the multiple objectives prescribed in Section 12.3.1, in which case the order of priority of the objectives determines the optimization procedure. The maximization of throughput is considered to be the most important, since it is a measure of averaged system performance. Load balancing is next important, since it is a measure of system performance under fluctuations. Fairness comes third. The scheme of CCM is equivalent to solving a three-step linear programming problem[130] involving the gate values $g_{ij}(t)$, which are the acceptance ratios for outgoing calls from node i to j in the control period t.

Step One: Maximize the throughput $\sum_{i,j} \lambda_{ij}(t) g_{ij}(t)$ subject to:

$$0 \leq g_{ij}(t) \leq 1, \tag{II.12.11}$$

$$\tau_0 \sum_j \lambda_{ij}(t) g_{ij}(t) + \tau_{0'} \sum_j \lambda_{ji}(t) g_{ji}(t) + \rho_{i,lo}(t) \leq \rho_{\max}, \quad 1 \leq i \leq N, \tag{II.12.12}$$

where $\rho_{i,lo}$ is the leftover load carried from the previous periods on node i. τ_0 is the average service time for outgoing calls arriving in the current period; $\tau_{0'}$ the corresponding service times for incoming calls. They are estimated by assuming the model in Table II.12.1. $\lambda_{ij}(t)$ is the outgoing call rate from node i to j, estimated for the control period t by averaging over a few previous periods. $\rho_{\max}$ is the predefined capacity threshold.

It turns out that the solution of the above inequalities is often degenerate. Removing the degeneracy enables us to optimize the secondary objectives of load balancing and fairness, which is done within the subspace of maximum throughput. Mathematically, this requires that all active constraints (equalities after the optimization) are preserved.

Removing the degeneracy is also important when CCM is used to generate examples for subsequent training of neural networks. Degeneracy means the teacher will prescribe different control actions for similar network situations. This is bad for supervised learning, since in this case, the student will only learn to output the mean value of the teacher's outputs. Hence, unambiguous examples should be provided.

Step Two: Optimize load balance by maximizing the vacancy parameter θ in the subspace of maximum throughput, where:

$$\tau_0 \sum_j \lambda_{ij}(t)g_{ij}(t) + \tau_{0'} \sum_j \lambda_{ji}(t)g_{ji}(t) + \rho_{i,lo}(t) + \theta \leq \rho_{\max}, \qquad (\text{II.12.13})$$

and each i denotes a non-full node in the subspace. Maximizing θ decreases the load of the most congested nodes. As a result, the traffic load is more evenly distributed among stations. If there is still degeneracy, the third optimization step is needed.

Step Three: Optimize fairness by maximizing the lower bound η in the subspace of maximum throughput and optimal load balance, where:

$$\eta \leq g_{ij}(t) \leq 1, \qquad (\text{II.12.14})$$

and each $g_{ij}(t)$ denotes an undetermined gate value in the previous optimization. Maximizing the lower bound η will avoid unfair rejection in some nodes. This step is repeated until all remaining degeneracies are lifted.

Understandably, this method is very time-consuming. The decision-making time grows as N^6 with the size N of the networks.

12.3.3   Neural Network Methods (NNM)

A neural network on a processor node receives input about the conditions of connected call processors and output corresponding control decisions about the gate values. It acquires this input-output mapping by a learning process using examples generated by CCM. It is difficult to train the neural networks properly using examples generated for a large range of traffic intensity, but on the other hand, training them at a fixed traffic intensity makes them inflexible to changes. Hence, as shown in Figure II.12.5 for each processor node, we build a group of neural networks, each member being a single layer perceptron trained by CCM using examples generated at a particular background traffic intensity. The final output is an interpolation of the outputs of all members using RBFs, which weight the outputs according to the similarity between background and real traffic intensities. This

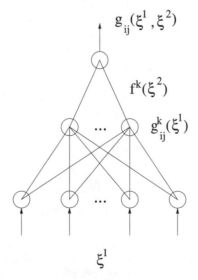

FIGURE II.12.5   The neural network for calculating the gate value g_{ij}. *Source*: Wu, S.I. and Wong, K.Y.M., *IEEE Trans. Neutral Networks*, 9, 6, 1998. With permission.

enables the neural controller to make a smooth fit to the desired control function, which is especially important during traffic upsurges.

This network architecture is similar to that of Stokbro et al.,[131] where each hidden unit produces as an output of a linear function of the inputs, and the final output is their average weighted by the RBFs. Our network differs from theirs in that the outputs of the hidden units are nonlinear sigmoid functions, and we save the effort of data clustering by taking advantage of the natural clusters according to their background traffic intensities. Of course, training for optimizing the RBF center could improve the performance further.

12.3.3.1 Training a Member of the Group of Neural Networks

For a neural controller associated with a node, the available information includes the measurements, within an updating period of all the outgoing and incoming call attempts, and the processing load of all nodes. Note that the processing load is the only global information fed into the neural controller.

To increase the learning efficiency of the neural networks, it is important to preprocess the inputs, so that they are most informative about the teacher control function, i.e., to make the control function as simple as possible. After exploring the geometry of the solution space in CCM for the most relevant parameters in locating the optimal point, we have chosen the following variables as the $2N - 1$ inputs to node i; detailed justification can be found in Wu and Wong:[28] (a) The first $N - 1$ inputs are $\min(\widetilde{\rho}_l(t)/\tau_0 \cdot \lambda_l(t), 1)$ for $l \neq i$, where $\widetilde{\rho}_l(t)$ is the load vacancy at node l as estimated by node i. (b) The Nth input is $\min(\widetilde{\rho}_l(t)/\tau_0'[\sum_l \lambda_{ij}(t)^2]^{1/2}, \sqrt{N})$. (c) The other $N - 1$ inputs are $\lambda_l(t)/[\sum_m \lambda_{im}(t)^2]^{1/2}$ for $l \neq i$.

The above inputs form a $2N - 1$ dimensional vector ξ^1 fed to each neural network in the group, each trained by a distinct training set of examples. The kth member outputs the gate values g_{ij}^k according to:

$$g_{ij}^k = f\left(\sum_{n=1}^{2N-1} J_{ijn}^k \xi_n^1 + J_{ij0}^k\right), \qquad (\text{II}.12.15)$$

where $f(h) = (1 + e^{-h})^{-1}$ is the sigmoid function. The coupling J_{ijn}^k and the bias J_{ij0}^k are obtained during the learning process by gradient descent minimization of an energy function:

$$E = \frac{1}{2}\sum_{k,\mu}(O_{ij}^{k,\mu} - g_{ij}^{k,\mu})^2, \qquad (\text{II}.12.16)$$

where $O_{ij}^{k,\mu}$ is the optimal decision of g_{ij} prescribed by the teacher, for example μ in the kth training set, and $g_{ij}^{k,\mu}$ is the output of the kth member of the group of neural networks.

12.3.3.2 Implementation of the Group of Neural Networks

Consider the part of the neural controller for calculating the gate value g_{ij}, as shown in Figure II.12.5 (the other parts have the same structure). The kth hidden unit is trained at a particular traffic intensity, and outputs the decision $g_{ij}^k(\xi^1)$ described in Section 12.3.3.1.

To weight the contribution of the kth output, we consider a $N - 1$ dimensional input vector ξ^2 which consists of the call rates $\lambda_{ij}^0(t)$, $j \neq i$. The weight $f^k(\xi^2)$ is the RBF given by:

$$f^k(\xi^2) = \frac{\exp[-(\xi^2 - \mu^k)^2/2\sigma_k^2]}{\sum_l \exp[-(\xi^2 - \mu^l)^2/2\sigma_l^2]}, \qquad (\text{II}.12.17)$$

where μ^k is the kth RBF center, and σ_k is the size of the RBF cluster. In our case, μ^k is the input vector ξ^2 averaged over the kth training set of examples, and describes the background traffic intensity. σ_k^2 is chosen to be the variance of the Poisson traffic at the kth RBF center.

TABLE II.12.2   Call Arrival Rates (per Hour) of Seven Nodes of the
Hong Kong Metropolitan Network under Normal Traffic Condition

	S_0	S_1	S_2	S_3	S_4	S_5	S_6
S_0	0	480	1070	1040	1640	280	670
S_1	360	0	220	320	390	240	300
S_2	900	400	0	2100	1550	450	520
S_3	700	410	2090	0	1020	270	410
S_4	1080	280	1300	970	0	380	400
S_5	250	220	290	170	230	0	210
S_6	500	260	490	430	450	230	0

Source: Wu, S.I. and Wong, K.Y.M., *IEEE Trans. Neutral Networks*, 9, 6, 1998. With permission.

The final output of the neural network is a combination of the weighted outputs of all hidden units, i.e.,

$$g_{ij}(\xi^1, \xi^2) = \sum_k f^k(\xi^2) g_{ij}^k(\xi^1). \tag{II.12.18}$$

Since the numerator of Equation II.12.17 is a decreasing function of the distance between the vector ξ^2 and μ^k, the RBF center nearest to ξ^2 has the largest weight. If ξ^2 moves between the RBF centers, their relative weights change continuously, hence providing a smooth interpolation of the control function.

12.3.4   Comparing the Performances of the Three Methods

To compare the above three methods, we perform simulations on part of the Hong Kong metropolitan network, which consists of seven fully connected switch stations, as shown in Figure II.12.4(a). The call arrival rates between different nodes under normal traffic condition are shown in Table II.12.2. In simulations, call attempts are generated according to the Poisson process, and accepted with probability given by the corresponding gate values. Taking into account hardware limitations, control speed, and statistical fluctuations, we choose the control period to be 5 sec. The accepted calls will queue in the buffer waiting for service. To account for the loss of customers who run out of patience after waiting too long, we assume a stochastic overflow process with a survival probability after waiting for t sec given by $p(t) = \min(1, \exp[-0.35(t-1)])$.

The RBF centers of the neural networks are chosen as 1–4, 6, and 8 multiples of the normal traffic intensity. To generate examples for neural network training by CCM algorithm, we simulate the traffic corresponding to each RBF center for more than 2×10^5 sec. The data of network scenarios and their associated globally optimal decisions are collected to train the neural controllers off-line.

12.3.4.1   Steady Throughput

Figure II.12.6 compares the throughputs of the network versus steady-state traffic intensities. The simulation for each case is done for 4000 sec. We see that the neural control performs comparably with the centralized teacher, and has a large improvement in throughput over the local control for a large range of traffic intensities.

12.3.4.2   Traffic Upsurges

Of particular interest to network management is the response of the system to traffic upsurges. In reality this occurs in such cases as phone-in programs, telebeting, and the hoisting of typhoon signals, when the amount of call attempts abruptly increases. It is expected that control schemes should respond as fast as possible to accommodate the changing traffic conditions.

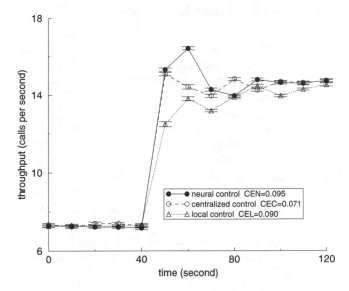

FIGURE II.12.6   Network throughput under constant traffic. The traffic intensities are measured in multiples of the normal rates. *Source*: Wu, S.I. and Wong, K.Y.M., *IEEE Trans. Neutral Networks*, 9, 6, 1998. With permission.

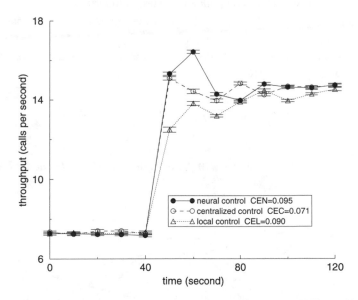

FIGURE II.12.7   Network throughput during a traffic upsurge on all nodes. The traffic intensities of all nodes increase to six times at $t = 40$ sec. CEL, CEC, and CEN are the average control errors of LCM, CCM, and NNM, respectively, within 50 sec after traffic increase. *Source*: Wu, S.I. and Wong, K.Y.M., *IEEE Trans. Neutral Networks*, 9, 6, 1998. With permission.

Figure II.12.7 shows how the system responds when the normal traffic intensity becomes sixfold at $t = 40$ sec. We see that NNM has a throughput higher than CCM (but with a slight compromise in control error), and they are both much better than LCM.

The neural controller significantly decreases the time for making decisions. For the network we simulated, it is about 10% of the CPU time of CCM. Hence, NNM can be implemented in real time.

12.4   Neural Networks for Fault Diagnosis

In telecommunications management, error messages are generated continuously. This makes it difficult to diagnose whether the system is normal or abnormal. Moreover, when a network breakdown takes place, error messages are generated at an enormous number, making it difficult to differentiate the primary sources and secondary consequences of the problem(s). Thus, it is desirable to have an efficient and reliable error message classifier.

Historically, intelligent techniques such as classification trees were used in analyzing system failures. Due to their hierarchical structures, classification trees are often too inflexible to deal with noisy and ambiguous features inherent in many diagnosis tasks. On the other hand, neural networks are good at providing probabilistic comparisons among the possible candidates for system failures. However, due to their flexibility, input data needs to be appropriately preprocessed before they can have maximal performance, as pointed out in Section 12.2.2.1. Naturally, it is instructive to consider whether these apparently different approaches, with their complementary advantages, can be hybridized to yield improvements over the individual systems. Here we report our work in Lau et al.,[63] showing that this is indeed valid for applications in error message classification.

12.4.1   A Hybrid Network Architecture

The hybrid classifier is composed of a rule-based hidden layer and a perceptron layer,[132] as shown in Figure II.12.8. The input layer contains N nodes receiving a binary vector $\mathbf{x}$ of input attributes $\{x_1, \ldots, x_N\}$. The hidden layer contains R nodes representing R classification rule vectors $\{\mathbf{r}^1, \ldots, \mathbf{r}^R\}$. Each rule $\mathbf{r}^j = (r_1^j, \ldots, r_N^j)$ has the same dimension as the input vector, but the component r_i^j can take the values 0, 1, or "don't care."

The output of a hidden node (y_j) is a matching function Φ between attributes in the input vector $\mathbf{x}$ and the rule vector $\mathbf{r}^j$. It is defined as:

$$y_j = \Phi\left(1 - \frac{H(\mathbf{x}, \mathbf{r}^j)}{D(\mathbf{r}^j)}\right), \qquad 1 \le j \le R \tag{II.12.19}$$

where $H(\mathbf{x}, \mathbf{r}^j)$ is the Hamming distance between $\mathbf{x}$ and $\mathbf{r}^j$, $D(\mathbf{r}^j)$ is the effective dimension of the rule $\mathbf{r}^j$, and the output of Φ is normalized in $[0, 1]$. In evaluating H and D, the "don't care" attributes

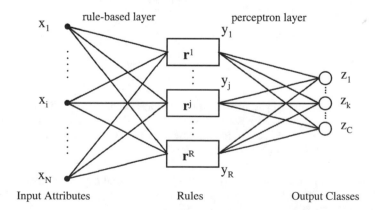

FIGURE II.12.8   The hybrid classifier network architecture. *Source*: Lau, H.C. et al., A hybrid expert system for error message classification, *Applications of Neutral Networks to Telecommunication* 2, Alspector, J., Goodman, R., and Brown, T.X., Lawrence Erlbaum, 1995, with permission.

in $\mathbf{r}^j$ and the corresponding attributes in $\mathbf{x}$ will be ignored. As a result, each hidden node will fire an output between 1 and 0, with perfectly matched and perfectly unmatched being the two extremes, respectively. y_{R+1} is set to 1 as a bias term for the perceptron.

The rule-based preprocessor aims at selecting the features of the input vector. With $R < N$, it also serves the purpose of dimensionality reduction to the network. The set of y's are then fed to an array of perceptrons labeled by the class variable k, where $k \in \{1, \ldots, C\}$. The output node activation z_k is given by:

$$z_k = \sum_{j=1}^{R+1} w_{kj} y_j. \qquad 1 \le k \le C \qquad \text{(II.12.20)}$$

It is an estimate of the corresponding class probability. Therefore:

$$c = \arg \max_{1 \le k \le C} \{z_k\} \qquad \text{(II.12.21)}$$

offers the first choice of our class prediction. Hence, this is often called a *winner-take-all* classifier. The second and third alternatives, etc., can be determined in a similar manner by finding the second and third largest argument in $\{z_k\}$.

12.4.2 Rule Extraction by CART

We now turn to the problem of finding the rules in the hidden layer. The rules are extracted from a classification and regression tree (CART).[133] When a training set is presented, the tree is grown by recursively finding splitting conditions until all terminal nodes have pure class membership.

Consider a branch node t and its left and right child t_L and t_R respectively, as in Figure II.12.9. Denote $p(i|t)$ as the conditional probability that the example belongs to class i, given that it stands in node i. Define the node impurity function by the Gini criterion:[133]

$$i(t) = \sum_j \sum_{i \neq j} p\left(i|t\right) p\left(j|t\right). \qquad \text{(II.12.22)}$$

An efficient discrimination is achieved when one selects the attribute that provides the greatest reduction in impurity among the examples. In other words, x_p is chosen to maximize:

$$\Delta i(x_p; t) = i(t) - i(t_L) p\left(t_L|t\right) - i(t_R) p\left(t_R|t\right), \qquad \text{(II.12.23)}$$

where $p(t_L|t)$ and $p(t_R|t)$ are the conditional probability that the example lands in node t_L and t_R, respectively, given that it lands in node t.

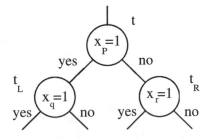

FIGURE II.12.9 A decision node of a classification tree. *Source*: Lau, H.C. et al., A hybrid expert system for error message classification, *Applications of Neutral Networks to Telecommunication 2*, Alspector, J., Goodman, R., and Brown, T.X., Lawrence Erlbaum, 1995, with permission.

After the tree is grown, it is pruned by minimizing the error complexity using a pruning factor which represents the cost per node. The number of rules (R) is thus kept below the input dimension N.

CART can be used independently for classification tasks, but here we use it to produce classification rules for subsequent neural network processing. The rule vectors $\mathbf{r}^j$ are generated by exhausting all routes of the hierarchical tree, setting up appropriate attributes in traversing the decision nodes of each route, and inserting "don't cares" when those attributes are not examined in the splitting criteria. These rules constitute the rule-based layer in Figure II.12.8, and for this purpose, the class outcome of each rule is not important at this stage.

12.4.3 Classification by Neural Network

The neural network classifier in the second layer is an extension of SLP introduced in Section 12.2.1.1 to the case of multi-class outputs. At each training step, an example with input vector $\mathbf{x}$ and output class ξ is selected randomly and fed to the network. Then the weights are modified according to a multi-class version of the perceptron learning algorithm:[134]

$$\Delta w_{kj} = 0, \qquad \text{if } z_k + \tau < z_\xi \tag{II.12.24}$$

$$\Delta w_{kj} = -\eta y_j, \qquad \text{if } z_k + \tau \geq z_\xi \tag{II.12.25}$$

$$\Delta w_{\xi j} = \mu \eta y_j, \tag{II.12.26}$$

where μ is the number of k's that accounts for the modification of Equations II.12.25, and η is the learning rate. A learning threshold τ is introduced to increase the stability of the network, because it is desirable that the predicted class c should be equal to the actual class ξ with high certainty. It ensures that weight updates will proceed when the desired node output z_ξ cannot exceed other outputs by the learning threshold τ. Learning is repeated until a satisfactory percentage of training examples are classified correctly. This algorithm has the advantage of fast convergence and fault tolerance.

12.4.4 Diagnosis of Error Messages

The classifier is tested on a set of error messages generated from a telephone exchange computer, indicating which circuit card is malfunctioning.[61] The training set consists of 442 samples and the test set of 112 samples. Each sample is in the format of a bit string consisting of an error vector of $N = 122$ bits and $C = 36$ possible classes.

Using CART alone, a decision tree is constructed and pruned. It achieves a best generalization rate of 69.6%. Using a multi-class neural network (NN) alone, the best generalization rate only reaches 61.6%. This is compared with the hybrid classifier by incorporating the perceptron with a preprocessing layer of CART rules, reducing the input dimension from 122 to 72. Using a linear function Φ for the matching criterion (Equation II.12.19), the resulting CART-NN hybrid network is found to have a performance boosted up to 74.1%. It is obviously better than the individual results of CART and NN. Table II.12.3 summarizes the classification results up to the first three choices. The hybridization of CART and NN yields a comparable performance with the Bayesian neural network.[61,62]

The study demonstrates the importance of preprocessing the data before feeding it to the neural network. In this specific application, there is a strong correlation inherent in the data. For example, there are concurrent error bits caused by the simultaneous breakdown of components. This probably accounts for the advantages of including rule-based approaches such as CART and the Bayesian neural network. Left alone, neural networks do not perform well when the input dimension is large and few training examples are available. Therefore, we use CART as a preprocessing layer of the multi-class perceptron for dimensionality reduction and feature extraction. Other conventional ways of preprocessing include vector quantization, self-organizing feature maps, radial basis functions, and principal component analysis.[110]

TABLE II.12.3   Classification Results of Various Techniques.

Choice	Backprop[61]	Multilayer Bayesian[61]	Higher Order Bayesian[62]	NN[63]	CART-NN Hybrid[63]
		Training Set Results (%)			
1st	91.2	80.8	86.7	92.8	82.6
1st—2nd	94.6	92.3	94.8	98.4	94.8
1st—3rd	95.1	96.1	97.1	99.5	98.2
Rest	4.9	3.9	2.9	0.5	1.8
		Testing Set Results (in %)			
1st	67.0	72.3	75.0	61.6	74.1
1st—2nd	75.9	82.1	80.4	75.9	83.0
1st—3rd	76.8	87.4	88.4	83.9	88.4
Rest	23.2	12.6	11.6	16.1	11.6

Note: The numbers indicate the successful rate classification when the specified choices are included.

Source: Lau, H.C. et al., A hybrid expert system for error message classification, *Applications of Neutral Networks to Telecommunication* 2, Alspector, J., Goodman, R., and Brown, T.X., Lawrence Erlbaum, Hillsdale, 1995, with permission.

On the other hand, rule-based approaches such as CART may generate irrevocable errors at an early stage of the hierarchy, and improvement of rules is hard to implement. Neural networks complement them by providing the necessary flexibility and fault tolerance to the generated rules.

12.5   A Data Model for Fault Diagnosis

We have demonstrated the advantages of the hybrid intelligent system in fault diagnosis. It illustrates the importance of preprocessing the data in using neural networks. However, it is misleading to conclude that fixed choices of cost functions and classifier architectures are always optimally applicable to all cases of diagnosis. Below, we consider an artificial data model introduced in Wong and Lau,[135] and study its effects on the behavior of different classifiers.

Most analyses of classification are based on models with rather uniform distribution of data noises. However, in many diagnostic tasks, data noises are much less uniform and highly correlated. Some symptoms are more essential to a given fault, and are therefore more informative than others, and some groups of symptoms have a high correlation in their occurrence. Indeed, this is the case in such a complex system as the telecommunications network.

To have a better understanding of these effects, we introduce the following *informator model* of nonuniform data, which resembles typical data for system faults in diagnostic classification tasks. The model is characterized by the presence of a minority of informative bits (or the "informators") among a background of less informative ones. Considering the limit of very high input dimensions, we will map out their regions of perfect and random generalization for comparison. While it is not surprising that the informative bits help classification, it is interesting to see the breakdown of conventional wisdom in some cases. For example, the Bayesian estimator is not always the best when examples are few. Implications to the choice of classifiers will be discussed.

12.5.1   The Informator Model of Data

We consider a data model with N input bits, which model the symptoms, and K output classes, which model the possible faults to be diagnosed. For an output class k, the ith input bit x_{ki} may

TABLE II.12.4   The Informator Model of Nonuniform Data

Input Bit Type	Informator	Strong Background	Weak Background
Error rate	p_0^ε	$\frac{3}{2}p_0$	$\frac{1}{2}p_0$
Frequency per input vector	C	$\frac{N-C}{2}$ (average)	$\frac{N-C}{2}$ (average)

either be 1, indicating that a symptom has occurred, or 0, indicating that the symptom is absent. The probability of symptom p_{ki} for an ith bit belonging to the kth class is assumed to be independent of each other. For each class k, there are two kinds of input bits:

1. There are C randomly chosen "informators," whose probability of occurrence p_{ki} has a typical magnitude of order p_c, and $C \ll N$.
2. All the other $N - C$ bits have a low probability of occurrence, i.e., $p_{ki} \sim p_0 \ll p_c$ and can be considered as background.

For convenience, we assume that the prior probabilities for all output classes are the same; for background bits, $p_{ki} = (1 \pm 1/2)p_0$ with probability $1/2$ respectively, whereas for the informators, $p_{ki} = p_c = p_0^\varepsilon$ with $0 < \varepsilon < 1$. So there are three types of symptoms: informators, and strong and weak backgrounds. All bits contain information about the output class, but the informators are more informative than the background bits. We will consider p_0 lying in the range $N^{-1} < p_0 < 1$ and will refer to p_0 as the *error rate*. The parameters of the model are summarized in Table II.12.4.

An example generated by the data model consists of the N input bits and the associated output class. To build a classifier, a set of P training examples per output class is provided. To test the performance of the resultant classifier, an example is drawn randomly from the data model, independent of the training set. The average probability that the example is classified correctly is the *generalization performance*.

The generalization behavior of the model depends on two factors. First, it depends on the degree of certainty that an informator associates with a given output class. Hence, we define $\bar{\varepsilon} \equiv 1 - \varepsilon$ as the *informator strength*. When $\bar{\varepsilon} = 0$, the informators are indistinguishable from the backgrounds. When $\bar{\varepsilon} = 1$, the informators occur with certainty for a given output class. Second, generalization depends on the number C of the informators in the inputs of an example. We consider cases that C scales as N^γ, where we define $\gamma \equiv \ln C / \ln N$ as the *informator frequency*. Below, we will consider four types of classifiers.

12.5.2   The Classifiers

12.5.2.1   The Bayesian Classifier

Suppose that from the training set, one observes that there are n_{ki} errors for the ith bit out of P examples of class k. The Bayesian probability of an output class k, given the input vector $\mathbf{x}$, is given by:

$$P\left(k|\mathbf{x}\right) = \frac{P\left(\mathbf{x}|k\right)P(k)}{P(\mathbf{x})} = P(k)\prod_i = \frac{P\left(x_i|k\right)}{P(x_i)}$$

$$P(k)\prod_i \frac{P\left(x_i|k\right)}{\sum_l P\left(x_i|l\right)P(l)}. \tag{II.12.27}$$

For binary inputs x_i, we can write:

$$P\left(x_i|k\right) = p_{ki}^{x_i}(1 - p_{ki})^{1-x_i}. \tag{II.12.28}$$

Since we have assumed that the prior probabilities of all classes are the same, we have:

$$P\left(k|\mathbf{x}\right) = \prod_i \frac{p_{ki}^{x_i}(1 - p_{ki})^{1-x_i}}{\sum_l p_{li}^{x_i}(1 - p_{li})^{1-x_i}}. \tag{II.12.29}$$

Hence the output k is given by:

$$F(\mathbf{x}) = \arg\max_k \left(\ln P\left(k|\mathbf{x}\right)\right) = \arg\max_k \left[\sum_i (x_i \ln p_{ki} + (1 - x_i) \ln (1 - p_{ki}))\right]. \tag{II.12.30}$$

Without further information about the prior distribution of p_{ki}, the Bayesian classifier estimates p_{ki} by the fraction n_{ki}/P. After collecting terms for the coefficients of x_i and the constant, the most probable output class $F(\mathbf{x})$ estimated by the Bayesian classifier is:

$$F(\mathbf{x}) = \arg\max_k \left[\sum_i P\left(S_i|k\right)\right] = \arg\max_k (z_k), \tag{II.12.31}$$

where z_k is the activation for output class k:

$$z_k = \sum_i w_{ki} x_i + w_{k0}, \tag{II.12.32}$$

and the weights and thresholds are given by:

$$w_{ki} = \ln \frac{n_{ki}}{P} - \ln \left(1 - \frac{n_{ki}}{P}\right), \tag{II.12.33}$$

$$w_{k0} = \sum_i \ln \left(1 - \frac{n_{ki}}{P}\right). \tag{II.12.34}$$

Zero values of n_{ki}/P or $1 - n_{ki}/P$ are replaced by a small number δ. In the limit of many examples, its behavior will approach that of the optimal classifier.

12.5.2.2 The Hebb Classifier

The Hebb classifier is equivalent to the maximum likelihood estimator prescribed by:

$$w_{ki} = \frac{1}{P} \sum_{\mu_k} x_i^{\mu_k} = \frac{n_{ki}}{P}, \tag{II.12.35}$$

where $x_i^{\mu_k}$ is the ith input of the μ_kth example belonging to class k. The Hebb classifier is particularly simple to implement. The corresponding cost function is different from the usual choice of the mean square error, and has the form:

$$E = -\frac{1}{P} \sum_{ki} \sum_{\mu_k} w_{ki} x_i^{\mu_k}. \tag{II.12.36}$$

(Strictly speaking, a regularization term of the form $\Sigma_{ki} w_{ki}^2/2$ should be added to prevent its divergence.)

12.5.2.3 The Perceptron Classifier

It learns the examples iteratively using the multi-class perceptron learning algorithm prescribed in Equations II.12.24–II.12.26. The corresponding cost function is:

$$E = \frac{1}{P} \sum_k \sum_{\mu_k} \sum_{l \neq k} \left(\tau - \sum_i (w_{ki} - w_{li}) x_i^{\mu_k}\right) \odot \left(\tau - \sum_i (w_{ki} - w_{li}) x_i^{\mu_k}\right), \tag{II.12.37}$$

where $\odot$ is the step function.

12.5.2.4 The CART-Perceptron Classifier

This is the rule-based hybrid network identical to the one used in error message classification in Section 12.4. The inputs are first matched with the CART rules, and the overlaps between the inputs and the rules are fed to the perceptron.

12.5.3 The Generalization Performance

The generalization performance f_g for output class 1 of the winner-take-all classifier is given by:

$$f_g = \left\langle \prod_{k>1} \Theta(z_1 - z_k) \right\rangle_{\mathbf{x}}, \tag{II.12.38}$$

where z_k is the activation of class k defined in Equation II.12.32, and the average is performed over input states $\mathbf{x}$ randomly generated from class 1. In the large N limit, the generalization behavior is conveniently studied in terms of the error rate exponent x and the training set size exponent y, respectively, defined by:

$$p_0 \sim N^x, \text{ or } x \equiv \frac{\ln p_0}{\ln N}, \tag{II.12.39}$$

$$P \sim N^y, \text{ or } y \equiv \frac{\ln P}{\ln N} \tag{II.12.40}$$

Using the permutation symmetry of the classes, $z_1 - z_k$ is a Gaussian distribution, with a mean of $M_1 - M_2$, a noise specific to an output class k with a variance $= (Q_2 - R_2)/N$, and a noise common to all output classes $k > 1$ with a variance $= (Q_1 - 2R_1 + R_2)/N$, where we have made use of the permutation symmetry to reduce the statistical parameters to the following six:[135]

$$M_1 = \langle z_1 \rangle, \quad M_2 = \langle z_k \rangle \text{ for } k > 1, \quad Q_1 = \langle z_1^2 \rangle - \langle z_1 \rangle^2, \quad Q_2 = \langle z_k^2 \rangle - \langle z_k \rangle^2 \text{ for } k > 1,$$

$$R_1 = \langle z_1 z_k \rangle - \langle z_1 \rangle \langle z_k \rangle \text{ for } k > 1, \quad R_2 = \langle z_k z_l \rangle - \langle z_k \rangle \langle z_l \rangle \text{ for } k > l > 1. \tag{II.12.41}$$

Hence we define the signal-to-noise ratio (SNR) as the ratio of the mean to the root of the total variance. In high dimensions N, SNR scales as N^E, and $E \equiv \ln \text{SNR}/\ln N$ is the SNR exponent. If $E > 0$, the classifier generalizes perfectly, i.e., $f_g = 1$; if $E < 0$, the classifier generalizes randomly, i.e., $f_g = 1/K$. At the boundary separating the two phases, $E = 0$ and f_g rises steeply in the large N limit. Thus, the generalization undergoes a phase transition. From Table II.12.4 we readily identify the following three regimes.

12.5.3.1 The White Regime

For few examples, namely $P \ll p_0^{-\varepsilon}$, or $y < -\varepsilon x$ on using Equations II.12.39 and II.12.40, all except a few n_{ki} are 0, and the informators and backgrounds are not distinguishable.

12.5.3.2 The Gray Regime

When $p_0^{-\varepsilon} \ll P \ll p_0^{-1}$, or $-\varepsilon x < y < -x$ on using Equations II.12.39 and II.12.40, n_{ki} for the background bits remain dominated by zeros but for the informators, $n_{ki} \gg 1$. Classification relies heavily on the informators.

12.5.3.3 The Black Regime

For sufficient examples, namely $P \gg p_0^{-1}$, or $y > -x$ on using Equations II.12.39 and II.12.40, $n_{ki} \gg 1$ for all bits. Both the informators and backgrounds contribute to classification, and the Bayesian probabilities can be estimated accurately.

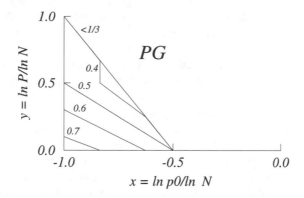

FIGURE II.12.10   Phase diagram for the Bayesian classifier at informator frequency $\gamma = 0.5$ for several values of informator strength $\bar{\varepsilon}$. Below and above the lines, the classifier is in the random generalization and perfect generalization (denoted by *PG*) phase respectively. *Source*: Wong, K.Y.M. and Lau, H.C., Neutral network classification of non-uniform data, *Progress in Neutral Information Processing*, vol. 1, Amari, S. et al., Eds., Springer, Singapore , 1996. With permission.

12.5.4   The Observed Trends

Figure II.12.10 shows the rich behavior in the phase diagrams for the Bayesian classifier when the informator strength $\bar{\varepsilon}$ varies at a given informator frequency γ. We observe the following trends.

12.5.4.1   Classification Eases with Error Rate

For a given error rate exponent x, there is a critical training set size, with exponent y, necessary for perfect generalization, which yields the phase lines. When the error rate p_0 (or its exponent x) increases, both the background bits and the informators carry more and more information, since the former become more and more populous, and the latter more and more certain. Hence the critical training set size exponent y decreases. When x is sufficiently large, classification is easy and the critical y reaches 0. A small training set size of the order N^0 is already sufficient for achieving perfect generalization.

12.5.4.2   Classification Eases with Informator Strength

When the informator strength $\bar{\varepsilon}$ is below $(1-\gamma)/(1+\gamma)$ (equals 1/3 in Figure II.12.10), the random generalization phase is maximally bounded by the phase line $2x + y + 1 = 0$. When the informator strength $\bar{\varepsilon}$ increases, classification becomes easier and the random generalization phase narrows. For informator strengths $\bar{\varepsilon}$ above $1 - \gamma/2$ (equals 3/4 in Figure II.12.10), the entire space has perfect generalization.

It is interesting to note the presence of a kink for $\bar{\varepsilon}$ lying between $(1-\gamma)/(1+\gamma)$ and $1-\gamma$ (the example of 0.4 in Figure II.12.10). It arises from the transition between the white and gray regimes. Indeed, the boundary $y = -\varepsilon x$ separating the two regimes passes through the kink, causing a discontinuity in the slope of the transition line between perfect and random generalization.

12.5.4.3   Backgrounds May Interfere or Assist Classification

In the gray regime, the probability for the occurrence of a symptom at a given background bit in the training set is of the order of $P p_0$. Dominantly, weights w_{ki} feeding out from these background bits have values $\ln(1/P)$. All the other background bits, recording no occurrence of symptoms in the training set, remain at the very negative value of $\ln \delta$. Since the number of background bits activated by a test example is of the order $p_0 N$, the contribution of the background bits to the activation z_k of an output class is of the order $P p_0^2 N \ln(1/P\delta)$, after subtracting the common background values of $\ln \delta$.

This should be compared with the contribution to the activation due to the informators. In the gray regime, dominantly, all informators record occurrence of symptoms in a fraction of p_0^ε of the training examples. Weights w_{ki} feeding from them are of the order $\ln(p_0^\varepsilon)$. When a test example is presented, the number of informators activated is of the order $p_0^\varepsilon C$. Hence, the contribution of informators to the activation z_k of an output class is of the order $p_0^\varepsilon C \ln(p_0^\varepsilon/\delta)$, after subtracting the common background values of $\ln \delta$.

When the background bits make a smaller contribution than the informators, generalization is performed by the information extracted from the informators, and background bits are only *interfering* with the task, since they send signals to *all* output classes. It is not surprising that classification becomes more difficult when the error rate p_0 (or its exponent x) increases. However, on further increase in p_0 or x, the background bits become numerous enough, so that their information traces from the background bits accumulate in the contribution to the activation of an output class. The background bits are *assisting* the classification, and the task becomes easier when p_0 or x increases.

We therefore conclude that there is a *role reversal* of the background bits from interference to assistance when p_0 or x increases. This happens when the contributions from the background bits and informators to the activation are comparable, namely, $Pp_0^2 N \sim p_0^\varepsilon C$, where the logarithmic terms can be neglected. In terms of the exponents, role reversal occurs at the line $(2-\varepsilon)x + y + 1 - \gamma = 0$. We expect a drop in generalization around this line.

This drop in generalization is most marked when $Pp_0^2 N \sim p_0^\varepsilon C \sim N^0$. In terms of the exponents, this occurs for certainty informators ($\bar{\varepsilon} = 1$) at the point $(x, y) = (-1/2, 0)$ when the number of informators scales as $N^0(\gamma = 0)$. As we will see, this accounts for the performance depression for finite values of N observed in simulations.

12.5.4.4 Weak Informators May Misinform

This is observed in the Hebb classifier, where generalization may become impossible if the error rates are too low, even for infinitely large training sets. This threshold error rate exists at intermediate informator strengths, namely for $(1 - \gamma)/3 < \bar{\varepsilon} < 1 - \gamma$. This is because the informators help classification when they are strong, and are irrelevant when they are weak. When their strength is intermediate, they are relevant but may misinform the classifier. We may say that $1 - \gamma < \bar{\varepsilon} < 1$ is the *true informator phase*, $(1 - \gamma)/3 < \bar{\varepsilon} < 1 - \gamma$ the *misinformator phase*, and $0 < \bar{\varepsilon} < (1 - \gamma)/3$ the *noninformator phase*. On the other hand, generalization in the Bayesian classifier is always possible for sufficiently large training sets, irrespective of the informator strength.

12.5.5 Simulations

12.5.5.1 No Informator

To study finite-size corrections to the large N limit, we perform Monte Carlo simulations of the classifiers using an input dimension of $N = 100$ and $K = 4$ output classes. For comparison, we first consider the predictions of the large N theory for the Bayesian and Hebb classifiers. With no informators ($C = 0$ or $\gamma \to -\infty$), the SNR exponent E of both classifiers becomes $2x + y + 1$ in the white and gray regimes, and $x + 1$ in the black regime. In both classifiers, the behavior changes from random to perfect generalization when x or y increases across the line $2x + y + 1 = 0$. The only exception is that at $x = -1$ in the black regime, f_g tends to 0.39 instead of 1, which is thus a singular line.

As shown in Figure II.12.11(a), for the case of no informators ($C = 0$) using the Hebb classifier, the region of poor generalization decreases with p_0, agreeing with the theory. The asymptotic generalization drops as x tends to -1, in agreement with the large N prediction that $x = -1$ is a singular line.

12.5.5.2 One Informator

For one informator ($C = 1$ or $\gamma = 0$) with certainty ($\bar{\varepsilon} = 1$), generalization of the Bayesian classifier is perfect in the entire space, but marginal along the singular line $2x + y + 1 = 0$, where f_g tends

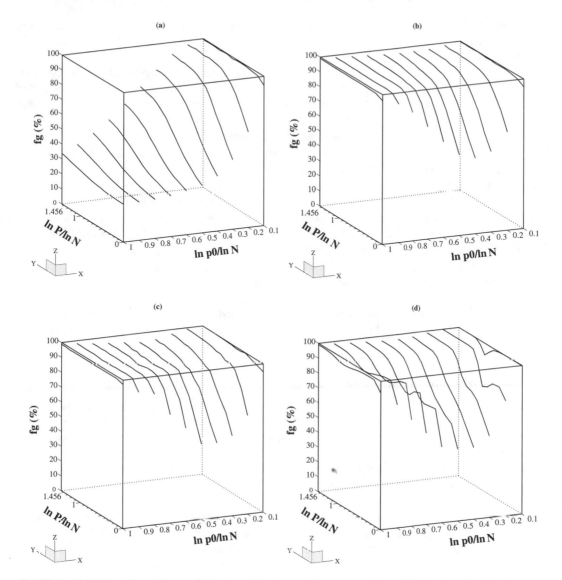

FIGURE II.12.11   Simulations of the generalization performance with no informators for (a) the Hebb classifier, and with one certain informator for (b) the Hebb classifier, (c) the perceptron classifier, and (d) the Bayesian classifier. *Source*: Wong, K.Y.M. and Lau, H.C., Neutral network classification of non-uniform data, *Progress in Neutral Information Processing*, vol. 1, Amari, S. et al., Eds., Springer, Singapore, 1996. With permission.

to 0.55, but 1 elsewhere. This dip in generalization is a consequence of the role reversal between the informators and backgrounds explained in Section 12.5.4.3. By comparison, marginal generalization in the Hebb classifier only exists at the singular point $(x, y) = (-1/2, 0)$, where f_g tends to 0.55, but to 1 elsewhere.

Figure II.12.11(b) shows the case with one informator ($C = 1$) classified by the Hebb classifier. Comparing with Figure II.12.11(a), the region of poor generalization is much reduced. This confirms that the informators provide a significant assistance to the classification task. The initial generalization drops near $x = -1/2$, reflecting the large N prediction that it is a singular point.

Figure II.12.11(c) shows the same case using the perceptron classifier. The generalization performance is similar to that of the Hebb classifier. A detailed comparison reveals that the asymptotic values improve slightly over the Hebbian results.

Figure II.12.11(d) shows the case with one informator using the Bayesian classifier. Comparing with Figure II.12.11(b–c), generalization is poorer for intermediate training set sizes in the present case, in agreement with the line of marginal generalization predicted by role reversal of the background bits in the large N theory.

For the CART-perceptron hybrid classifier, no analytical results are worked out. Turning to the simulation results in Figure II.12.12,[136] we see that in the case of no informators in Figure II.12.12(a), its performance is even worse than that of the perceptron classifier shown in Figure II.12.11(c). Since there are no informative bits in the model, CART takes no advantages of preprocessing the data.

The curves in Figure II.12.12(b) are interesting. For few background bits at small p_0, the classifier needs only a few training examples for perfect generalization, which is most satisfactory among the series of results presented here. It is because CART can extract the informator of the data and produce high quality rules for subsequent processing with the perceptron layer.

However, when p_0 increases further, the results are unsatisfactory, especially when the training examples are few. It is probably because the background bits are too active for CART to accurately extract the relevant informators.

We also perform simulations using uncertain informators with $\bar{\varepsilon} < 1$, and confirm that a task *lacking* informators with high certainty is not a suitable application of the CART-perceptron classifier.

12.5.6   Implications of Choice of Classifiers

By studying the phase diagram in the space of background error rate and training set size we have demonstrated, both analytically and simulationally, that the presence of informators makes the classification task easier. When the informators are not too strong, the data is relatively uniform. Hence, the Bayesian classifier performs better than other classifiers such as Hebb, as illustrated by the absence of the misinformator phase therein. However, the Bayesian classifier is not always optimal, as shown in both the theory and the large N limit, and simulations for finite size networks.

When the training examples are not sufficient, the Bayesian probabilities are not estimated accurately. This is especially valid in diagnostic models such as ours, in which the probability of occurrence differs largely between the informators and backgrounds.

On the other hand, a Hebb or a perceptron classifier is preferable in extracting the informator features with relatively few examples, though they are not necessarily optimal in the asymptotic limit of numerous examples.

We also confirm that the CART-perceptron hybrid classifier works best when there are strong informative bits and few background bits. This condition is exactly the same as that of fault diagnosis of the telephone system discussed in Section 12.4. Hence, this accounts for the success of the hybrid network approach to the problem. However, the same successful network may not be suitable in applications where strong informators are absent, or backgrounds are too active. The complementary dependence on the informator strength implies that a combination of all these classifiers may be useful in a wider range of applications.

12.6   Conclusion

We have surveyed the applications of neural networks in telecommunications systems, illustrating their usage with the two examples of overload control and fault diagnosis. While the technique has

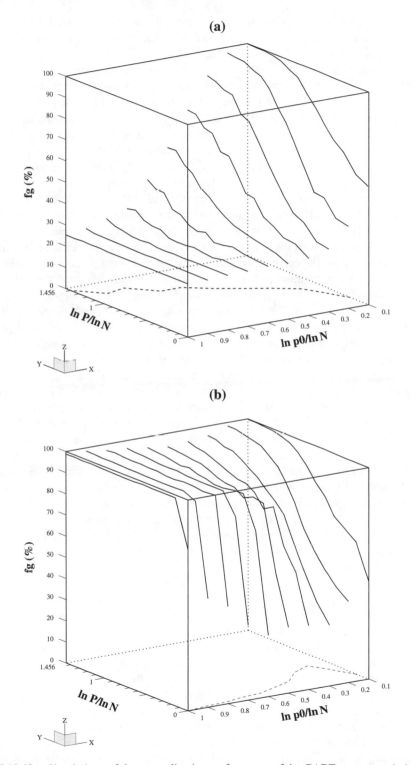

FIGURE II.12.12 Simulations of the generalization performance of the CART-perceptron hybrid classifier for (a) no informators, and (b) one certain informator. The dotted lines indicate the position where the generalization performance is at mid-value between the initial and asymptotic values. *Source*: Lau, H.C., Neural Network Classification Techniques for Diagnostic Problems, M.Phil. Thesis, 1995. With permission.

been extensively adopted in all levels of the system, the considerations involved in these applications are typical in many aspects, and exemplify the care needed to make them perform.

In the first example, we use a group of neural networks to implement a distributed overload control by learning the control actions of an optimal centralized teacher. This avoids two complications in the centralized control method, namely, the complex optimization which scales with the size of the network, and the overloading of the signaling network caused by the collection of network status information and the dissemination of control decisions. On the other hand, each of the neural network controllers learns to infer its own control action as part of the globally optimal decision, thus achieving a cooperative control even without global coordination. This is in contrast to the traditional local controllers, which base their decisions on strictly locally optimal objectives.

In the implementation of this concept, we have exercised care in a number of aspects. First, examples are generated by a teacher with globally optimal performance. The high quality of the examples is ensured by undertaking a sequence of linear programming for each network condition. Second, we have preprocessed the data to form inputs which are most relevant to the geometry of locating the optimal point in the solution space. Third, we have used the radial basis function architecture, so that each RBF center approximates the teacher control function within its range of loading level, and the combination provides a smooth interpolation among them all. Simulations show that the method is successful in both steady overload and traffic upsurges.

We have applied the same method to the dynamic routing of teletraffic in hierarchical networks.[24,137] First, a training set of globally optimal examples is obtained by a sequence of linear programming which aims at breaking the degeneracy of the optimal solution until no ambiguity is left. Then a group of neural networks, each located at an originating node, learn the teacher control function through the examples, and infer the optimal routing ratios as part of the globally optimal control, based on locally available information. The method yields blocking and crankback rates comparable to the centralized controller. Although we subsequently found a heuristic method which outperforms the neural controllers in hierarchical networks,[137] it is possible that the neural controllers may retain their advantages in networks with higher complexities. Further studies are needed.

In general, many control problems in telecommunications systems share the same features of a complicated centralized control vs. a sub-optimal local control. Examples are dynamic packet routing in computer networks, dynamic channel allocation in wireless networks, multiservice call admission control in wireless networks, and dynamic call admission control and routing in ATM networks. Neural network control provides an attractive solution to them.

In the second example, we use a hybrid classifier consisting of a rule-based preprocessing layer and a perceptron output layer. The rule-based layer plays the roles of feature extraction and dimensionality reduction, easing the classification task of the perceptron layer. That the hybrid network performs better than each individual component again illustrates the importance of preprocessing the data. We remark that the hybrid classifier works best when the components have complementary features. For example, we found that CART and Bayesian classifiers are incompatible, probably because their common decisive behavior offers no complementary advantages on hybridization.

We reason that a diagnostic problem differs from a typical pattern recognition problem in having a nonuniform distribution of information in the data. Some input bits are more informative or appear in high correlation, thus reducing the applicability of conventional classification techniques. Based on these characteristics, we further propose the informator model of diagnostic data.

Both analytical and simulational results show that the presence of informators makes the classification task easier, as evident from the reduction in the random generalization region. However, it is difficult to find a universal technique that can give superior performance on all problems, again illustrating the importance of carefully choosing the right classifier. For example, while the Bayesian classifier works perfectly in the asymptotic regime of many examples, it does not perform well in the regime of few examples, which does not allow the Bayesian probabilities to be estimated precisely. It also deteriorates quickly when there are correlations in the input bits.

When the case of insufficient training is taken into consideration, we find that the CART-perceptron classifier performs especially well when there are strong informative bits and less background noise in the data model, where the CART takes advantages of extracting the informative data for subsequent network processing.

For that reason, we conclude that different problems may require specialized techniques for good performance, and the use of classification trees as data preprocessing for a perceptron classifier is found applicable to diagnostic problems.

Acknowledgments

We thank the Research Grant Council of Hong Kong for partial support (grant no. HKUST6157/99P).

References

1. Hertz, J., Krogh, A., and Palmer, R.G., *Introduction to the Theory of Neural Computation*, Addison-Wesley, Redwood City, 1991.
2. Alspector, J., Goodman, R., and Brown, T.X., Eds., *Applications of Neural Networks to Telecommunications*, Lawrence Erlbaum, Hillsdale, 1993.
3. Alspector, J., Goodman, R., and Brown, T.X., Eds., *Applications of Neural Networks to Telecommunications 2*, Lawrence Erlbaum, Hillsdale, 1995.
4. Alspector, J., Goodman, R., and Brown, T.X., Eds., *Applications of Neural Networks to Telecommunications 3*, Lawrence Erlbaum, Hillsdale, 1997.
5. *IEEE Journal on Selected Areas in Communications*, vol. 15, no. 2, 1997.
6. *IEEE Journal on Selected Areas in Communications*, vol. 18, no. 2, 2000.
7. Bang, S.H., Sheu, B.J., and Choi, J., Programmable VLSI neural network processors for equalization of digital communication channels, 1–12 in Alspector et al.,[2] 1993.
8. Cid-Suerio, J. and Figueiras-Vidal, A.R., Improving conventional equalizers with neural networks, 20–26 in Alspector et al.,[2] 1993.
9. Brown, T.X., Neural networks for adaptive equalization, 27–33 in Alspector et al.,[2] 1993.
10. Meyer, M. and Pfeiffer, G., Multilayer perception-based equalizers applied to nonlinear channels, 188–195 in Alspector et al.,[2] 1993.
11. Sönmez, M.K. and Adali, T., Channel equalization by distribution learning: The least relative entropy algorithm, 218–224 in Alspector et al.,[2] 1993.
12. Bradley, M.J. and Mars, P., Analysis of recurrent networks as digital communication channel equalizer, 1–8 in Alspector et al.,[3] 1995.
13. Reay, D.S., Nonlinear channel equalization using associative memory neural networks, 17–24 in Alspector et al.,[3] 1995.
14. Jayakumar, A. and Alspector, J., Experimental analog neural network-based decision feedback equalizer for digital mobile radio, 33–40 in Alspector et al.,[3] 1995.
15. Gan, Q. et al., Equalisation of rapidly time-varying channels using an efficient RBF neural network, 224–231 in Alspector et al.,[4] 1997.
16. Dubossarsky, E., Osborn, T.R., and Reisenfeld, S., Equalization and the impulsive MOOSE: Fast adaptive signal recovery in very heavy tailed noise, 232–240 in Alspector et al.,[4] 1997.
17. Raivio, K., Henrikson, J., and Simula, O., Neural receiver structures based on self-organizing maps in nonlinear multipath channels, 241–247 in Alspector et al.,[4] 1997.
18. Brown, T.X., Neural Networks for Switching, *IEEE Comm. Mag.* 27, 72–80 1989.
19. Amin, S. and Gell, M., Constrained optimization for switching using neural networks, 106–111 in Alspector et al.,[2] 1993.

20. Park, Y.K., Cherkassky, V., and Lee, G., ATM cell scheduling for broadband switching systems by neural network, 112–118 in Alspector et al.,[2] 1993.

21. Park, Y.K. and Lee, G., NN-based ATM cell scheduling with queue length-based priority scheme, 261–269 in *IEEE J. Selected Areas in Common.*,[5] 1997.

22. Varma, A. and Antonucci, R., A neural-network controller for scheduling packet transmissions in a crossbar switch, 121–128 in Alspector et al.,[3] 1995.

23. Murgu, A., Adaptive flow control in multistage communications networks based on a sliding window learning algorithm, 112–120 in Alspector et al.,[3] 1995.

24. Lor, W.K.F. and Wong, K.Y.M., Decentralized neural dynamic routing in circuit-switched networks, 137–144 in Alspector et al.,[3] 1995.

25. Campbell, P. et al. Experiments with simple neural networks for real-time control, 165–178 in *IEEE J. Selected Areas in Common.*,[5] 1997.

26. Christiansen, A. et al., Neural networks for resource allocation in telecommunication networks, 265–273 in Alspector et al.,[4] 1997.

27. Wu, S. and Wong, K.Y.M., Overload control for distributed call processors using neural networks, 149–156 in Alspector et al.,[4] 1997.

28. Wu, S. and Wong, K.Y.M., Dynamic overload control for distributed call processors using the neural network method, *IEEE Trans. of Neural Networks* 9, 1377–1387, 1998.

29. de Vries, B. et al., Neural network speech enhancement for noise robust speech recognition , 9–16 in Alspector et al.,[3] 1995.

30. Frederickson, S. and Tarassenko, L., Text-independent speakers recognition using radial basis functions, 170–177 in Alspector et al.,[3] 1995.

31. Kasabov, N., Hybrid environments for building comprehensive AI and the task of speech recognition, 178–185 in Alspector et al.,[3] 1995.

32. Barnard, E. et al., Real-world speech recognition with neural networks, 186–193 in Alspector et al.,[3] 1995.

33. Yuang, M.C., Tien, P.L., and Liang, S.T., Intelligent video smoother for multimedia communications, 136–146 *IEEE J. Selected Areas in Common.*,[5] 1997.

34. Bustos, R.A. and Gedeon, T.D., Learning synonyms and related concepts in document collections, 202–209 in Alspector et al.,[3] 1995.

35. Gedeon, T.D. et al., Query word-concept clusters in a legal document collection, 189–197 in Alspector et al.,[4] 1997.

36. Liu, H. and Yun, D.Y.Y., Self-organizing finite state vector quantization for image coding, 176–182 in Alspector et al.,[2] 1993.

37. Chieuh, T.D., Tang, T.T., and Chen, L.G., Vector quantization using tree-structured self-organizing feature maps, 259–265 in Alspector et al.,[2] 1993.

38. Mekuria, F. and Fjällbrant, T., Neural networks for efficient adaptive vector quantization of signals, 218–225 in Alspector et al.,[3] 1995.

39. Carter, S., Frank, R.J., and Tansley, D.S.W., Clone detection in telecommunications software systems: a neural net approach, 273–280 in Alspector et al.,[2] 1993.

40. Barson, P. et al., Dynamic competitive learning applied to the clone detection problem, 234–241 in Alspector et al.,[3] 1995.

41. Connor, J.T., Prediction of access line growth, 232–238 in Alspector et al.,[2] 1993.

42. Giraud-Carrier, C. and Ward, M., Learning customer profiles to generate cash over the Internet, 165–170 in Alspector et al.,[4] 1997.

43. Mozer, M.C., Churn reduction in the wireless industry, *Advances in Neural Information Processing Systems* 12, S.A. Solla, T.K. Leen, K.-R. Müller, Eds., 935–941, MIT Press, Cambridge, 2000.

44. Engelbrecht, A.P. and Cloete, I., Dimensioning of telephone networks using a neural network as traffic distribution approximator, 72–79 in Alspector et al.,[3] 1995.

45. Zhang, C.X., Optimal traffic routing using self-organization principle, 225–231 in Alspector et al.,[2] 1993.

46. Lewis, L., Datta, U., and Sycamore, S., Intelligent capacity evaluation/planning with neural network clustering algorithms, 131–139 in Alspector et al.,[4] 1997.

47. Hoang, D.B., Neural networks for network topological design, 140–148 in Alspector et al.,[4] 1997.

48. Wang, G. and Ansari, N., Optimal broadcast scheduling in packet radio networks using mean field annealing, 250–260 in *IEEE J. Selected Areas in Common.*,[5] 1997.

49. Jagota, A., Scheduling problems in radio networks using Hopfield networks, 67–76 in Alspector et al.,[2] 1993.

50. Comellas, F. and Ozón, J., Graph coloring algorithms for assignment problems in radio networks, 49–56 in Alspector et al.,[3] 1995.

51. Berger, M.O., Fast channel assignment in cellular radio systems, 57–63 in Alspector et al.,[3] 1995.

52. Dixon, M.W., Bellgard, M.I., and Cole, G.R., A neural network algorithm to solve the routing problem in communication networks, 145–152 in Alspector et al.,[3] 1995.

53. Goodman, R.M., and Ambrose, B.E., Learning telephone network trunk reservation congestion control using neural networks, 258–264 in Alspector et al.,[3] 1995.

54. Fahmy, H.I., Develekos, G., and Douligeris, C., Application of neural networks and machine learning in network design, 226–237 in *IEEE. J. Selected Areas in commun.*,[5] 1975.

55. Hood, C.S. and Ji, C., An intelligent monitoring hierarchy for network management, 250–257 in Alspector et al.,[3] 1995.

56. Cortes, C., Jackel, L.D., and Chiang, W.P., Predicting failures of telecommunication paths: limits on learning machine accuracy imposed by data quality, 324–333 in Alspector et al.,[3] 1995.

57. Lewis, L. and Sycamore, S., Learning index rules and adaptation functions for a communications network fault resolution system, 281–287 in Alspector et al.,[2] 1993.

58. Collobert, M. and Collobert, D., A neural system to detect faulty components on complex boards in digital switches, 334–338 in Alspector et al.,[3] 1995.

59. Goodman, R. and Ambrose, B., Applications of learning techniques to network management, 34–44 in Alspector et al.,[2] 1993.

60. Chattell, A. and Brook, J.B., A neural network pre-processor for a fault diagnosis expert system, 297–305 in Alspector et al.,[2] 1993.

61. Holst, A. and Lansner, A., Diagnosis of technical equipment using a Bayesian neural network, 147–153 in Alspector et al.,[2] 1993.

62. Holst, A. and Lansner, A., A higher-order Bayesian neural network for classification and diagnosis, 347–354 in Alspector et al.,[3] 1995.

63. Lau, H.C. et al., A hybrid expert system for error message classification, 339–346 in Alspector et al.[3], 1995.

64. Sone, T., Using distributed neural networks to identify faults in switching systems, 288–296 in Alspector et al.,[2] 1993.

65. Sone, T., A strong combination of neural networks and deep reasoning in fault diagnosis, 355–362 in Alspector et al.,[3] 1995.

66. Leray, P., Gallinari, P., and Didelet, E., Local diagnosis for real-time network traffic management, 124–130 in Alspector et al.,[4] 1997.

67. Wietgrefe, H. et al., Using neural networks for alarm correlation in cellular phone networks, 248–255 in Alspector et al.,[4] 1997.

68. Yuhas, B.P., Toll-fraud detection, 239–244 in Alspector et al.,[2] 1993.

69. Connor, J.T., Brothers, L.B., and Alspector, J., Neural network detection of fraudulent calling card patterns, 363–370 in Alspector et al.,[3] 1995.

70. Field, S.D.H. and Hobson, P.W., Techniques for telecommunications fraud management, 107–115 in Alspector et al.,[4] 1997.

71. Junius, M. and Kennemann, O., Intelligent techniques for the GSM handover process, 41–48 in Alspector et al.,[3] 1995.

72. Biesterfeld, J., Ennigrou, E., and Jobmann, K., Neural networks for location prediction in mobile networks, 207–214 in Alspector et al.,[4] 1997.

73. Smith, K. and Palaniswami, M., Static and dynamic channel assignment using neural networks, 238–249 in *IEEE J. Selected Areas in commun.*,[5] 1997.

74. Singh, S. and Bertsekas, D., Reinforcement learning for dynamic channel allocation in cellular telephone systems, *Advances in Neural Information Processing Systems* 9, M.C. Mozer, M.I. Jordan, T. Petsche, Eds., 974–980, MIT Press, Cambridge, 1997.

75. Wilmes, E.J. and Erickson, K.T., Reinforcement learning and supervised learning control of dynamic channel allocation for mobile radio systems, 215–223 in Alspector et al.,[4] 1997.

76. Hiramatsu, A. ATM communications network control by neural networks, *IEEE Trans. Neural Networks*, 1, 122–140, 1990.

77. Hiramatsu, A. Integration of ATM call admission control and link capacity control by distributed neural networks, *IEEE J. Selected Areas in Commun.* 9, 1131–1138, 1991.

78. Youssef, S.A., Habib, I.W., and Sadaawi, T.N., A neurocomputing controller for bandwidth allocation in ATM Networks, *IEEE J. Selected Areas in Commun.*,[5] 191–199, 1997.

79. Brown, T.X., Adaptive access control applied to Ethernet data, in *Advances in Neural Information Processing Systems* 9, M.C. Mozer, M.I. Jordan, T. Petsche, Eds., 932–938, MIT Press, Cambridge, 1997.

80. Tham, C.K. and Soh, W.S., ATM connection admission control using modular neural networks, 71–78 in Alspector et al.,[4] 1997.

81. Zaremba, M.B. et al., Link bandwidth allocation in multiservice networks using neural technology, 64–71 in Alspector et al.,[3] 1995.

82. Nordström, E. and Carlström, J., A reinforcement learning scheme for adaptive link allocation in ATM networks, 88–95 in Alspector et al.,[3] 1995.

83. Gällmo, O. and Asplund, L., Reinforcement learning by construction of hypothetical targets, 300–307 in Alspector et al.,[3] 1995.

84. Marbach, P., Mihatsch, O., and Tsitiklis, J.N., Call admission control and routing in integrated services networks using neuro-dynamic programming, 197–208 in *IEEE J. Selected Areas in Commun.*,[6] 2000.

85. Tong, H. and Brown, T.X., Adaptive call admission control under quality of service constraints: a reinforcement learning solution, 209–221 in *IEEE J. Selected Areas in Commun.*,[6] 2000.

86. Carlström, J. and Nordström, E., Control of self-similar ATM call traffic by reinforcement learning, 54–62 in Alspector et al.,[4] 1997.

87. Estrella, A.D. et al., ATM traffic neural control: Multiservice call admission and policing function, 104–111 in Alspector et al.,[3] 1995.

88. Faragó, A. et al., Virtual lookahead—a new approach to train neural nets for solving on-line decision problems, 265–272 in Alspector et al.,[3] 1995.

89. Mahadevan, I. and Raghavendra, C.S., Admission control in ATM networks using fuzzy-ARTMAP, 79–87 in Alspector et al.,[4] 1997.

90. Liu, Y. and Douligeris, C., Rate regulation with feedback controller in ATM networks—a neural network approach, 200–208 in *IEEE J. Selected Areas in Commun.*,[5] 1997.

91. Pitsillides, A., Şekercioğlu, Y.A., and Ramamurphy, G., Effective control of traffic flow in ATM networks using fuzzy explicit rate marking (FERM), 209–225 in *IEEE J. Selected Areas in Commun.*,[5] 1997.

92. Murgu, A., Fuzzy mean flow estimation with neural networks for multistage ATM systems, 27–35 in Alspector et al.,[4] 1997.

93. Fan, Z. and Mars, P., Dynamic routing in ATM networks with effective bandwidth estimation by neural networks, 45–53 in Alspector et al.,[4] 1997.

94. Faragó, A. et al., Analog neural optimization for ATM resource management, 156–164 in *IEEE J. Selected Areas in Commun.*,[5] 1997.

95. Brown, T.X., Bandwidth dimensioning for data traffic, 88–96 in Alspector et al.,[4] 1997.

96. Edwards, T. et al., Traffic trends analysis using neural networks, 157–164 in Alspector et al.,[4] 1997.

97. Casilari, E. et al., Generation of ATM video traffic using neural networks, 19–26 in Alspector et al.,[4] 1997.

98. Casilari, E. et al., Model generation of aggregate ATM traffic using a neural control with accelerated self-scaling, 36–44 in Alspector et al.,[4] 1997.

99. Tarraf, A.A., Habib, I.W., and Sadaawi, T.N., Neural networks for ATM multimedia traffic prediction, 85–91 in Alspector et al.,[2] 1993.

100. Neves, J.E., de Almeida, L.B., and Leitão, M.J., ATM call control by neural networks, 210–217 in Alspector et al.,[2] 1993.

101. Gelenbe, E., Ghanwani, A., and Srinivasan, V., Improved neural heuristics for multicast routing, 147–155 in *IEEE J. Selected Areas in Commun.*,[5] 1997.

102. Ali, Z., Gafoor, A., and Lee, C.S.G., Media synchronization in multimedia web environment using a neuro-fuzzy framework, 168–183 in *IEEE J. Selected Areas in Commun.*,[6] 2000.

103. Davoli, F. and Maryni, P., A two-level stochastic approximation for admission control and bandwidth allocation, 222–233 in *IEEE J. Selected Areas in Commun.*,[6] 2000.

104. Chang, C.J. et al., Fuzzy/neural congestion control for integrated voice and data DS-CDMA/FRMA cellular networks, 283–293 in *IEEE J. Selected Areas in Commun.*,[6] 2000.

105. Cybenko, G., Approximation by superpositions of a sigmoid function, *Mathematics of Control, Signals and Systems* 2, 303 1989.

106. Boyan, J. and Littman, M.L., Packet routing in dynamically changing networks: a reinforcement learning approach, in *Adv. Neural Information Processing Systems* 6, J. Cowan, G. Tesauro, J. Alspector, Eds., 671–678, Morgan Kaufmann, San Francisco, 1994.

107. Choi, S. and Yeung, D.Y., Predictive Q-routing: a memory-based reinforcement learning approach to adaptive traffic control, *Advances in Neural Information Processing Systems* 8, D. Touretzky, M.C. Mozer, M.E. Hasselmo, Eds., 945–951, MIT Press, Cambridge, 1996.

108. Hérault, L., Dérou, D., and Gordon, M., New Q-routing approaches to adaptive traffic control, 274–281 in Alspector et al.,[4] 1997.

109. Brown, T.X., Low power wireless communication via reinforcement learning, *Adv. Neural Information Processing Systems* 12, S.A. Solla, T.K. Leen, K.-R. Mïler, Eds., 893–899, MIT Press, Cambridge, 2000.

110. Bishop, C.M., *Neural Networks for Pattern Recognition*, Clarendon Press, Oxford, 1995.

111. Amari, S., *Differential-Geometrical Methods in Statistics*, Springer-Verlag, New York, 1985.

112. Hopfield, J.J., Neural networks and physical systems with emergent computational abilities, *Proc. Natl. Acad. Sci. U.S.A.* 79, 2554–2558, 1982.

113. Rumelhart, D.E., Hinton, G.E., and Williams, R.J., Learning internal representations by error propagation, *Parallel Distributed Processing: Explorations in Microstructure of Cognition* 1, 318–362, MIT Press, Cambridge, 1988.

114. Moody, J. and Darken, C.J., Fast learning in networks of locally-tuned processing units, *Neural Computation* 1, 281–294, 1989.

115. Park, J. and Sandberg, I.W., Universal approximation using radial basis function networks, *Neural Computation* 3, 246–257, 1991.

116. Dempster, A.P., Laird, N.M., and Rubin, D.B., Maximum likelihood from incomplete data via the EM algorithm, *J. R. Statistical Society B* 39, 1–38, 1977.

117. Boser, B.E., Guyon, I.M., and Vapnik, V.N., A training algorithm for optimal margin classifier, *Proc. 5th ACM Workshop on Computational Learning Theory*, 144–152, 1992.

118. Vapnik, V., *The Nature of Statistical Learning Theory*, Springer-Verlag, New York, 1995.

119. Girosi, F., Jones, M., and Poggio, T., Regularization theory and neural network architectures, *Neural Computation* 10, 1455–1480, 1998.

120. Christianini, N. and Shawe-Taylor, J., *An Introduction to Support Vector Machines and Other Kernel-Based Methods*, Cambridge University Press, Cambridge, 2000.

121. Geman, S., Bienenstock, E., and Doursat, R., Neural networks and the bias/variance dilemma, *Neural Computation* 4, 1–58, 1992.

122. Bello, M.G., Enhanced training algorithms, and integrated training/architecture selection for multilayer perceptron networks, *IEEE Trans. Neural Networks* 3, 864–875, 1992.

123. Mozer, M.C. and Smolensky, P., Skeletonization: A technique for trimming the fat from a network via relevance assessment, *Advances in Neural Information Processing Systems* 1, 107–115, Morgan Kaufmann, San Mateo, 1989.

124. Amari, S., Natural gradient works efficiently in learning, *Neural Computation* 10, 252–276 1998.

125. W.H. Press et al., *Numerical Recipes in C: The Art of Scientific Computing* (2nd ed.), Cambridge University Press, Cambridge, 1992.

126. Hanselka, P., Oehlerich, J., and Wegmann, G., Adaptation of the overload regulation method stator to multiprocessor controls and simulation results, *ITC-12*, 395–401, 1989.

127. Manfield, D. et al., Overload control in a hierarchical switching system, *ITC-11*, 894–900, 1985.

128. Villen-Altamirano, M., Morales-Andres, G., and Bermejo-Saez,L., An overload control strategy for distributed control systems, *ITC-11*, 835–841, 1985.

129. Kaufman, J.S. and Kumar, A., Traffic overload control in a fully distributed switching environment, *ITC-12*, 386–394, 1989.

130. Best, M.J. and Ritter, K., *Linear Programming: Active Set Analysis and Computer Programs*, Prentice-Hall, Englewood Cliffs, 1985.

131. Stokbro, K., Umberger, D.K., and Hertz, J.A., Exploiting neurons with localized receptive fields to learn chaos, *Complex Syst.*, 4, 603–622, 1990.

132. Goodman, R.M. et al., Rule-based neural networks for classification and probability estimation, *Neural Computation* 4, 781–803, 1992.

133. Breiman, L. et al., *Classification and Regression Trees*, Wadsworth, Pacific Grove, 1984.

134. Sklansky, J. and Wassel, G.N., *Pattern Classifiers and Trainable Machines*, Springer-Verlag, New York, 1981.

135. Wong, K.Y.M. and Lau, H.C., Neural network classification of non-uniform data, Progress in *Neural Information Processing*, S. Amari, L. Xu, L.W. Chan, I. King, K.S. Leung, Eds., 242–246, Springer, Singapore, 1996.

136. Lau, H.C., Neural network classification techniques for diagnostic problems, M.Phil. Thesis, HKUST, 1995.

137. Au, W.K., Conventional and neurocomputational methods in teletraffic routing, M.Phil. Thesis, HKUST, 1999.

13

A Fuzzy Temporal Rule-Based Approach for the Design of Behaviors in Mobile Robotics

Manuel Mucientes
University of Santiago de Compostela

Roberto Iglesias
University of Santiago de Compostela

Carlos V. Regueiro
University of A Coruña

Alberto Bugarín
University of Santiago de Compostela

Senen Barro
University of Santiago de Compostela

Abstract. This chapter describes the application of a knowledge and reasoning representation model named Fuzzy Temporal Rules (FTRs) to the field of mobile robotics. This model makes it possible to explicitly incorporate time as a variable, thus allowing the evolution of variables in a temporal reference to be described. Specifically, two behaviors have been implemented on a Nomad 200 robot: wall following and avoidance of moving objects. In both cases it was shown how the use of FTRs not only enables us to determine the tendency of the different variables, but also to filter part of the sensor noise, as the input data are analyzed over temporal references. In this manner a more effective realization of both behaviors is obtained than by conventional fuzzy control.

13.1 Introduction

Currently, one of the principal fields of research in robotics is the development of techniques for the guidance of autonomous robots. Despite the great advances made in robotics in recent years, many complex unsolved problems still exist in this field. The greatest difficulties are due to the nature of the real world (environments difficult to model) and the uncertainty in these environments; knowledge about an environment is incomplete, uncertain, and approximated because the information supplied by the robot sensors is limited and not totally reliable. Also, the dynamism of the environment in which the robot is located is unpredictable.

There are two clearly differentiated solutions for developing a control architecture: reactive systems[1,2] and deliberative systems.[3] The former operate by sensing the environment and acting (stimulus–response) in a rapid manner, but without the capability for planning, while deliberative systems have to carry out a planning stage before acting. This renders them slow, and the environment may have changed significantly during this period and, thus, actions may already be inappropriate.

Hybrid architectures[4–6] are a half-way point between reactive and deliberative ones; a high-level deliberative strategy is followed when the environment is modeled and the plan to be followed is drawn up. At a low level, the execution layer is reactive, allowing it to act appropriately when faced with changes in the environment. The complexity of the execution layer increases as it must fulfill the different goals proposed by the planner and, at the same time, respond to any changes in the environment. One habitually used strategy to reduce complexity is to break down the execution stage into simple behaviors, such as avoiding obstacles, following a wall, going through a door. Thus, the planner supplies a series of goals that needs to be fulfilled, and different behaviors are activated to do so. The selection of the most suitable behaviors for each situation is implemented by means of an arbitration strategy, while the final order sent to the robot comes from the fusion of the commands of those selected behaviors.

In this chapter we describe two habitual behaviors in the field of mobile robotics (wall following and the avoidance of moving objects) in which a reasoning model called Fuzzy Temporal Rules (FTRs) has been employed. Fuzzy logic has proved to be a useful tool in the field of robotics,[7] as demonstrated in numerous studies carried out for guidance in real environments,[8,9] obstacle avoidance,[10] route planning,[11] and other tasks of interest. This is due, primarily, to fuzzy logic's capability for dealing with imprecise or uncertain measurements, which is of great interest in robotics, where sensorial readings carried out by robots in real environments always have a degree of imprecision associated to them. Despite this, fuzzy logic makes it possible to control a system even when its model is unknown. In this case, the system comprises the robot and the environment surrounding it. Dynamism is a common characteristic of real settings; there are moving obstacles, the environment may be altered by interaction with the robot, people, etc. so that completely modeling the system is an extremely complex task. The use of FTR makes it possible to endow conventional fuzzy control with the tools for carrying out temporal reasoning on system input variables. The significant advantages obtained are twofold: on one hand, it is possible to estimate the tendency of the input variables to examine the evolution of the system; on the other hand, it is possible to filter part of the noise associated to these variables (due fundamentally to erroneous readings of the ultrasound sensors).

The following section introduces fuzzy control, paying special attention to the advantages of fuzzy temporal control, while in Section 13.3 brief mention is made to the robot that is used (Nomad 200). In Section 13.4 wall-following behavior is explained in detail; Section 13.5 deals with behavior for avoiding moving objects; and conclusions are given in Section 13.6.

13.2   Fuzzy Control and Fuzzy Temporal Control

13.2.1   Fuzzy Control

One of the principal advantages of fuzzy control is its ability to incorporate knowledge and schemes of reasoning that are typically human into automation systems. This reasoning incorporates vagueness and imprecision into its description. This capability, inherent to fuzzy logic, for working with vague concepts makes it possible to reproduce experts' knowledge of a domain by means of fuzzy rules,

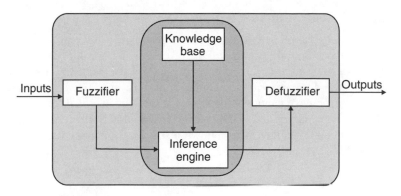

FIGURE II.13.1   Block diagram of a fuzzy control system.

principally when this is expressed by linguistic variables, the values of which cannot be defined precisely (such as, for example, *turn a little*).

Another interesting characteristic of fuzzy control is the lower degree of complexity in the mathematical expressions involved in the description of the reasoning process. This, added to the fact that a fuzzy controller works with a Knowledge Base (KB) comprising a set of more or less simple rules, makes it possible to implement parallel processing, giving rise to carrying out fuzzy inference within a calculation time that, in most of cases, makes its application in real-time problems possible.

Worthy of mention is the greater simplicity in tuning a fuzzy controller (it is relatively simple to verify which rules are being executed at any given moment), apart from the improved robustness due to a high degree of redundancy that allows the system a certain degree of immunity when faced with errors in the KB. Figure II.13.1 shows the typical architecture of a fuzzy controller.

A fuzzy rule base is generally made up of a set of IF–THEN rules that can be expressed as:

$$R^{(l)} : \text{IF } X_1^l \text{ is } A_1^l \text{ and } \dots \text{ and } X_M^l \text{ is } A_M^l, \text{ THEN } Y^l \text{ is } B^l \tag{II.13.1}$$

where $R^{(l)}, l = 1, 2, \dots, L$, is the l-th rule, $X_m^l, m = 1, 2, \dots, M$, and Y^l are linguistic variables of the antecedent and consequent parts, respectively, while A_m^l and B^l are linguistic values (labels) of these variables. One example of a fuzzy rule that follows this outline is

IF velocity is low and the collision time is medium, THEN turn moderately.

In this rule, *velocity, collision time*, and *turn* are linguistic variables, where *low, medium*, and *moderately* are their values, respectively. These fuzzy sets are defined on the universe of discourse U_m that is associated to each variable.

By means of the inference mechanism a mapping of the input fuzzy sets is carried out (those associated to the values described in the antecedent part of the rules). The most widely used inference mechanism in the bibliography is the compositional inference rule,[12] which interprets the process as the Sup-min composition of two relations: the relation induced by the simultaneous observation of the antecedents and the relation induced by the existence of the rules.

In the inference process, each rule is interpreted as a fuzzy implication that establishes a relation between the universes of discourse of the input and output variables:

$$R^{(l)} : A^l \rightarrow B^l \tag{II.13.2}$$

To determine the values of the control variables, we only need to take into account the values of the input variables at the current instant, which gives rise to a series of limitations, such as the impossibility of taking into account the evolution of the variables, or the fact that an erroneous measurement may produce an inadequate control output, as is explained in the following section.

13.2.2   Temporal Reasoning: FTR Model

In the majority of fuzzy logic applications, knowledge is modeled by nontemporal Fuzzy Knowledge Bases (FKBs), where the temporal dynamics of the process is not usually taken into consideration, with the exception of certain cases with variables defined for the purpose ("increase in distance," "accumulated error," etc.). In many real-time applications, this results in a strong restriction on the reasoning possibilities over the dynamics of the system, due to which models explicitly incorporating time as a variable have been proposed.[13–15] The model described in Barro et al.[16] and Bugarín et al.[17] has been adopted for our applications. The following structure for fuzzy temporal propositions is described:

$$X \text{ is } A \langle \text{ in } Q \text{ of } \rangle T \tag{II.13.3}$$

where X is a linguistic variable, A represents a linguistic value of X, T is a temporal reference or entity, and Q is a fuzzy quantifier.

The temporal entities T may represent both fuzzy temporal instants as well as fuzzy temporal intervals. In both cases membership functions are defined on a discrete set of values $\tau = \{\tau_0, \tau_1, \ldots, \tau_k, \ldots, \tau_{now}\}$, where each τ_k represents a precise temporal instant and τ_0 represents the origin.

We assume that the values of this set are evenly spaced, where $\Delta = \tau_j - \tau_{j-1}$ is the unit of time, whose size or granularity depends on the temporal dynamics of the application with which it is being dealt.

It is considered that all temporal distributions are defined in relation to the temporal point at which the proposition is evaluated, τ_{now} (current instant at each moment, since execution in real time is assumed), as can be seen in the example in Figure II.13.2. When the fuzzy temporal entity T represents an interval, it may require the fulfillment of "X is A" (spatial part of the proposition) for some point of interval T (a situation we refer to as *nonpersistence*), or require its fulfillment throughout the entire interval (*throughout T*: a situation of *persistence*), or that this should be fulfilled for some subinterval (*in the majority of T, in part of T*). An example of the first situation is the proposition "*velocity is high in the last three seconds*," and of the second, *velocity is high throughout the last three seconds*. In the first case, the existential quantifier *in* is included, and it is sufficient for the velocity to have been high at some temporal point which effectively belongs to *the last three seconds* for the compliance of the proposition to be considered high. In the second case (universal quantifier *throughout*), the fulfillment of this must be evaluated on all those temporal points that belong to the support* of *the last three seconds*.

The execution process of a FTR differs from that of a conventional fuzzy rule in the calculation of the degree of fulfillment (DOF), which now not only depends on the current measurement, but also on those corresponding to prior instants. The calculation of DOF is carried out in the model in the following manner: first, the degree of fulfillment of the spatial part of the proposition is calculated, which is defined as the spatial compatibility $sc(\tau_k)$:

$$sC(\tau_k) = \mu_A(X(\tau_k)), \ \tau_k \in SUPP_T \tag{II.13.4}$$

where μ_A is the spatial membership function associated to the value A of the proposition, and $X(\tau_k)$ is the value observed for the variable X at the temporal point τ_k (whose universe of discourse is U). The spatial compatibilities obtained are modulated by the temporal part of the proposition, so that in

---

*The support $SUPP_A$ of a membership function μ_A defined on a universe of discourse u is $SUPP_A = \{u \in U / \mu_A(u) > 0\}$.

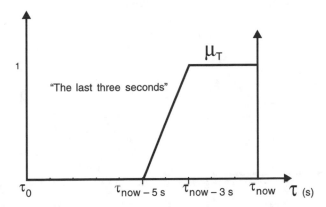

FIGURE II.13.2   Temporal reference μ_T associated with the expression *the last three seconds*.

all the three cases less weight is given to those temporal points of the support of T that do not form part of its core, to thus obtain the degree of fulfillment of the proposition:

- Non persistence: X is A in T

$$DOF = \bigvee_{\tau_k \in SUPP_T} sc(\tau_k) \wedge \mu_T(\tau_k) \qquad \text{(II.13.5)}$$

- Persistence: X is A throughout T

$$DOF = \bigwedge_{\tau_k \in SUPP_T} sc(\tau_k) \vee (1 - \mu_T(\tau_k)) \qquad \text{(II.13.6)}$$

- Intermediate case: X is A in Q of T

$$DOF = \mu_Q \left(\frac{\sum_{\tau_k \in SUPP_T} sc(\tau_k) \wedge \mu_T(\tau_k)}{\sum_{\tau_k \in SUPP_T} \mu_T(\tau_k)} \right) \qquad \text{(II.13.7)}$$

The operators $\wedge$ and $\vee$ are, respectively, the t-norm minimum and the t-conorm maximum, and μ_Q is the membership function associated to the linguistic quantifier Q. For a given history of values $X(\tau)$ and a linguistic value, depending on the temporal reference T and on the linguistic quantifier employed, it is possible to model different situations enabling the evaluation, in the most appropriate manner, of the different states in which the system to be controlled may find itself and to react to them suitably.

Figure II.13.3 shows an example of the *DOF* calculation for the propositions *velocity is high throughout the last seconds* and *velocity is high in the last seconds*. The process is as follows: first, the spatial compatibility is obtained for each instant τ_k (this spatial compatibility is calculated as the membership degree of $X(\tau_k)$ to the linguistic label high). Thus, five spatial compatibilities are obtained, one for each temporal instant belonging to the support of T. For the persistent type proposition (*throughout*), Equation II.13.6 is applied, which, in the example given, leads to a result of 0.2 (corresponding to $\tau_{now} - 2\Delta$) while for the nonpersistent type proposition, and applying Equation II.13.5, the final result obtained is 1.0 (corresponding to $\tau_{now} - 3\Delta$).

Figure II.13.4 shows some definitions for the membership functions μ_Q associated to temporal persistence quantifiers. This rule representation and reasoning model has been used in the two mobile robotics behaviors described in the following sections.

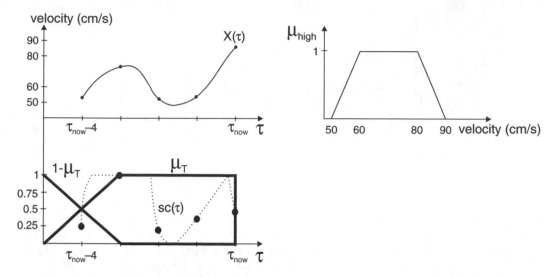

FIGURE II.13.3 Calculation of the DOF for the propositions *velocity is high throughout the last seconds* and *velocity is high in the last seconds*.

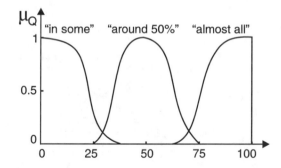

FIGURE II.13.4 Membership functions (μ_Q) of some temporal quantifiers.

13.3 Description of the Nomad 200 Robot

The robot on which the different controllers have been implemented is a Nomad 200 (Figure II.13.5). Table II.13.1 shows some of its principal technical characteristics.

The robot's base has odometric sensors that enable it to know its position, linear velocity, angular velocity, and the turret (upper part of the robot) and base angles. There are also four sensorial systems: bumper, sonar, infrared, and laser. The ultrasound ring (the only sensorial system used in the applications presented in the following sections) comprising 16 sensors supplies the input data requested by the system. The ultrasonic sensors have a range of approximately 6.5 m, and their layout is shown in Figure II.13.6.

To perform its maneuvers the Nomad 200 has a synchronous mechanism for handling its three motors: one for movement, another for turning the wheels, and a motor used to turn the turret independently from its base. It should be pointed out that the robot can turn on the spot, without displacement (null turning radius).

TABLE II.13.1   Characteristics of the
Nomad 200

Characteristic	Value
Diameter	53 cm
Height	112 cm
Weight	59 kg
Maximum translation speed	61 cm/s
Maximum linear acceleration	76 cm/s^2
Maximum angular velocity	45°/s

FIGURE II.13.5   Nomad 200 robot (Nomadic Technologies, Inc).

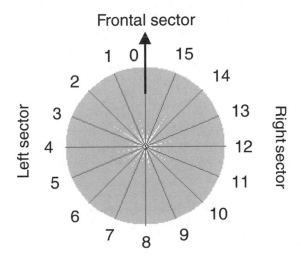

FIGURE II.13.6   Layout of the ultrasound sensors in a Nomad 200 robot. The arrow shows the direction in which the robot advances.

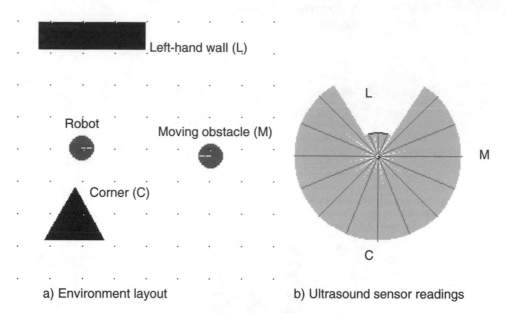

a) Environment layout   b) Ultrasound sensor readings

FIGURE II.13.7   Perception failure when the robot approaches an extremely open corner. It can be seen how moving obstacle M and corner C fail to be detected by the ultrasound sensors.

As is already well known, ultrasound sensors present a series of problems, which may produce a totally erroneous measurement of the distance to an obstacle. Some of these problems are, among others, "crosstalking" which consists of the reception of a pulse by a different sensor from the one that emitted it; specular reflection, which comes about when the incidence angle of the ultrasonic beam with respect to the normal to the wall is high (the smoother the surface of the wall onto which the pulse incides, the lesser the necessary angle). Due to this the reflected energy of the pulse does not return to the emitting sensor, or if it does so, it is after having bounced off other zones, and the measured distance will be erroneous. Last, the width of the beam that ultrasound sensors emit increases proportionally to distance, which leads to errors in the distances measured as well as low angular resolution. Because of this, it will not be possible to detect gaps or to receive information on corners.

Detection problems can be accentuated from the robot turning, above all with concave (closed) corners and, more importantly, in convex (open) corners, as is shown in Figure II.13.7. We can see in this figure how the robot detects the straight wall situated on its left, but the same is not true for the corner, which is undetected. If we opted only to group the sensors into sectors and to select the minimum distance within each sector we would find that no obstacle exists on the right, although the corner is actually located some 50 cm away. The same occurs with the moving obstacle situated in front of the robot, which is also not detected. This problem is compounded in real environments and occurs in a number of situations that are less extreme than the one shown in Figure II.13.7.

The next sections show how these detection and measurement errors can be handled to produce correct control actions for two implementations of behaviors that are usually described in robotics: wall following and avoidance of moving obstacles.

13.4   Application of FTRs in Wall Following

13.4.1   Introduction

One of the most frequently used behavior patterns in robot navigation is wall following.[18–21] This behavior is usually used when the robot needs to move between two given points. Thus, for example, a high-level order may indicate that the robot should move between two rooms in a building. The module that deals with planning would select the path to be followed and this would be executed by activating behavior patterns such as wall following and door crossing.

In this section we describe a controller for wall-following behavior that supplies the robot with suitable linear and angular velocity values for each given moment. The design has been modularized on the basis of two clearly differentiated objectives: first, to determine in which direction the robot should turn at each instant, so that it remains parallel to, and a certain distance from, the wall it is following; second, the maximum linear velocity at which the robot may move at each moment will need to be determined, without adversely affecting the behavior that is pursued. A good wall-following controller is characterized by three aspects: first, maintaining a suitable distance from the wall being followed; second, moving at a high velocity whenever the layout of the environment permits; and last, avoiding sharp movements, aiming for smooth and progressive turns and changes in velocity.

For all these reasons we have opted for implementing the linear velocity control by means of FTRs,[17] which enable us to collect the knowledge available in a suitable manner, as well as to consider the temporal evolution of the system variables. It is precisely this capacity for contemplating information of a temporal type which, as shall be seen, enables us to filter part of the large amount of sensorial noise and deal with possible erroneous perceptions of the environment due to unsuitable positioning of the robot. Furthermore, more continuous behavior patterns are obtained, which enable the maximum velocities to be selected by taking into consideration the layout of the environment at each instant, which also results in lower navigation time.

This approach differs significantly from others[22,23] in which the value of the control variables is habitually given according to the values of the input variables at a single instant. Other authors[21] have proposed as input to their control systems variables that reflect variations with respect to time derived from the distance and/or the orientation of the robot with respect to the wall, with the aim of antici-pating future positions of the robot, although these derivatives only reflect variations between two given instants and not throughout a particular period. On the contrary, when using FTRs we succeed in determining the evolution of a variable throughout a temporal interval, from which it is possible to anticipate future positions of the robot in the environment. Despite the computational requirements being higher when using this scheme, the reasoning model used allows execution of the knowledge base within real-time requirements, which in the case of Nomad 200 are three orders per second.

The other great advantage afforded by the temporal reasoning capability of the FTRs is filtering sensorial noise. A suitable implementation for wall-following behavior requires us to take into account the limitations characteristic of ultrasound sensors, which in some cases render the robot incapable of correctly perceiving the wall that is being followed. A number of strategies for tackling perception problems exist, the most common is the grouping together of the sensors into sectors. Within each sector a single measurement is obtained by means of different solutions: choosing the minimum distance,[22,24] employing the concepts of distributed and general perception,[20,21,25,26] applying a median filter (spatial filtering) and a Kalman filter (the measurements of the current and prior instants are taken into account),[27] etc. In all these cases, the fusion of information originating from sensors is still sensitive to erroneous measurements by one or more of the sensors.

In some studies,[20,25,26] this loss of perception of the environment is avoided by resorting to memorizing the parameters that identify the environmental situation in prior instants to be able to use them for calculating the position of the obstacles at the current instant in the event that the perception is not satisfactory.

13.4.2   Description of the Control System

A graphical description of the velocity control system for wall following is shown in Figure II.13.8. Linear velocity and angular velocity control blocks, respectively, calculate the values for the output

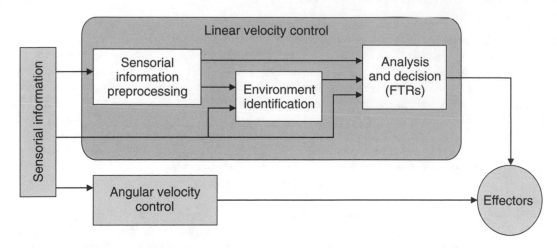

FIGURE II.13.8   Block diagram of the control system.

variables that control the movement of the robot. The calculation of the values of these control variables is carried out independently, in spite of the fact that they are directly linked through the turn radius. We now consider each block in more detail.

13.4.2.1   Angular Velocity Control

The angular velocity controller operates in a very simple manner. In a similar manner to that in Iglesias et al.,[28] information coming from the frontal and lateral sensors is conveniently processed to obtain a binary vector in which each component is associated with a particular direction with respect to the robot. Null components are associated with the presence of possible gaps or free spaces, into which it is possible for the robot to enter. On the contrary, the non-null components indicate which segments of the wall are closer to the robot.

This vector describes with a discrete set of 64 states all the situations in which the robot may find itself. A multilayer perceptron network was implemented for calculating the angular velocity that the robot should attain in each state. As a result in our case, the network is capable of generalizing and generating a suitable action for each one of the possible 64 states.

Other alternatives[22,29] could have been used for the design of this block, but the one described here was chosen due to its simplicity, and because it leads to robust behavior patterns. Moreover, it enables us to easily emphasize the capabilities of the linear velocity controller to guarantee that the robot move with a suitable turn radius, as well as the usefulness and interest of FTRs in the domain of mobile robotics.

13.4.3   Linear Velocity Control

Three modules make up the linear velocity control block (Figure II.13.8): sensorial *information preprocessing, environment identification,* and *analysis and decision.* To facilitate the process of tuning this knowledge and improving the final performance of the system, we consider it appropriate to distinguish between three mutually independent situations, labeled as straight wall, open corner, and closed corner.

Assuming that the robot attempts to follow a wall on its right, we represent any situation in which the robot has to turn to the right, with a low turn radius, and tracing an arc of approximately 90° or higher, as an *open corner* (O_i in Figure II.13.9). If the turn verifies similar conditions, but toward the left, we are in a *closed corner* situation (C in Figure II.13.9); any other situation is placed in the

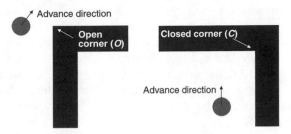

FIGURE II.13.9   Open and closed corner situations.

straight wall category. In the case of following a wall to the left, turns will be produced to the opposite direction.

The knowledge collected in the *analysis and decision* module, as well as the *modus operandi* of the processes included in the *environment identification* module, is based on two different levels of information abstraction: first, the measurements of distance originating from the ultrasound sensors are used. Second, we also use the information coming from the *sensorial information preprocessing module*, which, as will be seen subsequently, is suitable for losing sensitivity or the capability to discern among a number of different environmental configurations, which from a practical point of view can be considered identical.

13.4.3.1   Sensorial Information Preprocessing Module

This module carries out the task of abstracting the sensorial information provided by the robot's ultrasound sensors that perceive the wall being followed or the obstacles close to the robot's forward path. Self-organizing maps[30] are used to process this sensorial information. Instead of implementing one single network in which each neuron processes the information coming from all of the sensors, we use a number of one-dimensional networks, since this results in a superior performance and reduces the complexity of the problem. Each of these one-dimensional networks comprises 15 neurons, and receives measurements coming from three adjacent sensors. Since every measurement is processed by three different networks, nine are required for processing all the information coming from the eleven sensors that are taken into account.

It should be noted that every network learns to classify what the robot "perceives" in one given direction. In general, self-organizing maps can approximate the probability density function of a multidimensional pattern distribution. The network learns to recognize the most frequent patterns. In our case, the learning process was taken into account from a set of sensorial measurements obtained during the movement of the robot. Once the network is trained, its operation always occurs in the same manner; a competitive process takes place at every instant, which produces one winning neuron in each network. This winning neuron will be the one whose learned pattern is the most similar to the scenario the robot perceives in the direction corresponding to that network. It has been noted that for obstacles placed near the robot in one given direction, the winning neuron is usually one of the last neurons in the corresponding network, and vice versa. In this manner, it is possible to calculate a vector $\vec{P}$ in which each component identifies the winning neuron in each of the Kohonen networks. Each component of $\vec{P}$ codifies the global distance to the obstacles detected with a particular orientation to the robot.

13.4.3.2   Environment Identification Module

With this module we are interested in verifying in which of the three previously described situations (straight wall, open corner, closed corner) the robot finds itself at every instant.

To identify these situations whenever the robot is following a more or less straight wall, a reference system (characterized by presenting the y-axis, orientated, as far as is possible, in a parallel manner to the wall) should be constantly updated. In this reference system all points belonging to the wall being followed will be characterized by having the same value x_{ref} for the x-coordinate.

Thus, those sensors perceiving the same wall as the one being followed will be characterized by a coordinate x that is very similar to x_{ref}. In turn, when coordinates associated with the wall being followed stop being received, the vector $\vec{P}$ from the *sensorial preprocessing module* will aid in verifying whether this is due to the presence of an open or closed corner.

13.4.3.3 Analysis and Decision Module

FTRs in the *analysis and decision* module collect knowledge from the task expert in a suitable manner. It is also important for effectively filtering sensorial noise and anticipating future positions of the robot in the environment, as a result of the analysis of the evolution of the system during different temporal intervals. A set of rules describes the correct action to be performed for the three previously mentioned situations. This is described in detail in what follows.

13.4.3.3.1 *Straight Wall Situation*
The rules of the knowledge base for the straight wall situation have been designed to make the robot reach high velocities (up to 40 cm/s) when the distance to the wall of interest, in this case the right-hand one, is suitable (distances in the region of 50 cm are considered to be satisfactory). The robot is parallel to the wall, and furthermore, is closer to the right-hand wall than to the left-hand one. In any other situation the velocity will be reduced appropriately, adapting it to the circumstances in which the robot finds itself.

Assuming the robot follows the wall to the right, the most important variable for selecting the most suitable speed is the distance to the right-hand wall, defined as the minimum distance coming from the sensors in the right sector (Figure II.13.6).

The distance to the obstacles on the left must also influence the calculation of velocity (if the robot is in the center, closer to the right-hand wall, etc.). This information is collected by the variable *distances quotient*:

$$distances\ quotient = distance_{left}/distance_{right} \qquad (\text{II}.13.8)$$

where *distance$_{left}$* is calculated as the minimum distance of the left sector sensors. The distance to the right-hand wall often undergoes variations that do not lead to a change in the linguistic label. To obtain a smoother and more continuous behavior, a good quality controller should foresee and anticipate trend-changing situations. For example, it may happen that during a number of measurements the right-distance label continues to be the same, although the current value for this distance is changing (consistently increasing or decreasing). To also take this information into account, the value of the variable *trend of the right-hand distance* is calculated using the values of the right-hand distance in the current cycle (τ) and in the former one ($\tau - 1$) as:

$$Trend_{right\ distance} = distance_{right}(\tau) - distance_{right}(\tau - 1) \qquad (\text{II}.13.9)$$

Last, situations exist in which there is no appreciable variation in the distance to the wall of interest, but nevertheless, due to the proximity of the wall, or to the magnitude of the change in velocity, it is necessary to estimate the orientation of the robot with respect to the right-hand wall. For this reason, and using the vector $\vec{P}$ described in the *sensorial information preprocessing* module, a new variable is obtained that evaluates to what degree the robot is facing the wall being followed:

$$Orientation = max\{P_1, P_2, P_3\}/max\{P_4, P_5, P_6\} \qquad (\text{II}.13.10)$$

This variable is used as a complement to the trend of the right-hand distance, and more specifically when the latter is stable, to better describe the situation in which the robot finds itself. A low quotient indicates that the robot is not facing the wall of interest.

We now show an example of a rule from this knowledge base:

*IF the **distances quotient** is high approximately in the last three measurements AND the **right-hand distance** is medium approximately in the last three measurements AND the **right-hand distance** diminishes a little approximately in the last two measurements AND the **velocity** is high, THEN reduce the **velocity** a little.*

The temporal reasoning ability supplied by the FTRs has been exploited in several propositions from the antecedent part of the rules, as has been seen in the example given. The variable velocity proposition is a fuzzy nontemporal one, given that it is not necessary to carry out a filtering process to eliminate noise, and besides, it is not necessary to analyze the evolution of its values throughout a given temporal interval.

Conversely, the variables obtained from the readings of the ultrasonic sensors (right-hand distance, distances quotient, and trend of the right-hand distance) may indeed contain noise due to the reflections of the ultrasonic beams. Carrying out temporal processing of this information within the temporal reference framework ("the last three measurements," etc.) can avoid spurious erroneous data, which could cause unnecessary alterations in the velocity, enhancing the robustness of the behavior pattern. Furthermore, with the FTRs the evolution of the distance to the right-hand wall can be analyzed to determine its tendency within the temporal interval being analyzed. This means that attention only needs to be paid to those significant and minimally persistent variations of that variable. Thus, we succeed in smoothing out the behavior pattern, as the controller will anticipate future positions of the robot with respect to obstacles, thereby avoiding sharp changes of velocity.

Moreover, the use of FTRs provides flexible representation of knowledge, given that by using different quantifiers it is possible to adjust the percentage of points that need to verify the spatial part of the proposition. Thus, in the rule given as an example, the quantifier *approximately in* means that the degree of fulfillment of the proposition is proportional to the percentage of points in the temporal interval that fulfills the spatial part. In the same manner, the fulfillment of the spatial part could have been required in at least one point (quantifier *in*), at all points (*throughout*), or at a specified percentage of points of the temporal reference.

13.4.3.3.2 *Open Corner Situation*

This situation is defined as one in which the robot has to turn to the right, with a low turning radius, while tracing an arc of almost 90° or more (*O* in Figure II.13.9). This situation is truly complex, as aside from having to apply low velocities to obtain low-turning radius values, it is possible that at some given moments the robot's sensors cease to detect the right-hand wall correctly, even though some of them detect it occasionally (Figure II.13.7). FTRs are suitable tools for dealing with this problem. Even though in some temporal points perception may have been unsatisfactory, analysis of the variables in a temporal interval means that the system's response may be globally correct.

First, we assume that the robot needs to trace a circumference arc that is centered on the vertex of the corner, and with a radius of some 50 cm (the desired distance with respect to the wall). In this way it is possible to estimate what the linear velocity will be at each moment, taking into account the angular velocity at which the robot is moving: $V_{desired} = V_{angular} \times radius_{desired}$ where $radius_{desired} = 50$ cm. Given that $V_{desired}$ is only a value close to the desired one, and is taken as a reference, it is advisable, as a precaution so that the robot does not move too far away from the wall, that the linear velocity at which it moves initially be slightly lower than that one. Nevertheless, as the robot advances, a turning radius lower than the desired one may lead the robot to end up facing or too close to the wall. To avoid this, the velocity should be increased slightly, which will be reinforced by the increased probability that the open corner will end at any moment. For all these reasons, in the antecedent part of the rules the angle turned by the robot from the beginning of the corner is considered:

$$\text{angle}_{\text{turned}} = \text{angle}_{\text{initial}} - \text{angle}_{\text{current}} \qquad (\text{II}.13.11)$$

Finally, using the measurements obtained by the sensors in the right-hand section, it is possible to detect the presence of a nearby wall, which could be considered as the end of the open corner. Given that in the final instants it is possible for the robot to have moved with fairly unreliable information, it is necessary to verify that its orientation is suitable. For this reason in this situation, we once again take into account the *orientation* variable, as well as its tendency. These parameters will now be those that predominantly determine which behavior pattern the robot should adopt. Thus, for example, although in the most recent instants the orientation may have been suitable (robot not facing the wall) and the distance to the wall is correct, it will not be advisable to increase the linear velocity if the tendency of the orientation indicates that the robot is facing the wall, since such an increase would take it closer to the wall.

A typical rule from this knowledge base is:

> IF the **right-hand distance** is low approximately in the last four measurements AND the **left-hand distance** is high throughout the last two measurements AND the **angle turned** is medium or high AND the **robot** is not facing the wall approximately in the last four measurements AND the **orientation** is improving approximately in the last four measurements AND the **velocity is** medium, THEN increase the **velocity** a little.

This rule is activated typically in the final instants of an open corner, given that the right-hand wall is close, a fairly or very wide angle has been turned, the robot is not facing the right-hand wall, and furthermore, in the last four measurements the orientation with respect to the wall has improved.

Once again the use of FTRs enables us to carry out filtering of possible erroneous measurements in the distances. This filtering may be more or less demanding depending on the length of the selected temporal interval, and on the linguistic quantifier used (the membership functions μ_Q of the temporal persistence quantifiers used for this application are shown in Figure II.13.10). Thus, for the *right-hand distance*, the use of the quantifier *approximately in* evaluates the percentage of points within the temporal interval that fulfills the spatial part of the proposition, whereas the quantifier *throughout*, used for the left-hand distance, requires the fulfillment of the spatial part of the proposition in all the points of the temporal interval (in this case, this interval only covers two measurements), and as such, is much more restrictive. Finally, the orientation's evolution is evaluated throughout the last four measurements, to verify to what degree this tendency is being maintained.

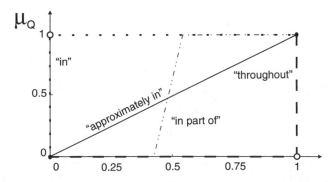

FIGURE II.13.10   Membership functions (μ_Q) of the temporal quantifiers used.

13.4.3.3.3 *Closed Corner Situation*

This situation is characterized by the robot turning to the left with a low turning radius (*C* in Figure II.13.9). As opposed to the other situations, the closed corner is distinguished due to the presence of a frontal object situated at the end of the straight wall that is being followed. For this reason, the frontal distance, calculated as the minimum distance from the frontal sensors, becomes one of the relevant variables that will determine the robot's behavior, even when it is also necessary to take into account the distances to the right and to the left.

To achieve a more precise control action when one of the distances has a low value, its trend will also be analyzed (trend is defined as the difference in distance between two successive cycles) to be able to detect tendencies in the evolution of the values of these variables.

While the robot is still a long way from the corner, its behavior will be no different from that shown in a straight wall situation; however, as it approaches the corner (there will be a reduction in the frontal distance) it has to significantly reduce velocity. Furthermore, the system takes into account that at a closed corner it is possible for the robot to have to turn up to 180°. This will depend on whether there is any obstacle situated to the left of the robot and close to the corner. These different situations can be identified by the fact that once the robot reduces speed and starts to turn, the wall placed on the left will now become a frontal wall. To deal with these situations the evolution of the quotient between the minimum distance to the left and the minimum frontal distance is taken into account, which indicates what type of closed corner the robot is faced with each time. Only when this quotient increases significantly will it be possible to consider that the corner has come to an end.

An example of this situation is found in the following rule:

*IF the **frontal distance** is low in part of the last four measurements AND the **frontal distance** is increasing approximately in the last four measurements AND the **quotient between the left-hand and frontal distances** is low in part of the last four measurements AND the **velocity** is low, THEN reduce the **velocity** a little.*

It could appear that the robot is starting to detect the straight wall and the closed corner is coming to an end, because *the frontal distance is increasing approximately in the last four measurements*, but given that *the quotient between the left-hand and frontal distances is low in part of the last four measurements*, it is advisable to continue reducing speed, given that there is a wall in the left-frontal zone. This situation is illustrated in Figure II.13.11, where it can be seen how, in spite of the frontal distance in (b) having increased with respect to the situation in (a), the quotient between the left and frontal distances is not high (the distance to the left is not significantly higher than to the front) and hence it is advisable to continue reducing speed slightly to avoid the wall situated in the left-frontal zone. In the case of no wall existing to the left, the quotient between both distances would be high, and thus the robot would start to accelerate, since the closed corner is about to end.

In this closed corner situation various walls are close to each other. It is highly possible, therefore, that due to the turning of the robot rebounding ultrasound signals may give rise to measurements somewhat higher than expected. For this reason temporal analysis of these distances in a wider time frame becomes essential to obtain measurement tendencies that have a high probability of containing noise, and also because in this situation the robot is moving at a notably lower speed. Moreover, the percentage of measurements that must verify the spatial part of the proposition is relaxed, including the quantifier *in part of* which requires the fulfillment of the spatial part in at least 50% of the points in the temporal interval.

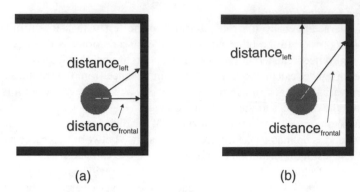

(a)   (b)

FIGURE II.13.11   Utilization of the quotient between the left and frontal distances.

TABLE II.13.2   Characteristics of the Linear Velocity Control System

	Straight wall	Open corner	Closed corner
Number of rules	108	140	65
Number of antecedents	5	7	8
Number of antecedents in which temporal reasoning is carried out	3	4	7

13.4.4 Results

To verify the system's operation a number of experimental tests have been carried out, in both real and simulated environments, with highly satisfactory results. Two examples implemented on the real robot* are now discussed.

Table II.13.2 shows some of the most relevant characteristics of the linear velocity control system. Despite the large number of rules needed for improving the behavior in any kind of situation, the execution time per cycle of the global control system is 0.22 sec, which allows ample margin for carrying out other tasks. Of the three situations in the environment, the open corner was the one that needed the highest number of rules, owing to its complexity. This is due to the large number of variables needed for its evaluation with all the rules. Conversely, for a closed corner situation, although the number of variables is high, they are not evaluated in all cases, and therefore the number of rules is substantially reduced.

A strict comparison of the results obtained with other control systems described in the bibliography is not possible; however, it is so in a qualitative manner. Thus we can emphasize that unlike Castellano, Attolico, and Distante[22] and García, Mandow, and López-Baldán,[23] our solution takes into account that the linear velocity of the robot may be variable, which enables navigation time to be reduced and velocity adapted to each situation. Furthermore, the results of the tests are given in environments in which the different situations (open corner, closed corner, and straight wall) are connected consecutively, thus increasing the complexity of the environment. The results given in García, Mandow, and López-Baldán[23] and Uribe, Urzelai, and Ezkerra[26] relate to each of these basic situations separately, and it is not possible to evaluate their global performance.

---

*Video recordings of some tests are available at http://www-gsi.dec.usc.es/areas/robotica_cas.htm.

In a field such as mobile robotics it is not viable to propose a standard "test bank" that would allow us to compare controllers in similar environments, and to consider the different dynamic characteristics of the various types of robots. For this reason, to carry out the results validation, we have chosen a set of examples sufficiently numerous and varied in the situations that have been considered. In this section we describe some of the most interesting cases. In the same manner, we include the values of certain objective parameters that endorse the global operation of the system (average velocity and average distances).

In the first example (Figure II.13.12) a complex, real environment is shown, with closed corners (C_i), open corners (O_i), gaps, doors, etc. To show the complexity of the proposed task, the same figure also shows the readings collected by the ultrasound sensors while the robot was in motion. On the other hand, by means of the robot's trajectory (represented by circles), information on its velocity at each instant (the higher the concentration of marks the lower the velocity) is implicitly included. The first point that should be emphasized is the control system's high degree of reliability and robustness, given that the average distance from the right-hand wall was 49 cm, which falls within the range of satisfactory distances. On the other hand, it can also be seen how the system is capable of satisfactorily resolving all situations, increasing velocity when the robot is following a straight wall (S_6) and reducing it when it approaches an open or closed corner. It is also possible to differentiate the reaction when confronted with an extremely closed corner (C_1) in which an about turn of 180° must be realized, or when it is not necessary to reduce speed so drastically because the is corner not so pronounced (C_3) and a turn of 90° is sufficient. The average velocity in this example was 23 cm/s.

Table II.13.3 shows some of the data measured in the different tests. It can be seen how the average velocity obtained in the example shown in Figure II.13.13 (example 2) is lower than the one obtained in the test shown in Figure II.13.12 (example 1) due to the complexity of the environment in the second example. The maximum and minimum velocities obtained in both environments are identical. It is noteworthy that in both cases, the robot, at some point, has to come to a halt to facilitate the turn, thus average velocities are not higher. With regard to obtained average distances to the right and left, the distances to the right are always close to 50 cm, while the average distances to the left are higher than to the right, and the values obtained in the different environments vary due to their different layouts.

One highly noteworthy aspect is the controller's ability to adapt linear velocity to the situation in which it finds itself, increasing or decreasing it, according to need. This aspect is evaluated in Figure II.13.14, which shows the evolution of velocity as opposed to time in the route shown in Figure II.13.12.

In the diagram, each environmental situation is labeled in the upper section, as well as the average velocity in each section (in cm/s). For straight walls the robot reaches and maintains the maximum

TABLE II.13.3   Data of the Two Examples Described in Figures II.13.11 and II.13.12

	Example 1	Example 2
Average right distance (cm)	49	48
Average left distance (cm)	171	202
Average velocity (cm/s)	23	17
Maximum velocity (cm/s)	44	43
Minimum velocity (cm/s)	0	0
Time spent	2 min 22 s	2 min 53 s

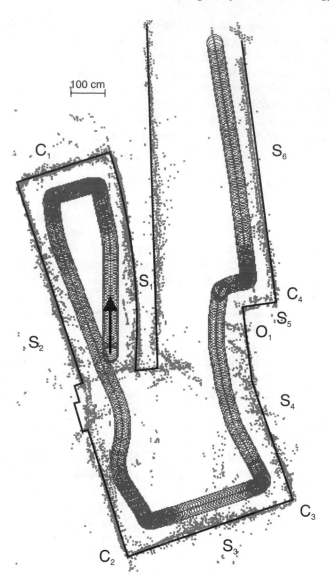

FIGURE II.13.12   Example 1. See text for description.

speed permitted (in the area of 40 cm/s) whenever the wall is sufficiently long (S_1, S_3, S_4, and S_6). In these situations the average velocity is clearly over 30 cm/s. In S_2 and S_5 these average velocities are not attained. In the latter case this is due to the short length of wall S_5. In the case of S_2, when the robot has attained velocities higher than 35 cm/s, available space narrows when going through a door 1.5 m wide. This gives rise to a decrease in velocity to 20 cm/s. Later, velocities over 35 cm/s are reached until, after the door has been crossed, the distance to the right-hand wall is greater than is desired, making it once again necessary to reduce speed. At closed corners the range of velocities is logically much lower than in the straight wall situation (average speeds close to 13 cm/s). In situations such as C_3, where the robot must turn less than 90°, the minimum velocity is 7 cm/s, but in very closed corners (180°) such as C_1 in some cycles the robot must stop to realize the turn, consequently reducing the final average velocity.

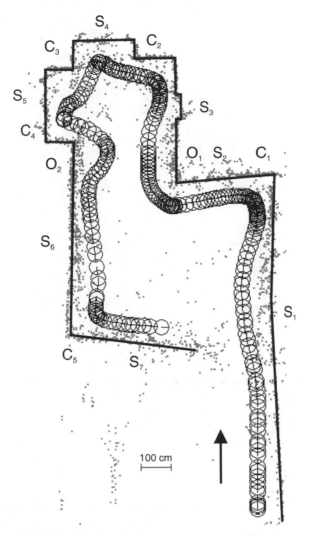

FIGURE II.13.13   Example 2. See text for description.

13.5   Application of FTRs in Avoidance of Moving Objects

13.5.1   Introduction

To the already complicated problem of autonomous robot guidance, we have to add in this case the presence of a moving object approaching the robot. Real-time operation is again one of the characteristics of this problem, since the robot has to be given orders in a rapid manner; if the difference between the time in which the measurements are taken and the moment in which the order is sent to the robot is great, the situation may have changed completely, and the response that was about to be carried out will not be adequate.

There are various approaches for tackling the problem of robot navigation in the presence of a moving obstacle. A number of studies deal with estimating the moving object's future positions. In this manner Elnagar and Gupta[31] use an autoregressive model, placing no restrictions on the trajectory of the obstacle. Chang and Song have[32] also used a neural network to this aim. Spence

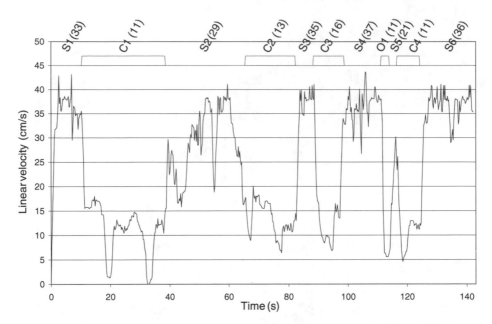

FIGURE II.13.14   Variation of linear velocity along the route shown in Figure II.13.12.

and Hutchinson[33] use a method based on attractive forces, depending on the distance to the goal intended for the robot to reach, and repulsive ones, due not only to the distance between the current state of the robot and the closest configuration in which a collision would occur, but also to the velocity of the moving object. On the other hand, Chakravarthy and Ghose[34] use an approach based on the concept of a collision cone, which allows the detection and avoidance of a collision between a robot and a moving object with an unknown trajectory, there being no constraint with regard to their forms. Gil de Lamadrid and Gini[35] have implemented a system for the monitoring of trajectories to be followed; there is a time limit of the arrival at the goal. The trajectories of the robot as well as of the moving objects are made up of linear segments along which they move at a constant speed. Tsoularis and Kambahmpati[36,37] also describe the avoidance of a moving obstacle, in this case with a uniform and rectilinear movement, resolving it in a geometrical manner. Finally, Garnier and Fraichard[38] and Pratihar, Deb, and Ghosh[39] have developed two fuzzy control systems. As antecedents to the rules, the former uses the velocity of the vehicle to be controlled and the distance between this and the moving object, measured by the different sensors, which also gives its relative position. The output commands are accelerations and changes in turning speed. Another module acts in conjunction with this one, which enables the vehicle to follow a trajectory without straying excessively when avoiding collision with the moving object. On the other hand, Pratihar, Deb, and Ghosh[39] uses the distance of the robot from the moving object and the relative angle that it forms, and the robot acts solely by turning. In both cases the moving objects move in a straight line at a constant velocity.

In all these approaches the robot acts according to the position of the moving object in the immediately preceding instants. In certain cases, this may lead to carrying out precipitated and inadequate actions. Our approach to the problem is to solve this by taking into account the history of more or less recent values of determined variables, which enable us to reflect the different scenarios through which the obstacle has been passing and, thus, verify what its trend is. In this way, one can deduce what the behavior of the robot should be, and take corresponding actions (modification of its speed and/or turning the robot) to obtain a behavior pattern in tune with recent situations. This system is robust as it permits the avoidance of collisions even when the moving object behaves in a totally unexpected manner. Furthermore, the moving obstacle is allowed to move freely. The need to evaluate past situations and previous values of the variables (which in many cases are fuzzy) and principally, to reason them out, has led us to incorporate FTRs.[40–42]

The representation of knowledge is modularized into three knowledge bases from which the different components of the control system are made independent for greater ease in tuning of the knowledge base, as well as achieving greater simplicity in each one of its components. The system enables the robot to operate with imprecise knowledge and takes into account the physical limitations of the environment in which the robot moves, obtaining satisfactory responses for the different situations analyzed by means of simulation software and in real environments.

13.5.2 Description of the Problem

As has been mentioned, movement of a robot in a dynamic environment is an extraordinarily complex problem. Besides avoiding collision with the moving object, the robot must move in an environment that may have fixed obstacles (walls, etc.) that are restrictions on the movement to be carried out to avoid the moving object. To this, one has to add the restrictions imposed by the robot's characteristics, such as the turn velocity, the linear acceleration, or the range of the sensors.

Unlike other cases, in our approach there is no limitation to the movement of the moving object; it is able to accelerate, brake, turn, etc. When the moving object is detected, the control system cuts in; its objective is for the robot, by means of accelerations and turns, to avoid the moving object approaching it. These robot movements will be limited by the fixed objects present in the environment. The restrictions to the problem are the width of the fixed object-free zone (which we refer to as a passage) and the additional condition imposed that the robot cannot move in reverse.

The aim of the control system is to obtain those control variables that are sent to the robot with each order: angular velocity and linear acceleration. To do this a series of steps is followed: first, the maneuver the moving object is intending to carry out (its trend) is estimated; the control system then selects the robot's behavior for this situation; and finally, the most adequate angular velocity and linear acceleration values are calculated in order to implement this behavior in the optimum manner.

For the input variables of the problem, we have all the parameters of the robot (current position, advance angle, translation velocity, and turn velocity) and the coordinates of the position of the moving object. From these the remaining necessary variables for the solution of the problem can be calculated.

Let $\vec{v}_{robot}$ be the velocity of the robot, $\vec{v}_{obstacle}$ the velocity of the moving object (calculated by simple kinematics based on the positional coordinates in two successive instants τ_l and τ_{l+1}), R_{robot} the radius of the robot, and $R_{obstacle}$ the radius of the moving object (it is assumed that both the robot and the moving object are circular, which does not lead to a loss in generality, see Figure II.13.15a). To be able to determine in a simpler manner the existence or nonexistence of a collision and where it will take place, we carry out a problem transformation,[43] which enables us to go from solving a cinematic problem between two nonpunctual objects to an equivalent static problem. In the equivalent transformed problem the velocity of the moving object is null and its size is $R = R_{robot} + R_{obstacle} + R_{security}$ and the robot is a punctual object with the velocity $\vec{v} = \vec{v}_{robot} - \vec{v}_{obstacle}$. A graphical representation of the transformed problem is shown in Figure II.13.15b.

$R_{security}$ is the minimum distance the robot is permitted to approach the moving object. This is established to maintain a safety margin, which, in any case, avoids a real collision between the object and the robot. Thus the collision test is reduced to verifying the intersection between the straight line given by the velocity of the robot relative to the moving object and the circumference that represents the moving object. In Figure II.13.15b there is an intersection point (collision point), at which a collision will occur. It was assumed in our problem that the robot and the moving object have the

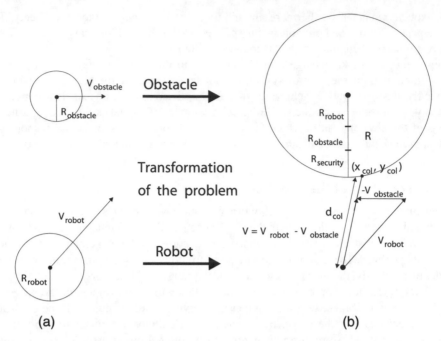

FIGURE II.13.15   Transformation of the original problem (a) in an equivalent one (b), where the robot is a punctual object.

same radius (approximately 25 cm), and the diameter of the robot was taken as the security radius, so that $R = 4 \times R_{robot} (R = 1 \text{ m})$.

Once the equivalent transformed problem has been proposed, a parameter (*noncollision index*) is defined that constantly evaluates the proximity of the current situation with respect to the collision situation.

The *noncollision index* (*nci*) is defined as:

$$
nci = \begin{cases} \frac{\sin \alpha}{|\sin \beta|} = \frac{d_c}{R} & \text{if } \alpha \in [-\frac{\pi}{2}, \frac{\pi}{2}] \\ -\frac{d_0}{R} & \text{if } \alpha \in [-\frac{\pi}{2}, -\pi] \\ \frac{d_0}{R} & \text{if } \alpha \in (\frac{\pi}{2}, \pi) \end{cases}
\tag{II.13.12}
$$

where, as can be seen in Figure II.13.16, d_0 is the distance between the robot and the moving object (in the transformed problem the moving object is represented by a circle with radius R; the coordinates of its center are taken as its position coordinates) and d_c is the distance between the moving object and the point with coordinates (x_c, y_c), that is given by the intersection of the straight line in which the robot is moving in the transformed problem (its direction is $\vec{v}$), and the radius of the moving object perpendicular to this straight line. The angle α is formed by the line that joins the robot and the moving object (straight line d_0) with the straight line $\vec{v}$, where:

$$
\sin \alpha = \frac{d_c}{d_0}
\tag{II.13.13}
$$

This angle α is measured with respect to the straight line d_0 and increases in a clockwise manner. Furthermore, angle β is the one formed by the straight line that is tangential to the circle with radius R (which represents the moving object in the transformed problem) and the straight line d_0:

$$
\sin \beta = \frac{R}{d_0}
\tag{II.13.14}
$$

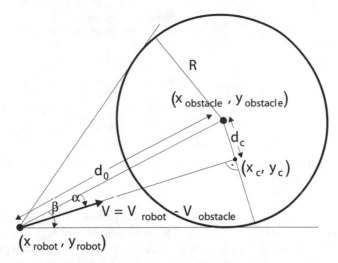

FIGURE II.13.16　Definition of the *noncollision index*.

The *nci* takes values between the interval $[\frac{-d_0}{R}, \frac{d_0}{R}]$, which reduces as the robot approaches the moving object (d_0 decreases in this case). As has been seen in Equation II.13.12, for $\alpha > \pi/2$ and $\alpha < -\pi/2$ the *nci* reaches its maximum and minimum values, respectively. These situations only happen when the robot is moving away from the moving object (in this case, no collision can occur). On the contrary, for values of the *nci* within the interval $[-1, +1]$ a collision situation exists. To obtain these values, angle α must be less than or equal to β (in absolute value), which indicates an intersection between the straight line given by the relative velocity of the robot with respect to the moving object and the circle with radius R. Positive values of the *nci* indicate that the robot is going to collide with the moving object on its left-hand side ($nci \leq 1$), or that it is going to pass before the moving object ($nci > 1$), while negative index values reflect a collision on the right-hand side of the robot ($nci \geq -1$) or that the robot has let the moving object pass ($nci < -1$).

The coordinates of the collision point (x_{col}, y_{col}) are given by the point at which the line $\vec{v}$ intersects with the circle representing the moving object (Figure II.13.15b). d_{col} is the distance separating the robot from this collision point—the point at which the robot will be placed after a certain amount of time (collision time) if the velocities of either the robot or the obstacle remain the same. At this point the robot will be at a distance $R_{security}$ from the moving object.

Variations in the value of *nci* and its temporal evaluation are of great interest for characterizing the dynamic behavior of the obstacle. Thus, for example, the *nci* value may increase because of a decrease in the obstacle's velocity, and it may decrease due to an increase in the obstacle's velocity.*

The *nci* is used as a basis for calculating new parameters related to the evolution of the moving object and/or the robot, since any change in the behavior of either of them will be reflected.

We now describe the intelligent control system in the following section. One aspect that is especially interesting at this point is the previously mentioned necessity to implement temporal reasoning on the evolution of the *nci*. By analyzing the past and present values of this variable, the moving object's current trend can be deduced in an intuitive manner. As an example, if an increase in the *nci* had been produced, but in the last few moments a decrease occurs, it is understood that the previous trend of the moving object to let the robot pass has changed, and has become that of passing

---

*This is true when the incidence occurs from the left side. For right-side incidences, a change in the *nci* must be made.

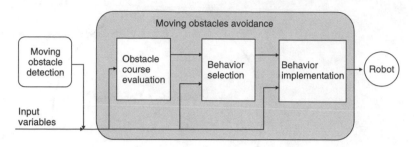

FIGURE II.13.17   Schematic diagram of the control system.

first. In real situations, it will be necessary to distinguish between true changes in the trend and sporadic movements of the obstacle in order to evaluate a situation as changing, a certain persistence or temporal maintenance is required in the new *nci* values. This need to bring temporal intervals into play and to analyze their occurrence in the variables' values has led us to use FTRs.

13.5.3 Description of the Control System

The control system was initially designed bearing in mind that the incidence of the moving object is produced from the left-hand side of the robot. Right-hand side incidences can be dealt with in an identical manner by simply transforming the problem.

Knowledge from the human expert, which has been fed into the system, has been structured into an FKB that has been modularized into three blocks. The objectives of this fragmentation of the FKB are, first, greater ease in tuning the knowledge base. In a nonmodularized knowledge base with a number of rules such as those of this FKB (a total of 117 rules) this process would be extremely complex. Another advantage is that the different blocks have a high degree of independence among themselves, hence modifications in one block do not influence the other blocks.

In order of execution (Figure II.13.17), the modules making up the knowledge base are:

- Obstacle course evaluation module. Its aim is to verify what movement strategy the obstacle is following (if it allows the robot to pass, if it wants to pass, or if it is not aware of the robot).
- Behavior selection module. The aim of this block is to decide on the optimum behavior that the robot should follow in light of the trend of the moving object.
- Behavior implementation module. This final module tries to obtain the angular velocity and linear acceleration with which the robot will most suitably implement the desired behavior for the current situation.

Before describing each of the modules of the FKB, all aspects related to the problem of moving obstacle detection by a real robot are presented.

13.5.4 Moving Obstacle Detection

Detection is based on real-time building of a map of the space occupancy from measurements from the ring of ultrasonic sensors. The degree of occupancy in each cell in the map is calculated by taking into account the number of times it is detected as being occupied and empty (count-based model).[44] This method comprises a very simple sensor model and an accumulation method. Each time a sonar echo is perceived, a value $v_t[i, j] = -1$ is assigned to the cells c_{ij} situated between the transmitter and the possible obstacle, and a value $v_t[i, j] = +1$ to the cells situated on the boundary sector of the sonar cone, independent of its position on it (Figure II.13.18). The accumulation method chosen

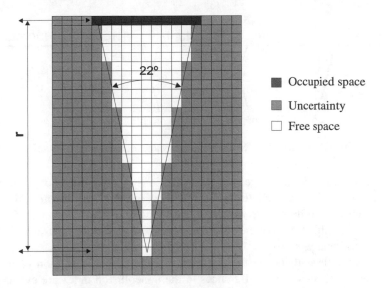

FIGURE II.13.18 Sensor model used, showing the distribution of free and occupied space with regard to one ultrasound measurement.

is the SUM model, in which the occupancy value of each cell $C[i, j]$ of the map is obtained by algebraically adding the different observations:

$$C_{t+1}[i, j] = C_t[i, j] + v_t[i, j] \qquad \text{(II.13.15)}$$

truncating to a maximum cell occupancy value. The initial value for all cells is 0, i.e., maximum uncertainty.

Detection of a moving obstacle is carried out by using the current measures from ultrasonic sensors and the existing map at the current instant. For all of the measurements provided by sensors, fulfillment of a number of heuristic conditions is requested to assume that a moving obstacle is detected by the sensors. These conditions are based on the following basic idea: a moving obstacle is assumed to be detected in a cell that is currently detected as occupied by a sensor, if that cell was previously detected as empty and is surrounded by empty cells.

In the following sections we examine each of the FKB modules. The description mainly focuses on the most relevant blocks that comprise the control system, rather than on a very detailed description of all the components (variables, processing blocks, etc.) in the controller.

13.5.5   Obstacle Course Evaluation Module

The objective of this module is to estimate the obstacle's movement tendencies, i.e., to attempt to characterize in which dynamic scenario the robot is placed. Evaluating this situation, the robot will assume that the object interfering with its trajectory is trying to pass before it or is letting it pass. In other cases it will not be able to estimate a clear trend in the object's movements.

The input variables for this block are the following: *collision time*, *collision status change*, and *nci trend*. Variable *collision time* estimates the time available before the robot enters into the obstacle's security radius. A low collision time supposes a sharp reaction on the part of the robot to avoid a collision which seems imminent, while a longer collision time enables it to observe the situation and act in a more gradual manner. As has been mentioned, besides indicating whether a collision will occur and what nature it will be, the *nci* reflects any changes in velocity (in module and/or direction) of the robot and the moving object.

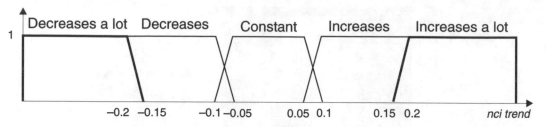

FIGURE II.13.19   Set of linguistic values for variable *nci trend*.

Variable *collision status change* is used to detect situations in which the robot passes from collision in one cycle, to not in collision in a later cycle, or vice versa. Knowing the *nci* values in these two cycles, makes it possible to determine whether the moving object wishes to pass first, or is letting the robot pass. The possible values of the *collision status change* variable considered in this problem are *decrease*, *neutral*, and *increase*.

The *nci trend* variable gathers, on the contrary, a more precise evolution of the trend of the moving object, in which successive differences in the noncollision index are evaluated. To do so, the difference between the values of the index in two successive instants are calculated, and then the mean is found between this difference and the difference in the previous cycle (to attempt to minimize errors due to imprecision in the measurements of the moving object's position):

$$nci\ trend = \frac{nci(\tau) - nci(\tau - 2)}{2} \tag{II.13.16}$$

The values used for this variable are *decreases a lot*, *decreases*, *constant*, *increases*, and *increases a lot* (Figure II.13.19).

The output variable for this block is the trend of the moving object, which estimates the behavior of the obstacle to then be able to act accordingly. In our case the trend is a crisp variable that may take the following values:

- *To give way*. The moving object intends to let the robot pass.
- *Indifferent*. This trend may be due to two reasons: on one hand, the moving object is moving in a random manner (braking, accelerating, or turning without any continuity in its movement) or because the moving object is not varying its speed (module direction).
- *To pass in front*. The moving object is attempting to pass first.

The rules of this knowledge base incorporate temporal reasoning and follow the FTRs model, since a correct evaluation of the object's movement must take previous situations into account, rather than a single summarizing average value. We now analyze the most noteworthy aspects of some representative examples.

One type of rule is:

*IF **collision time** is short AND **collision status change** is decrease in the last two seconds AND **nci trend** is not increasing throughout the last second, THEN **obstacle aim** is to pass in front.*

The meaning of this rule is that in a situation of relative proximity between the obstacle and the robot (*collision time is short*), it is assumed that the trend of the former is to pass in front if there has recently been at some point a decrease in the collision status change (this may occur when, for example, a noncollision situation with a high *nci* value changes into a collision situation – *collision status change decreases in the last two seconds*) and furthermore, even more recently, the *nci* has maintained its value or decreased (*nci trend is not increasing throughout the last second*). A strict decrease in the *nci* is not required, as this decrease has been produced implicitly since a change in the *collision status* has occurred.

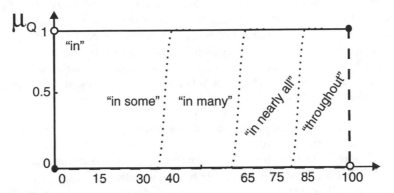

FIGURE II.13.20   Membership functions (μ_Q) of some of the temporal quantifiers used.

In general, if *collision status change* decreases, and the *nci* decreases or remains constant, the trend will indicate that the moving object intends to pass (for a *to give way* trend an increase in the *nci* is required), while if subsequent to the *collision status change* decrease the *nci* increases, the trend will be *indifferent*.

Another possible situation is no *collision status change*. In this case the trend will be *to give way* if the *nci* increases substantially (for a *to pass in front* trend a significant decrease in the *nci* would be required):

> IF **collision time** *is medium AND* **collision status change** *is neutral in the last three seconds AND* **nci trend** *is increasing a lot in the last three seconds, THEN* **obstacle aim** *is to give way.*

This rule analyzes whether for medium collision time a *collision status change* has not occurred at any point of the interval *the last three seconds*. In this case, a *to give way* trend is assigned to the moving object, if the *nci* has increased significantly at some point of this interval.

This increase may be given, for example, by a decrease in the moving object's velocity module. However, this decrease may take a more gradual form and have almost the same final effect (braking is not so sharp, thus the moving object will be closer to the robot). Rules like the one that follows have been introduced into the knowledge base to resolve this type of situation:

> IF **collision time** *is medium AND* **collision status change** *is neutral in the last three seconds AND* **nci trend** *is increasing in at least a few points of the last three seconds, THEN* **obstacle aim** *is to give way.*

The difference in this case is that the requirement for the increase in the *nci* is not as strict (the only requirement in this rule is that the nci should increase) so that the change in the index may be lower, but in this rule the increase is needed *in at least a few points of* the interval *the last three seconds* (*a few* represents approximately 30%). Definitions of some of the temporal quantifiers used for this application are shown in Figure II.13.20.

13.5.6   Behavior Selection Module

The objective of this block is to determine the type of behavior that should be adopted by a robot faced with the trend given by the moving obstacle's current situation. The robot implements an ample set of behavior types to try to resolve the multiple situations in which it may find itself. The input variables for this module are *collision time*, *trend*, and *limit situation*.

Limit situation is a crisp variable that indicates when the robot is in an extreme situation in which it will attempt to leave the trajectory of the moving object as quickly as possible. The conditions for the robot to be in a *limit situation* are two: it should be placed in the trajectory of the moving object,* and the incidence of the moving object should be *extremely frontal* or *extremely rear*.

As has been mentioned, a good number of possible behavior patterns exist. Moreover, given two equal behavior patterns, they do not necessarily have to be implemented in the same manner, rather this realization of behavior will depend on a series of variables; this is the task of the final block (behavior implementation module). The types of behavior that exist and the general descriptions of their implementations are:

- *To give way*. In this behavior pattern the robot lets the moving object pass by, and it does so braking and, sometimes, also turning.
- *To observe*. In this situation, the robot maintains its velocity (module and direction). This is normally because the trend of the moving object is not clear.
- *To pass in front*. Here, the robot attempts to pass before the moving object (for example, by turning and accelerating).

In addition to these basic behavior patterns, there is a specific behavior for *limit situations*. In these situations the trajectories of the robot and of the moving object are fairly parallel and it is essential for the robot to turn to move out of the path of the moving object. The behavior patterns for *limit situations* are characterized by trying to leave the path of the moving object as quickly as possible. To do so the robot will need to accelerate (as braking will not take it out of the moving object's path) and turn sharply (the turn is made in the direction with the greatest margin).

The most common rule used for the determination of the behavior is:

IF **collision time** is high AND the **obstacle's trend** is to give way, THEN the **robot's behavior** is to pass in front.

In this rule behavior heavily depends on the trend variable. If the trend is *to give way*, as a general norm, the behavior will be *to pass in front*, while if the trend is *to pass in front* the behavior will be *to give way*. For an *indifferent* trend the selected behavior will take into account the collision time. For high collision times the robot will act aggressively, and hence the corresponding behavior will be *to pass in front* while for low collision times the robot will act in a more conservative manner, implementing a *to give way* behavior. With medium collision times, the robot will adopt intermediate tactics, and the behavior pattern will be *observe*, due to which the robot will wait for future changes in the trend of the moving object.

In certain cases, the behavior patterns obtained in this block may be modified. This happens when the selected behavior cannot avoid collision (for example, the robot cannot turn, or it is in the trajectory of the obstacle and cannot brake, etc.). A set of criteria was implemented for these cases and decide the most favorable behavior pattern for the existing conditions.

13.5.7   Behavior Implementation Module

The aim of this block is to obtain, from a series of input data (*trajectory, behavior, collision time, robot's velocity, deviation,* and *incidence*) the angular velocity and the linear acceleration commands that will be sent to the robot. The crisp variable *trajectory* indicates whether the robot is in the path of the moving object or not.

---

*The trajectory of the moving obstacle is represented by a band, the width of which is equal to the diameter of the mobile object in the transformed problem.

Since the robot moves in the direction of a goal, any turn that makes the robot move toward the goal is labeled as favorable, while a turn that takes it away from this path is unfavorable. The variable *deviation* is defined for measuring how favorable a turn in a determined direction is. It may take the following values: *negative*, *null*, and *positive*. Negative deviations describe very favorable turns (movements toward the goal); null values will also allow the turn to be made, although this may produce a movement slightly away from the path. Positive deviations describe turns that will only be implemented in situations in which they are imperative for avoiding a collision.

The positive or negative sign of *deviation* is selected taking into account the following: if the robot is moving toward a point situated to the left of the goal, any turn to the right will be considered favorable since it will take it toward the goal, while a turn to the left will be considered unfavorable because it will take it away from the goal.

The behavior *to pass in front* will usually imply a turn to the right.* If it happens that the trajectory of the robot is placed to the right of the goal, the requested turning to the right will not be favorable (positive *deviation*). For the same behavior, if the trajectory of the robot is placed to the left of the goal, turning to the right is considered favorable and thus *deviation* is negative.

For *to give way* behavior, the turn is usually produced toward the left, so a turn in this direction is favorable (negative sign for the *deviation*) when the robot moves toward the right of the goal and unfavorable in the contrary case.

Variable *incidence* indicates whether the incidence of the obstacle is frontal, transversal, or from the rear. It is necessary to distinguish among these situations, as the optimal manner for avoiding collision is different. The *incidence* is defined as:

$$incidence = \frac{\pi/3 - angle}{\pi/3} \tag{II.13.17}$$

where *angle* $\in [0, \pi]$ is the angle formed between the velocity of the robot ($\vec{v}_{robot}$) and the relative velocity between the robot and the moving object ($\vec{v}$). The set of values for the variable is made up of: *rear*, *transversal*, *frontal*, *extremely frontal*, and *extremely rear*.

The 72 rules making up the FKB of this module enable us to implement different behavior patterns for each of the situations in a precise and adequate manner. The rules in this block can be grouped according to the behavior to be implemented and the type of incidence.

We now analyze a rule that represents those in which the behavior is *to pass in front* and the incidence is transversal. These rules are generally characterized by implementations of behavior that consist of right turns and accelerations. Thus, for example:

*IF the **robot's behavior** is to pass in front AND the **collision time** is medium AND the **robot's velocity** is medium AND the **deviation** is null AND the **incidence** is transversal, THEN increase **velocity** quite a lot AND **turn** a little.*

For high collision times, the reactions should be gentle (slight turns and accelerations) while for low collision times the system usually applies maximum turn and acceleration to avoid collision. In general, the aim is to avoid turns (to not move away from the trajectory the robot was following) except when these are favorable (negative deviations). For low collision times, this is not fulfilled (there is no other solution for avoiding the collision), and for positive deviations there will be a turn, although less intense than those implemented for negative or null deviations.

---

*Recall we are assuming that the incidence is from the left side.

TABLE II.13.4   Characteristics of the Tests Performed

Maximum Translation Velocity	61 cm/s
Range of distances between the robot and the obstacle	0.5 to 6 m
Range of collision times	1 to 30 s
Range of angles of incidence between the robot and the obstacle	0 to 2π
Types of movements of the obstacle	Accelerations, decelerations, and turns
Types of trajectories of the obstacle	Straight line, curved line, zigzag (and their combinations)

13.5.8   Analysis and Discussion of Results

Since the entire control system has been presented, we now analyze some representative examples of the robot's avoidance of a moving object. The system has been tested for a number of cases, simulated and real, to verify its validity and effectiveness. These tests have included the whole range of possible velocities, for both the robot and the moving object, as well as different angles of incidence (frontal, rear, left transversal, right transversal, etc.) as shown in Table II.13.4.

Furthermore, with the objective of making the simulations as realistic as possible, the tests were carried out with randomly introduced noise in the position of the moving object, to simulate the imprecision of the robot's real ultrasound sensors. This noise is a function of the distance between the robot and the moving object (the greater the distance, the higher the noise) and tests were carried out for a maximum 10% error.

Avoidance of the moving object was also made more difficult for the robot. To do so, when the robot started to implement a behavior pattern to avoid the collision, the moving object reacted by attempting to provoke a new collision situation.

Simulations have also been carried out in which the moving obstacle has a complex trajectory (for example, moving in a zigzag manner that takes up the entire passageway), or effecting a series of turns. The response was satisfactory in all cases, which demonstrates the robustness of the system as well as its reaction capabilities.

The examples we have chosen to illustrate the robot's behavior are given with graphical representations in which the trajectories of the moving object and the robot are described. Those of the former were chosen to show a selection of changes in the module and the direction of the velocity that face the robot with varied scenarios.

13.5.8.1   Example 1

In this first example, a moving object frontally approaches the robot, both moving at a velocity of 25 cm/s, in a situation in which at the beginning, there is not going to be a collision. Thus the robot maintains its speed in terms of module and direction (Figure II.13.21). When the moving object approaches to a distance of 380 cm from the robot, it starts to turn toward its left (Point A, Figure II.13.21) with an angular velocity of 30°/s, moving onto a collision course, and the control system starts to act. As a result of this turn, the *obstacle course evaluation* module correctly interprets that the moving object wishes to pass (the moving object has made a turn to its left). As the robot is now in the path of the moving object and the incidence is frontal, the behavior selected will be *limit situation*.

Under these circumstances the only possibility of avoiding collision is turning to the right. This behavior (*limit situation*) is characterized by high acceleration and turn, which will be more intense the lower the collision time is. In this situation the collision time is medium, and once the command has been sent to the robot, in the following cycle the robot is still on collision course (acceleration and turn were insufficient). The process is repeated, but this time with the collision time being

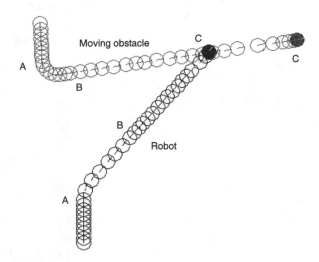

FIGURE II.13.21 Example 1. A, B, and C are the positions that the robot and the obstacle occupy in three time instants of the time interval represented.

medium-low, so that finally in the next cycle (with collision time already low), the robot succeeds in leaving the collision situation.

Once the collision has been avoided, the moving object continues turning until it reaches 45°, while the robot maintains a velocity of 61 cm/s. When the moving object reaches point B, it increases its velocity from 25 to 61 cm/s and once again the control system has to act because of the collision situation. Due to this acceleration, the *obstacle course evaluation* module indicates that the moving object is attempting to pass first, and the behavior selected is *to give way*. Its implementation for low collision time consists of braking sharply, taking the robot from a velocity of 61 cm/s to one of 36 cm/s, thus avoiding impact.

13.5.8.2 Example 2

In this second example a situation is described in which the moving object starts by letting the robot pass then changes strategy and passes first. The robot responds to these situations by simply varying its velocity (Figure II.13.22). At the start, the robot detects the moving object, and since it has no data stored (the moving object has entered within range of the robot's sensors for the first time), the tendency selected is *indifferent*. Because the collision time is high and the tendency of the moving object is *indifferent*, the robot adopts an aggressive behavior and attempts to pass first. The implementation of this behavior, for null *deviation* and medium robot velocity (25 cm/s), is a small acceleration. Due to this acceleration (which increases the robot's velocity from 25 to 33 cm/s) and because the moving object has started to turn to the right (point A), the *obstacle course evaluation* module now detects that the *nci* has increased substantially and the tendency will thus be *to give way*. The behavior of a robot faced with this attitude on the part of the moving object is to try to pass first, and it continues with the strategy begun in the previous cycle. Once again this behavior is implemented with a small increase in the robot's velocity, up to 38 cm/s, avoiding the collision.

A number of cycles later, the moving object changes its behavior by turning to its left and increasing velocity from 25 to 61 cm/s (B), which leads to a sharp decrease in the *nci*. Under these circumstances, the intention of the moving object is clearly to *pass first*. The *behavior selection* module selects *to give way*, and since the collision time is low, it acts by applying the maximum deceleration possible, decreasing the robot's velocity to 13 cm/s, thus avoiding collision.

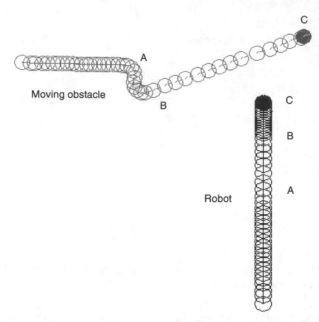

FIGURE II.13.22   Example 2. A, B, and C are the positions that the robot and the obstacle occupy in three time instants of the time interval represented.

13.5.8.3   Avoidance in Real Environments

In real environments, using people as moving obstacles, different tests have been carried out showing that the system is robust and reliable, despite the low precision the method for ultrasonic-based moving obstacle detection produces. A real example of collision avoidance* is shown in Figure II.13.23. Ultrasonic sensor measurements are represented by points, possible moving obstacles by ($\times$), and detected moving obstacles by ($+$). The solid line groups all detected moving obstacles that correspond to the same existing obstacle. In general, all possible obstacles correspond to detected ones (represented by ($*$), as happens for $M1$), although the final position may be calculated as the arithmetic mean of two detected positions.

The arrow on the right side of Figure II.13.23 represents an approximation of the real trace of the moving obstacle. At the beginning, the robot moves at a constant velocity (18 cm/s) and detects the moving obstacle (this happens when the obstacle is placed at $M1$ and the robot at $R1$), but takes no action because the trend of the obstacle is indifferent and the collision time is medium, so the robot will observe. Later, when the obstacle is placed at $M3$–$M4$, the *obstacle course evaluation* module detects that, because the obstacle turns to the left, its tendency is *to give way*. This is done by means of the following rule:

> *IF **collision time** is medium AND **collision status change** is increase in the last three seconds AND **nci trend** is not decreasing throughout the last two seconds, THEN **obstacle aim** is to give way.*

Due to this tendency, selected behavior is *to pass in front* and it is implemented by means of acceleration and turning to the left (since the obstacle's incidence is from right to left). This behavior is repeated for the next two operation cycles, until collision is avoided. To reach this, the robot

---

*A video recording of a similar test is available at http.//www-gsi.dec.usc.es/areas/robotica_cas.htm.

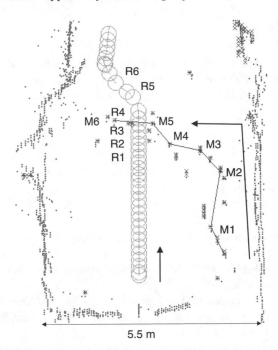

5.5 m

FIGURE II.13.23   Example of collision avoidance with a person who lets the robot pass.

needs to increase speed up to 45 cm/s and turn 40°. These values are reached two iterations after the first collision situation was detected, because robot's motor response is not instantaneous. One of the greatest difficulties encountered is that the obstacle's real position is often unknown, so the robot must use the last detected position. Thus, when the obstacle is placed at $M5$, another collision situation occurs after the change of obstacle position (1 m). This situation is solved by increasing speed and turning to the right, avoiding collision two cycles later. From final positions $R6$ (for the robot) and $M6$ (for obstacle), the robot resumes its trajectory toward the goal point, since the moving obstacle avoidance system does not take control anymore.

13.6   Conclusions

In this work we have described the application of fuzzy temporal controllers in the field of mobile robotics. Fuzzy logic is a highly useful tool in robotics, not only due to its reasoning capabilities with imprecise and/or uncertain values, but also for those situations in which the system model to be controlled is highly complex, or cannot be obtained.

In addition, FTRs exhibit the ability to carry out temporal reasoning by analyzing the evolution of variables throughout temporal references. In this manner, we succeed in determining the tendencies of variables and in filtering sensorial inputs. Hence, the control command not only depends on the values of the input variables at the current temporal point, but also takes into consideration previous values that may be highly significant in given situations.

The usefulness of this knowledge representation and reasoning model (FTRs) has been demonstrated with the implementation of two behaviors on a Nomad 200 robot: wall following and avoidance of moving objects. In both cases it has been shown, by means of tests in real environments, how due to their greater expressive capability the use of FTRs result in an increase in the reliability and robustness of these behaviors as well as their richness.

Acknowledgments

The authors wish to acknowledge the support from the Secretaría Xeral de I + D of the Xunta de Galicia through grant PGIDT99PXI20603A and from the European Commission and Spanish CICYT through grant 1FD97-0183.

References

1. Brooks, R.A., A robusted layered control system for a mobile robot, *IEEE Journal on Robotics and Automation*, RA-2, 24, 1986.
2. Arkin, R.C., Motor schema-based mobile robot navigation, *The International Journal of Robotics Research*, 8, 92, 1989.
3. Meystel, A., Autonomous mobile robots: Vehicles with cognitive control, vol. 1 of *Automation*, World Scientific, Singapore, 1991.
4. Konolige, K. et al., The Saphira architecture: A design for autonomy, *Journal of Experimental and Theoretical Artificial Intelligence*, 9, 215, 1997.
5. Gat, E., Three-layer architectures, in *Artificial Intelligence and Mobile Robots: Case Studies of Successful Robot Systems*, Kortenkamp, D., Bonasso, R.P., and Murphy, R., Eds., 195, AAAI Press/The MIT Press, Cambridge, 1998.
6. Arkin, R.C., Integrating behavioral, perceptual, and world knowledge in reactive navigation, *Robotics and Autonomous Systems*, 6, 105, 1990.
7. Saffiotti, A., The uses of fuzzy logic in autonomous robot navigation, *Soft Computing*, 1, 180, 1997.
8. Lee, E.T., Applying fuzzy logic to robot navigation, *Kybernetes*, 24, 38, 1995.
9. Toda, M. et al., Navigation method for a mobile robot via sonar-based crop row mapping and fuzzy logic control, *Journal of Agricultural Engineering Research*, 72, 299, 1999.
10. Althoefer, K. et al., Fuzzy navigation for robotic manipulators, *International Journal of Uncertainty, Fuzziness and Knowledge-Based Systems*, 6, 179, 1998.
11. de Lope, J., Maravall, D., and Zato, J.G., Topological modeling with fuzzy petri nets for autonomous mobile robots, in *Proceedings of the 11th International Conference on Industrial and Engineering Applications of Artificial Intelligence and Expert Systems (IEA-98-AIE)*, no. 1416 in Lecture Notes in Artificial Intelligence, Springer-Verlag, 1998, 290.
12. Zadeh, L.A., Outline of a new approach to the analysis of complex systems and decision-making approach, *IEEE Transactions on Systems, Man and Cybernetics*, 3, 28, 1973.
13. Chen, Z., Fuzzy temporal reasoning for process supervision, *Expert Systems*, 12, 123, 1995.
14. Más, O., Palá, P., and Miró, J.M., Razonamiento aproximado en presencia de conceptos temporales difusos, in *Actas del VI Congreso Español sobre Tecnologías y Lógica Fuzzy (ESTYLF '96)*, Oviedo (Spain), 1996, 185.
15. Qian, D.Q., Representation and use of imprecise temporal knowledge in dynamic systems, *Fuzzy Sets and Systems*, 50, 59, 1992.
16. Barro, S. et al., Petri nets for fuzzy reasoning on dynamic systems, in *Proceedings of the 7th IFSA World Congress*, Prague, 1997, 279.
17. Bugarín, A. et al., Reasoning with Fuzzy Temporal Rules on Petri Nets, *Fuzziness in Petri Nets*, vol. 22 of *Studies in Fuzziness and Soft Computing*, Physica-Verlag, Heidelberg, 1999, 174.
18. Mucientes, M. et al., Control de la velocidad de un robot móvil mediante reglas temporales borrosas, in *Actas del X Congreso Español sobre Tecnologías y Lógica Fuzzy (ESTYLF 2000)*, Sevilla (Spain), 2000, 127.
19. Mucientes, M. et al., A fuzzy temporal rule-based velocity controller for mobile robotics, *Fuzzy Sets and Systems* (special issue on fuzzy set techniques for intelligent robotic systems), accepted.

20. Arrúe, B.C. et al., Fuzzy behaviours combination to control a non-holonomic robot using virtual perception memory, in *Proceedings of the Sixth IEEE International Conference on Fuzzy Systems (Fuzz-IEEE '97)*, Barcelona (Spain), 1997, 1239.

21. Braunstingl, R., Mujika, J., and Uribe, J.P., A wall following robot with a fuzzy logic controller optimized by a genetic algorithm, in *Proceedings of the International Joint Conference of the Fourth IEEE International Conference on Fuzzy Systems and the Second International Fuzzy Engineering Symposium*, 5, Yokohama (Japan), 1995, 77.

22. Castellano, G., Attolico, G., and Distante, A., Automatic generation of fuzzy rules for reactive robot controllers, *Robotics and Autonomous Systems*, 22, 133, 1997.

23. García-Cerezo, A., Mandow, A., and López-Baldán, M.J., Fuzzy modelling operator navigation behaviours, in *Proceedings of the Sixth IEEE International Conference on Fuzzy Systems (Fuzz-IEEE '97)*, Barcelona (Spain), 1997, 1339.

24. Castellano, G. et al., Reactive navigation by fuzzy control, in *Proceedings of the Fifth IEEE International Conference on Fuzzy Systems (Fuzz-IEEE '96)*, 3, New Orleans (USA), 1996, 2143.

25. Arrúe, B.C. et al., Application of virtual perception memory to control a non-holonomic mobile robot, in *Proceedings of the 3rd IFAC Symposium on Intelligent Components and Instruments for Control Applications-SICICA '97*, Annecy (France), 1997, 537.

26. Uribe, J.P., Urzelai, J., and Ezkerra, M., Fuzzy controller for wall-following with a non-holonomous mobile robot, in *Proceedings of the Sixth IEEE International Conference on Fuzzy Systems (Fuzz-IEEE'97)*, Barcelona (Spain), 1997, 1361.

27. Ahrns, I. et al., Neural fuzzy techniques in sonar-based collision avoidance, in *Soft Computing for Intelligent Robotic Systems*, vol. 21 of *Studies in Fuzziness and Soft Computing*, Physica-Verlag, Heidelberg, 1998, 185.

28. Iglesias, R. et al., Implementation of a basic reactive behavior in mobile robotics through artificial neural networks, in *Proceedings of the International Work-Conference on Artificial Neural Networks (IWANN '97)*, no. 1240 in Lecture Notes in Computer Science, Springer-Verlag, Lanzarote (Spain), 1997, 1364.

29. Iglesias, R. et al., Supervised reinforcement learning: Application to a wall following behavior in a mobile robot, in *Tasks and Methods in Applied Artificial Intelligence*, vol. 2 of Lecture Notes in Artificial Intelligence, 1998, 300.

30. Kohonen, T., *Self-Organizing Maps*, Springer Verlag, Heidelberg, 1997.

31. Elnagar, A. and Gupta, K., Motion prediction of moving objects based on autoregressive model, *IEEE Transactions on Systems, Man and Cybernetics-Part A: Systems and Humans*, 28, 803, 1998.

32. Chang, C.C. and Song, K.T., Environment prediction for a mobile robot in a dynamic environment, *IEEE Transactions on Robotics and Automation*, 13, 862, 1997.

33. Spence, R. and Hutchinson, S., An integrated architecture for robot motion planning and control in the presence of obstacles with unknown trajectories, *IEEE Transactions on Systems, Man and Cybernetics*, 25, 100, 1995.

34. Chakravarthy, A. and Ghose, D., Obstacle avoidance in a dynamic environment: A collision cone approach, *IEEE Transactions on Systems, Man and Cybernetics-Part A: Systems and Humans*, 28, 562, 1998.

35. Gil de Lamadrid, J.F. and Gini, M.L., Path tracking through uncharted moving obstacles, *IEEE Transactions on Systems, Man and Cybernetics*, 20, 1408, 1990.

36. Tsoularis, A. and Kambahmpati, C., On-line planning for collision avoidance on the nominal path, *Journal of Intelligent Robotic Systems*, 21, 327, 1998.

37. Tsoularis, A. and Kambahmpati, C., Avoiding moving obstacles by deviation from a mobile robot's nominal path, *International Journal of Robotics Research*, 18, 454, 1999.

38. Garnier, P. and Fraichard, T., A fuzzy motion controller for a car-like vehicle, Tech. Rep. 3200, INRIA, 1997.

39. Pratihar, D.K., Deb, K., and Ghosh, A., A genetic-fuzzy approach for mobile robot navigation among moving obstacles, *International Journal of Approximate Reasoning*, 20, 145, 1999.
40. Mucientes, M. et al., Use of fuzzy temporal rules for avoidance of moving obstacles in mobile robotics, in *Proceedings of the 1999 Eusflat-Estylf Joint Conference*, Mallorca (Spain), 1999, 167.
41. Mucientes, M. et al., Avoidance of mobile obstacles in real environments, in *Proceedings of the IJCAI-2001 Workshop on Reasoning with Uncertainty in Robotics*, Fox, D. and Saffiotti, A., Eds., Seattle (USA), 2001, 69.
42. Mucientes, M. et al., Fuzzy temporal rules for mobile robot guidance in dynamic environments, *IEEE Transactions on Systems, Man and Cybernetics-Part C: Applications and Reviews*, 31, 391, 2001.
43. Fiorini, P., Robot Motion Planning among Moving Obstacles, Ph.D. thesis, University of California, Los Angeles, 1995.
44. Rodríguez, M. et al., Probabilistic and count methods in map building for autonomous mobile robots, in *Advances in Robot Learning*, vol. 1812 of Lecture Notes in Artificial Intelligence, Springer Verlag, Lausanne (Switzerland), 2000, 120.

Index

N

O

P